EXPLORING ANIMAL BEHAVIOR

Readings from *American Scientist*

EDITED BY
PAUL W. SHERMAN
CORNELL UNIVERSITY

AND
JOHN ALCOCK
ARIZONA STATE UNIVERSITY

Sinauer Associates, Inc., Publishers
Sunderland, Massachusetts

The Cover

Family of golden jackals (*Canis aureus*) in the Serengeti Plain.

Golden jackals form long-term pair bonds, and both parents provide food, protection, and care for their pups. Some pups stay and help to raise the next year's litter, their younger brothers and sisters.
Photograph courtesy of Patricia D. Moehlman.

Exploring Animal Behavior: Readings from *American Scientist*

Illustrations for the articles reprinted this collection were drawn by artists on the *American Scientist* staff, with the following exceptions: Figs. 4, 8–10, 13, and 14 in "Aerial defense tactics of flying insects" by Virge Kask. Figs. 2, 8, 10, and 12 in "The essence of royalty" by Elizabeth Carefoot. Figs. 1 and 5 in "How do planktonic larvae know where to settle?" and Fig. 1 in "Naked mole-rats" by Sally Black.

Library of Congress Cataloging-in-Publication Data
Exploring Animal Behavior
Readings from American Scientist
Edited by Paul W. Sherman and John Alcock
p. cm
Includes bibliographical references and index.
ISBN 0-87893-762-5 (paper)
1. Animal Behavior. I. Sherman, Paul W., 1949–. II. Alcock, John, 1942–.
III. American Scientist.
Ql751.6.E96 1993
591.51—dc20 93-1140
CIP

Printed in Hong Kong
8 7 6 5 4 3 2 1

Contents

Preface/Acknowledgments iv

Part I Doing Science: Integrity, Communication, and Controversy 1

Lewis M. Branscomb **Integrity in Science** 3

George D. Gopen and Judith A. Swan **The Science of Scientific Writing** 6

Sandra Ackerman ***American Scientist* Interviews: Barbara Smuts** 15

Sarah Blaffer Hrdy **Infanticide as a Primate Reproductive Strategy** 21

Richard Curtin and Phyllis Dolhinow **Primate Social Behavior in a Changing World** 31

Part II The Mechanisms of Behavior 39

Kay E. Holekamp and Paul W. Sherman **Why Male Ground Squirrels Disperse** 41

Donald R. Griffin **Animal Thinking** 49

Susan E. Nicol and Irving I. Gottesman **Clues to the Genetics and Neurobiology of Schizophrenia** 38

Meredith J. West and Andrew P. King **Mozart's Starling** 65

Mike May **Aerial Defense Tactics of Flying Insects** 74

M. Brock Fenton and James H. Fullard **Moth Hearing and the Feeding Strategies of Bats** 87

Carl Hirschie Johnson and J. Woodland Hastings **The Elusive Mechanism of the Circadian Clock** 97

Part III Evolutionary History and Behavioral Ecology 105

Bert Hölldobler and Edward O. Wilson **The Evolution of Communal Nest-Weaving in Ants** 108

Stephen Jay Gould **Evolution and the Triumph of Homology, or Why History Matters** 118

Mark L. Winston and Keith N. Slessor **The Essence of Royalty: Honey Bee Queen Pheromone** 128

Aileen N. C. Morse **How Do Planktonic Larvae Know Where to Settle?** 140

Kim Hill and A. Magdalena Hurtado **Hunter–Gatherers of the New World** 154

Thomas D. Seeley **The Ecology of Temperate and Tropical Honeybee Societies** 161

Part IV The Evolutionary Basis of Reproductive Behavior 170

Randy Thornhill and Darryl T. Gwynne **The Evolution of Sexual Differences in Insects** 172

William G. Eberhard **Animal Genitalia and Female Choice** 180

Robert R. Warner **Mating Behavior and Hermaphroditism in Coral Reef Fishes** 188

Douglas W. Mock, Hugh Drummond, and Christopher H. Stinson **Avian Siblicide** 197

Patricia D. Moehlman **Social Organization in Jackals** 209

Rodney L. Honeycutt **Naked Mole-Rats** 219

David M. Buss **Human Mate Selection** 230

Index 235

Preface

This book contains 25 articles that have appeared over the past 15 years in *American Scientist,* the journal of the scientific society Sigma Xi. We gathered these papers together to make them available as supplementary reading for students of animal behavior. We believe the articles can be profitably employed in classrooms in several ways, but especially as material for discussion and debate on key concepts. The articles also illustrate how behavioral scientists conduct research, providing greater depth of coverage than the generally brief textbook accounts of the same issues.

This anthology should be particularly useful for classes that use John Alcock's textbook, *Animal Behavior: An Evolutionary Approach,* because the articles are organized in a sequence complementary to that text. However, there is no reason why the collection cannot also enrich courses based on other textbooks. Indeed, these readings can stand alone as a sampler of the diversity of topics that constitute the modern study of animal behavior.

The material is organized into four sections: the first examines various aspects of science as a profession; the second focuses on investigations of the proximate mechanisms that underlie animal behavior; the third shifts to studies of the historical and adaptive bases of behavior; the final section illustrates how behavioral scientists analyze the possible reproductive consequences of behaviors. We provide each section with its own brief introduction.

We hope that students and teachers alike will enjoy and benefit from this collection of articles.

Acknowledgments

This book is a reality because of the goodwill and hard work of many individuals, most notably the authors of the articles, who graciously provided permission to reprint their work, and the many photographers and illustrators whose images brighten so many pages. Michelle Press, who was editor of *American Scientist* when most of the articles were first published, deserves our gratitude for her keen interest in the field of animal behavior, and for her efforts to see that behavior was prominently featured in the journal. The small size of this volume belies the number of details that had to be attended to in its production; Rosalind Reid, the current editor of *American Scientist,* and Lil Chappell, her editorial assistant, worked hard to see that everything was in order before we went to press. At Sinauer Associates, Peter Farley, Kathaleen Emerson, Joseph Vesely, and Christopher Small made sure that our two goals of a handsome book with a low price could be met without compromise. At home, our families kept our spirits from flagging. Thanks to all.

Paul W. Sherman
John Alcock

Part I
Doing Science: Integrity, Communication, and Controversy

We begin our collection with several articles that touch on some central features of what it means to be a scientist, especially how the "scientific method" is used. An interview with Barbara Smuts introduces us to a behavioral scientist who, at the time of the interview, was planning a new research project. When reading the interview, try to identify the questions that interest the researcher and the approach that she plans to use to try to answer them. Ask yourself what hypotheses (or causal explanations) she offers, and what her predictions (or expected observations) are. What tests does she plan involving the gathering of actual observations to be matched against the expected ones? Are Dr. Smuts's goals important ones? Is her career an attractive one to you?

A prominent feature of science is the competition between alternative hypotheses and their advocates. The interview with Smuts alludes to this phenomenon, which is further illustrated by the two following articles, by Sarah Hrdy, and by Richard Curtin and Phyllis Dolhinow, on infanticide in langurs and other primates. Readers can decide for themselves whether the advocacy approach, in which different persons champion different explanations, is superior, inferior, or equivalent to the method of multiple working hypotheses, in which one researcher formally tests a set of alternatives. Which of the two articles is more persuasive and why? How does progress in science arise from controversies of this sort? Is there a more sensible way to proceed when trying to evaluate competing explanations of some aspect of the natural world? If so, what is it and how could it be imposed on working researchers?

Lewis Branscomb discusses another element of science as an enterprise, which is the problem of whether tests of hypotheses as reported in scientific papers can be trusted. He argues that, although outright fraud is likely to be rare, self-deception is probably more common because of the desire of scientists to generate published accounts of their research and the correlated temptation to observe that which will yield publishable results. Given the pressures to produce research results, might not we expect that some persons will simply manufacture data in order to achieve the rewards that come from frequent publication? Blatant fraud by scientists has been detected on occasion. Are the detected cases really unusual or merely the tip of the iceberg? Readers of Branscomb's article may ask themselves what will prevent someone from simply inventing and then publishing a data set that provides strong support for, say, a hypothesis that infanticide in langurs is an adaptive tactic of males to eliminate future competitors for female mates. What is it about the nature of scientific competition and controversy that makes blatant fraud unlikely and that encourages the detection even of self-deception on the part of a "rival" researcher? Or are scientists engaged in another form of self-deception when they pat themselves on the back and speak of science as a self-correcting enterprise? How should the scientific enterprise be structured if "truth" is our goal? And what is "truth" anyway?

Publishing one's findings is one other central element of doing science, a point captured in the academic admonition "Publish or perish!" Writing about one's conclusions and the means by which they were reached enables scientists to communicate with a broad audience, which can then evaluate the message critically. Yet the training that most scientists receive rarely includes any formal instruction in how to communicate effectively, which may contribute to the widespread impression that scientific writing is generally turgid, close to incomprehensible, and no fun at all to read. Happily, there are numerous exceptions to this "rule," as the articles contained in this collection demonstrate. Moreover, useful formal instruction on how to write scientifically does exist. The analysis and advice on writing provided here by George Gopen and Judith Swan strikes us as being superbly helpful—not just for scientists but for anyone who wants to write in ways that readers will appreciate. A useful exercise might be to analyze and dissect a piece of writing—for example, an article in this collection—and, after pulling the writing apart, put it back together in improved form, taking advantage of the suggestions offered by Gopen and Swan.

Taken together, the articles in this first section provide insight into what it means to do science, an occupation whose usefulness to society depends on the integrity and writing skills of its practitioners as well as their ability to evaluate competing hypotheses, including the hypotheses they have advanced themselves.

Integrity in Science

Lewis M. Branscomb

Much of the problem of honor—or lack of honor—in science stems not from malice but from self-deception

In 1945 a physics graduate student at Harvard began a Ph.D. thesis project involving the use of molecular spectroscopy to determine the temperature of the atmosphere 1,000 km above the earth, at that time quite unknown. The Schumann-Runge bands of molecular oxygen had been observed as very weak emissions from the upper atmosphere. It was thought that they could be used as a thermometer, subject to verification in laboratory studies. But the bands had been observed only in absorption at very high pressures. Then in 1948 there appeared in *Nature* a report that stated that the Schumann-Runge bands had been observed in emission, excited at low pressure in a high-frequency discharge. The author also analyzed the molecular constants of the states involved (*1*).

Delighted to find from the literature that his thesis problem could be successfully attacked, the student set about reproducing the experiment described in *Nature*. After months of fruitless effort, he became suspicious that the results reported were in error and even that the photograph published with the text was not a picture of the Schumann-Runge spectrum at all. Indeed, it appeared that the results might have been fabricated from the proverbial whole cloth. In any case, six months of a predoctoral fellowship were lost, and another way to tackle the thesis problem had to be found.

I was that graduate student, and I have always felt sorry for the author of the article in *Nature*, who must have been under terrible pressure to show something for his efforts. I doubt that he had any intent to injure anyone, certainly not an unknown student thousands of miles away.

I believe that there are very few scientists who deliberately falsify their work, cheat on their colleagues, or steal from their students. On the other hand, I am afraid that a great many scientists deceive themselves from time to time in their treatment of data, gloss over problems involving systematic errors, or understate the contributions of others. These are the "honest mistakes" of science, the scientific equivalent of the "little white lies" of social discourse. But unlike polite society, which easily interprets those white lies, the scientific community has no way to protect itself from sloppy or deceptive literature except to learn whose work to suspect as unreliable. This is a tough sentence to pass on an otherwise talented scientist.

The pressures on young science faculty are often fierce, not so much from tenure committees or even from peers, but from within. A young untenured scientist has all his emotional eggs in one basket. He picks a research problem and invests a year or more in its pursuit. Getting a successful start is important to the opportunity to do research. A lifetime career hinges on nature's cooperation as well as his own diligence and ingenuity. As we are reminded on television, it is dangerous to trifle with Mother Nature. Scientists run that risk every day. It takes a very self-confident young scientist to laugh at Tom Lehrer's "Lobachevsky" without a twinge of fear.

The Sigma Xi project on Honor in Science must deal with the broader question of the integrity of scientists' behavior, not just with the morality of what is admittedly the more serious evil, deliberate cheating (*2–5*). Unless science students are thoroughly inculcated with the discipline of correct scientific process, they are in serious danger of being damaged by the temptation to take the easy road to apparent success. And outright cheating can best be contained if the standards in all disciplines are held at high levels.

When is an experiment complete?

The reader may feel that the rules are simple and easy to follow for those who care about the integrity of their work. That is not necessarily so. Take, for example, the problem of knowing when an experiment is finished and the results are ready to publish. In 1953, building on the work of Wade Fite and profiting from his help at a critical time, I succeeded in making the first laboratory measurement of the photodetachment cross section for a negatively charged atomic ion in vacuum (*6*). The absorption of light by the negative ion of hydrogen (H−) was believed by Rupert Wildt to dominate the opacity of the solar photosphere. Simply put, the temperature of the sun, and thus the wavelengths to which human eyes are sensitive, is determined by this cross section. No one knew how accurate the quantum calculation of this three-body problem might be.

In order to test the calculation, Stephen J. Smith and I undertook an experiment requiring an absolute measurement in a very complex crossed-beam apparatus. After several years of preparation, the experiment began

Lewis M. Branscomb is Vice President and Chief Scientist of the IBM *Corporation, President of Sigma Xi, and a past-president of the American Physical Society. He joined* IBM *in 1972 after a 21-year career at the National Bureau of Standards, of which he became director in 1969. He taught physics at several universities, was editor of* Reviews of Modern Physics, *and conceived and chaired for several years the Joint Institute for Laboratory Astrophysics at the University of Colorado. Dr. Branscomb was appointed by President Carter to the National Science Board in 1979 and was elected chairman the following year, serving until 1984. Address:* IBM *Corporation, Armonk, NY 10504.*

to yield data, and we made a reasonably diligent search for sources of systematic errors. The results differed from the quantum calculations by about 15%, a not unreasonable percentage considering the challenge of the three-body problem at the time. Stephen Smith and I were writing up the paper and making some final tests on the radiometric calibration system when the apparatus gave us a hint that something was amiss. We put the paper aside, tore the experiment down, and started over again on the calibrations. Three months later we had done everything necessary to quantify the limits of systematic error. Only then did we convert the results to cross-section units. We discovered to our utter amazement that the corrections we had introduced measured exactly 15%, bringing the experiment and the theory into an agreement so exact as to be clearly fortuitous.

At that point we were faced with a tough decision. What to do now? The experiment was finished. But we had thought it was finished once before. Were we in danger of stopping when we liked the answer? I realized then, as I have often said since, that Nature does not "know" what experiment a scientist is trying to do. "God loves the noise as much as the signal" (*7*). I decided to spend another three months looking for more sources of systematic error—a time exactly equal to the time we had spent on the last effort, which resulted in bringing experiment into agreement with theory. Fortunately, no additional sources could be found, and Steve and I felt we were ready to publish (*8*).

Perhaps this degree of conservatism is not necessary in every case, but it is certainly crucial in the case of absolute as opposed to relative measurements. The most severe requirement for such care is in the measurement of the fundamental constants of nature, a major interest of scientists at the National Bureau of Standards. Ever since the 1960s, scientists measuring atomic constants have adopted the policy of never reducing their data to final form (permitting comparison with the work of others) until all error analysis has been completed and the experiment is over.

Why, if the scientists are both honest and disciplined, is this necessary? Because the temptation to get a "good" (i.e., "safe" or "significant") result by stopping when the data pass through the desired coordinates is ever present. Some excellent scientists may have succumbed to the temptation. Back in the 1930s, for example, there was a long series of measurements of the universal constant of nature *c*, the speed of light in a vacuum. Following the pioneering measurement of Michelson and his co-workers, who used a rotating polygonal mirror to chop a beam of light passing between Mt. Wilson and Mt. Baldy in California, subsequent experimenters found more precise results using better equipment. In 1941, Birge's review of all the work concluded that the best weighted average of all the prior work was $c = 299{,}776 \pm 4$ km/sec (*9*).

Then came World War II. New technology and new people came into science. Very low frequency radio navigation (Loran) had been developed for military use, and electrical engineers realized that this system could be used to measure the speed of propagation of those 16 KHz waves in ways totally independent of the prewar optical methods. Within a few years, microwave cavity methods and free-space microwave interferometry gave consistent values with much higher precision. Froome found $c = 299{,}793 \pm 0.3$ km/sec (*10*). There had been a shift of 17 km/sec, yet the stated accuracy of most of the previous measurements was 4 km/sec or better.

In their review of this mystery, Cohen and DuMond concluded that "two things contributed strongly to mislead [Birge in 1941] and would have misled anyone else in the same circumstances. These were the great prestige of Michelson's name as an expert in the field, and the fact that . . . two measurements . . . in 1937 and 1941 agreed quite well with the Michelson-Pease-Pearson result" (*11*). Writing in 1957, Birge said: "In any highly precise experimental arrangement there are initially many instrumental difficulties that lead to numerical results far from the accepted value of the quantity being measured. . . . Accordingly, the investigator searches for the source or sources of such error, and continues searching until he gets a result close to the accepted value. Then he stops! . . . In this way one can account for the close agreement of several different results and also for the possibility that all of them are in error by an unexpectedly large amount" (*12*). Cohen and DuMond credit Peter Franken with labeling this tendency "intellectual phase locking."

Commitment to quality

What might be done to reduce these "honest mistakes," to support the quality and thus the integrity of science? It takes the concerted efforts of teachers and research mentors, of promotion and tenure committees, of journal editors and referees. Above all it takes renewed commitment by the working scientist.

Young scientists should understand all the subtle ways in which they can delude themselves in the design of observations and the interpretation of data and statistics. They should understand metrology and should know what tendencies to manipulate information are built into their digital signal processors. They should also get to know the algorithms used in their favorite computers, which may under certain circumstances give strange results. Above all they should be trained in the detection and control of systematic errors.

The responsibility of the gatekeepers of scientific careers, the tenure committees, deans, and laboratory directors, is a heavy one. To reward people solely on the basis of numbers of papers published is destructive of the quality of science. Publication is of course the conventional method of making one's work available for critical appraisal by one's peers, but it is not the only way. And while perhaps even a necessary way, it is most emphatically not sufficient.

Journal editors and referees are, of course, the stewards of scientific quality, and they face a very difficult task. No journal can afford to publish all the evidence required to support an author's experimental conclusions. But how can a referee approve publication, when information necessary to proof is missing? The traditional answer is that authors use a certain shorthand to refer to procedures used which are either common practice or documented elsewhere. The reader has to trust the author to invoke those procedures properly. Thus one's reputation for trustworthiness, call it intellectual integrity if not honesty, is crucial to a scientific career. Are young people entering the world of scientific research as aware of this as they should be?

The quality of science places another burden on the scientist: not only to ensure that his own work meets the highest standards, but to participate in both the peer review of primary literature and the authorship of reviews of areas of work in which he is competent. Maurice Goldhaber, when he was director of the Brookhaven National Laboratory, encouraged his staff to write scholarly reviews. He felt that the review literature was a special responsibility of scientists at national laboratories; his motto was, "A good review is the moral equivalent of teaching."

During the last two decades substantial organized efforts at professional reviews of the literature of physical science have been undertaken. Groups of research experts have undertaken critical evaluations of original literature, usually dealing with properties of matter and materials. The goal of such reviews is to increase the density of useful information in the literature. Information that is wrong is not useful. And information lacking evidence revealing whether it is right or wrong is scarcely more so. Quality control in original research is the responsibility of the individual, part of the duty, if not the honor, expected from each of us.

To make the literature worth reviewing, authors of original papers must give the reader quantitative estimates of the amount by which the values given may be in error, and scientific justification for their conclusions. Scientists must demand of others and of themselves a revival of sound scholarship, instead of the cream-skimming and large numbers of hastily written papers with which we are all too familiar.

Commitment to integrity

The broader view of honor in science that I have discussed here should help everyone understand that this is not someone else's problem and is not just the problem of fraud in science. Most of us will never encounter a piece of truly fraudulent research. But concerns about scientific integrity permeate every piece of research we do, every talk we hear, every paper we read. A revitalization of interest in scientific honesty and integrity could have an enormous benefit both to science and to the society we serve.

First of all, integrity is essential for the realization of the joy that exploring the world of science can and should bring to each of us. Beyond that, the integrity of science affects the way the public looks at the pronouncements of scientists and the seriousness with which it takes our warnings, whether they relate to acid rain, the loss of genetic materials from endangered species, or the possibilities for science to help solve the global problems facing mankind. The users of our results, the decision-makers who need our advice, will always press us to be more sure of ourselves than our data permit, for it would make their jobs easier. The pressures to take shortcuts in science come from outside, as well as inside, the community.

We must help the public understand the rules of scientific evidence, just as we insist on rules of judicial evidence in our courts. A precondition for success in this endeavor is to refine and apply those rules with great rigor in our own work and literature. The future of mankind hangs in no small measure on the integrity, and thus the credibility, of science.

References

1. L. Lal. 1948. *Nature* 161:477.
2. C. I. Jackson and J. W. Prados. 1983. *Am. Sci.* 71:462.
3. *Honor in Science.* 1984. Sigma Xi.
4. R. N. Hall. 1968. Gen. Elec. rep. no. 68-C-035.
5. R. P. Feynman. 1974. *Engineering and Science* 37(7):10.
6. L. M. Branscomb and W. L. Fite. 1954. *Phys. Rev.* 93:651.
7. L. M. Branscomb. 1980. *Phys. Today* 33(4):42.
8. L. M. Branscomb and S. J. Smith. 1955. *Phys. Rev.* 98:1028.
9. R. T. Birge. 1941. *Rep. Progr. Phys.* 8:90.
10. K. D. Froome. 1954. *Proc. Royal Soc. London* A223:195.
11. E. R. Cohen and J. DuMond. 1965. *Rev. Modern Phys.* 37:537.
12. R. T. Birge. 1957. *Nuovo Cimento,* supp. 6:39.

The Science of Scientific Writing

If the reader is to grasp what the writer means, the writer must understand what the reader needs

George D. Gopen and Judith A. Swan

Science is often hard to read. Most people assume that its difficulties are born out of necessity, out of the extreme complexity of scientific concepts, data and analysis. We argue here that complexity of thought need not lead to impenetrability of expression; we demonstrate a number of rhetorical principles that can produce clarity in communication without oversimplifying scientific issues. The results are substantive, not merely cosmetic: Improving the quality of writing actually improves the quality of thought.

The fundamental purpose of scientific discourse is not the mere presentation of information and thought, but rather its actual communication. It does not matter how pleased an author might be to have converted all the right data into sentences and paragraphs; it matters only whether a large majority of the reading audience accurately perceives what the author had in mind. Therefore, in order to understand how best to improve writing, we would do well to understand better how readers go about reading. Such an understanding has recently become available through work done in the fields of rhetoric, linguistics and cognitive psychology. It has helped to produce a methodology based on the concept of reader expectations.

Writing with the Reader in Mind: Expectation and Context

Readers do not simply read; they interpret. Any piece of prose, no matter how short, may "mean" in 10 (or more) different ways to 10 different readers. This methodology of reader expectations is founded on the recognition that readers make many of their most important interpretive decisions about the substance of prose based on clues they receive from its structure.

This interplay between substance and structure can be demonstrated by something as basic as a simple table. Let us say that in tracking the temperature of a liquid over a period of time, an investigator takes measurements every three minutes and records a list of temperatures. Those data could be presented by a number of written structures. Here are two possibilities:

t (time) = 15', T (temperature) = 32°; t = 0', T = 25°; t = 6', T = 29°; t = 3', T= 27°; t=12', T = 32°; t = 9', T = 31°

time (min)	temperature (°C)
0	25
3	27
6	29
9	31
12	32
15	32

Precisely the same information appears in both formats, yet most readers find the second easier to interpret. It may be that the very familiarity of the tabular structure makes it easier to use. But, more significantly, the structure of the second table provides the reader with an easily perceived context (time) in which the significant piece of information (temperature) can be interpreted. The contextual material appears on the left in a pattern that produces an expectation of regularity; the interesting results appear on the right in a less obvious pattern, the discovery of which is the point of the table.

If the two sides of this simple table are reversed, it becomes much harder to read.

temperature (°C)	time (min)
25	0
27	3
29	6
31	9
32	12
32	15

Since we read from left to right, we prefer the context on the left, where it can more effectively familiarize the reader. We prefer the new, important information on the right, since its job is to intrigue the reader.

Information is interpreted more easily and more uniformly if it is placed where most readers expect to find it. These needs and expectations of readers affect the inter-

George D. Gopen is associate professor of English and Director of Writing Programs at Duke University. He holds a Ph.D. in English from Harvard University and a J.D. from Harvard Law School. Judith A. Swan teaches scientific writing at Princeton University. Her Ph.D., which is in biochemistry, was earned at the Massachusetts Institute of Technology. Address for Gopen: 307 Allen Building, Duke University, Durham, NC 27706.

pretation not only of tables and illustrations but also of prose itself. Readers have relatively fixed expectations about where in the structure of prose they will encounter particular items of its substance. If writers can become consciously aware of these locations, they can better control the degrees of recognition and emphasis a reader will give to the various pieces of information being presented. Good writers are intuitively aware of these expectations; that is why their prose has what we call "shape."

This underlying concept of reader expectation is perhaps most immediately evident at the level of the largest units of discourse. (A unit of discourse is defined as anything with a beginning and an end: a clause, a sentence, a section, an article, etc.) A research article, for example, is generally divided into recognizable sections, sometimes labeled Introduction, Experimental Methods, Results and Discussion. When the sections are confused—when too much experimental detail is found in the Results section, or when discussion and results intermingle—readers are often equally confused. In smaller units of discourse the functional divisions are not so explicitly labeled, but readers have definite expectations all the same, and they search for certain information in particular places. If these structural expectations are continually violated, readers are forced to divert energy from understanding the content of a passage to unraveling its structure. As the complexity of the content increases moderately, the possibility of misinterpretation or noninterpretation increases dramatically.

We present here some results of applying this methodology to research reports in the scientific literature. We have taken several passages from research articles (either published or accepted for publication) and have suggested ways of rewriting them by applying principles derived from the study of reader expectations. We have not sought to transform the passages into "plain English" for the use of the general public; we have neither decreased the jargon nor diluted the science. We have striven not for simplification but for clarification.

Reader Expectations for the Structure of Prose

Here is our first example of scientific prose, in its original form:

> The smallest of the URF's (URFA6L), a 207-nucleotide (nt) reading frame overlapping out of phase the NH_2-terminal portion of the adenosinetriphosphatase (ATPase) subunit 6 gene has been identified as the animal equivalent of the recently discovered yeast H^+-ATPase subunit 8 gene. The functional significance of the other URF's has been, on the contrary, elusive. Recently, however, immunoprecipitation experiments with antibodies to purified, rotenone-sensitive NADH-ubiquinone oxido-reductase [hereafter referred to as respiratory chain NADH dehydrogenase or complex I] from bovine heart, as well as enzyme fractionation studies, have indicated that six human URF's (that is, URF1, URF2, URF3, URF4, URF4L, and URF5, hereafter referred to as ND1, ND2, ND3, ND4, ND4L, and ND5) encode subunits of complex I. This is a large complex that also contains many subunits synthesized in the cytoplasm.*

Ask any ten people why this paragraph is hard to read, and nine are sure to mention the technical vocabulary; several will also suggest that it requires specialized background knowledge. Those problems turn out to be only a small part of the difficulty. Here is the passage again, with the difficult words temporarily lifted:

> The smallest of the URF's, an [A], has been identified as a [B] subunit 8 gene. The functional significance of the other URF's has been, on the contrary, elusive. Recently, however, [C] experiments, as well as [D] studies, have indicated that six human URF's [1-6] encode subunits of Complex I. This is a large complex that also contains many subunits synthesized in the cytoplasm.

It may now be easier to survive the journey through the prose, but the passage is still difficult. Any number of questions present themselves: What has the first sentence of the passage to do with the last sentence? Does the third sentence contradict what we have been told in the second sentence? Is the functional significance of URF's still "elusive"? Will this passage lead us to further discussion about URF's, or about Complex I, or both?

Information is interpreted more easily and more uniformly if it is placed where most readers expect to find it.

Knowing a little about the subject matter does not clear up all the confusion. The intended audience of this passage would probably possess at least two items of essential technical information: first, "URF" stands for "Uninterrupted Reading Frame," which describes a segment of DNA organized in such a way that it could encode a protein, although no such protein product has yet been identified; second, both ATPase and NADH oxido-reductase are enzyme complexes central to energy metabolism. Although this information may provide some sense of comfort, it does little to answer the interpretive questions that need answering. It seems the reader is hindered by more than just the scientific jargon.

To get at the problem, we need to articulate something about how readers go about reading. We proceed to the first of several reader expectations.

Subject-Verb Separation

Look again at the first sentence of the passage cited above. It is relatively long, 42 words; but that turns out not to be the main cause of its burdensome complexity. Long sentences need not be difficult to read; they are only difficult to write. We have seen sentences of over 100 words that flow

*The full paragraph includes one more sentence: "Support for such functional identification of the URF products has come from the finding that the purified rotenone-sensitive NADH dehydrogenase from *Neurospora crassa* contains several subunits synthesized within the mitochondria, and from the observation that the stopper mutant of *Neurospora crassa*, whose mtDNA lacks two genes homologous to URF2 and URF3, has no functional complex I." We have omitted this sentence both because the passage is long enough as is and because it raises no additional structural issues.

easily and persuasively toward their clearly demarcated destination. Those well-wrought serpents all had something in common: Their structure presented information to readers in the order the readers needed and expected it.

The first sentence of our example passage does just the opposite: it burdens and obstructs the reader, because of an all-too-common structural defect. Note that the grammatical subject ("the smallest") is separated from its verb ("has been identified") by 23 words, more than half the sentence.

Beginning with the exciting material and ending with a lack of luster often leaves us disappointed and destroys our sense of momentum.

Readers expect a grammatical subject to be followed immediately by the verb. Anything of length that intervenes between subject and verb is read as an interruption, and therefore as something of lesser importance.

The reader's expectation stems from a pressing need for syntactic resolution, fulfilled only by the arrival of the verb. Without the verb, we do not know what the subject is doing, or what the sentence is all about. As a result, the reader focuses attention on the arrival of the verb and resists recognizing anything in the interrupting material as being of primary importance. The longer the interruption lasts, the more likely it becomes that the "interruptive" material actually contains important information; but its structural location will continue to brand it as merely interruptive. Unfortunately, the reader will not discover its true value until too late—until the sentence has ended without having produced anything of much value outside of that subject-verb interruption.

In this first sentence of the paragraph, the relative importance of the intervening material is difficult to evaluate. The material might conceivably be quite significant, in which case the writer should have positioned it to reveal that importance. Here is one way to incorporate it into the sentence structure:

> The smallest of the URF's is URFA6L, a 207-nucleotide (nt) reading frame overlapping out of phase the NH_2-terminal portion of the adenosinetriphosphatase (ATPase) subunit 6 gene; it has been identified as the animal equivalent of the recently discovered yeast H^+-ATPase subunit 8 gene.

On the other hand, the intervening material might be a mere aside that diverts attention from more important ideas; in that case the writer should have deleted it, allowing the prose to drive more directly toward its significant point:

> The smallest of the URF's (URFA6L) has been identified as the animal equivalent of the recently discovered yeast H^+-ATPase subunit 8 gene.

Only the author could tell us which of these revisions more accurately reflects his intentions.

These revisions lead us to a second set of reader expectations. Each unit of discourse, no matter what the size, is expected to serve a single function, to make a single point. In the case of a sentence, the point is expected to appear in a specific place reserved for emphasis.

The Stress Position

It is a linguistic commonplace that readers naturally emphasize the material that arrives at the end of a sentence. We refer to that location as a "stress position." If a writer is consciously aware of this tendency, she can arrange for the emphatic information to appear at the moment the reader is naturally exerting the greatest reading emphasis. As a result, the chances greatly increase that reader and writer will perceive the same material as being worthy of primary emphasis. The very structure of the sentence thus helps persuade the reader of the relative values of the sentence's contents.

The inclination to direct more energy to that which arrives last in a sentence seems to correspond to the way we work at tasks through time. We tend to take something like a "mental breath" as we begin to read each new sentence, thereby summoning the tension with which we pay attention to the unfolding of the syntax. As we recognize that the sentence is drawing toward its conclusion, we begin to exhale that mental breath. The exhalation produces a sense of emphasis. Moreover, we delight in being rewarded at the end of a labor with something that makes the ongoing effort worthwhile. Beginning with the exciting material and ending with a lack of luster often leaves us disappointed and destroys our sense of momentum. We do not start with the strawberry shortcake and work our way up to the broccoli.

When the writer puts the emphatic material of a sentence in any place other than the stress position, one of two things can happen; both are bad. First, the reader might find the stress position occupied by material that clearly is not worthy of emphasis. In this case, the reader must discern, without any additional structural clue, what else in the sentence may be the most likely candidate for emphasis. There are no secondary structural indications to fall back upon. In sentences that are long, dense or sophisticated, chances soar that the reader will not interpret the prose precisely as the writer intended. The second possibility is even worse: The reader may find the stress position occupied by something that does appear capable of receiving emphasis, even though the writer did not intend to give it any stress. In that case, the reader is highly likely to emphasize this imposter material, and the writer will have lost an important opportunity to influence the reader's interpretive process.

The stress position can change in size from sentence to sentence. Sometimes it consists of a single word; sometimes it extends to several lines. The definitive factor is this: The stress position coincides with the moment of syntactic closure. A reader has reached the beginning of the stress position when she knows there is nothing left in the clause or sentence but the material presently being read. Thus a whole list, numbered and indented, can occupy the stress position of a sentence if it has been clearly announced as being all that remains of that sentence. Each member of that list, in turn, may have its own internal stress position, since each member may produce its own syntactic closure.

Within a sentence, secondary stress positions can be

formed by the appearance of a properly used colon or semicolon; by grammatical convention, the material preceding these punctuation marks must be able to stand by itself as a complete sentence. Thus, sentences can be extended effortlessly to dozens of words, as long as there is a medial syntactic closure for every piece of new, stress-worthy information along the way. One of our revisions of the initial sentence can serve as an example:

> The smallest of the URF's is URFA6L, a 207-nucleotide (nt) reading frame overlapping out of phase the NH_2-terminal portion of the adenosinetriphosphatase (ATPase) subunit 6 gene; it has been identified as the animal equivalent of the recently discovered yeast H^+-ATPase subunit 8 gene.

By using a semicolon, we created a second stress position to accommodate a second piece of information that seemed to require emphasis.

We now have three rhetorical principles based on reader expectations: First, grammatical subjects should be followed as soon as possible by their verbs; second, every unit of discourse, no matter the size, should serve a single function or make a single point; and, third, information intended to be emphasized should appear at points of syntactic closure. Using these principles, we can begin to unravel the problems of our example prose.

Note the subject-verb separation in the 62-word third sentence of the original passage:

> Recently, however, immunoprecipitation experiments with antibodies to purified, rotenone-sensitive NADH-ubiquinone oxido-reductase [hereafter referred to as respiratory chain NADH dehydrogenase or complex I] from bovine heart, as well as enzyme fractionation studies, have indicated that six human URF's (that is, URF1, URF2, URF3, URF4, URF4L, and URF5, hereafter referred to as ND1, ND2, ND3, ND4, ND4L, and ND5) encode subunits of complex I.

After encountering the subject ("experiments"), the reader must wade through 27 words (including three hyphenated compound words, a parenthetical interruption and an "as well as" phrase) before alighting on the highly uninformative and disappointingly anticlimactic verb ("have indicated"). Without a moment to recover, the reader is handed a "that" clause in which the new subject ("six human URF's") is separated from its verb ("encode") by yet another 20 words.

If we applied the three principles we have developed to the rest of the sentences of the example, we could generate a great many revised versions of each. These revisions might differ significantly from one another in the way their structures indicate to the reader the various weights and balances to be given to the information. Had the author placed all stress-worthy material in stress positions, we as a reading community would have been far more likely to interpret these sentences uniformly.

We couch this discussion in terms of "likelihood" because we believe that meaning is not inherent in discourse by itself; "meaning" requires the combined participation of text and reader. All sentences are infinitely interpretable, given an infinite number of interpreters. As communities of readers, however, we tend to work out tacit agreements as to what kinds of meaning are most likely to be extracted from certain articulations. We cannot succeed in making even a single sentence mean one and only one thing; we can only increase the odds that a large majority of readers will tend to interpret our discourse according to our intentions. Such success will follow from authors becoming more consciously aware of the various reader expectations presented here.

We cannot succeed in making even a single sentence mean one and only one thing; we can only increase the odds that a large majority of readers will tend to interpret our discourse according to our intentions.

Here is one set of revisionary decisions we made for the example:

> The smallest of the URF's, URFA6L, has been identified as the animal equivalent of the recently discovered yeast H^+-ATPase subunit 8 gene; but the functional significance of other URF's has been more elusive. Recently, however, several human URF's have been shown to encode subunits of rotenone-sensitive NADH-ubiquinone oxido-reductase. This is a large complex that also contains many subunits synthesized in the cytoplasm; it will be referred to hereafter as respiratory chain NADH dehydrogenase or complex I. Six subunits of Complex I were shown by enzyme fractionation studies and immunoprecipitation experiments to be encoded by six human URF's (URF1, URF2, URF3, URF4, URF4L, and URF5); these URF's will be referred to subsequently as ND1, ND2, ND3, ND4, ND4L, and ND5.

Sheer length was neither the problem nor the solution. The revised version is not noticeably shorter than the original; nevertheless, it is significantly easier to interpret. We have indeed deleted certain words, but not on the basis of wordiness or excess length. (See especially the last sentence of our revision.)

When is a sentence too long? The creators of readability formulas would have us believe there exists some fixed number of words (the favorite is 29) past which a sentence is too hard to read. We disagree. We have seen 10-word sentences that are virtually impenetrable and, as we mentioned above, 100-word sentences that flow effortlessly to their points of resolution. In place of the word-limit concept, we offer the following definition: A sentence is too long when it has more viable candidates for stress positions than there are stress positions available. Without the stress position's locational clue that its material is intended to be emphasized, readers are left too much to their own devices in deciding just what else in a sentence might be considered important.

In revising the example passage, we made certain decisions about what to omit and what to emphasize. We put subjects and verbs together to lessen the reader's syntactic burdens; we put the material we believed worthy of emphasis in stress positions; and we discarded material for which we could not discern significant connections. In doing so, we have produced a clearer passage—but not one that necessarily reflects the author's intentions; it reflects only our interpretation of the author's intentions. The more problematic the structure, the less likely it becomes that a grand majority of readers will perceive the discourse in exactly the way the author intended.

It is probable that many of our readers—and perhaps even the authors—will disagree with some of our choices. If so, that disagreement underscores our point: The original failed to communicate its ideas and their connections clearly. If we happened to have interpreted the passage as you did, then we can make a different point: No one should have to work as hard as we did to unearth the content of a single passage of this length.

The information that begins a sentence establishes for the reader a perspective for viewing the sentence as a unit.

The Topic Position

To summarize the principles connected with the stress position, we have the proverbial wisdom, "Save the best for last." To summarize the principles connected with the other end of the sentence, which we will call the topic position, we have its proverbial contradiction, "First things first." In the stress position the reader needs and expects closure and fulfillment; in the topic position the reader needs and expects perspective and context. With so much of reading comprehension affected by what shows up in the topic position, it behooves a writer to control what appears at the beginning of sentences with great care.

The information that begins a sentence establishes for the reader a perspective for viewing the sentence as a unit: Readers expect a unit of discourse to be a story about whoever shows up first. "Bees disperse pollen" and "Pollen is dispersed by bees" are two different but equally respectable sentences about the same facts. The first tells us something about bees; the second tells us something about pollen. The passivity of the second sentence does not by itself impair its quality; in fact, "Pollen is dispersed by bees" is the superior sentence if it appears in a paragraph that intends to tell us a continuing story about pollen. Pollen's story at that moment is a passive one.

Readers also expect the material occupying the topic position to provide them with linkage (looking backward) and context (looking forward). The information in the topic position prepares the reader for upcoming material by connecting it backward to the previous discussion. Although linkage and context can derive from several sources, they stem primarily from material that the reader has already encountered within this particular piece of discourse. We refer to this familiar, previously introduced material as "old information." Conversely, material making its first appearance in a discourse is "new information." When new information is important enough to receive emphasis, it functions best in the stress position.

When old information consistently arrives in the topic position, it helps readers to construct the logical flow of the argument: It focuses attention on one particular strand of the discussion, both harkening backward and leaning forward. In contrast, if the topic position is constantly occupied by material that fails to establish linkage and context, readers will have difficulty perceiving both the connection to the previous sentence and the projected role of the new sentence in the development of the paragraph as a whole.

Here is a second example of scientific prose that we shall attempt to improve in subsequent discussion:

> Large earthquakes along a given fault segment do not occur at random intervals because it takes time to accumulate the strain energy for the rupture. The rates at which tectonic plates move and accumulate strain at their boundaries are approximately uniform. Therefore, in first approximation, one may expect that large ruptures of the same fault segment will occur at approximately constant time intervals. If subsequent mainshocks have different amounts of slip across the fault, then the recurrence time may vary, and the basic idea of periodic mainshocks must be modified. For great plate boundary ruptures the length and slip often vary by a factor of 2. Along the southern segment of the San Andreas fault the recurrence interval is 145 years with variations of several decades. The smaller the standard deviation of the average recurrence interval, the more specific could be the long term prediction of a future mainshock.

This is the kind of passage that in subtle ways can make readers feel badly about themselves. The individual sentences give the impression of being intelligently fashioned: They are not especially long or convoluted; their vocabulary is appropriately professional but not beyond the ken of educated general readers; and they are free of grammatical and dictional errors. On first reading, however, many of us arrive at the paragraph's end without a clear sense of where we have been or where we are going. When that happens, we tend to berate ourselves for not having paid close enough attention. In reality, the fault lies not with us, but with the author.

We can distill the problem by looking closely at the information in each sentence's topic position:

Large earthquakes
The rates
Therefore... one
subsequent mainshocks
great plate boundary ruptures
the southern segment of the San Andreas fault
the smaller the standard deviation...

Much of this information is making its first appearance in this paragraph—in precisely the spot where the reader looks for old, familiar information. As a result, the focus of the story constantly shifts. Given just the material in the topic positions, no two readers would be likely to construct exactly the same story for the paragraph as a whole.

If we try to piece together the relationship of each sentence to its neighbors, we notice that certain bits of old information keep reappearing. We hear a good deal about the recurrence time between earthquakes: The first sentence introduces the concept of nonrandom intervals between earthquakes; the second sentence tells us that recurrence rates due to the movement of tectonic plates are more or less uniform; the third sentence adds that the recurrence rate of major earthquakes should also be somewhat predictable; the fourth sentence adds that recurrence rates vary with some conditions; the fifth sentence adds information about one particular variation; the sixth sentence adds a recurrence-rate example from California; and the last sentence tells us something about how recurrence rates can be described statistically. This refrain of "recurrence intervals" constitutes the major string of old information in the paragraph. Unfortunately, it rarely appears at the beginning of sentences, where it would help us maintain our focus on its continuing story.

In reading, as in most experiences, we appreciate the opportunity to become familiar with a new environment before having to function in it. Writing that continually begins sentences with new information and ends with old information forbids both the sense of comfort and orientation at the start and the sense of fulfilling arrival at the end. It misleads the reader as to whose story is being told; it burdens the reader with new information that must be carried further into the sentence before it can be connected to the discussion; and it creates ambiguity as to which material the writer intended the reader to emphasize. All of these distractions require that readers expend a disproportionate amount of energy to unravel the structure of the prose, leaving less energy available for perceiving content.

We can begin to revise the example by ensuring the following for each sentence:

1. The backward-linking old information appears in the topic position.
2. The person, thing or concept whose story it is appears in the topic position.
3. The new, emphasis-worthy information appears in the stress position.

Once again, if our decisions concerning the relative values of specific information differ from yours, we can all blame the author, who failed to make his intentions apparent. Here first is a list of what we perceived to be the new, emphatic material in each sentence:

time to accumulate strain energy along a fault
approximately uniform
large ruptures of the same fault
different amounts of slip
vary by a factor of 2
variations of several decades
predictions of future mainshock

Now, based on these assumptions about what deserves stress, here is our proposed revision:

> Large earthquakes along a given fault segment do not occur at random intervals because it takes time to accumulate the strain energy for the rupture. The rates at which tectonic plates move and accumulate strain at their boundaries are roughly uniform. Therefore, nearly constant time intervals (at first approximation) would be expected between large ruptures of the same fault segment. [However?], the recurrence time may vary; the basic idea of periodic mainshocks may need to be modified if subsequent mainshocks have different amounts of slip across the fault. [Indeed?], the length and slip of great plate boundary ruptures often vary by a factor of 2. [For example?], the recurrence interval along the southern segment of the San Andreas fault is 145 years with variations of several decades. The smaller the standard deviation of the average recurrence interval, the more specific could be the long term prediction of a future mainshock.

Many problems that had existed in the original have now surfaced for the first time. Is the reason earthquakes do not occur at random intervals stated in the first sentence or in the second? Are the suggested choices of "however," "indeed," and "for example" the right ones to express the connections at those points? (All these connections were left unarticulated in the original paragraph.) If "for example" is an inaccurate transitional phrase, then exactly how does the San Andreas fault example connect to ruptures that "vary by a factor of 2"? Is the author arguing that recurrence rates must vary because fault movements often vary? Or is the author preparing us for a discussion of how in spite of such variance we might still be able to predict earthquakes? This last question remains unanswered because the final sentence leaves behind earthquakes that recur at variable intervals and switches instead to earthquakes that recur regularly. Given that this is the first paragraph of the article, which type of earthquake will the article most likely proceed to discuss? In sum, we are now aware of how much the paragraph had not communicated to us on first reading. We can see that most of our difficulty was owing not to any deficiency in our reading skills but rather to the author's lack of comprehension of our structural needs as readers.

In our experience, the misplacement of old and new information turns out to be the No. 1 problem in American professional writing today.

In our experience, the misplacement of old and new information turns out to be the No. 1 problem in American professional writing today. The source of the problem is not hard to discover: Most writers produce prose linearly (from left to right) and through time. As they begin to formulate a sentence, often their primary anxiety is to capture the important new thought before it escapes. Quite naturally they rush to record that new information on paper, after which they can produce at their leisure the contextualizing material that links back to the previous discourse. Writers who do this consistently are attending more to their own need for unburdening themselves of their information than to the reader's need for receiving the material. The methodology of reader expectations articulates the reader's needs explicitly, thereby making writers consciously aware of structural problems and ways to solve them.

Put in the topic position the old information that links backward; put in the stress position the new information you want the reader to emphasize.

A note of clarification: Many people hearing this structural advice tend to oversimplify it to the following rule: "Put the old information in the topic position and the new information in the stress position." No such rule is possible. Since by definition all information is either old or new, the space between the topic position and the stress position must also be filled with old and new information. Therefore the principle (not rule) should be stated as follows: "Put in the topic position the old information that links backward; put in the stress position the new information you want the reader to emphasize."

Perceiving Logical Gaps

When old information does not appear at all in a sentence, whether in the topic position or elsewhere, readers are left to construct the logical linkage by themselves. Often this happens when the connections are so clear in the writer's mind that they seem unnecessary to state; at those moments, writers underestimate the difficulties and ambiguities inherent in the reading process. Our third example attempts to demonstrate how paying attention to the placement of old and new information can reveal where a writer has neglected to articulate essential connections.

> The enthalpy of hydrogen bond formation between the nucleoside bases 2'deoxyguanosine (dG) and 2'deoxycytidine (dC) has been determined by direct measurement. dG and dC were derivatized at the 5' and 3' hydroxyls with triisopropylsilyl groups to obtain solubility of the nucleosides in non-aqueous solvents and to prevent the ribose hydroxyls from forming hydrogen bonds. From isoperibolic titration measurements, the enthalpy of dC:dG base pair formation is –6.65 ± 0.32 kcal/mol.

Although part of the difficulty of reading this passage may stem from its abundance of specialized technical terms, a great deal more of the difficulty can be attributed to its structural problems. These problems are now familiar: We are not sure at all times whose story is being told; in the first sentence the subject and verb are widely separated; the second sentence has only one stress position but two or three pieces of information that are probably worthy of emphasis—"solubility... solvents," "prevent... from forming hydrogen bonds" and perhaps "triisopropylsilyl groups." These perceptions suggest the following revision tactics:

1. Invert the first sentence, so that (*a*) the subject-verb-complement connection is unbroken, and (*b*) "dG" and "dC" are introduced in the stress position as new and interesting information. (Note that inverting the sentence requires stating who made the measurement; since the authors performed the first direct measurement, recognizing their agency in the topic position may well be appropriate.)
2. Since "dG" and "dC" become the old information in the second sentence, keep them up front in the topic position.
3. Since "triisopropylsilyl groups" is new and important information here, create for it a stress position.
4. "Triisopropylsilyl groups" then becomes the old information of the clause in which its effects are described; place it in the topic position of this clause.
5. Alert the reader to expect the arrival of two distinct effects by using the flag word "both." "Both" notifies the reader that two pieces of new information will arrive in a single stress position.

Here is a partial revision based on these decisions:

> We have directly measured the enthalpy of hydrogen bond formation between the nucleoside bases 2'deoxyguanosine (dG) and 2'deoxycytidine (dC). dG and dC were derivatized at the 5' and 3' hydroxyls with triisopropylsilyl groups; these groups serve both to solubilize the nucleosides in non-aqueous solvents and to prevent the ribose hydroxyls from forming hydrogen bonds. From isoperibolic titration measurements, the enthalpy of dC:dG base pair formation is –6.65 ± 0.32 kcal/mol.

The outlines of the experiment are now becoming visible, but there is still a major logical gap. After reading the second sentence, we expect to hear more about the two effects that were important enough to merit placement in its stress position. Our expectations are frustrated, however, when those effects are not mentioned in the next sentence: "From isoperibolic titration measurements, the enthalpy of dC:dG base pair formation is –6.65 ± 0.32 kcal/mol." The authors have neglected to explain the relationship between the derivatization they performed (in the second sentence) and the measurements they made (in the third sentence). Ironically, that is the point they most wished to make here.

At this juncture, particularly astute readers who are chemists might draw upon their specialized knowledge, silently supplying the missing connection. Other readers are left in the dark. Here is one version of what we think the authors meant to say, with two additional sentences supplied from a knowledge of nucleic acid chemistry:

> We have directly measured the enthalpy of hydrogen bond formation between the nucleoside bases 2'deoxyguanosine (dG) and 2'deoxycytidine (dC). dG and dC were derivatized at the 5' and 3' hydroxyls with triisopropylsilyl groups; these groups serve both to solubilize the nucleosides in non-aqueous solvents and to prevent the ribose hydroxyls from forming hydrogen bonds. Consequently, when the derivatized nucleosides are dissolved in non-aqueous solvents, hydrogen bonds form almost exclusively between the bases. Since the interbase hydrogen bonds are the only bonds to form upon mixing, their enthalpy of formation can be determined directly by measuring the enthalpy of mixing. From our isoperibolic titration measurements, the enthalpy of dG:dC base pair formation is –6.65 ± 0.32 kcal/mol.

Each sentence now proceeds logically from its predecessor. We never have to wander too far into a sentence without being told where we are and what former strands of discourse are being continued. And the "measurements" of the last sentence has now become old information, reaching back to the "measured directly" of the preceding sentence. (It also fulfills the promise of the "we have directly measured" with which the paragraph began.) By following our knowledge of reader expectations, we have been able to spot discontinuities, to suggest strategies for bridging gaps, and to rearrange the structure of the prose, thereby increasing the accessibility of the scientific content.

Locating the Action

Our final example adds another major reader expectation to the list.

> Transcription of the 5S RNA genes in the egg extract is TFIIIA-dependent. This is surprising, because the concentration of TFIIIA is the same as in the oocyte nuclear extract. The other transcription factors and RNA polymerase III are presumed to be in excess over available TFIIIA, because tRNA genes are transcribed in the egg extract. The addition of egg extract to the oocyte nuclear extract has two effects on transcription efficiency. First, there is a general inhibition of transcription that can be alleviated in part by supplementation with high concentrations of RNA polymerase III. Second, egg extract destabilizes transcription complexes formed with oocyte but not somatic 5S RNA genes.

The barriers to comprehension in this passage are so many that it may appear difficult to know where to start revising. Fortunately, it does not matter where we start, since attending to any one structural problem eventually leads us to all the others.

We can spot one source of difficulty by looking at the topic positions of the sentences: We cannot tell whose story the passage is. The story's focus (that is, the occupant of the topic position) changes in every sentence. If we search for repeated old information in hope of settling on a good candidate for several of the topic positions, we find all too much of it: egg extract, TFIIIA, oocyte extract, RNA polymerase III, 5S RNA, and transcription. All of these reappear at various points, but none announces itself clearly as our primary focus. It appears that the passage is trying to tell several stories simultaneously, allowing none to dominate.

We are unable to decide among these stories because the author has not told us what to do with all this information. We know who the players are, but we are ignorant of the actions they are presumed to perform. This violates yet another important reader expectation: Readers expect the action of a sentence to be articulated by the verb.

Here is a list of the verbs in the example paragraph:

is
is... is
are presumed to be
are transcribed
has
is... can be alleviated
destabilizes

The list gives us too few clues as to what actions actually take place in the passage. If the actions are not to be found in the verbs, then we as readers have no secondary structural clues for where to locate them. Each of us has to make a personal interpretive guess; the writer no longer controls the reader's interpretive act.

Worse still, in this passage the important actions never

As critical scientific readers, we would like to concentrate our energy on whether the experiments prove the hypotheses.

appear. Based on our best understanding of this material, the verbs that connect these players are "limit" and "inhibit." If we express those actions as verbs and place the most frequently occurring information—"egg extract" and "TFIIIA"—in the topic position whenever possible,* we can generate the following revision:

> In the egg extract, the availability of TFIIIA limits transcription of the 5S RNA genes. This is surprising because the same concentration of TFIIIA does not limit transcription in the oocyte nuclear extract. In the egg extract, transcription is not limited by RNA polymerase or other factors because transcription of tRNA genes indicates that these factors are in excess over available TFIIIA. When added to the nuclear extract, the egg extract affected the efficiency of transcription in two ways. First, it inhibited transcription generally; this inhibition could be alleviated in part by supplementing the mixture with high concentrations of RNA polymerase III. Second, the egg extract destabilized transcription complexes formed by oocyte but not by somatic 5S genes.

As a story about "egg extract," this passage still leaves something to be desired. But at least now we can recognize that the author has not explained the connection between "limit" and "inhibit." This unarticulated connection seems to us to contain both of her hypotheses: First, that the limitation on transcription is caused by an inhibitor of TFIIIA present in the egg extract; and, second, that the action of that inhibitor can be detected by adding the egg extract to the oocyte extract and examining the effects on transcription. As critical scientific readers, we would like to concentrate our energy on whether the experiments prove the hypotheses. We cannot begin to do so if we are left in doubt as to what those hypotheses might be—and if we are using most of our energy to discern the structure of the prose rather than its substance.

Writing and the Scientific Process

We began this article by arguing that complex thoughts expressed in impenetrable prose can be rendered accessible and clear without minimizing any of their complexity. Our

*We have chosen these two pieces of old information as the controlling contexts for the passage. That choice was neither arbitrary nor born of logical necessity; it was simply an act of interpretation. All readers make exactly that kind of choice in the reading of every sentence. The fewer the structural clues to interpretation given by the author, the more variable the resulting interpretations will tend to be.

examples of scientific writing have ranged from the merely cloudy to the virtually opaque; yet all of them could be made significantly more comprehensible by observing the following structural principles:

1. Follow a grammatical subject as soon as possible with its verb.
2. Place in the stress position the "new information" you want the reader to emphasize.
3. Place the person or thing whose "story" a sentence is telling at the beginning of the sentence, in the topic position.
4. Place appropriate "old information" (material already stated in the discourse) in the topic position for linkage backward and contextualization forward.
5. Articulate the action of every clause or sentence in its verb.
6. In general, provide context for your reader before asking that reader to consider anything new.
7. In general, try to ensure that the relative emphases of the substance coincide with the relative expectations for emphasis raised by the structure.

None of these reader-expectation principles should be considered "rules." Slavish adherence to them will succeed no better than has slavish adherence to avoiding split infinitives or to using the active voice instead of the passive. There can be no fixed algorithm for good writing, for two reasons. First, too many reader expectations are functioning at any given moment for structural decisions to remain clear and easily activated. Second, any reader expectation can be violated to good effect. Our best stylists turn out to be our most skillful violators; but in order to carry this off, they must fulfill expectations most of the time, causing the violations to be perceived as exceptional moments, worthy of note.

It may seem obvious that a scientific document is incomplete without the interpretation of the writer; it may not be so obvious that the document cannot "exist" without the interpretation of each reader.

A writer's personal style is the sum of all the structural choices that person tends to make when facing the challenges of creating discourse. Writers who fail to put new information in the stress position of many sentences in one document are likely to repeat that unhelpful structural pattern in all other documents. But for the very reason that writers tend to be consistent in making such choices, they can learn to improve their writing style; they can permanently reverse those habitual structural decisions that mislead or burden readers.

We have argued that the substance of thought and the expression of thought are so inextricably intertwined that changes in either will affect the quality of the other. Note that only the first of our examples (the paragraph about URF's) could be revised on the basis of the methodology to reveal a nearly finished passage. In all the other examples, revision revealed existing conceptual gaps and other problems that had been submerged in the originals by dysfunctional structures. Filling the gaps required the addition of extra material. In revising each of these examples, we arrived at a point where we could proceed no further without either supplying connections between ideas or eliminating some existing material altogether. (Writers who use reader-expectation principles on their own prose will not have to conjecture or infer; they know what the prose is intended to convey.) Having begun by analyzing the structure of the prose, we were led eventually to reinvestigate the substance of the science.

The substance of science comprises more than the discovery and recording of data; it extends crucially to include the act of interpretation. It may seem obvious that a scientific document is incomplete without the interpretation of the writer; it may not be so obvious that the document cannot "exist" without the interpretation of each reader. In other words, writers cannot "merely" record data, even if they try. In any recording or articulation, no matter how haphazard or confused, each word resides in one or more distinct structural locations. The resulting structure, even more than the meanings of individual words, significantly influences the reader during the act of interpretation. The question then becomes whether the structure created by the writer (intentionally or not) helps or hinders the reader in the process of interpreting the scientific writing.

The writing principles we have suggested here make conscious for the writer some of the interpretive clues readers derive from structures. Armed with this awareness, the writer can achieve far greater control (although never complete control) of the reader's interpretive process. As a concomitant function, the principles simultaneously offer the writer a fresh re-entry to the thought process that produced the science. In real and important ways, the structure of the prose becomes the structure of the scientific argument. Improving either one will improve the other.

The methodology described in this article originated in the linguistic work of Joseph M. Williams of the University of Chicago, Gregory G. Colomb of the Georgia Institute of Technology and George D. Gopen. Some of the materials presented here were discussed and developed in faculty writing workshops held at the Duke University Medical School.

Bibliography

Williams, Joseph M. 1988. *Style: Ten Lessons in Clarity and Grace.* Scott, Foresman, & Co.

Colomb, Gregory G., and Joseph M. Williams. 1985. Perceiving structure in professional prose: a multiply determined experience. In *Writing in Non-Academic Settings*, eds. Lee Odell and Dixie Goswami. Guilford Press, pp. 87–128.

Gopen, George D. 1987. Let the buyer in ordinary course of business beware: suggestions for revising the language of the Uniform Commercial Code. *University of Chicago Law Review* 54:1178–1214.

Gopen, George D. 1990. *The Common Sense of Writing: Teaching Writing from the Reader's Perspective.* To be published.

American Scientist Interviews

Barbara Smuts

*The work of Barbara Smuts is rich in surprising associations. An example is her joint appointment at the University of Michigan as associate professor of both anthropology and psychology—an unusual combination, but understandable for someone who studies the behavior of primates, our closest living relatives. In her observations of olive baboons (*Papio cynocephalus anubis*) in eastern Africa, Smuts has been one of a handful of researchers who first recognized the existence of friendships among nonhuman primates. Her efforts in developing objective criteria by which these could be documented and compared have contributed greatly toward the redefining of friendships and other social relationships as objects of serious scientific study.*

Unexpectedly, in an interview with American Scientist, *Smuts mentions that her next field trip will take her to Western Australia—to study the social relationships of dolphins. As she explains the reasoning behind this new project, her excitement is infectious. Because certain forms of social behavior appear peculiar to dolphins and primates alone among all the mammals, they may be linked to another trait found in both groups: a disproportionately large brain. The search is on for the origins of intelligence.*

Smuts credits her early interest in primates to the pioneering work of Jane Goodall, whose sharp and sympathetic observations of chimpanzees in the Gombe Stream Reserve, Tanzania, brought millions of people their first appreciation of nonhuman primates in the wild. As an undergraduate at Harvard University, Smuts was strongly influenced by evolutionary biologist Robert L. Trivers and anthropologist Irven DeVore. As a graduate student she worked with Jane Goodall and David A. Hamburg at Stanford University and began research for her doctoral dissertation on the chimpanzees of Gombe; the reserve became closed to non-Tanzanian researchers in 1975, however, and she shifted her attention to a troop of about 150 olive baboons frequenting the Eburru Cliffs of Kenya, about 160 km northwest of Nairobi. She found that she soon developed "a fascination with baboons and a conviction that their behavior was complex, multifaceted, and worthy of intense scrutiny." Smuts recently talked with American Scientist *about what that scrutiny has revealed so far and where it may lead her next.*

You have been studying an intriguing topic: the evolution of intimate relationships. What sorts of material do you use for this?

I try to pull together material from a lot of different sources: from primate studies in the lab and in the wild and, more recently, from work on dolphins.

I didn't expect to be hearing—

About dolphins, I know. After we've talked about primates for a while, I'll tell you why I think it makes sense for primatologists to broaden their research to include dolphins.

Okay. Let's start with the primates. You and other researchers have recently been looking at how females choose mates—rather than considering mating as determined entirely by competition among males, with the female passively accepting the outcome. Is female choice a new notion in observations of primate behavior?

The importance of female choice has been recognized theoretically since Darwin, but there was very little empirical work on it until the 1960s. I think attention shifted toward the idea of female choice for a variety of reasons. The feminist movement definitely had something to do with it, as did the influx of women scientists into the field; but also the revitalization of evolutionary theory that began with W. D. Hamilton's paper on kin selection in 1963, G. C. Williams's book on natural selection in 1966, and Robert Trivers's article on sexual selection and parental investment in 1972. Trivers emphasized female choice, and that paper has formed the theoretical backbone for everything in the area of mate choice and competition for mates since then. So a number of influences converged.

It also had a lot to do with the fact that by the late 1970s there were several primate studies that had been going on for ten or fifteen years, and people had finally been out there long enough to begin to focus on some of the more subtle aspects of social behavior. Female choice is just one example of that.

I suppose anybody could recognize the spectacle of two males fighting, but to observe female choice you have to know who everybody is, who they were with before, and who they seem to be avoiding.

Precisely. And because of competition among males, females often have to be quite subtle about how they express their preferences.

Also, males in many species, including baboons, do sometimes use aggression as a way to exert leverage on females. So females have to develop counterstrategies that don't depend on brute force, since they're so much smaller. Those include making friends with other males, relying on female relatives as allies, and controlling benefits that males can't take by force—such as grooming (a male can't force a female to groom him) and copulations (if a female isn't interested in mating, all she has to do is sit down).

You have said elsewhere that protection from aggression seems to be one of the main advantages of friendship for the female. Are there other benefits of forming a friendship?

In addition to protecting the female, the male usually forms a very close bond with her infant. So, indirectly, whatever benefits the infant receives the female will receive too. These include protection and a variety of other advantages, such as access to favored food sources, because the adult male will let the little infant feed in a small area very close to him where he won't let any other animals. And on several occasions I've seen a male run to an infant and carry it away when there was an alarm; so it adds another individual in the troop who will go to an infant's aid if need be. It looks as if these relationships between the male and the infant persist for many years, so the infant continues to derive some of these benefits as a juvenile and perhaps even as an adolescent.

Are there costs that the female—or, for that matter, the male—incurs for these friendships?

Yes, I think there are two kinds of costs. The first is the time that's involved in developing and maintaining the relationship—and it does take a lot of time and energy. The second has to do with how one male baboon, in challenging another, will sometimes pick on a female friend in order to provoke an interaction with another male who has been ignoring him. One of the main strategies that male baboons use to reduce the risks of aggression is to ignore a lot of challenges that they receive from other males. On several occasions I have seen a male who had been trying repeatedly to provoke a rival go and attack the rival's female friend. So the females sometimes get caught in the middle.

Are infants ever used in that way?

No, because baboons—especially females—are extremely sensitive to any threat to an infant, particularly if that threat comes from a male who hasn't been in the troop for very long. This makes sense, since we know that infanticide does occur; and it does seem to be perpetuated by males who have moved into the troop recently.

What happens instead is that, in a tense situation with a rival, a male who has a friendship with a female will sometimes invite her infant to climb onto him, which the infant very often willingly does. And once he has the infant clinging to him, the tension tends to abate; the other male's level of aggression decreases. They use the infant as a sort of buffer.

Smuts's book Sex and Friendship in Baboons *was published in 1985 by Aldine. In 1987 she was an editor, with D. L. Cheney, R. M. Seyfarth, R. W. Wrangham, and T. T. Struhsaker, of* Primate Societies, *published by the University of Chicago Press.*

That seems to indicate a lot of trust on the infant's part. What is the basis for this bond—is the male usually the infant's father?

Well, no. What I have argued, based on the data from my study, is that paternity is *not* the primary factor in the forming of these relationships. There were many cases in which a male was very unlikely to be the father of an infant, but if he had a strong friendship with the mother before the infant was born, he was likely to form a friendly bond with the infant. In contrast, we had other cases in which the male probably was the father but did not have a friendship with the mother; and in those cases he was unlikely to form one of these bonds with the infant. Sometimes the male was a likely father as well as a friend, and then of course he generally formed such a bond.

So it seems that it's friendship with the mother, rather than paternity, that is the primary determinant. And that's surprising at first, because primatologists have long thought there was no reason for a male to invest in an infant that wasn't likely to share his genes. But

Smuts takes time out for a midday rest with her subjects near the Eburru Cliffs, Kenya.

what seems increasingly evident, at least in primates, is that there are circumstances under which males invest in infants in order to influence their future relationship with the mother. One could think of it as long-term mating effort, rather than parental investment.

A striking description from your book is that of tiny infant baboons at play around a large adult male who is resting. What actually is going at such moments?

This is a fairly typical scene. The male is resting, and several infants are playing in his vicinity; they run right over him, or use his body as a springboard during their chases, or jump up and down on him. It's very striking, because male baboons are so paradoxical. They're so huge and they have these incredible canines, and they do kill infants under certain circumstances; females and young are very attuned to the males in their troop, because they are potentially dangerous. And yet, within these special relationships, you get a great degree of familiarity and ease.

Did you ever see infants playing near a male that was not a friend of their mother?

Near, but not on. Infants are very clear about which males will be lenient in that way and which ones won't.

And does a session like this go on for just a few minutes?

Usually just a few minutes; and then the infants will get distracted by something else and go running off, just like human children.

When you first were documenting friendship, was it a struggle to get colleagues to take your observations seriously?

Two adult friends, male and female, sleep together on the cliffs.

No, actually it wasn't. I think that's in part because I realized from the beginning that it would not be sufficient for me to leave the field and just assert that these relationships existed. I worked hard to come up with objective measures of friendship that could be replicated, such as the amount of time animals spend together, the frequency of grooming, and so on.

When I first noticed these relationships, before I had analyzed the data, it seemed to me there were two possibilities. One was that, given a female baboon with 18 males in her troop, you would naturally expect her to be closer to some of those males than to others. And it might be that if you looked at her proximity scores or her grooming scores, you'd find a continuous distribution—in which case you could simply define friends as males who were at the high-scoring end of that distribution.

But the other possibility was that friendships really were discrete phenomena. That's one of the main questions I asked of my data, and they were very clear in that regard. The males weren't on a continuum; almost all the males were at one end of, say, proximity or grooming, and then there would be one, two, or three who stood out with a very high score, and no one in between. So that indicated to me that it was a real phenomenon, and that we were recognizing something that was meaningful to the baboons themselves.

It's important to emphasize that several other researchers observing baboon troops had also observed these long-term friendships, and Robert Seyfarth and Jeanne Altmann had also provided quantitative data on them. So that of course helped as well.

It's hard to imagine what friendship could be based on, if not personality. Do baboons have personality as we understand it?

It depends on who you ask. In general, people in this field have shied away from talking about personality and emotions—in print. And yet, when primatologists get together, either in a camp at night or at a conference, it's the personalities that they talk about.

Let me give you an example of how this might be studied. Shirley Strum and other people, including myself, have emphasized that male baboons have a number of fairly subtle social strategies that they use in competition with one another. And I had the sense, when I was observing males, that the ones who were most successful in mating with females tended to be the ones who had a calm demeanor and who were able to maintain their cool in tense situations. But that's a subjective impression; the challenge was to come up with a way to investigate that scientifically.

Traditionally, an approach of two individ-

An adult male holds the hand of a daughter of his female friend.

uals is recorded only if something transpires between them; if one animal just walks past another one, it very often doesn't get written down and it doesn't get analyzed. But what I did, when I was following a male and another male came within a meter of him, was to record exactly what happened between them—whether they looked at each other, whether the male that I was watching stopped feeding briefly, glanced nervously at the other guy, got up and walked away, or just kept on with what he was doing. And it turned out that there was wide individual variation among the males in this measure—and that the males who ignored other males the most, and who were ignored the most, were the ones having the highest mating success.

I've argued that being ignored and ignoring others is one indication of a personality that is relatively at ease. And that in turn benefits a male, because it decreases the frequency with which he gets involved in aggressive interactions that have no real payoff. The best strategy for a male baboon is to be able to keep his cool 99% of the time and only fight when it's really worth it. But most males aren't able to do that.

This was a long digression, but it's an example of how you can begin to get at something that you might call personality. There are all sorts of interesting things that you can begin to do once you open up this Pandora's box. What you have to do is be even more rigorous about your methods of recording and analyzing data than you normally are.

How do the baboons spend most of their time?

They spend most of the day wandering through open grassland foraging for food, but they always take some time out to rest and groom and socialize. And that's when you can really see the special relationships, much more clearly than when they're foraging and everybody's spread out, paying attention to what they're eating.

The time that their social relationships are clearest, though, is on the cliffs at night—who sleeps with whom.

Is that where we get that expression?

I guess so! But it was impossible for me to get those data on a regular basis, because of the way the cliffs are arranged—some were just inaccessible. But I say sometimes, only half jokingly, if I could have had information on who slept with whom every night, I wouldn't have needed any other information to tell you the entire social organization of the baboon troop.

The baboon, Papio cynocephalus, *occurs in four subspecies: the olive baboon* (P. cynocephalus anubis) *inhabits East Africa, as does the yellow baboon* (P. cynocephalus cynocephalus); *the chacma baboon* (P. cynocephalus ursinus) *is found in southern Africa, and the guinea baboon* (P. cynocephalus papio) *in West Africa.*

A separate species, the hamadryas baboon (P. hamadryas), *is found in Ethiopia and Saudi Arabia.*

What would you like to explore next?

I'd like to go back and focus on the fine-grained, day-to-day dynamics of how a relationship is formed.

What you really need to do it right is two observers. Say you picked a pair that you thought were going about forming a relationship; you'd have one person look at the female and another look at the male simultaneously, because the signals that are being exchanged can be extremely subtle, and if you are focusing on one animal you won't necessarily see what the other is doing. The ideal thing would be to have videos going on both of them at the same time, and then go back to the lab. You'd have the videos hooked up on the same clock, so that you could really look at the synchronization.

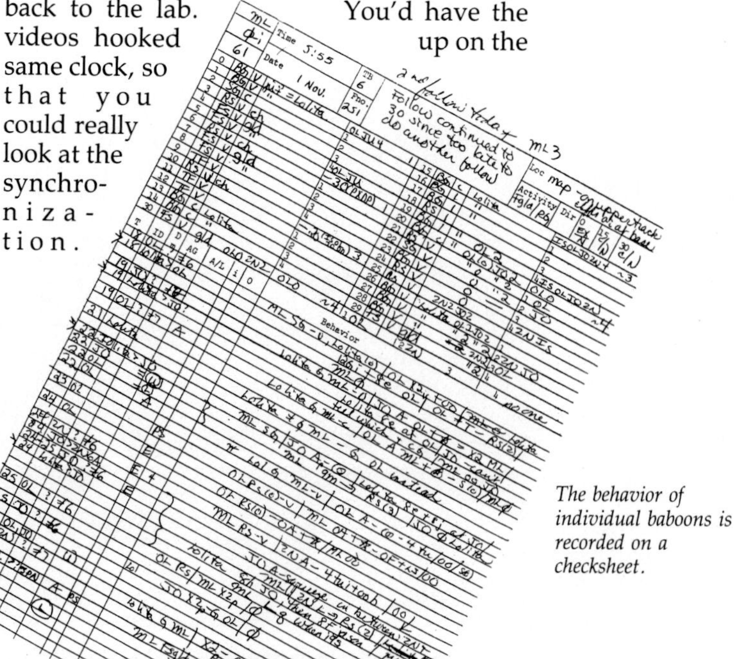

The behavior of individual baboons is recorded on a checksheet.

Male dolphins that have formed a social alliance, or coalition, swim together and rise to the surface in precise synchrony.

The kind of video with the time always running underneath it?

Yes—we're getting into my fantasies now.

And then you could come back and play it on a split screen.

Yes. You could make a wonderful educational film out of it. That would really be the way to convince people that there are meaningful emotional exchanges going on between these animals that are highly reminiscent of what goes on among humans. A picture is worth a thousand words—especially a moving picture.

1972: B.A. (Social anthropology), Harvard University
1981: Lecturer in Anthropology, Harvard University
1982: Ph.D. (Neuro- and biobehavioral sciences), Stanford University
1983: Fellow, Center for Advanced Study in the Behavioral Sciences, Stanford, CA
1984: Assistant Professor of Psychology and Anthropology, University of Michigan
1988: Associate Professor, University of Michigan

In an article in Science, *you mentioned "the intriguing hypothesis that primate intelligence, including our own, originally evolved to solve the challenge of interacting with one another." Are you still thinking along those lines?*

Well, let's say that it's one interesting hypothesis in this area. We don't know enough.

You're not pushing it especially.

No. I'm not pushing it. I am interested in exploring it further—and that's actually where the dolphins come in.

The dolphins have evolved totally independent of nonhuman primates for over 60 million years. They're particularly interesting because, along with the monkeys and apes and humans, dolphins show the largest brains in relation to body size. And work with captive dolphins indicates that they form very close relationships with one another; they're socially complex, flexible, adaptable animals, and that's also true of primates.

Data from two sites—the one in Western Australia that I'm becoming involved in, and one in Florida, run by Randy Wells—indicate some really striking parallels in social organization between bottlenose dolphins and chimpanzees. They both live in fission-fusion societies—that is, individuals travel around in small parties of variable size whose composition is constantly changing. A second parallel is the significance of long-term male coalitions in reproductive competition, now being documented in dolphins by Richard Connor. And a third is that in both chimps and dolphins the females tend to travel in smaller, core areas, whereas the males tend to range more widely. Those three characteristics are found together, as far as we know, *only* in chimpanzees and bottlenose dolphins. I'm not saying there aren't any other animals who show these particular features, but they're not common among mammals.

These parallels in brain evolution and social organization suggest a possible link between the evolution of a large brain and cognitive capacities, and certain aspects of social life. There's not enough known to establish that that is the case, but there's enough, I think, to make comparative studies of apes and the dolphin family particularly interesting.

It sounds as if these two kinds of animals could almost have been drawn up by somebody wanting to test the possible associations between social structure and intelligence. Because in terms of their environment, you can't get much further apart than chimpanzees and dolphins. And in terms of the morphology of the animal itself—

Right. They're so different. That's why it's so interesting, because it may narrow down our search for the essential common elements in the evolution of intelligence. If these similarities in social organization hold up when we know more about dolphins, then the next step will be to determine whether the social similarities reflect similar ecological pressures, by studying the distribution and availability of dolphins' food, which we don't know very much about.

In studying dolphins, are primatologists working with marine biologists?

Well, it seems to be in the air. Just in the last couple of weeks I've heard of two other primatologists who are now studying dolphins. This has not been typical in the past; close observation of individually recognized wild dolphins is really just beginning. And we know from work with primates that once you start to do this, all sorts of surprises are in store. That's when you really begin to understand what's going on—

when you know individuals, when you can keep going back year after year and trace their relationships.

Work with whales and dolphins, methodologically, is where primatology was fifteen or twenty years ago. So in the next five to ten years we're likely to see an explosion of new, unexpected results. And it may happen faster than it did in primatology, because all the lessons that have been learned from studying terrestrial mammals can be applied in the work with whales and dolphins; they won't have to reinvent the wheel.

But they'll have to take the wheel and. . . .

Modify it for the water.

Do you see that as a much smaller problem than developing those methods in the first place? In the past you have often made use of focal observations, for example—following an individual for some length of time and closely observing his or her behavior. How would that be done with a dolphin?

I don't think it's going to be that difficult to modify the methods for the more accessible coastal species. The real challenge is going to be the pelagic species, the ones that are way out there in the ocean and traveling over huge distances. It's going to require the use of some new technology, or some kind of radioing—not a radio collar, but some kind of radio telemetry.

A final question, this one about conservation. Are primates in the wild really endangered? And should we be concerned?

Yes, and yes. It's turning out that nonhuman primates have more than we'd ever imagined to tell us about our own social relationships, the evolution of the capacity for friendship, its functional significance in nature, and much more besides. And I think that alters our relationships with them in two ways. One, it makes them even more important as subjects for scientific research. Two, it makes our affinity to them more dramatic. The more we learn about them, the more it appears that they are clearly sentient beings, with emotional capacities and intimate bonds with one another that are very reminiscent of humans.

So there's a double reason for the preservation of primate species: both their value as objects of scientific study and their intrinsic worth as sentient beings. And nonhuman primates are severely endangered all over the world—because their primary habitat is the tropical forest, which of course is the fastest disappearing ecosystem of all. If present trends continue, there won't be any great apes in the wild in another twenty years. I think that's something every educated person should know. Try to imagine a world in which apes exist only in zoos. And apes don't breed well in zoos, so that means possibly imagining a world in which apes don't exist at all. Your grandchildren might know them only through photographs.

Do you see any promising lines of action?

There are a number of effective conservation organizations that are focusing their efforts on tropical forests right now, and there are two encouraging developments. One is the purchasing of land for reserves; a good example is Costa Rica, where money has been raised to allow the government to buy private land, to add to their national parks. They have a very good record of protecting the parks once they're established, but they're too poor to be able to buy up this land on their own. But that's one way that people with money can help.

Dolphin greets researcher.

The other trend, and the one that's really critical, is that ways must be found to incorporate the local people into the running of the parks. And the parks need to be set up in such a way that the local people benefit from their existence. Otherwise, it's the humans versus the animals; you're stuck with this irresolvable moral dilemma. I think there has been a real shift in the conservation movement in the last few years toward finally incorporating that human element into the planning. That's something that people should look out for if they want to support a conservation organization: whether the organizations are including the human element in their proposals. Otherwise, you don't have a long-term solution.

In a continuing series of interviews with young scientists, managing editor Sandra Ackerman talked with Barbara Smuts for American Scientist.

Sarah Blaffer Hrdy

Infanticide as a Primate Reproductive Strategy

Conflict is basic to all creatures that reproduce sexually, because the genotypes, and hence self-interests, of consorts are necessarily nonidentical. Infanticide among langurs illustrates an extreme form of this conflict

The Hanuman langur, *Presbytis entellus,* is the most versatile member of a far-flung subfamily of African and Asian leaf-eating monkeys known as Colobines. Langurs are traditionally classified as arboreal, but these elegant monkeys are built like greyhounds and can cover distances on the ground with speed and agility. Far more omnivorous than "leaf-eater" implies, Hanuman langurs feed on fully mature leaves, leaf flush, seeds, sap, fruit, insect pupae, and whatever delicacies might be fed them or left unguarded by local people. In forests, langurs spend much of their days in trees, but near open areas the adaptable Hanuman descends to the ground to feed and groom and may spend as much as 80 percent of daytime there. Monkeys are considered sacred by Hindus. This tolerance and their flexibility of diet and locomotion combine to make the Hanuman langur the most widespread primate other than man on the vast subcontinent of India. Ranging from as high as 400 meters in the Himalayas down to sea level, and living in habitats that grade from moist montane forest to semidesert, this flexible Colobine occurs in pockets and in connected swaths from Nepal, down through India, to the island country of Sri Lanka.

Sarah Blaffer Hrdy received her Ph.D. from Harvard in 1975 and was appointed a lecturer in biological anthropology there. Five years of research on langurs are chronicled in her forthcoming book, The Langurs of Abu: Male and Female Strategies of Reproduction *(Harvard Univ. Press). Currently she is doing research on monogamous primates.*
Dr. Hrdy wishes to acknowledge her debt to the community of langur fieldworkers, most especially to P. (Jay) Dolhinow, S. M. Mohnot, and Y. Sugiyama, and to other primatologists, J. Fleagle, D. Fossey, G. Hausfater, S. Kitchener, J. Oates, T. Struhsaker, R. Tilson, and K. Wolf, who allowed her access to unpublished findings. D. Hrdy, J. Seger, and R. Trivers made valuable comments on the manuscript.
Dr. Blaffer Hrdy is also author of The Black-man of Zinacantan *(Univ. of Texas Press, 1972), an analysis of myths of Maya-speaking people. Address: Department of Anthropology, Harvard University, Cambridge, MA 02138.*

The stable core of langur social organization is overlapping generations of close female relatives who spend their entire lives in the same matrilineally inherited 40 hectare plot of land. Troops have an average of 25 individuals, including as many as three or more adult males, but more often only one fully adult male is present. Whereas females remain in the same range and in the company of the same other females throughout their lives, males typically leave their natal troop or are driven out by other males prior to maturity. Loose males join with other males (in some cases brothers or cousins) in a nomadic existence. These all-male bands, containing anywhere from two to 60 or more juvenile and fully adult males, traverse the ranges of a number of female lineages. They will not return again to troop life unless as adults they are successful in invading a bisexual troop and usurping resident males.

With the exception of male invasions, langur troops are closed social units. Troops are spaced out in separate ranges with some areas of overlap between them. When troops meet at the borders of their ranges, both males and females participate in defending their territory. Males are especially active, relying on a wide repertoire of impressive audiovisual displays, such as whooping, canine grinding, and daring leaps that create a swaying turmoil in the treetops. Despite chases and lunges, the apparent aggressiveness of intertroop encounters is largely bravado and almost never results in injuries. Serious fighting among langurs is largely confined to the business of defending troops against invading males; invasions are the only encounters in which males have actually been seen to inflict injuries on one another.

Because of the close association between man and langurs in a part of the world where monkeys are considered sacred, the earliest published accounts of their behavior date back before the time of Darwin and provide us with extraordinary descriptions of langur males battling among themselves for access to females and of females going to great lengths to defend their own destinies. In the 1836 issue of the *Bengal Sporting Magazine,* for example, we are told that in langur society, males compete for females and "the strongest usurps the sole office of perpetuating his species" (Hughes 1884). Another account (see also Hughes 1884) was written by a Victorian naturalist who witnessed invading males attack and kill a resident male followed by a counterattack against the invaders by resident females, who—if we are to believe the account—castrated and mortally wounded one of the invaders:

> In April 1882, when encamped at the village of Singpur . . . my attention was attracted to a restless gathering of Hanumans. . . . Two opposing troops [were] engaged in demonstrations of an unfriendly character. Two males of one troop . . . and one of another—a splendid looking fellow of stalwart proportions—were walking round and displaying their teeth. . . . It was some time—at least a quarter of an hour—before actual hostilities took place, when, having got within striking distance, the two monkeys made a rush at their adversary. I saw their arms

Figure 1. Aggressive behavior in encounters between troops of Hanuman langurs is largely bluff. Here, a spread-eagled, open-mouthed male hurtles dangerously through space. In fact, he landed nowhere near his opponent, the familiar resident male from the neighboring troop. (Photos by D. B. Hrdy and author.)

and teeth going viciously, and then the throat of one of the aggressors was ripped right open and he lay dying.

He had done some damage however before going under having wounded his opponent in the shoulder.... I fancy the tide of victory would have been in [this male's favor] had the odds against him not been reinforced by the advance of two females. ... Each flung herself upon him, and though he fought his enemies gallantly, one of the females succeeded in seizing him in the most sacred portion of his person, and depriving him of his most essential appendages. This stayed all power of defense, and the poor fellow hurried to the shelter of a tree where leaning against the trunk, he moaned occasionally, hung his head, and gave every sign that his course was nearly run. ... Before the morning he was dead.

Social pathology hypothesis

Despite the vivid accounts of langur aggression set down by early naturalists, one of the first steps of modern primatology was to put aside these anecdotes so that the fledgling science of primatology could be laid on a purely factual foundation. By the late 1950s the modern era of primate studies, launched primarily by social scientists, had begun. The early workers were profoundly influenced by current social theory and in particular by the work of Radcliffe-Brown, who believed that any healthy society had to be a "fundamentally integrated social structure" and that in such a society every class of individuals would have a role to play in the life of the group in order to ensure its survival.

In 1959, Phyllis Jay went out from the University of Chicago to the Indian forest of Orcha (Fig. 2) and to Kaukori, a village on the heavily cultivated Gangetic plain. Jay found among North Indian langurs a remarkably peaceful society. She reported that relations among adult male langurs were relaxed, dominance relatively unimportant, and aggressive threats and fighting exceedingly uncommon (1963 diss., 1965). All troop members, she wrote, were functioning so as to maintain the fabric of the social structure. Because of the overriding conviction that primates behave as they do for the good of their group, the early naturalists' descriptions were dismissed as "anecdotal, often bizarre, certainly not typical behavior" (Jay 1963 diss.).

Nevertheless, a second study turned up findings that forced reconsideration of the question of langur aggressiveness. In 1963, a team of Japanese primatologists led by Yukimaru Sugiyama were tracking langurs through the teak forests near Dharwar, South India, when they witnessed a band of seven langur males drive out the leader of a bisexual troop, after which one male from among the invaders usurped control and remained in sole possession of the troop. Within days of this takeover, all six infants in the troop were bitten to death by the new male. Curiously, and contrary to all previous reports concerning the solicitude of langur mothers (who have been known to carry the corpse of a dead infant for days), mothers whose infants were wounded by the usurping male abandoned them (Sugiyama 1967).

It was difficult to explain such behavior in terms of group survival and of a "fundamentally integrated" social structure. To circumvent this problem, it was suggested that there was something abnormal about the langurs of Dharwar and that their extreme aggressiveness was somehow pathological. In fact, if Jay's Kaukori study—the only other one available at that time—was taken as the norm, there *was* something unusual about Dharwar: langurs there were living in an area of rapid deforestation and of environmental disruption. Population densities (84–133 langurs per km^2) were some 30 times higher than the very low density recorded at Kaukori (3 per km^2).

Almost concurrently, John Calhoun (1962), at the National Institutes of Health, was studying the effects of crowding on the behavior of rats. He demonstrated that when the animals were crowded, normal rat social conventions broke down. The rats sank into a "pathological" state characterized by excessively high infant mortality due to inadequate maternal care, infanticide, and cannibalism. Comparisons between Calhoun's rats and the langurs of Dharwar were inevitable. A number of explanations were offered as to why langur infants were killed, and the social pathology hypothesis figured prominently among them. It was suggested that infanticide was a product of crowding (Sugiyama 1967; Eisenberg et al. 1972) and as such a mechanism for population control (Rudran 1973; Kummer et al. 1974). Alternatively, it was suggested that the behavior had no adaptive value (Bygott 1972) or that it was "dysgenic" (Warren 1967). Functional explanations for infanticide included the idea that males were somehow displacing aggression built up by the "simultaneous sexual excitement and enragement" of the new leader (Mohnot 1971) or that the male attacked infants in order to strengthen his "social bonds" with females in his new troop (Sugiyama 1965). All of these explanations derived from the basic assumption that under normal conditions animals act so as to maintain, not disrupt, the prevailing social structure.

Only one of the early explanations focused on the possible advantages of infanticide for the animal actually responsible for the act—the male. In 1967, Sugiyama suggested that the male attacked infants to avoid the two- to three-year delay in female sexual receptivity while she continued to nurse her offspring. This argument has been expanded into the more general sexual selection hypothesis that will be offered here.

It was to find out whether crowding really was at issue in infant-killing and desertion that in 1971 I first went to India. By the time I arrived, there was a new report of infanticide, this time from the desert region near Jodhpur, far to the north of Dharwar at a location where the Indian primatologist S. M. Mohnot had been studying langurs for several years. Already it seemed possible that infanticide was a more widespread and normal behavior than the social pathology hypothesis suggested.

From Jodhpur, I traveled southwestward to Mt. Abu. For 1,503 hours during five annual two- to three-month study periods between 1971 and 1975 I monitored political changes in five troops of langurs in and around the town. In the following section I will summarize the evidence—based largely on work at Abu, but drawing also on the detailed observations of Y. Sugiyama and S. M.

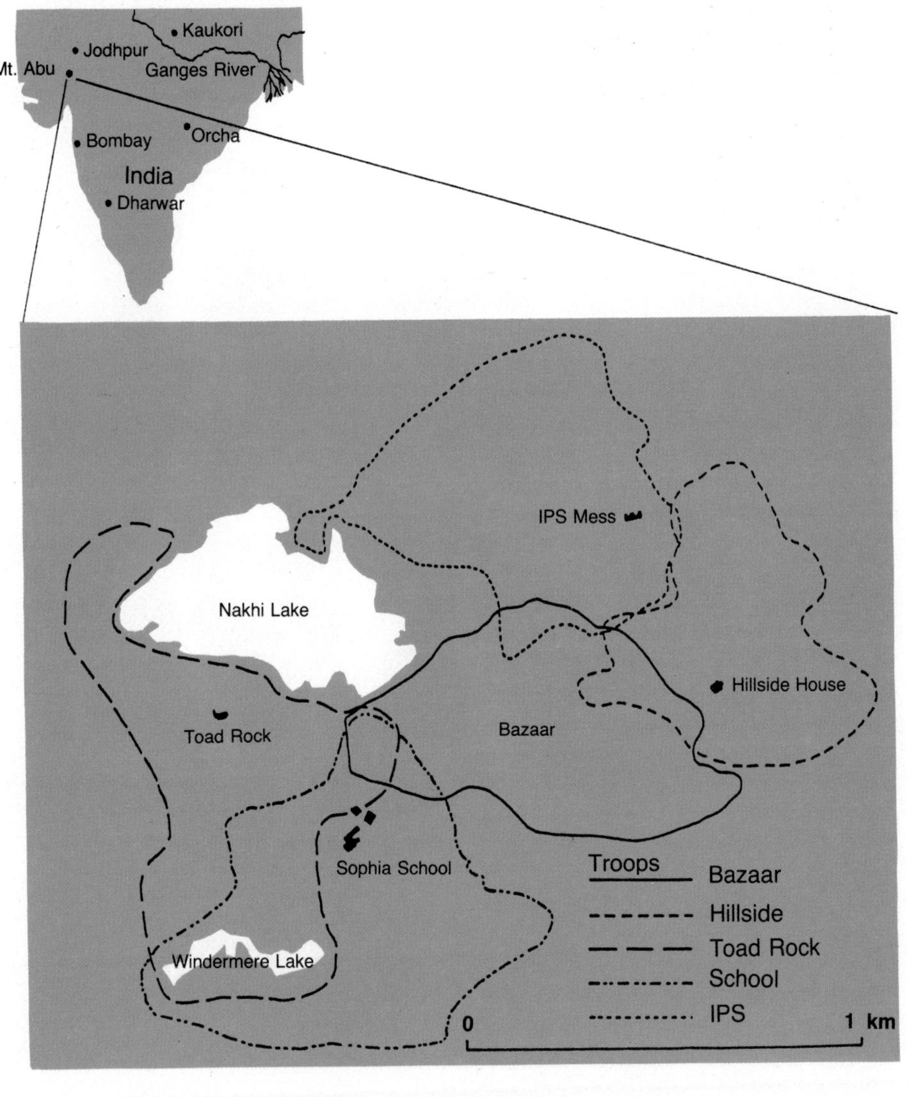

Figure 2. Over a 5-year period, the author studied five troops of Hanuman langurs in and around Mt. Abu, one of several sites on the Indian subcontinent where these widespread monkeys have been investigated.

Mohnot—that led me to reject my initial crowding hypothesis in favor of the theory that infanticide is adaptive behavior, extremely advantageous for the males who succeed at it.

The langurs of Abu

The forested hillsides of Mt. Abu rise steeply from the parched Rajasthani plains. The town itself is an Indian pilgrimage and tourist center 1,300 m above sea level. My study concentrated on five troops in the vicinity of the town, but I will focus here on just two of these: the small Hillside troop and its neighbor, the Bazaar troop, whose name derives from the fact that these langurs spent a portion of almost every day scavenging in the bazaar (see Fig. 2).

In June 1971 the Hillside troop contained one adult male, seven adult females, six infants, and one juvenile male. In August of that year, Mug was replaced by a new male, Shifty Leftless—named for a bite-sized chunk missing from his left ear. At the time of the takeover, one adult female and all six infants disappeared from the troop. Soon after, mothers who had lost infants came into estrus and solicited the new male. Local inhabitants witnessed the killing of two infants by an adult male. Each killing took place at a site well within the range of the Hillside troop; in fact, one occurred at a location used exclusively by that group. It seemed highly probable that the missing infants had been killed, and that the usurping male Shifty was the culprit. (These events are discussed in greater detail in Blaffer Hrdy 1974 and 1975 diss.)

On my return to Abu in June 1972, I was surprised to find that the same male, Shifty, had now transferred to the neighboring Bazaar troop. In 1971, Bazaar troop had contained three adult males, ten subadult and adult females, five infants, and four juveniles. Three of these infants were now missing. The killing of one had been observed by a local amateur ornithologist who lived beside the bazaar. The three Bazaar troop males remained in the vicinity of their former troop; the second-ranking of these bore a deep wound in his right shoulder.

During 1972, Mug took advantage of Shifty's absence to return to his former troop. At this time Hillside troop consisted of the same six adult females and their four new infants. Two females, an older, one-armed female called Pawless and a very old female named Sol, had no infants. Although Mug was able to return to his troop for extended visits, whenever Shifty left Bazaar troop on reconnaissance to Hillside troop, Mug fled. On at least eight occasions, Mug left the troop abruptly just as the more dominant Shifty arrived, or else the "interloper" was actually chased by Shifty. Typically, Shifty's visits to Hillside troop were brief, but if one of the Hillside females was in estrus he might remain for as long as eight hours before returning to Bazaar troop.

During the periods Mug was able to spend with his former harem he made repeated attacks on infants that had been born since his loss of control. On at least nine occasions in 1972, Mug actually assaulted the infants he was stalking. Each time one or both childless females intervened to thwart his attack. Despite their heroic intervention, on three occasions the infant was wounded. During this same period, other animals in the troop were never wounded by the male. When the same male, Mug, had been present in the troop in 1971, he had not attacked infants. Similarly, during Shifty's visits to the Hillside troop in 1971, his demeanor toward infants was aloof but never hostile. Whereas Hillside mothers were very

Figure 3. This Hillside infant was conceived in 1971, during the time that Shifty was the troop's resident male, and was later killed by an adult male langur, probably Mug. The age of langur infants can be determined with some precision: between the third and fifth months of life, the all-black natal coat changes to cream color, starting with the top of the head and a little white goatee.

Figure 4. Juveniles and subadults threaten and lunge at two females from another troop. The adult male looks on calmly but does not participate.

restrictive with their infants when Mug was present, gathering them up and moving away whenever he approached, these same mothers were quite casual around Shifty. Infants could be seen clambering about and playing within inches of Shifty without their mothers' taking notice.

In 1973, Mug was joined by a band of five males. Nevertheless, the double usurper Shifty could still chase out all six males whenever he visited the Hillside troop. A daughter born to Pawless during the period when both Shifty and Mug were vying for control of Hillside troop was assaulted on several occasions by the five newer invaders; the infant eventually disappeared and was presumed dead.

By 1974, Mug was once again in sole possession of the Hillside troop and holding his own against Shifty. When the Hillside and Bazaar troops met, Mug remained with his harem. On several occasions, the newly staunch Mug confronted Shifty and in one instance grappled with him briefly before retreating behind females in the Hillside troop. Mug resolutely chased away members of a male band who attempted to enter his troop. By 1975, Mug's star had risen.

When I returned to Abu in March of that year, Shifty was no longer with the Bazaar troop. In his place was Mug. It was not known what had become of the extraordinary old male with the bite out of his left ear. Perhaps he died or moved on to another troop, or perhaps he was at last usurped by his longtime antagonist Mug.

Mug's former position in the Hillside troop was filled by a young adult male called Righty Ear. Righty (with a missing half-moon out of his right ear) was one of the five males who had joined Mug in the Hillside troop two years previously. Since that time, Righty had passed in and out of the troop's range, traveling with other males but not (so far as I knew) attempting to enter the troop. Righty's "waiting game" apparently paid off that March, when he came into sole possession of the Hillside troop. But, as in the case of his predecessors, Hillside troop was only a stepping stone: in April 1975, Righty replaced Mug as the leader of Bazaar troop.

The first indication I had of Righty's arrival in Bazaar troop was a report from local inhabitants that an adult male langur had killed an infant. On the following day when I investigated this report, the young adult male with the unmistakable half-moon out of his right ear was present in Bazaar troop; Mug was nowhere to be found. An elderly langur mother still carried about the mauled corpse of her infant; by the following day, she had abandoned it. Righty subsequently made more than 50 different assaults on mothers carrying infants. Nevertheless, only one other infant disappeared. Five infants in the Bazaar troop remained unharmed when my observations terminated on June 20.

After Righty switched from Hillside to Bazaar troop, there followed some nine or more weeks during which the Hillside females had no resident male except for brief visits from Righty. Whenever the two troops met at their common border, Hillside females sought out Righty Ear and lingered beside him. These females were fiercely rebuffed by resident females in the Bazaar troop. Hostility of Bazaar troop females toward "trespassers" from Righty's previous harem prevented a merger of the two. The troops were still separate when Harvard biologist James Malcolm visited Abu in October 1975, but the vacuum in Hillside troop had been filled by a new male, christened Slash-neck for the deep gash in his neck.

The evolution of infanticide

Over a period of five years, then, political histories of the Hillside and Bazaar troops were linked by a succession of shared usurpers. First Shifty, then Mug, and finally Righty switched from the small and apparently rather vulnerable Hillside troop to the larger Bazaar troop (Fig. 5). Possibly the shifts were motivated by the greater number of reproductively active females in Bazaar troop. Between 1971 and 1975, at least four different males usurped control of Hillside troop. Infant mortality in this troop between 1971 and 1974 reached 83 percent, and extinction of the troop loomed as a real possibility. In contrast, during the same period, another troop at Abu, the School troop, was exceedingly stable, retaining the same male throughout.

Figure 5. The vicissitudes of male tenure in two troops of langurs are charted during the months the author spent observing the troops at intervals during 1971–75. Observations of infants missing, killed, or assaulted coincided with tenure shifts (as shown in italics in the chart).

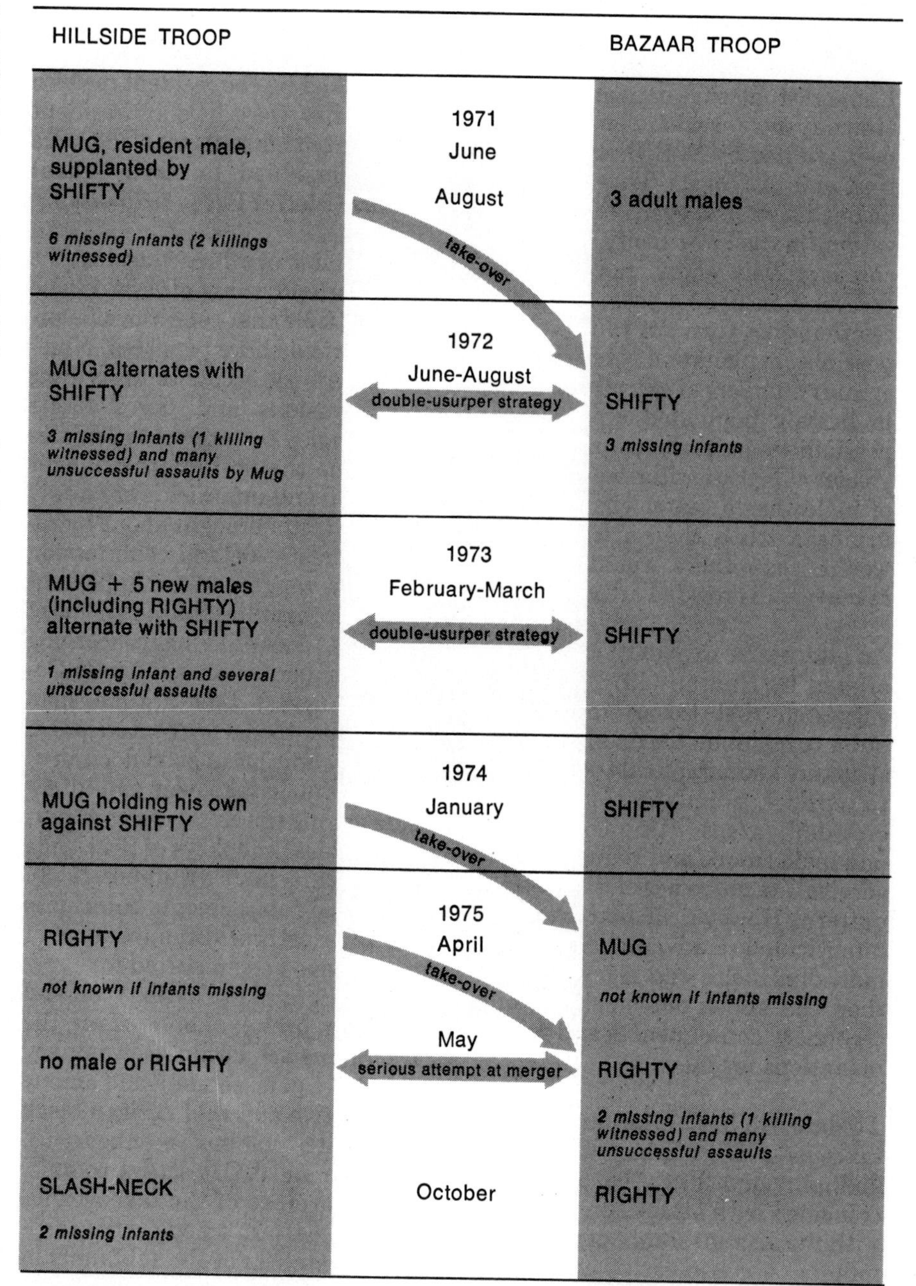

Combining all troop studies, the average male tenure at Abu was 27.5 months, a figure astonishingly close to the average tenure of 27 months calculated by Sugiyama for Dharwar (1967).

The short average duration of male tenure among langurs underlies the most crucial counterargument against the social pathology hypothesis: the extent to which adult males appear to gain from killing infants. Given that the tenure of a usurper is likely to be short, he would benefit from telescoping as much of his females' reproductive career as possible into the brief period during which he has access to them. By eliminating infants sired by a competitor, the usurping male hastens the mothers' return to sexual receptivity; on average, a mother whose infant is killed will become sexually receptive again within eight days of the death. In other words, infanticide permits an incoming male to use his short reign more efficiently than if he allowed unweaned infants present in the troop at his entrance to survive, to continue to suckle, and thus to delay the mother's next conception.

In three troops at Dharwar and Abu for which we have reliable information on subsequent births, 70 percent of the females who lost infants gave birth again within 6 to 8 months of the death of their infants, on average—just over one langur gestation period later. In the harsh desert environment of Jodhpur, however, the postinfanticide birth interval was much longer, up to 27 months.

Once infant-killing began, a usurper would be penalized for *not* committing infanticide. If a male failed to kill infants upon taking over a troop, and instead waited for those infants already in the troop to be weaned before he inseminated their mothers, then his infants would still be unweaned and hence vulnerable when the next usurper (presumably an infanticidal male) entered.

Other variations in the social system might likewise be expected to select for changes in male behavior. For example, if the rate of takeovers were speeded up and then held constant over time, male tolerance toward weaned immatures might be drastically altered. With a faster rate of takeover, it would be unlikely that one male could remain in control of a troop long enough for immature females to reach menarche and to give birth to an infant that would in turn grow old enough to survive the next takeover. Immature females, then, would be worth no more to the usurper than young males would be, and they might compete with the productive females of his harem for resources. Under these circumstances, it would behoove a usurper to drive out immatures of both sexes. This is precisely what occurs among a related langur species, *Presbytis senex*, living at very high densities (as high as 215 animals per km^2) at Horton Plains in Sri Lanka (Rudran 1973). The ousted females travel with former male troopmates in mixed-sex bands.

Up to this point, I have not dealt with the apparent correlation between male takeovers and high population density. At both Dharwar (84–133 langurs per km^2) and Abu (50 per km^2), population densities are relatively high. In the desert region near Jodhpur, langurs have vast open areas available to them but tend to cluster about waterholes and garden spots. Infanticide has been reported at all three locations, but it has been recorded for none of the areas with low densities (at Jay's Orcha and Kaukori study sites or at any of three Himalayan sites where langurs have been studied by N. Bishop, the Curtins, and C. Vogel). This finding is inconclusive, however, since observations in the low-density areas were comparatively short, ranging from several months to a year. If the correlation does turn out to be valid, a possible explanation may be the greater numbers of extratroop males in heavily populated areas. If the possibilities for male recruitment are greater at high densities, and if a band of males has a better chance of usurping a troop than a single male would, then there would be more takeovers in crowded areas.

An alternative explanation has been offered by Rudran (1973), who has suggested that takeovers occur in order to maintain the one-male troop structure and infanticide occurs so as to curtail population growth in crowded areas. Unquestionably, one-male troops and reduced infant survival are outcomes of the takeover pattern. However, if takeovers and infanticide are advantageous to the individual males who engage in them, then the above outcomes are only secondary consequences and not explanations for them.

To date, we have information on 15 takeovers, 5 at Dharwar, one at Jodhpur, and 9 at Abu. At least 9 coincided with attacks on infants or with the disappearance of unweaned infants. A conservative estimate of the number of infants who have disappeared at the time of takeovers is 39. The important point (and this is the second line of evidence against the social pathology hypothesis) is that attacks on infants have been observed *only* when males enter the breeding system from outside—even if, as in the case of Mug, they have been only temporarily outside it. Such males are unlikely to be the progenitors of their victims. In contrast to what is considered "pathological" behavior, attacks on infants were highly goal-directed. An important area of future research will be learning exactly what means a langur male has at his disposal for discriminating infants probably his own from those probably sired by some other male. Quite possibly, males are evaluating past consort relations with the mother (Blaffer Hrdy 1976). Interestingly, infants kidnapped by females from neighboring troops were not attacked by the resident male so long as they were held by resident females from his own troop and were not accompanied by their (alien) mothers (Blaffer Hrdy, 1975 diss.).

The third line of evidence against the social pathology hypothesis is the length of time that conditions favoring infanticide have persisted. Nineteenth-century accounts describing male invasions and fierce fights among males for access to females undermine the position that langur aggression and infanticide are newly acquired traits brought about by recent deforestation and compression of langur ranges. More important (and this constitutes the fourth line of evidence), recent findings concerning other members of the subfamily Colobinae suggest that a time span much longer than a few centuries is at issue. In addition to good documentation for male takeovers and infanticide among the closely related purple-faced leaf-monkeys of Sri Lanka (*Presbytis senex*) (Rudran 1973), adult male replacements coinciding with the death or disappearance of infants have been reported for *Presbytis cristata* of Malaysia (Wolf and Fleagle, in press); *P. potenziani,* the rare Mentawei Island leaf-monkey (R. Tilson, pers. comm.); and among both captive and wild African black and white colobus monkeys (S. Kitchener and J. Oates, pers. comm.). This recurrence of the takeover/infanticide pattern among widely separated members of the subfamily in Africa, India, and Southeast Asia argues strongly for its antiquity. Though the possibility of environmental convergence cannot be ruled out, the case of phylogenetic inheritance of these traits among geographically disparate relatives is a compelling one. Far from being recent responses to crowded conditions, it appears that a predisposition to male takeovers and infanticide has been part of the colobine repertoire since Pliocene times, some ten million or more years ago, when the split between the African and Asian forms occurred.

Beyond the Colobines

But the tale of infanticide does not stop with the Colobines. In what may be the most startling finding by primatologists in recent years, we are discovering that the gentle souls we claim as our near relatives in the animal world are by and large an extraordinarily murderous lot. It is apparent now that the events witnessed at Abu and Dharwar are not aberrations. Increased observation of primates had led to an increase in the number of species in which adult males are known to attack and kill infants—and, occasionally, each other. Although murder is uncommon, cases of adults fighting to the death have been reported for rhesus, pig-tailed, and Japanese macaques, baboons, and chimpanzees, as well as Hanuman langurs.

At the time of this writing, infanticide, either observed or inferred from the disappearance of infants at times when males have usurped new females, has been reported for more than a dozen species of primates. Every major group of primates, including the prosimians, the New and Old World monkeys, apes, and man, is represented.

Not all these reports parallel the pattern of events recorded for Hanuman langurs, but many are disturbingly similar: males attack infants when they come into possession of females who are accompanied by offspring sired by another male. Typically, these are unfamiliar females. Perhaps the clearest illustration of the potential importance of previous acquaintance is provided by an experiment with caged crab-eating macaques (Thompson 1967). Here, infanticide was the unexpected outcome in a cage study on the effects of familiarity or lack of it in relations between male and female *Macaca fascicularis.* When paired with his accustomed companion and her infant, the adult male displayed typical behavior, mounting the female briefly and then casually exploring his surroundings. He entirely ignored the infant. Paired with an unfamiliar mother-infant pair, the male responded quite differently. After a brief attempt at mounting, the male attacked the infant as it lay clutched to its mother's belly. When the mother tried to escape, the male pinned her to the ground and gnawed the infant, making three different punctures in its brain with his canines.

Two suspected cases of infanticide

Figure 6. A female langur holding a newborn infant takes food from a priest of Shiva who lives in one of the sacred caves in the hillsides surrounding Mt. Abu.

among wild hamadryas baboons were occasioned by human manipulation. In the course of capture-and-release experimentation on the process of harem formation among *Papio hamadryas* of Ethiopia, two mothers were switched to new one-male units. In one case the infant was missing a day later; in the other, the infant was seen dead, "its skull pierced and its thighs lacerated by large canine teeth." The witnessed killing of two hamadryas infants at the Zurich Zoo just after their mothers changed "owners" adds plausibility to the inference that the wild infants were similarly murdered (Kummer et al. 1974).

Less contrived perhaps is the following account of chimpanzees from the Gombe Stream Reserve in Tanzania, where a young British researcher, David Bygott, happened to be following a band of five male chimpanzees when they encountered a strange female whom "in hundreds of hours of field observation," Bygott had never seen before. This female and her infant were immediately and intensely attacked by the males. For a few moments, the screaming mass of chimps disappeared from Bygott's view. When he relocated them, the strange female had disappeared and one of the males held a struggling infant. "Its nose was bleeding as though from a blow, and [the male], holding the infant's legs, intermittently beat its head against a branch. After 3 minutes, he began to eat the flesh from the thighs of the infant which stopped struggling and calling" (Bygott 1972). In contrast with normal chimp predation, this cannibalized corpse was nibbled by several males but never consumed.

Dian Fossey's remarkable decade-long study of wild mountain gorillas in central Africa provides what may be the most dramatic instances of adult male invaders mauling infants. For several days, a lone "silverback" (or fully mature) male had been following a harem of gorillas, presumably in quest of females. At last, he made his move, penetrating the group with a "violent charging run." A primiparous female who had given birth to an infant on the previous night countered his charge by running at him. Halting within arm's reach of the male, she stood bipedally to beat her chest. The male struck her ventrally exposed body region where her newly born infant was clinging. Immediately following this blow, a "thin wail" was heard from the dying infant. On two other occasions, Fossey witnessed silverbacks kill infants belonging to primiparous mothers. In the best documented of these cases, the mother subsequently copulated with the male who had killed her infant (Fossey 1974; pers. comm.). To date, of the killings witnessed, only first-born gorilla infants have been seen to be victims. This could be owing to maternal inexperience, or, as I believe is more probable, to Fossey's finding that in gorilla society only *young* females routinely change social units. Since an older mother would in all likelihood not join a usurper anyway, he would rarely benefit from killing her infant.

Isolated instances of infanticide by adult males have also been reported for various prosimians: among free-ranging Barbary (*Macaca sylvana*) and rhesus macaques (*M. mulatta*) (Burton 1972; Carpenter 1942) and among wild *Cercopithecus ascanius*, the red-tailed monkeys of Africa (T. Struhsaker, in press). Infanticide is suspected among wild chacma baboons (*Papio ursinus*) (Saayman 1971); wild howler monkeys (*Alouatta*) of South America (Collias and Southwick 1952); and among caged squirrel monkeys (*Saimiri*) (Bowden et al. 1967).

The explanation for infanticide need not be the same in every case, but the parallels with the well-documented langur pattern are striking. According to the explanatory hypothesis offered here for langurs, infant-killing is a reproductive strategy whereby the usurping male increases his own reproductive success at the expense of the former leader (presumably the father of the infant killed), the mother, and the infant. If this model applies, the primatewide phenomenon of infanticide might be viewed as yet another outcome of the process Darwin termed *sexual selection:* any struggle between individuals of one sex (typically males) for reproductive access to the other sex, in which the result is not death to the unsuccessful competitor, but few or no offspring (Trivers 1972). Crucial to the evolution of infanticide are, first, a nonseasonal and flexible female reproductive physiology such that it is both feasible and advantageous for a mother to ovulate again soon after the death of her infant and, second, competition between males such that tenure of access to females is on average short.

Female counterstrategies

Confronted with a population of males competing among themselves, often with adverse consequences for females and their offspring, one would expect natural selection to favor those females most inclined and best able to protect their interests. When an alien langur male invades a troop, he may be chased away and harassed by resident females as well as by the resident male. After a new male takes over, females may form temporary alliances to prevent him from killing their infants (e.g. Sol and Pawless's combined front against the infanticidal Mug).

Females are often able to delay infanticide. Less often are they able to prevent it. Pitted against a male who has the option to try again and again until he finally succeeds, females have poor odds. For this reason, one of the best counterinfanticide tactics may be a peculiar form of female deceit. Almost invariably, langur males have attacked infants sired by some other male; a male who attacked his own offspring would rapidly be selected against. It may be significant then that at Dharwar, Jodhpur, and Abu, pregnant females confronted with a usurper displayed the traditional langur estrous signals: the female presents her rump to the male and frenetically shudders her head. These females mated with the usurper even though they could not possibly have been ovulating at the time. Postconception estrus in this context may serve to confuse the issue of paternity.

After birth, an infant's survival is best ensured if its mother is able to associate with the father, or at least with a male who "considers" himself the father or who acts like one—in short, a male who tolerates her infant. In at least three instances at Abu, females with unweaned infants left recently usurped troops to spend time in the vicinity of males that on the basis of other evidence I suspected of having fathered their offspring.

If all else fails and her infant is attacked and wounded, a mother may continue to care for it, or abandon it. In several cases at Dharwar and Jodhpur, mothers abandoned their murdered infants soon after or even before death (Sugiyama 1967; Mohnot 1971). Rudran has suggested that the mother abandons her infant for fear of injury to herself and "because an adult female is presumably more valuable than an infant to the troop" (1973). It is far more likely, however, that desertion reflects a practical evaluation of what *this* infant's chances are weighed against the probability that her next infant will survive.

Under some circumstances a mother may opt to abandon an unwounded infant. In a single case from Abu, a female in a recently usurped troop who had been traveling apart from the troop (presumably to avoid the new male's assaults) left her partially weaned infant in the company of another mother and returned to the main body of the troop alone. If this was in fact an attempt to save her infant by deserting it, the ploy failed when the babysitter herself returned to the troop, some time later, bringing both infants with her. Nevertheless, both infants did survive the takeover.

Despite the various tactics that a female may employ to counter males, infanticide was the single greatest source of infant mortality at Abu. The plight of these females raises a perplexing question: How has this situation come about? Langur males contribute little to the rearing of offspring; apart from insemination, females have little use for males *except* to protect them from other langur males who might otherwise invade the troop and kill infants. Why then should females tolerate males at all, suffering subjection to the tyranny of warring polygynists? On the vast time scale of evolution, alternatives have been open to the female since the dawn of Colobines. Large body size, muscle mass, and saber-sharp canines might just as well have been selected among females as among males. Why should females weigh only 12 kg, on average, and not the 18 kg that males routinely do? Alternatively, female relatives could ally themselves to a much greater extent than they do. The combined 36 kg of three females operating as a united front against an infanticidal male surely should prevail. Infanticide depends for its evolutionary feasibility on the prior female adaptation of conceiving again as soon as possible after the death of an infant. If females failed to ovulate after a male killed their infants, or if they "refused" to copulate with an infanticide, the trait would be eliminated from the population.

The facts that females do not grow so large as males, that they do not selflessly ally themselves to one another, and that they do not boycott infanticides, suggest that counterselection is at work. Once again, the pitfall is intrasexual competition—this time competition among females themselves for representation in the next generation's gene pool. Whereas head-on competition between males for access to females selects for males who are as big and as strong (or stronger) than their opponents, a female who "opted" for large size in order to fight off males might not be so well-adapted for her dual role of ecological survivor and childbearer. An over-sized female might produce fewer offspring than her smaller cousin. In time, the smaller cousin's progeny would prevail.

Intrasexual competition is mitigated by the close genetic relatedness between female troop members, but it is by no means eliminated. A female in her reproductive prime who altruistically defended her kin, in spite of the cost to herself, might be less fit than her cousin who sat on the sidelines. Finally, if infanticide really is advantageous behavior for males, a female who sexually boycotted infanticides would do so to the detriment of her male progeny. Her sons would suffer in competition with the offspring of nondiscriminating mothers.

For generations langur females have possessed the means to control their own destinies. Caught in an evolutionary trap, they have never been able to use them.

References

Blaffer Hrdy, S. 1974. Male-male competition and infanticide among the langurs (*Presbytis entellus*) of Abu, Rajasthan. *Folia. Primat.* 22:19–58.

———. Male and female strategies of repoduction among the langurs of Abu. 1975 diss., Harvard University.

———. 1976. The care and exploitation of nonhuman primate infants by conspecifics other than the mother. In *Advances in the Study of Behavior 6,* ed. J. Rosenblatt, R. Hinde, C. Beer, and E. Shaw. Academic Press.

———. In press. *The Langurs of Abu.* Harvard Univ. Press.

Bowden, D., P. Winter, and D. Ploog. 1967. Pregnancy and delivery behavior of the squirrel monkey (*Saimiri sciureus*) and other primates. *Folia Primat.* 5:1–42.

Burton, F. 1972. The integration of biology and behavior in the socialization of *Macaca sylvana* of Gibraltar. In *Primate Socialization,* ed. F. Poirier. Random House.

Bygott, D. 1972. Cannibalism among wild chimpanzees. *Nature* 238:410–11.

Calhoun, J. 1962. Population density and social pathology. *Sci. Am.* 206:139–48.

Carpenter, C. R. 1942. Societies of monkeys and apes. *Biol. Symposia* 8:177–204.

Collias, N., and C. H. Southwick. 1952. A field study of population density and social organization in howling monkeys. *Proc. of the Amer. Phil. Soc.* 96:143–56.

Eisenberg, J. F., N. A. Muckenhirn, and R. Rudran. 1972. The relation between ecology and social structure in primates. *Science* 176:863–74.

Fossey, D. 1974. Development of the mountain gorilla (*Gorilla gorilla beringei*) through the first thirty-six months. Paper presented at Berg Wartenstein symposium no. 62, The Behavior of the Great Apes, Wenner-Gren Foundation for Anthropological Research.

Hughes, T. H. 1884. An incident in the habits of *Semnopithecus entellus,* the common Indian Hanuman monkey. *Proc. Asiatic Soc. of Bengal,* pp. 147–50.

Jay, P. The social behavior of the langur monkey. 1963 diss., University of Chicago.

———. 1965. The common langur of North India. In *Primate Behavior,* ed. I. DeVore. Holt, Rinehart and Winston.

Kummer, H., W. Gotz, and W. Angst. 1974. Triadic differentiation: An inhibitory process protecting pairbonds in baboons. *Behaviour* 49:62.

Mohnot, S. M. 1971. Some aspects of social change and infant-killing in the Hanuman langur, *Presbytis entellus* (Primates: Cercopithecinae) in Western India. *Mammalia* 35:175–98.

Rudran, R. 1973. Adult male replacement in one-male troops of purple-faced langurs (*Presbytis senex senex*) and its effects on population structure. *Folia Primat.* 19: 166–92.

Saayman, G. S. 1971. Behaviour of the adult males in a troop of free-ranging chacma baboons (*Papio ursinus*). *Folia Primat.* 15: 36–57.

Struhsaker, T. In press. Infanticide in the redtail monkey (*Cereopithecus ascanius schmidti*). In *Proceedings of the Sixth Congress of International Primatological Society.* Academic Press.

Sugiyama, Y. 1965. On the social change of Hanuman langurs (*Presbytis entellus*) in their natural conditions. *Primates* 6:381–417.

———. 1967. Social organization of Hanuman langurs. In *Social Communication among Primates,* ed. S. Altmann. Univ. of Chicago Press.

Thompson, N. S. 1967. Primate infanticide: A note and request for information. *Laboratory Primate Newsletter* 6(3):18–19.

Trivers, R. L. 1972. Parental investment and sexual selection. In *Sexual Selection and the Descent of Man 1871–1971,* ed. B. Campbell, pp. 136–79. Aldine.

Warren, J. M. 1967. Discussion of social dynamics. In *Social Communication Among Primates,* ed. S. Altmann. Univ. of Chicago Press.

Wolf, K., and J. Fleagle. In press. Adult male replacement in a group of silvered leafmonkeys (*Presbytis cristata*) at Kuala Selangor, Malaysia. *Primates.*

Richard Curtin
Phyllis Dolhinow

Primate Social Behavior in a Changing World

Human alteration of the environment may be pushing the gray langur monkey of India beyond the limits of its adaptability

The gray langur monkey (*Presbytis entellus*) finds itself at present in the midst of heated controversy. Infant mortality in one of the langur troops at Mt. Abu, in northwest India (Fig. 1), rose to an astonishing 83% over a period of four years, and the causes for this striking figure immediately commanded the attention of students of primate behavior. It was assumed by some researchers that the Abu monkeys, and other langurs living at crowded sites, were typical of the species (Mohnot 1971; Sugiyama 1965, 1966, 1967; Hrdy 1974, 1977), and Hrdy has proposed an evolutionary model that includes males routinely killing infants as an adaptation to competition for females. Our observations of langurs at the less crowded sites of Orcha and Kaukori (Dolhinow 1972) and at Junbesi (Curtin 1975 diss.) have forced us to question this interpretation and to ask instead: How could such levels of infant death arise from the more widespread, generally peaceful pattern of behavior associated with the gray langurs' notable ecological and evolutionary success?

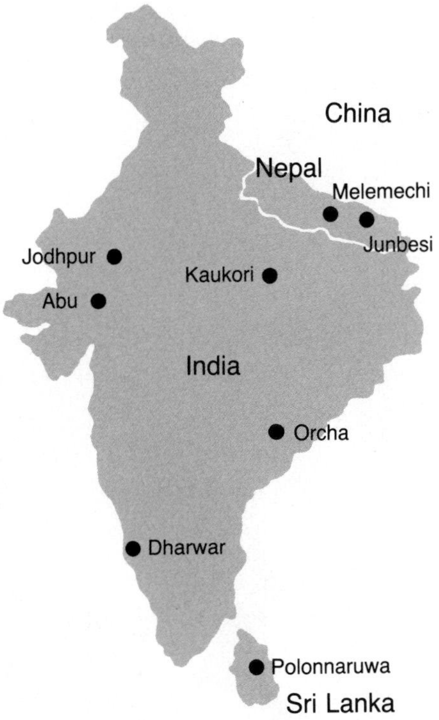

Figure 1. Gray langurs have been studied at a variety of sites in India, Nepal, and Sri Lanka, ranging from heavily populated areas to relatively unspoiled forest.

Langurs are members of the subfamily Colobinae, the Old World monkeys that possess digestive tracts which allow them to eat mature leaves. Primate studies have historically focused on the other subfamily—the Cercopithecinae, which includes such animals as baboons and rhesus macaques. Perhaps more important, colobines have been neglected because most of them are forest dwellers, and early studies concentrated on animals that could be easily seen. However, field researchers of the 1960s and 1970s have been less intimidated by the difficulties of working in forest habitats, and several excellent studies of forest leaf-eaters have been completed or are under way.

Gray langurs live from Nepal to Sri Lanka in an array of habitats ranging from tropical jungle to desert margin, and from undisturbed forest to bustling village. Their behavior varies greatly over this range, but they have now been well studied at a number of sites scattered over the Indian subcontinent, and the limits of their behavioral variation have emerged clearly. The behavior of langurs at Orcha and Junbesi appears to represent a pattern of social organization that probably occurs over the species' whole range and persists under all but the most stressful environmental conditions.

Typical of the gray langur social pattern are multiple infant caretakers and male rivalry that can be dramatic but that does not normally disrupt the orderly rearing of the young. Langur troops include relatively stable cores of adult females plus adult and subadult males whose membership in the group is somewhat less constant. Males can and do move from troop to troop, and although relations among adult males differ from one site to another, competition for females is intense enough in some areas to bring about changes in troop composition. At Orcha, serious fighting among males was never observed during a year's study, but at Junbesi, fighting among males, particularly during the peak mating season, led to animals being driven from the troop, at least temporarily. These populations can be said to have

Richard Curtin, who has studied langur monkeys in the Nepal Himalaya for twenty months, received his Ph.D. in 1975 from the University of California, Berkeley, where he is currently a Research Associate and Lecturer and does research on captive langurs. During his field work he was supported by an NIH Traineeship.
Phyllis Dolhinow, Professor of Anthropology at Berkeley, spent three and one half years in India studying langur monkeys and rhesus macaques. She has also worked in East Africa. Since 1972 she has done research on langur monkeys in a captive colony at the University of California. Her recent research concentrates on langur development, attachment, caretaking and adoption, and the effects of mother-loss. Address: Department of Anthropology, University of California, Berkeley, CA 94720.

a "multi-male troop pattern," and are to be distinguished from populations in which troops typically have only one adult male, and a very high percentage of males are not members of reproductive troops.

The multi-male troop pattern is as widespread as the species itself. It has been found in all three continuous long-term studies of gray langurs living where human influence on ecological conditions has been slight: at Orcha, where leopards and tigers still prowled the virgin forest; in minimally disturbed oak forest in central Nepal (Bishop 1975 diss.); and at Junbesi, where a marginal habitat and substantial predation helped limit population density to a single langur every square kilometer. The pattern has also persisted under quite different circumstances: Ripley (1967) reported it from Sri Lanka where langurs lived at a population density fifty times that at Junbesi, and the langurs at Kaukori maintained their stable multi-male troop in an area where 98% of the land was occupied by villages or under cultivation (see Fig. 2). This remarkable adaptability of langur social organization carries with it a high degree of behavioral variability, but the wealth of studies now complete allows us to describe, in general terms, the social life a langur is likely to lead.

Figure 2. A juvenile gray langur begs a bite of mango from her mother, who ignores her. Although the Kaukori troop, of which they are members, takes a toll of both unripe and ripe fruit from the mango orchards, an abundance of fruit remains on the large trees.

Normal social life

From the time a newborn presence is noticed in a group, it becomes a focus of great interest for young and old females alike (Fig. 3). The infant is conspicuous because the bare pink skin of its hands, feet, face, and disproportionately large and convoluted ears contrasts sharply with its sparse dark brown fur. The newborn may be passed among waiting females more than fifty times in four hours on the first day of life. It may be away from its mother for hours, although normally she can retrieve it when she desires, and most transfers are calm and without peril to the infant. Many females allow the infant to nurse, even though they are not lactating or may never have had an infant.

This high frequency of passing the young infant from female to female lasts only a few weeks. By the end of the second month of life the infant is very active and ably goes where and when it wishes. In the meanwhile, it has had many hours of contact with most of the females in the troop and has developed strong preferences for certain of them.

Weaning, at the end of the first year, is seldom a harsh or traumatic experience. Inconsistent maternal rejection often prolongs weaning, and occasionally an adult with a neonate is observed being followed closely by her last infant. At the time of weaning, a yearling may adopt another adult female to serve as its mother. Although adoptions seldom have been witnessed in the field, probably because of difficulties of observation, they occur frequently in captivity and may be stimulated by even brief removals of the mother from the group. This is, of course, assuming that substitute caretakers are available—as they normally are.

Male and female paths of development diverge soon after the first year of life. Play is important for both sexes, but the immature female also shows interest in neonates. By the time she is two, she takes an active part in attempting to take, hold, and carry new infants. At first, her efforts are hopelessly clumsy, and her ineptness, as reflected in the infant's struggles and cries, stimulates older females to retrieve it. She gradually acquires skill and becomes a fair or even a competent caretaker well before she becomes a mother. There is a great individual variability in motivation to perform caretaking activities among females of all ages. A woefully inadequate juvenile may become a splendidly able mother!

Furthermore, adult females may treat successive infants in very different manners. The degree of interest in the infants of other females varies greatly among adults, and definite preferences are shown for certain infants by some females. The reasons for these preferences and for changes in caretaking styles of individual females are being investigated, because to the human observer it is not at all apparent why the variations exist. The vast majority of attentions directed toward infants can best be described as adequate and protective. Infants too are far from a homogeneous lot: some are strugglers and squealers, while others seem passive, seldom showing discomfort.

Adult females live an orderly existence. There are social changes as, for example, when males enter and leave the group, but the paramount markers of the passage of time are a succession of infants. A female has her first infant when she is 4 to 5 years old, and her ability to produce viable offspring improves as she ages. In the wild, females usually give birth at intervals of 18 to 24 months. There is variability in peak birth seasons in different areas of the Indian subcontinent, but most areas are characterized by seasonal regularity of births. When a female is in estrus, she solicits and mates with several males if the group is large enough, often showing a preference for one, who may not always be the alpha male. Complete copulations may occur during pregnancy but are by far more frequent during the female's mid-cycle estrous periods.

Precise information on the improvement of reproductive capability with aging is being gathered in a captive colony at Berkeley, where ages are known. By 9 years a female is bearing regularly, and evidence from both field and colony suggests strongly that she continues to do so for many years. No doubt, if she survives long enough, she will eventually experience a diminution of reproductive ability, but very aged females are rarely observed to survive the rigors of feral existence. Estimating ages in the field is exceedingly difficult, and indicators such as worn teeth, wrinkles, bags under the eyes, or generally poor health and emaciation are extremely unreliable.

Adult female social ranks are seldom clearly apparent or linearly ordered, and neither age nor reproductive history is a good predictor of social rank. The older females are by no means usually the highest in social authority, nor do they automatically yield to younger females. The scales of power tip with the events of the reproductive cycle such as sexual receptivity, pregnancy, delivery, and lactation. Fluctuations occur within periods of months as well as over years, and a female may move up or down relative to other females, only to change her position in the hierarchy after some months. Body size does not determine status once females are mature.

Male rivalry

Juvenile males, in contrast to females of the same age, spend most of their time in rough play and rarely show interest in newborn infants. As a male grows he tends to become somewhat peripheralized to the troop because his boisterousness annoys adults. A young male may leave the group when he is sexually mature, and a few do even before that. An exit may be hastened or stimulated by tensions arising between him and the adult males. Such departures are not always permanent, and reentries as well as entries into new troops have been recorded. The adult males in a troop can usually be ranked in a linear hierarchy of status, but there is no clear correlation between size or age and position in the hierarchy.

Rivalry with other males is inevitably an extremely important factor in a male langur's life. Since it is one of the most significant variables in langur behavior, and since it is probably at the root of the infant disappearances reported from Abu and other sites, its expression within the multi-male troop demands close scrutiny.

One end of the male-rivalry spectrum is shown by the relaxed behavior and absence of fighting in the troop at Orcha. Further along the spectrum, rivalry was far more apparent among the Kaukori langurs, where a stable dominance hierarchy was easily discernible among the troop's six adult males. Prolonged harassment with only occasional outbreaks of fighting accompanied the gradual exchange of the two top male positions. Notably, this dominance shift took place *within* the troop; after the exchange was effected, the defeated male assumed a stable, albeit lower, rank within the hierarchy.

In contrast to Kaukori, male rivalry was dramatic among the Junbesi langurs, and its effects were the most striking so far reported within the multi-male troop pattern. Relations among the adult males were tense throughout the 16 months of observation. During the first winter of the study, one of the troop's four original males repeatedly left after skirmishes with the other three males. These departures were temporary and are best described as peripheralizations, since the male did not leave the troop's home range, which was so large that he could stay within it and rarely be in close contact with the troop. He rejoined the troop in late winter but was frequently driven from it during the peak mating season in summer. He was again able to remain within the home range, and as the mating season ended he was gradually able to shorten his enforced distance from the troop. He was one of the two adult males in the troop when the study ended.

The other two males had left the troop permanently after repeated aggressive exchanges and violent fights at the beginning of the summer. One of them was later discovered in a different troop some distance away. A similar gradual process had apparently led to his successful entry, and he was a well-assimilated member of the multi-male troop.

Relations among males within the multi-male troop can thus follow several patterns: males can coexist amicably, with little apparent aggression, as at Orcha. Rivalry can be resolved within a troop by threat, fighting, or temporary peripheralization, or it can lead more dramatically to a male's permanent exclusion from his troop. Males can also enter strange troops or reenter their own with very little fighting. This is a remarkable range of behavior, and at present it is difficult to identify or explain the critical elements that cause the variation.

We do not know, for instance, what causes some males to accept a lowered position in the dominance ranking and why others elect to leave the troop. We do not know how some males successfully manage to enter strange troops while others, as witnessed at Kaukori, are actively rebuffed. One generalization, however, is possible. The patterns of behavior that allow males to remain in or to enter troops in the face of male rivalry take time, and, perhaps most important, they take *space*. The dominance shift at Kaukori took 2½ weeks, and the reentry of the Junbesi male into his troop took even longer. At first the male in the troop threatened the other when he was 1 km away, but as the mating season ended, his threats became less intense, and the distance gradually decreased. When troops are compressed into a restricted space that forces them to interact frequently and aggressively over limited food, water, and places of safety, or when they are harassed by humans, male rivalry may produce new and maladaptive results.

Behavior at crowded sites

The langurs at Abu (Hrdy 1974, 1977) and those watched by Mohnot (1971) at Jodhpur and by Sugiyama (1965, 1966, 1967) and his colleagues at Dharwar shared a pattern of behavior that falls outside the broad range so far discussed. Male behavior was much less variable. These investigators reported that the expression of male rivalry extended far beyond threat and fighting to include the routine killing of infants sired by competing males. This pattern differed further in being associated with only very specific ecological conditions. The habitats at all three sites were greatly disturbed by human activities, and langur population densities were either absolutely or effectively extremely high. The Jodhpur langurs lived on the edge of the desert and spent most of their time in very limited areas near water sources or gardens. Much of the 40-hectare home ranges of some of the Abu troops lay in the town itself. The forest near Dharwar had recently been cleared and the langurs concentrated in what little remained, with the result that their population density reached 134 langurs per km^2.

Troops at these sites typically contained only one adult male, and strange males did not enter the troops by a gradual process similar to that observed at Junbesi. Nontroop males were therefore numerous, and these joined together in male groups of as many as 60 members. Both male groups and lone males attacked bisexual troops. In successful attacks—*male takeovers*—they drove out the male originally in the troop, and a *usurper* became the troop's new leader. Several infants disappeared during some of these takeovers, and both Sugiyama and Hrdy assumed that the usurping males killed the missing infants. Both observers noted that some of the infants' mothers soon became pregnant, and Hrdy elaborated a model in which infant killing is an evolved behavioral trait of male langurs, selected for because it renders victims' mothers quickly available for insemination. We will review the events observed at Dharwar, Jodhpur, and Abu before turning to the model.

The circumstances surrounding the disappearances of some 40 infant langurs at these sites are known in varying degrees. Infant disappearance following a male takeover was first noted at Dharwar in 1962. Eyewitness accounts of the takeover of two troops at Dharwar are available, and Sugiyama's (1965) description of the takeover of the 30th troop there is perhaps the most detailed. The downfall of Z, the original male leader of the 30th troop, began when three of his females, one in estrus and one accompanied by her infant, approached and actually entered an outside male group. Four days later, following repeated fighting in which Z suffered serious wounds, L, the largest member of the male group, was established as the lone male of the "new" 30th troop. Fighting among L, Z, the rest of the male group, and the 30th troop females and juveniles continued sporadically for 15 days after the start of the takeover, and during this period two infants disappeared. Another infant survived the protracted battling with what appeared to be a canine bite wound. After six days, still alive, it was abandoned by its mother.

Despite the continued fighting, L was quickly able to consolidate his position, and ten days after the start of the takeover, he was leader of the new troop. As he gradually stopped attacking the females, they stopped running away screaming when he approached.

Figure 3. A group of females huddle around an infant in a meadow above Junbesi. Their troop, on the flanks of 6,959-meter Numbur peak, lives at higher elevations than any other langurs in this Himalayan region.

This reduction of tension was apparently somewhat illusory, because four more infants disappeared in the month and a half after the last observed contact between the 30th troop and either Z or other members of the male group. Like the earlier disappearances and wounding of infants, three of these new disappearances were discovered in the morning at the beginning of daily observation periods. The fourth death followed the only male attack on an infant that was actually observed at Dharwar: L attacked a mother-infant pair and bit the infant near the base of the tail.

Three other takeovers at Dharwar are mentioned by Sugiyama (1967), and four or five infants reportedly disappeared during one of these. A more complete account is available of the experimentally induced takeover of the 2nd troop at Dharwar (Sugiyama 1966). Investigators removed the male leader from the troop, and males from two adjacent troops then contested control over it. Events did not differ drastically from those already observed during the takeover of the 30th troop. Four infants disappeared, two of them after receiving wounds adjudged the result of langur canine bites. Unfortunately, neither the attacks of the two wounded infants nor the events leading to the disappearance of the other two were witnessed.

The next reports of infant disappearance followed three cases of infanticide at Jodhpur (Mohnot 1971). In another somewhat artificial removal of a troop leader, the male and 70 other members of an 82-member one-male troop died of disease or poisoning. An outside male group attacked the survivors, and one of this male group established himself as the new leader. Mohnot's report includes his eyewitness account of this male killing three infants within a month of his takeover. A fourth infant disappeared during this time, and a fifth vanished six months later.

The most recent reports of infant disappearance and male takeover come from Abu, where Hrdy (1974, 1977) observed langurs for five 2-to-3-month periods between June 1971 and June 1975. On the basis of her own observations, from reports of Abu townspeople, and from changes in troop membership, she inferred that at least nine takeovers of the Abu B-6 and B-3 troops occurred during this 4-year period. At least 17 infants disappeared from the two troops, and infant mortality in the B-6 troop between 1971 and 1974 reached an astonishing 83%. The events surrounding these dramatic levels of infant death remain somewhat obscure. Hrdy witnessed male attacks on infants after three of the takeovers, but although one such attack led to serious injury, none inflicted fatal wounds. Only the reports of casual bystanders linked male langurs to any infant deaths.

When the evidence is reviewed carefully, the immediate causes of infant death remain perhaps the least understood aspect of the takeover/disappearance phenomenon. There is no question that males have definitely injured infants following some takeovers; however, the motivations behind their aggression are far from clear. Hrdy described the male assaults on mother-infant pairs that she saw as "highly goal-directed and organized" attacks specifically aimed at the infants, but Mohnot attributed the cases of actual infanticide that he witnessed to the male's "simultaneous sexual excitement and enragement." It is noteworthy that at least two of the three Jodhpur infanticides occurred during battles between the attacking male and members of his old male group.

It is quite probable that the instance Hrdy witnessed in which an infant was seriously injured may have occurred in a similar contest, because the male from a nearby troop chased the infant's attacker only moments after the assault. At Dharwar, as at Jodhpur, male attacks on females were frequent, particularly when the females approached male groups. This behavior raises serious doubts as to whether *infants* are the targets of the attacks; after the takeover of the 30th troop at Dharwar, the new leader continued to attack at least four females even after they had lost their infants.

The role of females in the takeover/disappearance complex is also far from understood. The central question is whether the deaths of the infants hastened the subsequent pregnancies of their mothers or whether the females might have given birth when they did even had they *not* lost their infants.

A great increase in sexual behavior typically followed takeovers, but data on subsequent pregnancies are available in only four instances. Langurs have a gestation period of approximately 6½ months. Following the takeovers of the 30th and 2nd troops at Dharwar, and the 1972 takeover of the B-3 troop at Abu, 70% (11 of 15) of the females who lost infants gave birth within 8 months. At Jodhpur, however, none of the 4 females who lost infants had given birth even a year after the takeover.

Even where births have followed infant disappearances, the significance of the timing is difficult to evaluate. Data available from the infanticide sites indicate that the age of the missing infants is a crucial factor. At the time of their death the infants at Jodhpur were, at most, 3 months old, and the mother of a 2-week-old infant experimentally removed from the 2nd Dharwar troop had not given birth even a year after the infant's loss. In marked contrast, the pregnancies begun just after the takeovers at Dharwar and Abu followed the disappearances of infants aged 6 months to a year. These relatively advanced ages lead inevitably to the question of how far apart langur births are when infants survive. Although we do not have a definite answer, estimates of langur birth intervals in the wild usually range from 1½ to 2 years. At the Berkeley colony, langur mothers with surviving infants give birth to subsequent offspring after an average interval of 16.3 months (Vonder Haar Laws, ms.), and data from Junbesi indicate that similar intervals occur in the wild.

Such short birth intervals mean that the infant disappearances at Dharwar and Abu in fact may not have drastically altered normal female reproductive cycles. Females who became pregnant after the loss of their infants may have done so simply because of the passage of time, their receptivity perhaps accelerating only slightly as a result of the sudden presence of new males. Data on the behavior of females obtained immediately prior to takeovers are too few to support any definitive statements, but it is interesting that the mother of an infant who subsequently disappeared approached the male group—with the infant—at the very start of the takeover of the 30th troop at Dharwar. Future research will reveal whether changes in female behavior—such as those induced by females' attraction to male groups—have an important role in precipitating takeovers.

Sugiyama (1965, 1966) reported that female sexual activity was noticeable within hours after the invasion of a troop by a male group. He surmised that the sudden appearance of the new males might be effective in inducing receptivity and explained the desertion of wounded infants by their mothers as a result of the mothers' already developing bonds with new males, and thus their loss of motivation to care for injured offspring.

After reviewing the evidence, we adjudge that it is at present not at all clear that female receptivity *follows* infant disappearance. If birth intervals are short, and strange males can indeed induce estrus in the mothers of older infants, the weaning infants' disappearance can more easily be considered to follow from the unfortunate combination of newly estrous females and sexually active and competitive males at a time when the infants are at a particularly vulnerable age.

Explaining infant death

The data from the three "infanticide" sites are too equivocal to guide interpreters toward any single conclusion. Hrdy's recent model of infanticide as a reproductive strategy carries the interpretation of the data further than any other hypothesis, but to do so it requires many assumptions.

According to the model, the history of a langur troop is cyclic. Adult males tolerate their own sons, and thus one-male troops may grow to include several adult males belonging to different generations. However, takeover by outside males is a normal process that checks development into a multi-male troop. Takeovers are most frequently successful in densely populated areas because of the high numbers of nontroop males. Most troops in such areas contain only one adult male, and the length of time that he remains in the troop is short.

Infanticide is explained as an evolutionary adaptation to this brief male tenure. When a male enters a troop after a successful takeover, the unweaned infants already in the troop are unrelated to him; because their presence postpones their mothers' fertility, they are obstacles to his reproductive success, and he systematically kills them. Natural selection is viewed as favoring male behavior that will remove the infants, and infanticide is seen as an easy solution.

Selective pressures favoring infanticide are especially powerful where there are many takeovers. First, situations in which males enter new troops—opportunities for "worthwhile" infanticide—are frequent. Second, and more important, a male can "expect" that he himself will soon be driven from his new troop, and his promptness in fathering infants is crucial to his evolutionary success. If killing their infants brings mothers into estrus, infanticide evolves essentially as an extension of male rivalry (and competition between male and female) into the past. An infanticidal male enjoys accelerated and lengthened access to receptive females at their expense, that of their infants, and that of the male who preceded him. The genes of the infanticidal male, rather than those of the more tolerant deposed male, are transmitted to the next generation and incorporated into the gene pool.

Under this model, infanticide can drastically affect the reproductive success of langurs of either sex, and many behaviors function to control or mitigate its effects. Not only must male langurs kill the progeny of their competitors, they must also *not* injure their own. Males moved back and forth quite frequently at Abu, and they sometimes usurped troops that probably contained their own offspring, and thus the model proposes a mechanism that enables males to avoid attacking their own children. Interestingly, the mechanism essentially concerns females, not infants: a male does not attack a female-infant pair if he was in consort with the female at the time of the infant's conception. The troubling questions raised by the gray langur's cooperative mothering and the possibility of adoption remain unanswered.

Since loss of her infant is always detrimental to a female, female counterstrategies against infanticide are thought to be even more elaborate. These strategies range from her interference with a male's "paternity testing" to her ability to select the lesser of two evils. According to the model, a female who has already conceived will again become receptive after a takeover in order to deceive the usurping male into sparing her infant as though it were also his own. Females know when to cut their losses; a mother will desert her wounded infant in the face of a male attack after "practical evaluation of what *this* infant's chances are, weighed against the probability that her next infant will survive." A female finally mates with the killer of her infant because "a female who sexually boycotted infanticides would do so to the detriment of her male progeny. Her sons would suffer in competition with the offspring of nondiscriminating mothers." (Hrdy 1977:48–49).

Hrdy's model does not necessarily impute the langurs' decisions to conscious processes, but langurs are credited with remarkable abilities. Before committing infanticide, a usurping male in effect judges his potential victim's age and counts back one langur gestation period—about 6½ months from his birth date. Armed then with his target's date of conception—18 months in the past in the case of a yearling infant—he guards against eroding his own reproductive success by remembering whether he was in consort with the infant's caretaker at the crucial time. Females make similar calculations; before abandoning her infant, a mother evaluates its cost to her so far in terms of time and energy, the cost of its defense, the likelihood of its survival, and the cost and chances of a new infant.

This is an elegant and remarkably thorough model, but if we recall what is actually *known* of the events at Dharwar, Jodhpur, and Abu, it is clear that the model explains not the results of observation but the products of assumption. The primary assumption concerns the occurrence of infanticide itself. As we have seen, from all of India there has been clear-cut evidence of only four langur infants being killed by males. This is despite the fact that infanticide is a cause of infant death that would be likely to be seen. Visibility at Dharwar, Jodhpur, and Abu was apparently good; langurs are diurnal, and infanticide presumably takes place in the daylight; and, since habituated animals were studied, the presence of observers presumably would not deter the behavior. In spite of this, the model rests on the assumption that infanticide was the leading cause of death among the langurs at the three sites. The disappearance of 39 infants became evidence of the regular occurrence of infanticide.

The assumption does not stop at the cause of infant death; the identity and motivations of the males who committed the purported infanticides are also crucial to the model. Usurping males killed the infants in "highly goal-directed and organized attacks" which are largely the result of their particular genetic endowments. When an infant langur disappears in a situation marked by restricted space, violent fighting and chasing among males and females, and frequent harassment by men and dogs, we are asked to attribute his death to deliberate attack by a member of his own species and to name a specific male as his attacker.

The second major assumption concerns the effects of the purported infanticides on the reproductive behavior of the victims' mothers. Some females appear to have become receptive before takeovers; some have become so following their exposure to strange, usurping males but prior to the disappearance of their infants; and some, indeed, have come into estrus following the deaths of their infants. The available data seem to indicate that in some cases a mother's return to estrus can endanger her infant, but the model asks us to assume one direct relation: because the infant is killed, the mother comes into estrus and is impregnated.

These two assumptions are necessary to impart an easily explainable *order* to the chaotic events at Dharwar, Jodhpur, and Abu. A third assumption is needed to justify an *evolutionary* rationale for "infanticide as a reproductive strategy": the takeovers at these three sites represent normal langur behavior over the centuries. It is supposed that takeovers do occur at locations such as Orcha, Kaukori, and Junbesi, but are infrequent because of low numbers of nontroop males. This rarity explains why takeovers are never observed and why multi-male troops persist at those sites.

The evolutionary context

Our studies lead us to argue from the opposite direction. The pattern that, we believe, represents the typical and evolutionarily adaptive behavior of gray langurs includes male rivalry but not any egregious interference with female reproduction or the killing of infants, and does not have any systematic effect on troop structure. Males contest for status directly among themselves, by means of threat or combat. The occasional expulsion of males from troops during conflict is balanced both by their frequent return and the ability of strange males sometimes to enter troops without major disruption. The latter counters any tendency toward a pattern of one-male troops and reduces the numbers of males living outside troops.

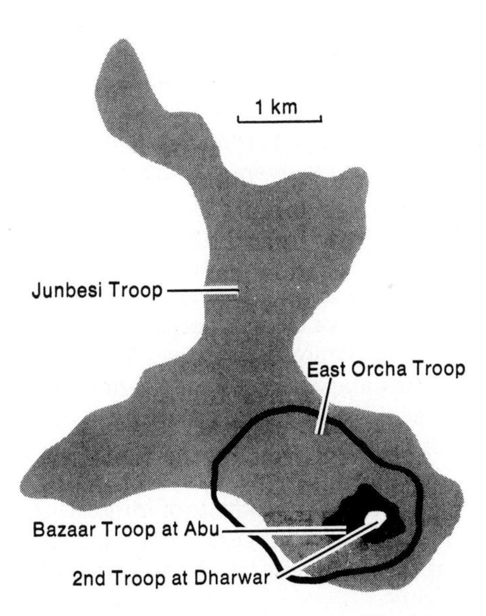

Figure 4. A comparison of the ranges of four langur troops illustrates variation in range size, which seems to influence troop structure. Troops at Junbesi and Orcha contained several adult males, but those at Abu and Dharwar contained only one, and new males could enter only after driving out the old.

This pattern has been reported from almost every ecological condition in which the gray langur lives; it has been found wherever langurs have been studied in habitats that remained close to the aboriginal condition; and it has persisted in spite of greatly varying population densities. In sharp contrast, the one-male troop/male takeover pattern has been observed *only* where langur population densities are high, and where human influence on the ecology is immense. This close correspondence of the specialized pattern with distinctive ecological conditions raises the possibility that the pattern arises directly as an effect of the conditions, and the conditions have not been prevalent for long.

The crowding associated with the pattern of male takeover has almost certainly arisen because of habitat destruction and predator removal in historical—not evolutionary—time. Agriculture and human handouts helped support the langurs at Abu and Jodhpur in their constricted, suburban home ranges, and at Dharwar recent logging has concentrated the langurs in what remains of the forest. Predator removal is another factor of undoubted importance that has been dramatic only since the development of agriculture and urbanization. Schaller's (1967) demonstration that the large Indian cats prey substantially on langurs, as well as the low langur densities at Orcha, Melemchi, and Junbesi, where predator populations remain effectively undisturbed, indicate that predation normally exercises an important restraint on langur population sizes. Habitat disruption and faunal impoverishment even approaching present levels began only after the advent of the Mughal Empire (Mukherjee 1974); both the introduction of firearms and an increase in poaching after Indian independence have since led to acceleration of the destruction of Indian wildlife.

Small home ranges stand out as a major characteristic of the "infanticide" sites, particularly when they are compared to similar sites elsewhere (Fig. 4). The 40-hectare home ranges at Abu should be compared to the 770- and 1,200-hectare ranges at Kaukori and Junbesi, respectively, and the 16.8-hectare ranges in the Dharwar forest should similarly be contrasted with the 380- and 220-hectare ranges at Orcha and Melemchi.

Since langur crowding is almost certainly a very recent phenomenon in terms of evolutionary time, it is appropriate to search for the origins of the male-takeover pattern in the effects of crowding on those patterns of behavior that probably *did* exist in the langur's evolutionary past. Although, as we have seen, there are powerful male rivalries within the widespread multi-male troop pattern, such rivalries are balanced by processes that allow outside males to enter troops. These balancing processes, however, require time and space. Crowding removes the space requisite for the resolution of male conflict within troops, or the peaceful entry of strange males into the troops. The levels of human harassment typical of crowded sites effectively remove the requisite time; the gradual reduction of tension between males is impossible if the animals are being chased frequently.

Clearly, it will be difficult for multi-male troops to exist in crowded habitats. Crowding increases male tension by enforcing close proximity between animals, and by increasing competition for food—particularly where humans feed the animals—or for receptive females. Crowding also escalates the consequences of male agonistic behavior. Any serious fighting will easily drive defeated males beyond home range or territorial boundaries and into direct confrontation with other troops. A neighboring troop is extremely unlikely to tolerate a precipitate entry into *its* range, particularly since it has probably had previous territorial encounters with the intruders. The defeated male will probably be unable to remain in the area; an interaction that might lead merely to temporary peripheralization in an uncrowded habitat will drive him permanently from his troop in a crowded area. High density increases the frequency of aggression and greatly magnifies its effect; losers become nontroop males.

The argument from here to the pattern of male takeover observed at Dharwar, Jodhpur, and Abu is short. The exiles form male groups, and, again because of crowding, contacts between these male groups and the troops are frequent. Since the male groups are numerically superior to the lone males in reproductive troops, takeover attempts are often successful.

A pattern of one-male troops and male takeovers does not represent normal langur behavior either today or in the evolutionary past. Neither does the one-male troop/male-takeover complex represent normal langur social organization; it is instead a symptom of the failure of the multi-male troop structure to endure environmental conditions far different from those in which it evolved.

This argument leads to a simple explanation of infant mortality at Dharwar, Jodhpur, and Abu. Infanticide cannot be an evolutionary adaptation to a takeover pattern that is itself only an artifact of recent environmental pressures. Infant mortality at the three sites need have no evolutionary significance beyond the obviously detrimental. The observed events need not be interpreted as evidence of intricate but hidden patterns; they are, more simply, the indicators of chaos that they appear to be. Male takeovers involve fighting, chasing, and display among both males and females, approach of male groups by mothers carrying infants, and male attacks on mother-infant pairs. Any of these creates danger for infants. Removing the need to ascribe evolutionary function to infant death makes it unnecessary to assume that the 35 out of 39 deaths at the three sites which resulted from unknown causes were all products of infanticide by adult male langurs possessing the capacity to discern their own offspring.

The relationship between infant death and female reproductive cycles represents a similar situation. Pregnant females sometimes exhibit estrous cycles, just as they do in undisturbed habitats and in captivity, and females under attack sometimes abandon their infants. Again, there is no reason to extract from the observed variability of behavior either elegant "female counterstrategies" or the single rule that estrus is caused by infant loss.

The evolution of langur behavior took hundreds of thousands of years, and we can be confident that langurs were much as they are today long before *Homo sapiens*' rise to eminence. In the last few thousand years, at most, civilization has transformed India, and there now are langurs living under conditions that simply did not exist ten thousand years ago. To study the evolution of langur behavior, we must consider carefully the whole range of behavior that exists today. Behaviors that appear everywhere, or almost everywhere, and that exist in habitats little altered by man are those most likely to have survived from the evolutionary past. Behaviors that occur only at selected sites and only under novel environmental conditions are quite likely to be artifacts of modern conditions. These new patterns of behavior are of great interest—not because they reveal the course of langur evolution, but because they demonstrate the disturbing results when even a remarkably adaptable species is pushed beyond the range of its flexibility.

References

Bishop, N. H. Social behavior of langur monkeys (*Presbytis entellus*) in a high altitude environment. 1975 diss., Univ. of California, Berkeley.

Curtin, R. A. Socioecology of the common langur, *Presbytis entellus*, in the Nepal Himalaya. 1975 diss., Univ. of California, Berkeley.

Dolhinow, P. 1972. The north Indian langur. In *Primate Patterns*, ed. P. Dolhinow. Holt, Rinehart and Winston.

Hrdy, S. B. 1974. Male-male competition and infanticide among the langurs (*Presbytis entellus*) of Abu, Rajasthan. *Folia Primat.* 22:19–58.

———. 1977. Infanticide as a primate reproductive strategy. *Am. Sci.* 65:40–49.

Mohnot, S. M. 1971. Some aspects of social change and infant-killing in the Hanuman langur, *Presbytis entellus* (Primates: Cercopithecidae) in Western India. *Mammalia* 35:175–98.

Mukherjee, A. K. 1974. Some examples of recent faunal impoverishment and regression. In *Ecology and Biogeography in India*, ed. M. Mani. The Hague: D. W. Junk.

Ripley, S. 1967. Intertroop encounters among Ceylon gray langurs. In *Social Communication among Primates*, ed. S. Altmann. Univ. of Chicago Press.

Schaller, G. B. 1967. *The Deer and the Tiger: A Study of Wildlife in India*. Univ. of Chicago Press.

Sugiyama, Y. 1965. On the social change of Hanuman langurs (*Presbytis entellus*) in their natural condition. *Primates* 6:381–417.

———. 1966. An artificial social change in a Hanuman langur troop. *Primates* 7:41–72.

———. 1967. Social organization of Hanuman langurs. In *Social Communication among Primates*, ed. S. Altmann. Univ. of Chicago Press.

Vonder Haar Laws, J. Female langur monkey (*Presbytis entellus*) reproduction and sexual behavior: Records and analyses from captive social groups. Manuscript.

Part II
The Mechanisms of Behavior

The study of animal behavior involves research at different levels of analysis. On the one hand, there are questions about the internal machinery of behavior, the proximate mechanisms within animals that make it possible for individuals to do things. On the other hand, there are questions about how these mechanisms and the behaviors they control came about over evolutionary time.

Kay Holekamp and Paul Sherman set the stage for the organization of the articles in this section by showing how the proximate and evolutionary approaches to animal behavior can be subdivided and yet integrated. They do so in the context of a concrete example, the dispersal behavior of Belding's ground squirrels, arguing that all studies of this animal's behavior (and that of any other species as well) can be categorized as research into (1) the underlying developmental or (2) physiological mechanisms responsible for the behavior, *or* as research on (3) the evolutionary history or (4) fitness consequences of the behavior. What kinds of academic disagreements are prevented by adopting such a four-part scheme? What would happen, for example, if the hypothesis that male dispersal is caused by male hormones was pitted against the hypothesis that dispersal is caused by selection for avoidance of inbreeding?

There have been challenges to the view that four fundamental questions underlie the entire field of behavioral research. Holekamp and Sherman say that one adaptive reason why male Belding's ground squirrels disperse is because the behavior advances individual fitness (reproductive success) by reducing the chances of father–daughter and sister–brother incest. Someone might argue, however, that the fitness consequences of a behavior can hardly be invoked as a *cause* of that behavior when it is actually an *effect* of the behavior. Readers may ask whether Holekamp and Sherman provide the basis for an effective response to this criticism.

The other articles in this second section were written by persons interested primarily in the mechanisms underlying behavior, although their analyses at this level do not prevent them from also considering the evolutionary causes of the behavior they are studying. For example, Donald Griffin's article on animal thinking raises the possibility that many animals other than ourselves possess consciousness mechanisms (i.e., physiological systems in the brain) that are fundamentally similar to our own. He encourages researchers to try to get inside the nervous systems of other animals to determine what they are "aware" of. If we could, our understanding of animal behavior would be greatly advanced, according to Griffin. How would it change your view of "lower animals" if you knew that an animal's decisions about performing a particular behavior were conscious as opposed to unconscious? If crickets (or fish or snakes) could think, would you feel differently about the need to treat them humanely when conducting scientific experiments with them? Griffin points out that many scientists resist the notion that animals other than ourselves are consciously aware of certain things. He speaks out strongly against this attitude, but readers may wish to ask why so many of Griffin's fellow scientists, after careful consideration of Griffin's position, still question the utility of trying to examine consciousness in other animals.

Griffin discusses the possible fitness consequences of consciousness mechanisms, so his article also integrates two different levels of analysis. He proposes potential reproductive benefits that animals might gain by having conscious thoughts in some situations. But are there any reproductive costs associates with this ability? If so, what will determine the spread of the genes underlying the development of the brain systems required to produce consciousness in a killdeer, a sea otter, or any other animal, ourselves included? What sorts of information are conscious and which are unconscious, and how might knowing this help us understand the costs and benefits of consciousness in ourselves and other creatures?

The way the human brain operates determines whether we behave "normally" or are afflicted by a mental illness, such as schizophrenia, that very much affects our consciousness. Susan Nicol and Irving Gottesman discuss the role that genetic differences among individuals play in the development of schizophrenia. They report on studies testing the general hypothesis that genetic differences among people underlie variations in the way human brain cells communicate with each other, and conclude that some patterns of brain cell communication result in schizophrenia in some individuals. Their article provides a superb example of how causal explanations can be evaluated by testing key predictions derived from alternative hypotheses. Can you recreate the scientific process followed by Nicol and Gottesman based on information presented in this article? For example, the data in their Figure 3 provide evidence with respect to what prediction and what hypothesis? What conclusion follows unequivocally from this test?

Some scientists interested in human behavior have used studies of the sort described by Nicol and Gottesman to claim that there may be a genetic basis for the differences between people's behavior, a key condition required for the evolution of behavior by natural selection. But critics of this position have not been persuaded by the discovery that there is a genetic basis for behavioral disorders like schizophrenia. Readers might consider whether or not a case of this sort is compelling with respect to the possibility that human behavior is the product of natural selection. If not, why not?

Meredith West and Andrew King explore the developmental basis of a "normal" behavior (bird song) in another kind of organism (the starling). According to modern developmental theory, all behavior is the ontogenetic product of a gene–environment interaction. Developmental theory, however, does not specify what elements of the environment will have what effects for a particular species or ontogenetic stage. Starlings exhibit considerable skill in vocal mimicry, even to the point of being able to mimic human speech. How do the developmental mechanisms of starlings permit this to happen? In their delightful article, West and King describe the key experiments that provide an answer to this question, and in so doing, they clear up an entertaining mystery involving Mozart and his pet starling. If West and King are correct, what kinds of birds will be immune to social influences on the development of their vocalizations? Does the evidence on how starling song develops suggest testable hypotheses about how humans acquire a language?

Mike May moves from the developmental to the physiological level of analysis as he explores the mechanisms that permit certain night-flying insects to detect and escape from bats. His work builds on the classic studies of this subject by Kenneth Roeder. A particularly appealing element of the article is May's reconstruction of his thoughts as he designed and conducted his doctoral research, where readers can see a young researcher's own ontogeny and how one experiment led to another.

In the course of his article, May notes, "Not all useful observations come as a result of premeditated experimental design; sometimes it's useful just to play with the equipment." (The same point is raised by Winston and Slessor in the next section, where they describe their discovery of a way to test how certain chemicals affect the behavior of worker honeybees.) What about this issue? Do you believe that May had no hypothesis at all in mind when he made certain observations while playing around with a cricket and his experimental equipment? Is it possible to make scientific progress without using the scientific method? Does the method always require premeditation?

Because some nocturnal insects use bat detection devices and evasive responses, natural selection favors bat predators that can counter their prey's anti-bat tactics. The article by Brock Fenton and James Fullard is included here because it complements so nicely the research by May and others on the insect side of the predator–prey equation. Fenton and Fullard examine the wonderful diversity in foraging behavior exhibited by bat predators and the interactions between bats and their potential victims. What level of analysis do Fenton and Fullard employ in their research? Why do different bats use different types of sound production, and how do the bats that don't echolocate at all ever find food? What is the likely evolutionary basis for the immense diversity in bat calls and insect responses?

Carl Johnson and Woodland Hastings conclude this section with an exploration of a mechanism that has proven very hard to understand, or to even find: the system responsible for the circadian rhythms (roughly 24–hour cycles that can express themselves independently of environmental cues) that occur in most organisms. The fact that even a single-celled alga can exhibit clock-driven circadian rhythms tells us something about the difficulty of identifying where a physiological clock might be found in humans or fruit flies with their millions of cells. The study of circadian rhythms and the biological clocks that underlie them has led investigators to explore the operation of everything from genes to nerve cells, but Johnson and Hastings focus on the biochemical events that may contribute to the operation of a biological clock. Their article demonstrates how challenging it can be to get to the heart of a mysterious mechanism, or even to find out where it is located amid the complex internal machinery of multicelled organisms. At what level of analysis would you place research on the biochemical bases of behavior? Where does this kind of work fit into the Holekamp–Sherman organizational scheme? Why do so many very different kinds of organisms possess circadian rhythms? And why do circadian mechanisms underlying behavior apparently occur in so many different places within organisms? What kinds of creatures can you predict will lack circadian mechanisms and rhythms altogether?

The articles in this section combine to illustrate that many different avenues exist for the study of behavioral mechanisms. Moreover, the articles show that studies of mechanism and evolution are interrelated, because an understanding of precisely how a physiological element works helps us understand what natural selection designed it to do. And by considering the ultimate, evolutionary aspects of behavior, we can better know what mechanisms to look for in a given species if individuals are to reproduce successfully. Figuring out how behavioral mechanisms work and why they have evolved is the essence of the modern study of animal behavior.

Why Male Ground Squirrels Disperse

Kay E. Holekamp
Paul W. Sherman

When they are about two months old, male Belding's ground squirrels *(Spermophilus beldingi)* leave the burrow where they were born, never to return. Their sisters behave quite differently, remaining near home throughout their lives. Why do juvenile males, and only males, disperse? This deceptively simple question, which has intrigued us for more than a decade *(1, 2)*, has led us to investigate evolutionary, ecological, ontogenetic, and mechanistic explanations. Only recently have answers begun to emerge.

A multilevel analysis helps us to understand why male and not female Belding's ground squirrels leave the area where they were born

Dispersal, defined as a complete and permanent emigration from an individual's home range, occurs sometime in the life cycle of nearly all organisms. There are two major types: breeding dispersal, the movement of adults between reproductive episodes, and natal dispersal, the emigration of young from their birthplace *(3, 4)*. Natal dispersal occurs in virtually all birds and mammals prior to first reproduction. In most mammals, young males emigrate while their sisters remain near home (the females are said to be philopatric); in birds, the reverse occurs *(4–6)*. Although naturalists have long been aware of these patterns, attempts to understand their causal bases have been hindered by both practical and theoretical problems. The former stem from difficulties of monitoring dispersal by free-living animals, and of quantifying the advantages and disadvantages of emigration *(6)*. The latter stem from failure to distinguish the two types of dispersal, and from confusion among immediate and long-term explanations for each type.

We begin with a discussion of the latter point and develop the idea that natal dispersal, like other behaviors and phenotypic attributes, can be understood from multiple, complementary perspectives. Separating these levels of analysis helps organize hypotheses about cause and effect in biology *(7)*. In the case of natal dispersal, this approach can minimize misunderstandings in terminology and allow for clearer focus on the issues of interest.

Questions of the general form "Why does animal A exhibit trait X?" have always caused confusion among biologists. And even today, the literature is full of examples. The nature-nurture controversy, which arose over the question of whether behaviors are innate or acquired through experience, is a classic case *(8)*. After two decades of spirited but inconclusive argument in the nature-nurture debate, it became apparent to Mayr *(9)* and Tinbergen *(10)* that a lack of consensus was caused by the failure to realize that such questions could be analyzed from multiple perspectives.

In 1961, Mayr proposed that causal explanations in biology be grouped into proximate and ultimate categories. Proximate factors operate in the day-to-day lives of individuals, whereas ultimate factors encompass births and deaths of many generations or even entire taxa. Pursuing this theme in 1963, Tinbergen further subdivided each of Mayr's categories. He noted that complete proximate explanations of any behavior involve elucidating both its ontogeny in individuals and its underlying physiological mechanisms. Ultimate explanations require understanding both the evolutionary origins of the behavior and the behavior's effects on reproduction. The former involves inferring the phylogenetic history of the behavior, and the latter requires comparing the fitness consequences of present-day behavioral variants.

There are two key implications of the Mayr-Tinbergen framework. First, competition among alternative hypotheses occurs within and not between the four analytical levels. Second, at least four "correct" answers to any question about causality are possible, because explanations at one level of analysis complement rather than supersede those at another. Deciding which explanations are most interesting or satisfying is largely a matter of training and taste; debating the issue is usually fruitless *(7)*.

With the Mayr-Tinbergen framework in mind, let us turn to the question of natal dispersal in ground squir-

Kay Holekamp is a research scientist in the Department of Ornithology and Mammalogy at the California Academy of Sciences. She received a B. A. in psychology in 1973 from Smith College, and a Ph.D. in 1983 from the University of California, Berkeley. From 1983 to 1985 she studied reproductive endocrinology as a postdoctoral fellow at the University of California, Santa Cruz. She is currently observing mother-infant interactions and the development of social behaviors in hyenas in Kenya. Paul Sherman is an associate professor of animal behavior at Cornell. He received a B. A. in biology in 1971 from Stanford, and a Ph.D. in zoology in 1976 from the University of Michigan. Following a postdoctoral appointment at the University of California, Berkeley (1976–78), he joined the psychology faculty there. He moved to his present position at Cornell in 1980, and is currently studying the behaviors of naked mole-rats, Idaho ground squirrels, and wood ducks. Address for Dr. Sherman: Section of Neurobiology and Behavior, Seeley G. Mudd Hall, Cornell University, Ithaca, NY 14853.

Figure 1. A female Belding's ground squirrel *(Spermophilus beldingi)* sits with two of her pups in the central Sierra Nevada of California. The pups are about four weeks old, and have recently emerged above ground. At about six or seven weeks of age, male ground squirrels begin to disperse; young females always remain near home. The causes of male dispersal in ground squirrels and many other mammals are complex, but can be explained by using a multilevel analytical approach in which four categories of causal factors are considered separately. (Photo by Cynthia Kagarise Sherman.)

rels. Following analyses of why natal dispersal occurs from each of the four analytical perspectives, we attempt an integration and a synthesis. Our studies reaffirm the usefulness of levels of analysis in determining biological causality.

From 1974 through 1985 we studied three populations of *S. beldingi* near Yosemite National Park in the Sierra Nevada of California (Figs. 1 and 2). In each population, the animals were above ground for only four or five months during the spring and summer; during the rest of the year they hibernated *(1, 2)*. Females bore a single litter of five to seven young per season, and reared them without assistance from males. Most females began to breed as one-year-olds, but males did not mate until they were at least two. Females lived about twice as long as males, both on average (four versus two years) and at the maximum (thirteen versus seven years) *(11)*.

During each field season ground squirrels were trapped alive, weighed, and examined every two to three weeks. About 5,300 different ground squirrels were handled. The animals were marked individually and observed unobtrusively through binoculars for nearly 6,000 hours. Natal dispersal behavior was measured by a combination of direct observations, livetrapping, radio telemetry, and identification of animals killed on nearby roads *(12)*. The day on which each emigrant was last seen within its mother's home range was defined as its date of dispersal. Only those juveniles that were actually seen after leaving their birthplace were classified as dispersers.

Observations of marked pups revealed that natal dispersal was a gradual process, visually resembling the fissioning of an amoeba (see Fig. 3). Young first emerged from their natal burrow and ceased nursing when they were about four weeks old. Two or three weeks later some youngsters began making daily excursions away from, and evening returns to, the natal burrow. Eventually these young stopped returning, restricting their activities entirely to the new home range; by definition, dispersal had occurred.

As shown in Figure 4, natal dispersal is clearly a sexually dimorphic behavior. In our studies, every one of over 300 surviving males dispersed by the end of its second summer; a large majority (92%) dispersed before their first hibernation, by the age of about 16 weeks. In contrast, only 5% of over 250 females recaptured as two-year-olds had dispersed from their mother's home range. The universality of natal dispersal by males suggested no plasticity in its occurrence; however, there was variation among individuals in the age at which dispersal occurred.

During the summer following their birth, males that

Figure 2. *S. beldingi* **in the central Sierra Nevada are found above ground only four or five months of the year, during the spring and summer; they hibernate during the rest of the year. The group above is emerging from an underground burrow. Female adults bear litters of five to seven young each year and rear them in underground burrows without assistance from males. (Photo by George D. Lepp)**

had dispersed as juveniles often moved again, always farther from their birthplace (Fig. 4). Yearling males were last found before hibernation an average of 170 m from their natal burrow, whereas yearling females moved on average only 25 m from home in the same time period. As two-year-olds, males mated at locations that were on average ten times farther from their natal burrows than the mating locations of females *(13)*.

By the time they were two years old, male *S. beldingi* had attained adult body size. In the early spring they collected on low ridges beneath which females typically hibernated. As snow melted and females emerged, the males established small mating territories. Only the most physically dominant males—especially the old, heavy ones—retained territories throughout the three-week mating period. Although dominant males usually copulated with multiple females, the majority of males rarely mated. After mating, the most polygynous males again dispersed. They typically settled far from the places where they had mated; indeed, their new home ranges usually did not include their mating territories. Less successful males tended not to move, and they attempted to mate the following season in the same area where they were previously unsuccessful.

Females were all quite sedentary. After mating on a ridge top close to her hibernation burrow, each female dug a new nest burrow or refurbished an old one—sometimes her own natal burrow. There she reared her pups. As a result of philopatry, females spent their lives surrounded by and interacting with female relatives. Close kin cooperated to maintain and defend nesting territories and to warn each other when predators approached *(13, 14)*. Natal philopatry has facilitated similar nepotism, or favoring of kin, among females in many other species of ground-dwelling sciurid rodents *(15)*.

Physiological mechanisms

We began our analysis of natal dispersal in *S. beldingi* by considering physiological mechanisms. Of the two broad categories of such mechanisms, neuronal and hormonal, we were most interested in the latter. Gonadal steroids can influence the development of a specific behavior in two general ways: through organizational effects, which are the result of hormone action, in utero or immediately postpartum, on tissues destined to control the behavior, and through activational effects, which result from the direct actions of hormones on target tissues at the time the behavior is expressed *(16)*. We suspected that gonadal steroids might mediate natal dispersal, and so we tested for organizational versus activational effects of androgens.

Under the activational hypothesis, levels of circulating androgens should be elevated in juvenile males at the time of natal dispersal. Conversely, in the absence of androgens, males should not disperse. To test this, we studied male pups born and reared in the laboratory. Blood samples were drawn every few weeks for four months *(17)*. We also conducted a field experiment: soon after weaning but prior to natal dispersal, a number of juvenile males and females were gonadectomized; sham operations were performed on a smaller sample of each sex. After surgery, these juveniles were released into their natal burrow and subsequent dispersal behavior was monitored.

Castration was found to have little effect on natal dispersal. Although castrated males and those subjected to sham operations dispersed a few days later than untreated males, probably because of the trauma of surgery, castration did not significantly reduce the fraction that dispersed. Likewise, removal of ovaries did not increase the likelihood of dispersal by juvenile females. Finally, radioimmunoassays revealed only traces of testosterone in the blood of lab-reared juvenile males throughout their first four months, and no increase in circulating androgens was detected at the age when natal dispersal typically occurs (7–10 weeks).

Sex and body mass together were the most consistent predictors of dispersal status

Under the organizational hypothesis, exposing perinatal or neonatal females to androgens should masculinize subsequent behavior, including natal dispersal. We tested this idea by capturing pregnant females and housing them at a field camp until they gave birth. Soon after parturition, female pups were injected with a small amount of testosterone propionate dissolved in oil; a control group was given oil only. After treatment, the pups and their mothers were taken back to the field, where the mothers found suitable empty burrows and successfully reared their young.

Twelve of the female pups treated with androgens were located when they were at least 60 days old, and

75% of them had dispersed *(17)*. The distances they had traveled and their dispersal paths closely resembled those of juvenile males. By comparison, only 8% of untreated juvenile females in the same study area had dispersed by day 60, whereas 60% of juvenile males from the transplanted litters and 74% of males from unmanipulated litters born in the same area had dispersed by day 60.

It is possible that transplantation and not treatment with androgens caused the juvenile females in our experiment to disperse; unfortunately, we were unable to test this because none of the transplanted females treated with only oil were recovered. However, transplantation did not seem to affect the behavior of the juvenile males in the experiment. Also, other behavioral evidence linked natal dispersal in the females with androgen treatment. For example, treated juvenile females did not differ significantly from untreated juvenile males of the same age, but did differ from control females with respect to several indices of locomotor and social behavior. Androgen treatment masculinized much of the behavior of juvenile females, apparently including the propensity to disperse.

These results, which suggest an organizational role for steroids in sexual differentiation of *S. beldingi*, are consistent with those from studies of many other vertebrates *(18)*. In mammals, females are homogametic (XX) and males are heterogametic (XY), whereas in birds the situation is reversed. In each taxon, natal dispersal occurs primarily in the heterogametic sex. In both birds and mammals, sex-typical adult behavior in the homogametic sex can often be reversed by perinatal exposure to the gonadal steroid normally secreted at a particular developmental stage by the heterogametic sex. These considerations suggest that natal dispersal in mammals and birds has a common underlying mechanism, namely the organizational effects of gonadal steroids on the heterogametic sex.

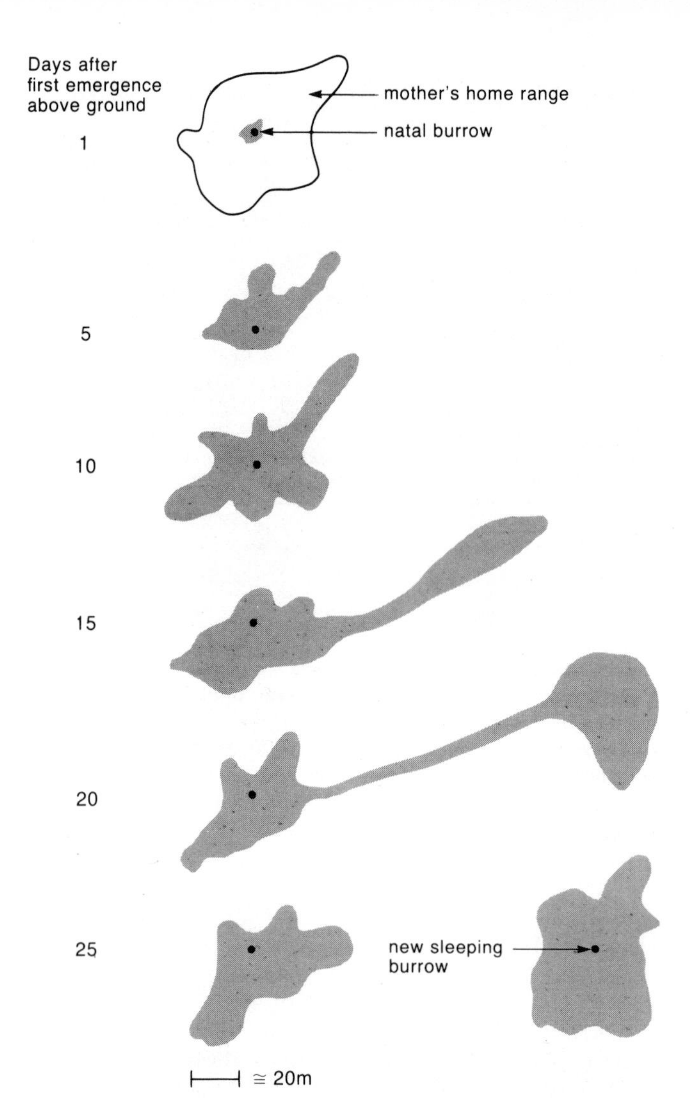

Figure 3. The process by which *S. beldingi* males disperse visually resembles the fissioning of an amoeba. When a male first emerges from the natal burrow, at an age of about four weeks, his daily range of movement is restricted to the immediate vicinity of the burrow. He soon enlarges that range into an amorphous shape, the boundaries of which are established by topographic features or the presence of other animals. By about the 15th day above ground, his range has surpassed the scope of his mother's home range. At this time he may spend long periods far from the natal burrow, yet he will return home at nightfall. Near the 25th day, when he is roughly seven weeks old, he will cease returning at dusk, thereby accomplishing dispersal. (After ref. *35*.)

Ontogenetic processes

Natal dispersal might be triggered during development by changes in either the animal's internal or external environment. We tested two hypotheses about external factors. First, natal dispersal might be caused by aggression directed at juveniles by members of their own species. Under this hypothesis, prior to or at the time of dispersal, the frequency or severity of agonistic behavior between adults and juvenile males should increase. However, observations revealed that adults neither attacked nor chased juvenile males more frequently or vigorously than juvenile females *(19)*, and there was no increase in aggression toward juvenile males at the time of dispersal. Moreover, there were no differences between juvenile males and females in the number and severity of wounds inflicted by other ground squirrels. Thus the data offered no support for the social aggression hypothesis.

A second hypothesis is that natal dispersal occurred because juvenile males attempted to avoid their littermates (current and future competitors) or their mother *(20)*. For a large number of litters, we found no significant relationship between litter size or sex ratio and dispersal behavior *(2, 19)*. Males who dispersed during their natal summer were not from especially large or small litters, or predominantly male or female litters. Also, the timing of juvenile male dispersal depended neither on the mother's age nor on whether the mother was present or deceased. Thus the ontogeny of natal dispersal was apparently not linked to either of the exogenous (external) influences usually invoked to account for it.

In view of these results, we suspected that natal dispersal was triggered by endogenous (internal) factors. In particular, we hypothesized that males might stay home until they attained sufficient size or energy reserves to permit survival during the rigors of emigration. This ontogenetic-switch hypothesis predicts that juvenile

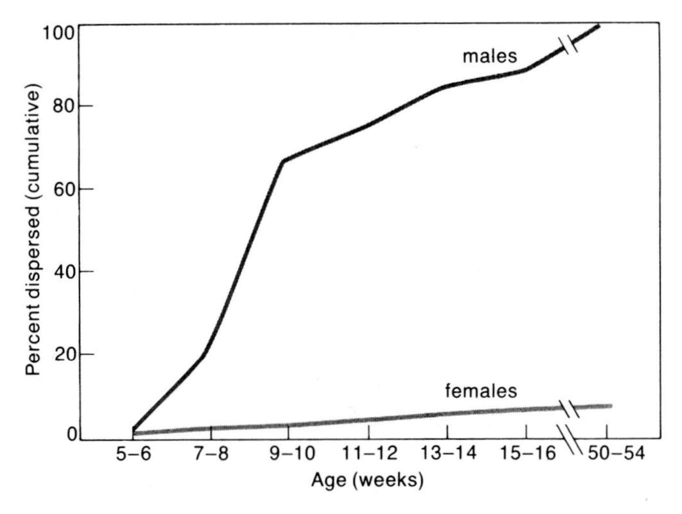

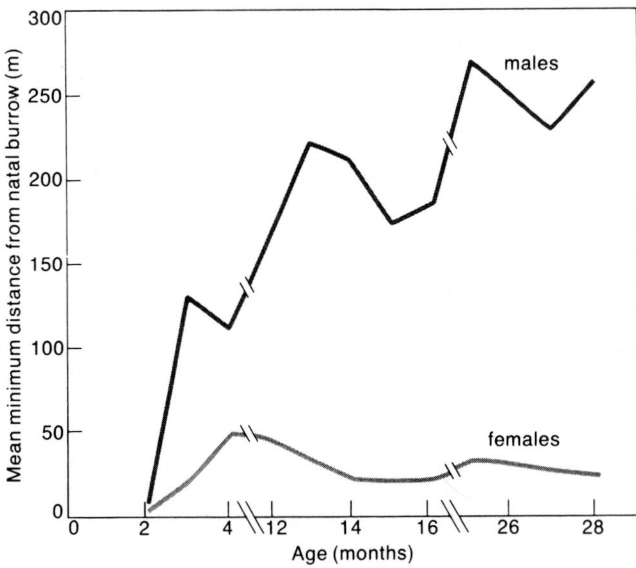

Figure 4. Although a small fraction of female *S. beldingi* disperse, the behavior is very evidently male-biased. The majority of male pups disperse by the 10th week; by about the 54th week, all males have dispersed *(above)*. Although many other mammals exhibit male-biased natal dispersal, *S. beldingi* is unusual in that all males eventually disperse. Males also move considerably farther from their natal burrows than do females, and they continue to move away from home throughout their first three years *(below)*. (After refs. *2* and *12*.)

males will disperse when they attain a threshold body mass and that dispersers should be heavier, or exhibit different patterns of weight gain, than predispersal males of equivalent ages.

Our data were consistent with the ontogenetic-switch hypothesis. Emigration dates were correlated with the time at which males reached a minimum body weight of about 125 g, as shown in Figure 5. Emigrant juveniles were significantly heavier than male pups that had not yet dispersed. Most males attained the threshold weight during their natal summer, and dispersed then. Only the smallest males, who did not put on sufficient weight in the first summer, overwintered in their natal area. All these males dispersed the following season once they had become heavy enough.

Sex and body mass together were the most consistent predictors of dispersal status. Occasionally, however, predispersal and immigrant juvenile males with body weights exceeding the threshold were captured in the same area. This observation suggested that something closely associated with body weight, such as fat stores, may be the actual dispersal trigger.

Behavioral changes also accompanied natal dispersal. The frequencies of movement and distances moved per unit time by juvenile males were found to be greater than those of females, and these behaviors peaked at the time of dispersal. Relative to juvenile females, juvenile males also spent significantly more time climbing and digging and exploring nonfamilial burrows and novel objects—for example, a folding footstool; they also reemerged from a burrow into which they had been frightened much sooner than did females. These observations of spontaneous ontogenetic changes in the behavior of young males reinforced the hypothesis that endogenous factors triggered natal dispersal.

Effects on fitness

Natal dispersal might enable juvenile males to avoid fitness costs associated with life in the natal area and might allow them to obtain benefits elsewhere *(6)*. Possible disadvantages of remaining at home include shortages of food or burrows *(21)*, ectoparasite infestations or diseases, competition with older males for mates *(5, 22)*, and nuclear family incest *(4, 23, 24)*. We examined each of these hypotheses as functional explanations for natal dispersal in *S. beldingi*.

If natal dispersal occurs because of food shortages, then juveniles whose natal burrow is surrounded by abundant food should be more philopatric than those from food-poor areas; immigration to food-rich areas should exceed emigration from them; dispersing individuals should be in poorer condition (perhaps weigh less) than males of the same age residing at home; and, based on the strong sexual dimorphism in natal dispersal, food requirements of young males and females should differ.

Detailed observations revealed that juvenile males and females ate similar amounts of the same plants and at similar rates. Juvenile males spent only slightly more time foraging than did juvenile females. The diets and foraging behaviors of males that had not yet dispersed and males that had immigrated to that same area were indistinguishable. As discussed previously, dispersing males were significantly heavier than predispersal males, a result contrary to that predicted in the scenario of emigration because of lack of food. Finally, juvenile male immigration equaled emigration every year. This is important because preferred foods were unevenly distributed within and among populations *(1, 2)*. Evidence consistently suggested no link between immediate food shortages and natal dispersal.

A second reason for natal dispersal might be to locate a nest burrow. Ground squirrels depend on burrows for safety from predators, as places to spend the night, and as nests in which to hibernate *(25)*. Given the sexual dimorphism in natal dispersal, this hypothesis predicts differences between males and females in the type or location of habitable burrows and implies that dispersers should emigrate from areas of high population density or low burrow quality to areas where unoccupied holes of high quality are available. To test this idea we monitored population density each week and counted burrow entrances in the territories of lactating females. We found that neither the probability of juvenile male dispersal nor its timing was significantly related to population density or burrow availability near home, and that dispersers did not settle in areas of higher burrow density.

The only unusual aspect is that every male eventually leaves home

Another cause of natal dispersal might be ectoparasite infestation. If parasites build up in the natal nest and if juvenile males are more affected by them than are juvenile females, then males in particular might emigrate to avoid them. We examined this hypothesis indirectly, by counting the number of fleas and ticks on every captured juvenile. We found low levels of ectoparasitism throughout the animals' natal summer, and no consistent differences between infestations in males and females prior to or at the time of dispersal.

Do juvenile males disperse to avoid future competition with older males for sexual access to females? Because males always emigrated, it was not possible to determine if dispersers experienced less severe mate competition than hypothetical nondispersers. However, the mate competition hypothesis was examined indirectly by comparing, at sites where males were born and on ridge tops where those males mated two or more years later, three parameters: the density of breeding adult males, the mean number of fights adults engaged in for each successful copulation, and the mean daily ratio of breeding males to receptive females. We found no significant differences in any of these parameters, suggesting that dispersing males did not find better access to females than they would have if they had remained at home.

Do juvenile males disperse to avoid future nuclear family incest? A test of this hypothesis requires comparing the reproductive consequences of various degrees of inbreeding *(26, 27)*. However, of more than 500 copulations observed, none occurred between close kin; therefore we could not directly test this hypothesis. Nonetheless, the nonrandom movements of males away from the natal area clearly resulted in complete avoidance of kin as mates (Fig. 4). Furthermore, during post-breeding dispersal, the highly polygynous males moved farthest. Under this hypothesis, the polygynous males who had sired many female pups in an area would have the most to gain by emigrating. Under the mate-competition hypothesis, successful males would be expected to stay put, while unsuccessful males might gain by dispersing. The observed pattern is thus most consistent with avoiding inbreeding.

Belding's ground squirrels are not unusual in the rarity of close inbreeding. Consanguineous mating is minimized in most mammals and birds *(23, 24, 28, 29)*, often via the mechanism of sex-specific natal dispersal. But why are males the dispersive sex in mammals generally and ground squirrels particularly? The answer probably relates to a sexual asymmetry in the significance

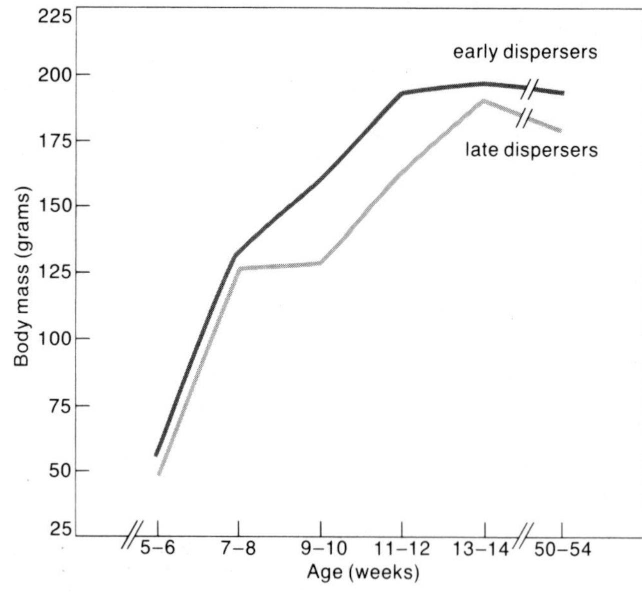

Figure 5. Weight gain among juvenile *S. beldingi* has been positively correlated with the onset of dispersal. Early dispersers (most males) left home at 7–10 weeks of age; late dispersers, in contrast, remained at home until they were 11–14 weeks old. Dispersal seems to occur when a threshold body mass of 125–150 g is attained. (After ref. *19*.)

of the location and quality of burrows for procreation *(6, 30)*. The depth and dryness of nest burrows, their proximity to food, and their degree of protection from both inter- and intraspecific predators are vital to pup survival *(31)*. The significance of the burrow, in turn, favors females who seek out and defend high-quality nest sites and who remain in them from year to year *(25)*. The quality of a nursery burrow is of negligible significance to nonparental males. To avoid predators and inclement weather, and to forage, males can move frequently without jeopardizing the survival of their young. Thus the sexual bias in natal dispersal might occur because inbreeding is harmful to both sexes and males incur lower procreative costs by leaving home.

Sexual selection could reinforce a sex-bias in natal dispersal generated by incest avoidance. If consanguineous mating is indeed harmful, then the philopatric females should prefer to mate with unrelated (unfamiliar) males. A reproductive advantage should therefore accrue to males that seek and locate unfamiliar females *(32)*.

Evolutionary origins

The fourth component of our investigation of natal dispersal was an attempt to infer evolutionary origins. A first hypothesis was that the male bias in natal dispersal arose in an evolutionary ancestor of *S. beldingi* as a developmental error (for example, in the timing of hormone secretion) or as a by-product of natural selection on males for the high levels of activity associated with finding mates and defending mating territories. Alternatively, perhaps natal dispersal was favored directly by selection, for example, as a mechanism to avoid inbreeding, throughout the evolutionary history of *S. beldingi*.

One way to evaluate these alternatives is to consider

Why do juvenile male Belding's ground squirrels disperse? Answers have been found at each of four levels of analysis.

Level of analysis	Summary of findings
Physiological mechanisms	Dispersal by juvenile males is apparently caused by organizational effects of male gonadal steroid hormones. As a result, juvenile males are more curious, less fearful, and more active than juvenile females.
Ontogenetic processes	Dispersal is triggered by attainment of a particular body mass (or amount of stored fat). Attainment of this mass or composition apparently also initiates a suite of locomotory and investigative behaviors among males.
Effects on fitness	Juvenile males probably disperse to reduce chances of nuclear family incest.
Evolutionary origins	Strong male biases in natal dispersal characterize all ground squirrel species, other ground-dwelling sciurid rodents, and mammals in general. The consistency and ubiquity of the behavior suggest that it has been selected for directly across mammalian lineages.

the taxonomic distribution of male-biased natal dispersal. If selection has consistently and directly favored dispersal by juvenile males, then phylogenetic relatives of *S. beldingi* should share this trait to a greater degree than if it were a hormonally mediated side effect or developmental error. This is because any hormonal link between adult male sexual activities and dispersal by juveniles two years previously could presumably be broken by mutation in some species through evolutionary time. This, in turn, would lead to a spotty taxonomic distribution of the behavior if it were neutral for fitness.

Members of the squirrel family first appeared in the fossil record 35 to 40 million years ago; thus they are one of the most ancient of extant rodent families *(33)*. Belding's ground squirrel is one of 32 species in the genus *Spermophilus*; this genus is more closely related to marmots and prairie dogs than to tree squirrels *(34)*. Strongly male-biased natal dispersal occurs in all 12 *Spermophilus* species that have been studied in this regard *(5, 15, 35)*. Male-biased natal dispersal patterns are also the rule in marmots *(35)* and prairie dogs *(36)*. The dispersal behavior of *S. beldingi* is therefore probably a conservative rather than a derived trait; in other words, it is likely quite ancient.

As far as we know, the only unusual aspect of natal dispersal in *S. beldingi* is that every male eventually leaves home, whereas in a few other species a tiny fraction of males are philopatric. Male-biased natal dispersal is widespread among mammals *(4–6, 30, 32, 37)*, suggesting that this behavior may predate the appearance of the squirrel family. The ubiquity of natal dispersal seems more consistent with the hypothesis that it has been favored directly by natural selection in various lineages than that it originated as a mistake or a correlated response to selection for some other male attribute and is maintained by phylogenetic inertia rather than adaptive value.

Synthesis

Our data reveal that there are at least four types of answers to the question of why juvenile male Belding's ground squirrels disperse (see the box). These answers complement rather than supersede each other. Clearly, however, the causal variables we have identified within each analytical level do not operate in isolation, and it seems appropriate to consider how they may interrelate.

During embryogenesis, sex chromosomes cause the formation of testes in male *S. beldingi*. The gonads secrete a pulse of androgens before birth, which, we hypothesize, sets up an ontogenetic switch, presumably by modifying the morphology or behavior of neurons or nuclei in the brain. When juvenile males have accumulated sufficient weight or fat stores, the switch turns on. The young males then boldly explore their environment, making increasingly longer forays away from home. The timing of dispersal by each individual may be influenced by any environmental factor that accelerates or delays arrival at the dispersal threshold (for example, food abundance or scarcity). The main cost of natal dispersal is probably mortality during emigration; the main benefits are likely related to reduced inbreeding and optimal outbreeding. Male biases in natal dispersal occur consistently across modern mammalian taxa *(37)*, suggesting an evolutionary history of natural selection favoring such behavior directly, and a taxon-wide consistency of function.

By employing the levels-of-analysis framework for developing and testing hypotheses, we have come to appreciate the complexity of what at first appeared to be a simple behavior. We suspect that our explanations for the proximate and ultimate causes of natal dispersal in *S. beldingi* will be applicable to other species. Perhaps equally important, our study illustrates that there can be multiple correct answers to questions of causality in behavioral biology *(38)*. The usefulness of the levels-of-analysis approach is thereby reemphasized.

References

1. P. W. Sherman. 1976. Natural selection among some group-living organisms. Ph.D. thesis, Univ. of Michigan.
2. K. E. Holekamp. 1983. Proximal mechanisms of natal dispersal in Belding's ground squirrels *(Spermophilus beldingi beldingi)*. Ph.D. thesis, Univ. of California.
3. W. Z. Lidicker, Jr. 1975. The role of dispersal in the demography of small mammals. In *Small Mammals: Their Productivity and Population Dynamics*, ed. F. B. Golley, K. Petruscewicz, and C. Ryszkowski, pp. 103–28. Cambridge Univ. Press.
4. P. J. Greenwood. 1980. Mating systems, philopatry and dispersal in birds and mammals. *Animal Behav.* 28:1140–62.
5. F. S. Dobson. 1982. Competition for mates and predominant juvenile male dispersal in mammals. *Animal Behav.* 30:1183–92.
6. A. E. Pusey. 1987. Sex-biased dispersal and inbreeding avoidance in birds and mammals. *Trends in Ecol. and Evol.* 2:295–99.
7. P. W. Sherman. 1988. The levels of analysis. *Animal Behav.* 36: 616–19.
8. D. S. Lehrman. 1970. Semantic and conceptual issues in the

nature-nurture problem. In *Development and Evolution of Behavior,* ed. L. R. Aronson, E. Tobach, D. S. Lehrman, and J. S. Rosenblatt, pp. 17–52. W. H. Freeman.

9. E. Mayr. 1961. Cause and effect in biology. *Science* 134:1501–06.
10. N. Tinbergen. 1963. On aims and methods of ethology. *Zeitschrift für Tierpsychologie* 20:410–33.
11. P. W. Sherman and M. L. Morton. 1984. Demography of Belding's ground squirrels. *Ecology* 65:1617–28.
12. K. E. Holekamp. 1984a. Natal dispersal in Belding's ground squirrels *(Spermophilus beldingi). Behav. Ecol. Sociobiol.* 16:21–30.
13. P. W. Sherman. 1977. Nepotism and the evolution of alarm calls. *Science* 197:1246–53.
14. P. W. Sherman. 1981a. Kinship, demography, and Belding's ground squirrel nepotism. *Behav. Ecol. Sociobiol.* 8:251–59.
15. G. R. Michener. 1983. Kin identification, matriarchies, and the evolution of sociality in ground-dwelling sciurids. In *Recent Advances in the Study of Mammalian Behavior,* ed. J. F. Eisenberg and D. G. Kleiman, pp. 528–72. Am. Soc. Mammal.
16. C. H. Phoenix, R. W. Goy, A. A. Gerall, and W. C. Young. 1959. Organizing action of prenatally administered testosterone propionate on the tissues mediating mating behavior in the female guinea pig. *Endocrinology* 65: 369–82.
17. K. E. Holekamp, L. Smale, H. B. Simpson, and N. A. Holekamp. 1984. Hormonal influences on natal dispersal in free-living Belding's ground squirrels *(Spermophilus beldingi). Hormones and Behavior* 18:465–83.
18. E. Adkins-Regan. 1981. Early organizational effects of hormones: An evolutionary perspective. In *Neuroendocrinology of Reproduction,* ed. N. T. Adler, pp. 159–228. Plenum Press.
19. K. E. Holekamp. 1986. Proximal causes of natal dispersal in Belding's ground squirrels *(Spermophilus beldingi). Ecol. Monogr.* 56: 365–91.
20. S. Pfeifer. 1982. Disappearance and dispersal of *Spermophilus elegans* juveniles in relation to behavior. *Behav. Ecol. Sociobiol.* 10:237–43.
21. F. S. Dobson. 1979. An experimental study of dispersal in the California ground squirrel. *Ecology* 60:1103–09.
22. J. Moore and R. Ali. 1984. Are dispersal and inbreeding avoidance related? *Animal Behav.* 32:94–112.
23. A. E. Pusey and C. Packer. 1987. The evolution of sex-biased dispersal in lions. *Behaviour* 101:275–310.
24. A. Cockburn, M. P. Scott, and D. J. Scotts. 1985. Inbreeding avoidance and male-biased natal dispersal in *Antechinus* spp. (Marsupialia: Dasyuridae). *Animal Behav.* 33:908–15.
25. J. A. King. 1984. Historical ventilations on a prairie dog town. In *The Biology of Ground-dwelling Squirrels,* ed. J. O. Murie and G. R. Michener, pp. 447–56. Univ. of Nebraska Press.
26. W. M. Shields. 1982. *Philopatry, Inbreeding, and the Evolution of Sex.* State Univ. of New York Press.
27. P. J. Greenwood, P. H. Harvey, and C. M. Perrins. 1978. Inbreeding and dispersal in the great tit. *Nature* 271:52–54.
28. J. L. Hoogland. 1982. Prairie dogs avoid extreme inbreeding. *Science* 215:1639–41.
29. K. Ralls, P. H. Harvey, and A. M. Lyles. 1986. Inbreeding in natural populations of birds and mammals. In *Conservation Biology: The Science of Scarcity and Diversity,* ed. M. E. Soulé, pp. 35–56. Sinauer.
30. P. M. Waser and W. T. Jones. 1983. Natal philopatry among solitary mammals. *Q. Rev. Biol.* 58:355–90.
31. P. W. Sherman. 1981b. Reproductive competition and infanticide in Belding's ground squirrels and other animals. In *Natural Selection and Social Behavior,* ed. R. D. Alexander and D. W. Tinkle, pp. 311–31. Chiron Press.
32. A. E. Pusey and C. Packer. 1986. Dispersal and philopatry. In *Primate Societies,* ed. B. B. Smuts, D. L. Cheney, R. M. Seyfarth, R. W. Wrangham, and T. T. Struhsaker, pp. 250–66. Univ. of Chicago Press.
33. W. P. Luckett and L. J. Hartenberger, eds. 1985. *Evolutionary Relationships among Rodents.* Plenum Press.
34. D. J. Hafner. 1984. Evolutionary relationships of the nearctic Sciuridae. In *The Biology of Ground-dwelling Squirrels,* ed. J. O. Murie and G. R. Michener, pp. 3–23. Univ. of Nebraska Press.
35. K. E. Holekamp. 1984b. Dispersal in ground-dwelling sciurids. In *The Biology of Ground-dwelling Squirrels,* ed. J. O. Murie and G. R. Michener, pp. 297–320. Univ. of Nebraska Press.
36. M. G. Garrett and W. L. Franklin. 1988. Behavioral ecology of dispersal in the black-tailed prairie dog. *J. Mammal.* 69:236–50.
37. B. D. Chepko-Sade and Z. T. Halpin, eds. 1987. *Mammalian Dispersal Patterns.* Univ. of Chicago Press.
38. P. W. Sherman. 1989. The clitoris debate and the levels of analysis. *Animal Behav.* 37:697–98.

Animal Thinking

Donald R. Griffin

Ethologists are once again investigating the possibility that animals have conscious awareness

What is it like to be an animal? What do monkeys, dolphins, crows, sunfishes, bees, and ants think about? Or do nonhuman animals experience any thoughts and subjective feelings at all? Aside from Lorenz (1963) and Hediger (1947, 1968, 1980) very few ethologists have discussed animal thoughts and feelings. While seldom denying their existence dogmatically, they emphasize that it is extremely difficult, perhaps impossible, to learn anything at all about the subjective experiences of another species. But the difficulties do not justify a refusal to face up to the issue. As Savory (1959) put the matter, "Of course to interpret the thoughts, or their equivalent, which determine an animal's behaviour is difficult, but this is no reason for not making the attempt to do so. If it were not difficult, there would be very little interest in the study of animal behaviour, and very few books about it" (p. 78).

Most biologists and psychologists tend, explicitly or implicitly, to treat most of the world's animals as mechanisms, complex mechanisms to be sure, but unthinking robots nonetheless. Mechanical devices are usually considered to be incapable of conscious thought or subjective feeling, although it is currently popular to ascribe mental experiences to computer systems. John (in Thatcher and John 1977), among others, has equated consciousness with a sort of internal feedback whereby information about one part of a pattern of information flow acts on another part. This may be a necessary condition for conscious thinking, but it is also an aspect of many physiological processes that operate without any conscious awareness on our part.

Many comparative psychologists seem petrified by the notion of animal consciousness. Historically, the science of psychology has been reacting for fifty years or more against earlier attempts to understand the workings of the human mind by introspective self-examination—trying to learn how we think by thinking about our thoughts. This effort led to confusing and contradictory results, so in frustration experimental psychologists largely abandoned the effort to understand human consciousness, replacing introspection with objective experiments. While experiments have been very helpful in analyzing learning and other human abilities, the rejection of any concern with consciousness and subjective feelings has gone so far that many psychologists virtually deny their existence or at least their accessibility to scientific analysis.

In one rather extreme form of this denial, Harnad (1982) has argued that only after the functioning of our brains has determined what we will do does an illusion of conscious awareness arise, along with the mistaken belief that we have made a choice or had control over our behavior. The psychologists who thus belittle and ignore human consciousness can scarcely be expected to tell us much about subjective thoughts and feelings of animals. If we cannot gather any verifiable data about our own thoughts and feelings, the argument has run, how can we hope to learn anything about those of other species?

A long-overdue corrective reaction to this extreme antimentalism is well under way. To a wide range of scholars, and indeed to virtually the whole world outside of narrow scientific circles, it has always been self-evident that human thoughts and feelings are real and important (see, for example, MacKenzie 1977 and Whiteley 1973). This is not to underestimate the difficulties that arise when one attempts to gather objective evidence about other people's feelings and thoughts, even those one knows best. But it really is absurd to deny the existence and importance of mental experiences just because they are difficult to study.

Why do so many psychologists appear to ignore a central area of their subject matter when most other branches of science refrain from such self-inflicted paralysis? The usual contemporary answer to such a question is that a relatively new sort of cognitive psychology has developed during the past twenty or thirty years, based in large part on the analysis of human and animal behavior in terms of information-processing (reviewed in Norman 1981). Analogies to computer programs play a large part in this approach, and many cognitive psychologists draw their inspiration from the success of computer systems, feeling that certain types of programs can serve as instructive models of human thinking. Words that used to be reserved for conscious human beings are now commonly used to describe the impressive accomplishments of computers. Despite the

Donald R. Griffin, a professor at The Rockefeller University, is an authority on animal physiology and behavior, best known for his work on echolocation in bats and other animals. He is the author of numerous articles and books on ethology and comparative physiology, including Listening in the Dark *(1958),* Echoes of Bats and Men *(1959),* Bird Migration *(1964), and* The Question of Animal Awareness *(1976). The present article is adapted by permission of the publisher from* Animal Thinking, *published in April 1984 by Harvard University Press. Address: The Rockefeller University, 1230 York Avenue, New York, NY 10021.*

Figure 1. A chimpanzee, having selected a suitably shaped small branch, strips it of twigs and leaves, transforming it into a serviceable tool for capturing termites. The chimpanzee will then walk to a termite nest, often at a considerable distance, and probe it with the stick for termites. Such behavior, which differs radically from other activities of chimpanzees, seems difficult to understand unless the animal is consciously thinking about gathering termites while preparing the probe. (Photograph by Linda Koebner, Bruce Coleman Inc.)

optimism of computer enthusiasts, however, it is highly unlikely that any computer system can spontaneously generate subjective mental experience (Boden 1977; Dreyfus 1979; Baker 1981).

Conspicuously absent from most of contemporary cognitive psychology is any serious attention to conscious thoughts or subjective feelings. For example, Wasserman (1983) defends cognitive psychology to his fellow behaviorists by arguing that it is not subjective and mentalistic. Analyzing people as though they were computers may be useful as an initial, limited approach, just as physiologists began their analysis of the functioning of hearts by drawing analogies to mechanical pumps. But it is important to recognize the limitations inherent in this approach; it suffers from the danger of leading us into what Savory (1959) called by the apt but unfortunately tongue-twisting name of "the synechdochaic fallacy." This means the confusion of a part of something with the whole, or as Savory put it, "the error of nothing but." Information-processing is doubtless a necessary condition for mental experience, but is it sufficient? Human minds do more than process information; they think and feel. We experience beliefs, desires, fears, expectations, and many other subjective mental states.

Many cognitive psychologists imply that a computer system that could process information exactly as the human brain does would duplicate all essential elements of thinking and feeling; others simply feel that subjective experience is beyond the reach of scientific investigation. Perhaps the issue will someday be put to an empirical test, but the extent and the complexity of information-processing in our brains is so great that available procedures can detect only a tiny fraction of it, and even if it could be monitored in full detail, we do not know whether any computer system could duplicate it.

The difference between conscious and nonconscious states is a significant one, yet most scientists concerned with animal behavior have felt that looking for consciousness in animals would be a futile anachronism. This defeatist attitude is based in part on convincing evidence that we do a great deal of problem-solving, decision-making, and other kinds of information-processing without any consciousness of what is going on. Harnad (1982) bases his belief that human consciousness is merely an illusion on the fact that we are conscious of only the tip of the iceberg of information-processing in our brains. Indeed the ratio of conscious to unconscious brain activity is probably even smaller than the density ratio of ice to water. The intellectual excitement of this discovery has obscured the obvious fact that we are conscious some of the time, and we certainly do experience many sorts of thoughts and feelings that are very important to us and our companions. If the choice were open, would anyone prefer a lifelong state of sleepwalking?

What behavior suggests conscious thinking?

Just what is it about some kinds of behavior that leads us to feel that it is accompanied by conscious thinking? Comparative psychologists and biologists worried about this question extensively around the turn of this century. No clear and generally accepted answers emerged from their thoughtful efforts, and this is one reason why the behavioristic movement came to dominate psychology.

Complexity is often taken as evidence that some behavior is guided by conscious thinking. But complexity is a slippery attribute. One might think that simply running away from a frightening stimulus was a rather simple response, yet if we make a detailed description of every muscle contraction during turning and running away, the behavior becomes extremely complex. But, one might object, this complexity involves the physiology of locomotion; what is simple is the direction in which the animal moves. If we then ask what sensory and central nervous mechanisms cause the animal to move in this direction, the matter again becomes complex. Does the animal continuously listen to the danger signal and push more or less hard with its right or left legs in order to keep the signal directly behind it? Or does it head directly toward some landmark? If the latter,

Figure 2. The assassin bug at the top has camouflaged itself chemically and tactilely by gluing bits of a termite nest all over its body. In this way it is able to capture a termite at the opening of the nest without alarming the soldier termites. After sucking out the termite's semifluid organs, the assassin bug jiggles the empty exoskeleton in front of the nest opening in order to attract another termite worker, which will normally attempt to consume or dispose of the corpse. When a second termite worker seizes the first, it is then captured and consumed itself, as shown in the photograph below, and the process may be repeated continuously many times by the same assassin bug. The extraordinary complexity and coordination of these actions strongly suggest conscious thought, even though the assassin bug's central nervous system is very small. (Photographs by Raymond A. Mendez.)

how does it coordinate vision and locomotion? Again one might say that the direction of motion is simple, and it is irrelevant to worry about the complexities of the physiological mechanisms involved.

But how is this simple direction "away from the danger" represented within the animal's central nervous system? Does the animal employ the concepts of *away from* and *danger*? If so, how are such concepts established? Even though we cannot answer the question in neurophysiological terms, it is clear that running away from something is a far simpler behavior than, say, the construction of a bird's nest. Conversely, even the locomotor motions of a caterpillar that will move toward a light with a machine-like consistency hour after hour are not simple when examined in detail. What is simple is the abstract notion of *toward* or *away*, but the mechanistic interpretation of animal behavior tends to deny that the animal could think in terms of even such a simple abstraction.

One very important attribute of animal behavior that seems intuitively to suggest conscious thinking is its adaptability to changing circumstances. If an animal repeats some action in the same way regardless of the results, we assume that a rigid physiological mechanism is at work, especially if the behavior is ineffective or harmful to the animal. When a moth flies again and again at a bright light or burns itself in an open flame, it is difficult to imagine that the moth is thinking, although one can suppose that it is acting on some thoughtful but misguided scheme. When members of our own species do things that are self-damaging or even suicidal, we do not conclude that their behavior is the result of a mechanical reflex. But to explain the moth flying into the flame as thoughtful but misguided seems far less plausible than the usual interpretation that such insects automatically fly toward a bright light, which leads them to their death in the special situation where the brightest light is an open flame.

Conversely, if an animal manages to obtain food by a complex series of actions that it has never performed before, intentional thinking seems more plausible than rigid automatism. For example, Japanese macaques learned a new way to separate grain from inedible material by throwing the mixture into the water; the kernels of grain would float while the inorganic sand and other particles tended to sink (Kawai 1965). These new types of food handling were first devised by a few monkeys, then were gradually acquired by other members of their social group through observational learning.

Behavioral versatility came into play in a spectacular fashion in the 1930s when two species of tits discovered that milk bottles delivered to British doorsteps could be a source of food (Fisher and Hinde 1949; Hinde and Fisher 1951). At that time milk-bottle tops were made of soft metal foil, and the milk was not homogenized, so the cream rose to the top of the bottle. One or more birds discovered that the same type of behavior used to get at insects hidden under tree bark could also be used to get cream from milk bottles. The people whose milk was disturbed immediately noticed it, and careful studies were made of the gradual spread of this behavior throughout much of England. A change in the technology of covering milk bottles eventually ended the whole business, but meanwhile thousands of birds had learned, almost certainly through observation, to exploit a newly available food source.

Connected patterns of behavior

Another criterion upon which we tend to rely in inferring conscious thinking is the element of interactive steps in a relatively long sequence of appropriate behavior patterns. Effective and versatile behavior often entails many steps, each one modified according to the results of the previous actions. In such a complex sequence the animal must pay attention not only to the immediate stimuli, but also to information obtained in the past. Psychologists once postulated that complex behavior can be understood as a chain of rigid reflexes, the outcome of one serving as stimulus for the next. Students of insect behavior have generally accepted this explanation for such complex activities as the construction of elaborate shelters or prey-catching devices, ranging from the underwater nets spun by certain caddis-fly larvae to the magnificent webs of spiders. But the steps an animal takes often vary, depending on the results of the previous behavior and on many influences from the near or distant past. The choice of *which* past events to attend to may be facilitated by conscious selection from a broad spectrum of memories.

Figure 3. With an apparent deliberateness that suggests intentional thinking, nesting killdeer will conspicuously lead a potential predator away from their nest or young, adjusting the speed of their abnormally awkward movements away from the nest so as to remain within sight but out of reach of the intruder. Quite often the killdeer acts as if it is injured, as shown here, a behavior that would make it more attractive to a predator. The killdeer will sometimes employ a very different tactic in response to approaching cattle, which may trample on eggs or nestlings but will not eat them; the birds will then stand close to the nest and spread their wings in a conspicuous display that usually causes the cattle to step aside. (Photograph by Noble S. Proctor.)

An outstanding example of such sequences of interactive behaviors is the use of probes by chimpanzees to gather termites from their mounds (Goodall 1968, 1971). The chimpanzee prepares a probe by selecting a suitable branch, pulling off its leaves and side branches, breaking the stick to the right length, carrying it—often for several minutes—to a termite mound, and then probing into the openings used by the termites (see Fig. 1). If the hole yields nothing, the chimpanzee moves to another one. Even after the tool has been prepared, its use is far from stereotyped. When curious scientists try to imitate the chimpanzees' techniques, they find it rather difficult and seldom gather as many termites. It is especially interesting that the young chimpanzees seem to learn this use of tools by watching their mothers or other members of their social group. Youngsters have been observed making crude and relatively ineffectual attempts to prepare and use their own termite probes; the termite "fishing" of chimpanzees gives every evidence of being learned.

Examples can be found among insects of long, complex sequences of behaviors that are as suggestive of thought as the chimpanzee's use of a termite probe. McMahan (1982) has discovered that in tropical rain forests one species of assassin bug, a predatory insect, uses two effective tricks, illustrated in Figure 2, to capture the workers of termite colonies. The bug glues small bits of the outer layers of a termite nest to its head, back, and sides. Then it stands near an opening to the termite colony. The bits of termite nest on the assassin bug apparently smell and perhaps feel familiar to the termites, so no alarm signals are emitted, which otherwise would attract the well-armed members of the soldier caste that attack intruders. Although the assassin bug's actions often attract soldier termites, its camouflage seems to prevent them from recognizing it as an intruder, and they return to the nest. This chemical and tactile camouflage allows the assassin bug to reach into the opening and capture a termite worker, which it kills and consumes by sucking out all the semifluid internal organs, leaving only the exoskeleton.

Such camouflage-assisted prey capture is remarkable enough, but the next step is even more thought-provoking. The assassin bug pushes the empty exoskeleton of its victim into the nest opening and jiggles it gently. Another termite worker seizes the corpse as part of a normal behavior pattern of devouring the body of a dead sibling or carrying the corpse away for disposal. The assassin bug pulls the exoskeleton of the first victim out with the second worker attached. This one is eaten and its empty exoskeleton used in another "fishing" effort. In one case an assassin bug was observed to thus devour thirty-one termites before moving away with a fully distended abdomen.

When chimpanzees fashion sticks to probe for termites, their actions are considered among the most convincing cases of intentional behavior yet described for nonhuman animals. When McMahan discovers assassin bugs carrying out an almost equally elaborate feeding behavior, must we assume that the insect is only a genetically programmed robot incapable of understanding what it does? Perhaps we should be ready to infer conscious thinking whenever any animal shows such ingenious behavior, regardless of its taxonomic group and our preconceived notions about limitations of animal consciousness.

An example even more suggestive of thoughtful behavior among insects is the so-called "dance language" of honeybees, which was discovered by the remarkably brilliant and original experiments of von Frisch (reviewed by von Frisch 1967, 1972). One significant reaction to von Frisch's discovery was that of Jung (1973). Late in his life he wrote that although he had believed insects were merely reflex automata:

this view has recently been challenged by the researches of Karl von Frisch. . . . Bees not only tell their comrades, by means of a peculiar sort of dance, that they have found a feeding-place, but they also indicate its direction and distance, thus enabling beginners to fly to it directly. This kind of message is no different in principle from information conveyed by a human being. In the latter case we would certainly regard such behavior as a conscious and intentional act and can hardly imagine how anyone could prove in a court of law that it had taken place unconsciously. . . . We are . . . faced with the fact that the ganglionic system apparently achieves exactly the same result as our cerebral cortex. Nor is there any proof that bees are unconscious.

In many cases the networks of informative events that suggest animal thinking are sufficiently complicated that we are not sure what the animal is doing even when we know most of the relevant facts. Consider, for example, how certain ground-nesting birds, such as the killdeer or piping plovers, lead predators away from their nests or young. At a considerable distance, long before an approaching large intruder, such as a person or other mammal, can see the cryptically colored bird or its eggs, the plover may stand up and walk slowly to a point a few meters from the nest. Then the bird may flutter slowly but conspicuously away from the nest, staying relatively close to the intruder. It almost always makes loud piping or peeping sounds similar to those a bird makes when disturbed or mildly irritated.

It is common for the bird to hold its tail or wing in an abnormal position as it moves. Often the tail almost drags on the ground, and the wings are slightly extended, sometimes one more than the other, strongly suggesting some weakness or injury. After running a few meters, the bird may flop about on the ground, extending one or both wings, as if injured. This behavior, shown in Figure 3, is often called the "broken-wing display," and it requires considerable effort for an observer to believe that the bird is really quite healthy. Predators are extremely sensitive to minor differences in the gait and demeanor of potential prey and are much more likely to attack animals that are behaving abnormally.

Throughout most of this predator-distraction behavior, the bird watches the intruder. Typically it does not move in a straight line and stops from time to time. If the intruder approaches, the bird moves farther ahead. If not, the bird usually flies back closer to the intruder and repeats the behavior. The bird will allow the intruder to approach quite close, sometimes within two meters, but it always moves just fast enough and far enough to avoid capture. Typically the bird continues the injury simulation while leading the intruder some distance away from the nest or young. Finally, however, it flies away rapidly, usually in the same direction, then circling back to the general vicinity, though seldom to the exact spot, where the eggs or young are located. One Wilson's plover, a close relative of the killdeer and piping plover, led me more than 300 meters along a sandy beach before flying off.

Killdeer have been observed to use very different tactics when their nests are approached by cattle, which may trample on eggs or nestlings but will not eat them. Rather than moving away from the nest and fluttering as though injured, the birds stand close to the nest and spread their wings in a conspicuous display that usually causes the cattle to step aside (Skutch 1976).

Adaptations to novelty

One further consideration can help refine the criteria for determining the presence of conscious thought. We can easily change back and forth between thinking consciously about our own behavior and not doing so. When we are learning some new task such as swimming, riding a bicycle, driving an automobile, flying an airplane, operating a vacuum cleaner, caring for our teeth by some new technique recommended by a dentist, or any of the large number of actions we did not formerly know how to do, we think about it in considerable detail. But once the behavior is thoroughly mastered, we give no conscious thought to the details that once required close attention.

This change can also be reversed, as when we make the effort to think consciously about some commonplace and customary activity we have been carrying out for some time. For example, suppose you are asked about the pattern of your breathing, to which you normally give no thought whatsoever. But you can easily take the trouble to keep track of how often you inhale and exhale, how deeply, and what other activities accompany different patterns of breathing. You can find out that it is extremely difficult to speak while inhaling, so talking continuously requires rapid inhalation and slower exhalation. This and other examples that will readily come to mind if one asks the appropriate questions show that we can bring into conscious focus activities that usually go on quite unconsciously.

The fact that our own consciousness can be turned on and off with respect to particular activities tells us that in at least one species it is not true that certain behavior patterns are always carried out consciously while others never are. It is reasonable to guess that this is true also for other species. Well-learned behavior patterns may not require the same degree of conscious attention as those the animal is learning how to perform. This in turn means that conscious awareness is more likely when the activity is novel and challenging; striking and unexpected events are more likely to produce conscious awareness.

Thus it seems likely that a widely applicable, if not all-inclusive, criterion of conscious awareness in animals is *versatile adaptability of behavior to changing circumstances and challenges.* If the animal does much the same things regardless of the state of its environment or the behavior of other animals nearby, we are less inclined to judge that it is thinking about its circumstances or what it is doing. Consciously motivated behavior is more plausibly inferred when an animal behaves appropriately in a novel and perhaps surprising situation that requires specific actions not called for under ordinary circumstances. This is a special case of versatility, of course, but the rarity of the challenge combined with the appropriateness and effectiveness of the response are important indicators of thoughtful actions.

For example, Janes (1976) observed nesting ravens make an enterprising use of rocks. He had been closely observing ten raven nests in Oregon, eight of which were near the top of rocky cliffs. At one of these nests two ravens flew in and out of a vertical crack that extended from top to bottom of a twenty-meter cliff. Janes and a companion climbed up the crevice and inspected the six nearly fledged nestlings. As they started down,

Figure 4. When sea otters cannot open shellfish with their claws or teeth, they often employ a stone for the purpose, smashing it against the shells. The apparently conscious, intentional nature of the behavior is further indicated by the fact that an otter carefully selects a stone of suitable size and weight and often carries the stone under its armpit for considerable periods. (Photograph by William F. Bryan.)

both parents flew at them repeatedly, calling loudly, then landed at the top of the cliff, still calling. One of the ravens then picked up small rocks in its bill and dropped them at the human intruders. Several of the rocks showed markings where they had been partly buried in the soil, so the birds presumably had pried them loose. Only seven rocks were dropped, but the raven seemed to be seeking other loose ones and apparently stopped only because no more suitable rocks were available.

While many birds make vigorous efforts to defend their nests and young from intruders, often flying at people who come too close, regurgitating or defecating on them, and occasionally striking them with their bills, rock throwing is most unusual. Nor do ravens pry out rocks and drop them in other situations. It is difficult to avoid the inference that this quite intelligent and adaptable bird was anxious to chase the human intruders away from its nest and decided that dropping rocks might be effective.

There are limits to the amount of novelty with which a species can cope successfully, and this range of versatility is one of the most significant measures of mental adaptability. This discussion of adaptable versatility as a criterion of consciousness implies that conscious thinking occurs only during learned behavior, but we should be cautious in accepting this belief as a rigid doctrine.

Another aspect of conscious thinking is anticipation and intentional planning of an action with conscious awareness of its likely results. An impressive example is the use of small stones by sea otters to detach and open shellfish (Kenyon 1969). These intelligent aquatic carnivores feed mostly on sea urchins and mollusks. The sea otter must dive to the bottom and pry the mollusk loose with claws or teeth, but some shells, especially abalones, are tightly attached to the rocks and have shells that are too tough to be loosened in this fashion. The otter will search for a suitable stone, which it carries while diving, then uses the stone to hammer the shellfish loose, holding its breath all the while.

The otter usually eats while floating on its back, as shown in Figure 4. If it cannot get at the fleshy animal inside the shell, it will hold the shell against its chest with one paw and pound it with the stone. The otter often tucks a good stone under an armpit as it swims or dives. Although otters do not alter the shapes of the stones, they do select ones of suitable size and weight and often keep them for considerable periods. The otters use tools only in areas where sufficient food cannot be obtained by other methods. In some areas only the young and very old sea otters use stones; vigorous adults can dislodge the shellfish with their unaided claws or teeth. Thus it is far from a simple stereotyped behavior pattern, but one that is used only when it is helpful. Sea otters sometimes use floating beer bottles to hammer open shells. Since the bottles float, they need not be stored under the otter's armpit.

Anticipation and planning are of course impossible to observe directly in another person or animal, but indications of their likelihood are often observable. As early as the 1930s Lorenz studied the intention movements of birds (Lorenz 1971), and other ethologists have noted that these movements, small-scale preliminaries to major actions such as flying, often serve as signals to others of the same species. Although Lorenz interpreted the movements as indications that the bird was planning and preparing to fly, the term *intention movement* has been quietly dropped from ethology in recent years. I suspect this is because the behavioristic ethologists fear that the term has mentalistic implications. Earlier ethologists such as Daanje (1951) described a wide variety of intention movements in many kinds of animals, but their interest was in whether the movements had gradually become specialized communication signals in the course of behavioral evolution. The possibility that intention movements indicate the animal's conscious intention has been totally neglected by ethologists during their behavioristic phase, but we may hope that the revival of scientific interest in animal thinking will lead cognitive ethologists to study whether such movements are accompanied by conscious intentions.

Animal communication

The very fact that intention movements so often evolve into communicative signals may reflect a close linkage between thinking and the intentional communication of thoughts from one conscious animal to another. These considerations lead us directly to a recognition that because communicative behavior, especially among social animals, often seems to convey thoughts and feelings from one animal to another, it can tell us something about animal thinking: it can be an important "window" on the minds of animals.

Human communication is hardly limited to formal language; nonverbal communication of mood or intentions also plays a large and increasingly recognized role in human affairs. We make inferences about people's feelings and thoughts, especially those of very young children, from many kinds of communication, verbal and nonverbal; we should similarly use all available evidence in exploring the possibility of thoughts or feelings in other species. When animals live in a group and depend on each other for food, shelter, warning of dangers, or help in raising the young, they need to be able to judge correctly the moods and intentions of their companions. This extends to animals of other species as well, especially predators or prey. It is important for the animal to know whether a predator is likely to attack or whether the prey is so alert and likely to escape that a chase is not worth the effort. Communication may either inform or misinform, but in either case it can reveal something about the conscious thinking of the communicator.

Vervet monkeys, for example, have at least three different categories of alarm calls, which were described by Struhsaker (1967) after extensive periods of observation. He found that when a leopard or other large carnivorous mammal approached, the monkeys gave one type of alarm call; quite a different call was used at the sight of a martial eagle, one of the few flying predators that captures vervet monkeys. A third type of alarm call was given when a large snake approached the group. This degree of differentiation of alarm calls is not unique, although it has been described in only a few kinds of animals. For example, ground squirrels of western North America use different types of calls when frightened by a ground predator or by a predatory bird such as a hawk (Owings and Leger 1980).

The question is whether the vervet monkey's three types of alarm calls convey to other monkeys information about the type of predator. Such information is important because the animal's defensive tactics are different in the three cases. When a leopard or other large carnivore approaches, the monkeys climb into trees. But leopards are good climbers, so the monkeys can escape them only by climbing out onto the smallest branches, which are too weak to support a leopard. When the monkeys see a martial eagle, they move into thick vegetation close to a tree trunk or at ground level. Thus the tactics that help them escape from a leopard make them highly vulnerable to a martial eagle, and vice versa. In response to the threat of a large snake they stand on their hind legs and look around to locate the snake, then simply move away from it, either along the ground or by climbing into a tree.

To answer this question, Seyfarth, Cheney, and Marler (1980a, b) conducted some carefully controlled playback experiments under natural conditions in East Africa. From a concealed loudspeaker, they played tape recordings of vervet alarm calls and found that the playbacks of the three calls did indeed elicit the appropriate responses. The monkeys responded to the leopard alarm call by climbing into the nearest tree; the martial eagle alarm caused them to dive into thick vegetation; and the python alarm produced the typical behavior of standing on the hind legs and looking all around for the nonexistent snake.

Inclusive behaviorists—that is, psychologists interested only in contingencies of reinforcement during an individual's lifetime, and ethologists or behavioral ecologists solely concerned with the effects of natural selection on behavior—insist on limiting themselves to stating that an animal benefits from accurate information about what the other animal will probably do. But within a mutually interdependent social group, an individual can often anticipate a companion's behavior most easily by empathic appreciation of his mental state. The inclusive behaviorists will object that all we need postulate is behavior appropriately matched to the probabilities of the companions *behaving* in this way or that—all based on contingencies of reinforcement learned from previous situations or transmitted genetically.

But empathy may well be a more efficient way to gauge a companion's disposition than elaborate formulas describing the contingencies of reinforcement. All the animal may need to know is that another is aggressive, affectionate, desirous of companionship, or in some other common emotional state. Judging that he is aggressive may suffice to predict, economically and parsimoniously, a wide range of behavior patterns depending on the circumstances. Neo-Skinnerian inclusive behaviorists may be correct in saying that this empathy came about by learning, for example, the signals that mean a companion is aggressive. But our focus is on the animal's possible thoughts and feelings, and for this purpose the immediate situation is just as important as the history of its origin.

Humphrey (1976) has extended an earlier suggestion by Jolly (1966) that consciousness arose in primate evolution when societies developed to the stage where it became crucially important for each member of the group to understand the feelings, intentions, and thoughts of others. When animals live in complex social groupings, where each one is crucially dependent on cooperative interactions with the others, they need to be "natural psychologists," as Humphrey puts it. They need to have internal models of the behavior of their companions, to feel with them, and thus to think consciously about what the other one must be thinking or feeling.

Following this line of thought, we might distinguish between the animals' interactions with some feature of the physical environment or with plants, and their interactions with other reacting animals, usually their own species, but also predators and prey. Although Humphrey has so far restricted his criterion of consciousness to our own ancestors within the past few million years, it could apply with equal or even greater force to other animals that live in mutually interdependent social groups.

All this adds up to the simple idea that when animals communicate to one another they may be conveying something about their thoughts or feelings. If so, eavesdropping on the communicative signals they exchange may provide us with a practicable source of data about their mental experiences. When animals devote elaborate and specifically adjusted activities to communication, each animal responding to messages from its companion, it seems rather likely that both sender and receiver are consciously aware of the content of these messages.

The adaptive economy of conscious thinking

The natural world often presents animals with complex challenges best met by behavior that can be rapidly adapted to changing circumstances. Environmental conditions vary so much that for an animal's brain to have programmed specifications for optimal behavior in all situations would require an impossibly lengthy instruction book. Whether such instructions stem from the animal's DNA or from learning and environmental influences within its own lifetime, providing for all likely contingencies would require a wasteful volume of specific directions. Concepts and generalizations, on the other hand, are compact and efficient. An instructive analogy is provided by the hundreds of pages of official rules for a familiar game such as baseball. Once the general principles of the game are understood, however, quite simple thinking suffices to tell even a small boy approximately what each player should do in most game situations.

Of course, simply thinking about various alternative actions is not enough; successful coping with the challenges of life requires that thinking be relatively rapid and that it lead both to reasonably accurate decisions and to their effective execution. Thinking may be economical without being easy or simple, but consideration of the likely results of doing this or that is far more efficient than blindly trying every alternative. If an animal thinks about what it might do, even in very simple terms, it can choose the actions that promise to have desirable consequences. If it can anticipate probable events, even if only a little way into the future, it can avoid wasted effort. More important still is being able to avoid dangerous mistakes. To paraphrase Popper (1972), a foolish impulse can die in the animal's mind rather than lead it to needless suicide.

I have suggested that conscious thinking is economical, but many contemporary scientists counter that the problems mentioned above can be solved equally well by unconscious information-processing. It is quite true that skilled motor behavior often involves complex, rapid, and efficient reactions. Walking over rough ground or through thick vegetation entails numerous adjustments of the balanced contraction and relaxation of several sets of opposed muscles. Our brains and spinal cords modulate the action of our muscles according to whether the ground is high or low or whether the vegetation resists bending as we clamber over it. Little, if any, of this process involves conscious thought, and yet it is far more complex than a direct reaction to any single stimulus.

We perform innumerable complex actions rapidly, skillfully, and efficiently without conscious thought. From this evidence many have argued that an animal does not need to think consciously to weigh the costs and benefits of various activities. Yet when we acquire a new skill, we have to pay careful conscious attention to details not yet mastered. Insofar as this analogy to our own situation is valid, it seems plausible that when an animal faces new and difficult challenges, and when the stakes are high—often literally a matter of life and death—conscious evaluation may have real advantages.

Inclusive behaviorists often find it more plausible to suppose that an animal's behavior is more efficient if it is automatic and uncomplicated by conscious thinking. It has been argued that the vacillation and uncertainty involved in conscious comparison of alternatives would slow an animal's reactions in a maladaptive fashion. But when the spectrum of possible challenges is broad, with a large number of environmental or social factors to be considered, conscious mental imagery, explicit anticipation of likely outcomes, and simple thoughts about them are likely to achieve better results than thoughtless reaction. Of course, this is one of the many areas where we have no certain guides on which to rely. And yet, as a working hypothesis, it is attractive to suppose that if an animal can consciously anticipate and choose the most promising of various alternatives, it is likely to succeed more often than an animal that cannot or does not think about what it is doing.

References

Baker, L. R. 1981. Why computers can't act. *Am. Philos. Q.* 18:157–63.

Boden, M. A., ed. 1977. *Artificial Intelligence and Natural Man.* Basic Books.

Daanje, A. 1951. On the locomotory movements in birds and the intention movements derived from them. *Behaviour* 3:48–98.

Dreyfus, H. L. 1979. *What Computers Can't Do: The Limits of Artificial Intelligence*, rev. ed. Harper and Row.

Fisher, J., and R. A. Hinde. 1949. The opening of milk-bottles by birds. *Brit. Birds* 42:347–57.

Frisch, K. von. 1967. *The Dance Language and Orientation of Bees*. Harvard Univ. Press.

______. 1972. *Bees, Their Vision, Chemical Senses and Language*, 2nd ed. Cornell Univ. Press.

Goodall, J. van Lawick. 1968. Behaviour of free-living chimpanzees of the Gombe Stream area. *Anim. Beh. Monogr.* 1:165–311.

______. 1971. *In the Shadow of Man*. Houghton Mifflin.

Harnad, S. 1982. Consciousness: An afterthought. *Cog. Brain Theory* 5:29–47.

Hediger, H. 1947. Ist das tierliche Bewusstsein unerforschbar? *Behaviour* 1:130–37.

______. 1968. *The Psychology of Animals in Zoos and Circuses*. Dover.

______. 1980. *Tiere verstehen, Erkenntnisse eines Tierpsychologien*. Munich: Kindler.

Hinde, R. A., and J. Fisher. 1951. Further observations on the opening of milk bottles by birds. *Brit. Birds.* 44:393–96.

Humphrey, N. K. 1976. The social function of intellect. In *Growing Points in Ethology*, ed. P. P. G. Bateson and R. A. Hinde. Cambridge Univ. Press.

Janes, S. W. 1976. The apparent use of rocks by a raven in nest defense. *Condor* 78:409.

Jolly, A. 1966. Lemur social behavior and primate intelligence. *Science* 153:501–06.

Jung, C. G. 1973. *Synchronicity, a Causal Connecting Principle*. Princeton Univ. Press.

Kawai, M. 1965. Newly acquired pre-cultural behavior of the natural troop of Japanese monkeys on Koshima Islet. *Primates* 6:1–30.

Kenyon, K. W. 1969. *The Sea Otter in the Eastern Pacific Ocean*. North American Fauna, no. 68. US Bureau of Sport Fisheries and Wildlife.

Lorenz, K. 1963. Haben Tiere ein subjectives Erleben? *Jahr. Techn. Hochs. München*. Eng. trans., Do animals undergo subjective experience? In *Studies in Animal and Human Behavior*, vol. 2. Harvard Univ. Press.

______. 1971. *Studies in Animal and Human Behavior*, vol. 2. Harvard Univ. Press.

MacKenzie, B. D. 1977. *Behaviorism and the Limits of Scientific Method*. London: Routledge and Kegan Paul.

McMahan, E. A. 1982. Bait-and-capture strategy of termite-eating assassin bug. *Insectes Sociaux* 29:346–51.

Norman, D. A., ed. 1981. *Perspectives on Cognitive Science*. Hillsdale, N. J.: Erlbaum.

Owings, D. H., and D. W. Leger. 1980. Chatter vocalizations of California ground squirrels: Predator- and social-role specificity. *Z. Tierpsychol.* 54:163–84.

Popper, K. R. 1972. *Objective Knowledge*. Oxford Univ. Press.

Savory, T. H. 1959. *Instinctive Living, a Study of Invertebrate Behaviour*. London: Pergamon.

Seyfarth, R. M., D. L. Cheney, and P. Marler. 1980a. Monkey responses to three different alarm calls: Evidence for predator classification and semantic communication. *Science* 210:801–03.

______. 1980b. Vervet monkey alarm calls: Semantic communication in a free-ranging primate. *Anim. Beh.* 28:1070–94.

Skutch, A. F. 1976. *Parent Birds and Their Young*. Univ. of Texas Press.

Struhsaker, T. T. 1967. *The Red Colobus Monkey*. Univ. of Chicago Press.

Thatcher, R. W., and E. R. John. 1977. *Foundations of Cognitive Processes*. Hillsdale, N. J.: Erlbaum.

Wasserman, E. A. 1983. Is cognitive psychology behavioral? *Psychol. Record.* 33:6–11.

Whiteley, C. H. 1973. *Mind in Action, an Essay in Philosophical Psychology*. Oxford Univ. Press.

Susan E. Nicol
Irving I. Gottesman

Clues to the Genetics and Neurobiology of Schizophrenia

Uncertainty about the etiology of schizophrenia persists, but compelling evidence implicates genes and links dopamine to the disorder

Cases of schizophrenia, one of the common forms of major psychiatric illness, have been reported almost since the beginning of recorded history, yet the cause of the disorder is still unknown. Our interpretation of the accumulated evidence from studies of families—both biological and adopted—of schizophrenics is that patients have a genetic predisposition to the disorder but that this predisposition by itself is not sufficient for the development of schizophrenia. This article can mention only some of the major findings from clinical genetics and promising leads in neurobiological research on schizophrenia; more detailed information can be found in the references, particularly in Gottesman and Shields (1982).

In discussing the neurobiology of schizophrenia, we should keep in mind the following findings from genetic and epidemiologic work on the disorder. First, no environmental causes have been discovered that will invariably, or even with moderate probability, produce schizophrenia in persons who are not related to a schizophrenic. The disorder occurs in both industrialized and undeveloped societies, and although within large urban communities the prevalence of schizophrenia rises dramatically in the lower classes of society, this is true in general because individuals predisposed to the disorder tend to move downward through the social classes even before the onset of the illness (Goldberg and Morrison 1963).

Second, the risk of schizophrenia to the relatives of schizophrenics increases markedly with the degree of genetic relatedness (Bleuler 1978) even if the relatives have not shared a specific environment with the patient (Fig. 1). An identical twin of a schizophrenic is at least 3 times as likely as a fraternal twin to develop schizophrenia, and some 35–60 times as likely as an unrelated person from the same general population. However, fewer than half of the identical twins of schizophrenics in recent studies have schizophrenia themselves, although they share all their genes with schizophrenics, which demonstrates unequivocally the importance of environmental factors. The incomplete expression of genotypes, on the other hand, also plays a role, since the normal identical twins from discordant pairs transmit schizophrenia to their offspring at the same high rate as their schizophrenic twins do (Gottesman and Shields 1982).

Identical twins of schizophrenics are about as likely to develop schizophrenia whether they were reared apart from or together with their schizophrenic twins since childhood. Raising identical twins in different homes is a very rare event—only 12 pairs involving schizophrenia have been authenticated; it may not be possible, therefore, to generalize the information yielded by these cases.

Data from other research strategies confirm the importance of genetics. For example, children of schizophrenics adopted by nonrelatives early in life still develop schizophrenia as adults at rates that are considerably higher than those of the population at large, and that sometimes are as high as the rates for children reared by their own schizophrenic parents (Heston 1966; Rosenthal et al. 1968). The adoptive parents and siblings of adoptees who became schizophrenics do not have significantly elevated rates of schizophrenia, but the schizophrenics' biological relatives do have high rates (Kety et al. 1976). Furthermore, children of normal parents adopted by a nonrelative who became schizophrenic after the adoption do not show a significantly increased rate of schizophrenia (Wender et al. 1974).

Being a twin does not increase the risk for schizophrenia. Gender is not generally relevant in schizophrenia: the risk of developing the disorder among half-siblings of schizophrenic adoptees does not depend on whether the siblings share a father or mother with the schizophrenic (Kety et al. 1976), the offspring of male schizophrenics are as often schizophrenic as those of female schizophrenics, female identical twins of schizophrenics are not significantly more likely to develop the disorder than male identical twins, and opposite-sex fraternal twins of schizophrenics have the same risk as same-sex fraternal twins in recent studies. The one exception is that male schizophrenics, on average, develop the disorder earlier in

Susan E. Nicol is a clinical psychologist in the Department of Psychiatry of the Hennepin County Medical Center in Minneapolis. She obtained her Ph.D. in behavioral genetics and neuroscience at the University of Minnesota in 1972. Her research interests include quantifying the diathesis-stressor model of mental illness and investigating the biochemical genetics of psychiatric disorders. Irving I. Gottesman has been Professor of Psychiatric Genetics at the Washington University School of Medicine since 1980. He received his Ph.D. from the University of Minnesota in 1960 and has taught there as well as at Harvard University and the University of North Carolina. His research focuses on the genetic epidemiology of mental disorders.
Preparation of this paper was made possible in part by federal grants MH 31302 and AA 13539.
Address for Dr. Gottesman: Department of Psychiatry, Washington University School of Medicine, St. Louis, MO 63110.

life than female schizophrenics, but there are as many female as male schizophrenics by the end of the risk period, at age 65.

The observed risks for schizophrenia, however, are not compatible with any simple Mendelian genetic models, since they also vary with the severity of the schizophrenic's illness, the number of other relatives already affected by the disorder, and, in the case of the patient's offspring, the other parent's mental health (O'Rourke et al. 1982). The risks are compatible with a polygenic model of transmission similar to that used to study other common genetic diseases and congenital malformations, which involves the idea of a threshold and a range of contributory factors, but theories that a single major genetic locus is responsible for some schizophrenia cannot be discarded completely (McGue et al., in press).

Because no laboratory test can be equated with a schizophrenia genotype, it is not now possible to predict which individuals of those known to be at risk will develop the disorder. Hence, ambiguity and uncertainty haunt attempts to test specific models of genetic transmission. To complicate matters further, some psychoses that resemble schizophrenia are actually different conditions. These occur after head injuries or epileptic seizures, and the patients' relatives have the same low risk of schizophrenia as the population at large. Childhood psychoses that appear before puberty do not seem to be genetically related to schizophrenia.

Environmental factors continue to be important after the onset of schizophrenia. Excessively critical or overly intrusive behavior on the part of relatives of schizophrenics in remission increases their rate of relapse. Social intervention, therefore, should complement drug treatments of schizophrenics.

Biochemical genetics

In theorizing about the biochemical genetics of schizophrenia, it is important to begin by considering what we know about other genetic diseases. Even with Huntington chorea, a relatively simple autosomal Mendelian disease that is dominant and has complete penetrance, we cannot specify the exact nature of the genetic disorder at the molecular level. We infer the existence of an abnormal gene, but we are not yet able to isolate it by restriction enzymes. We cannot account for the variable age at onset of the disease, which covers the entire life span, and we cannot detect which of the offspring of affected individuals are destined to develop the disease, although 50% of them do so. For some 70% of other genetic disorders whose mode of transmission has been clearly identified as that of recessive inheritance, which we know to be associated with an enzyme deficiency, we cannot identify the enzyme (McKusick 1983). Although the pattern of inheritance of color blindness, one of the oldest known genetic disorders, was understood in the early 1700s, it was not assigned to the X chromosome until 1911 (Morgan 1911). In the 1980s we are still dependent on a behavioral test to identify the individuals who are color-blind, and we know little about the disorder's biological basis.

We know even less about the biochemical genetics of multifactorial disorders (Fraser 1981). However, the genes associated with these conditions are not qualitatively different from those underlying Mendelian traits at the molecular level: both groups of genes are subject to the same rules of inheritance because they are chromosomal and thus segregate, show dominance, can be suppressed or enhanced by other genes, and interact with their environment. Although from the beginning of this century geneticists have identified specific loci involved in polygenic conditions and have located them on specific chromosomes by linkage with major genes, these feats were accomplished with genetically tractable organisms such as wheat and fruit flies, not with humans.

Impressive recent advances in the biochemical genetics of common human illnesses including heart disease, diabetes, and hypertension have been made through the study of low-density lipoprotein receptors (e.g., Goldstein and Brown 1982), insulin receptors (e.g., Jarett 1979), and cation permeability, but genetic differences in receptor structure or function are only beginning to be explored as biochemical techniques such as high specific activity radioligands become available. Major genes partially responsible for diabetes and lupus erythematosus, however, have been identified (e.g., Fraser 1981).

Where does one begin the search for the biochemical differ-

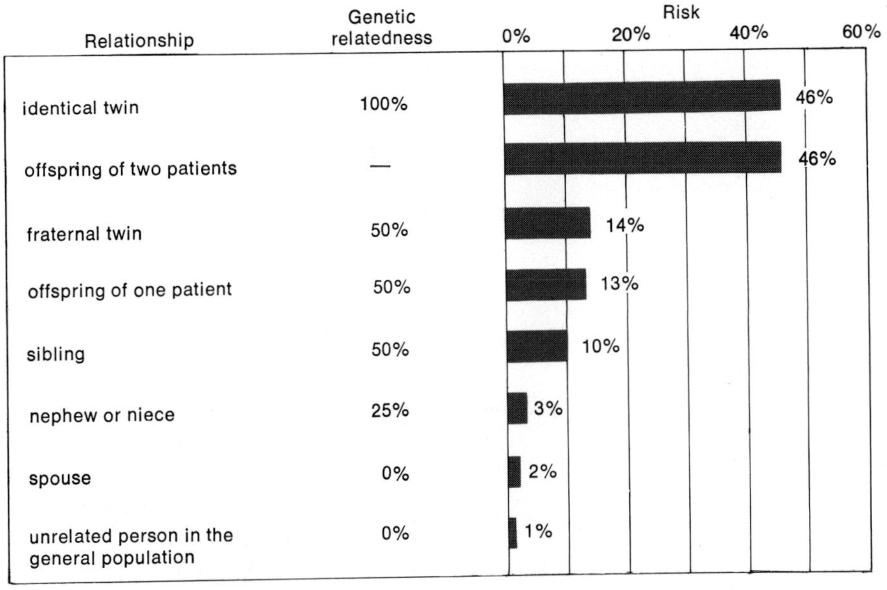

Relationship	Genetic relatedness	Risk
identical twin	100%	46%
offspring of two patients	—	46%
fraternal twin	50%	14%
offspring of one patient	50%	13%
sibling	50%	10%
nephew or niece	25%	3%
spouse	0%	2%
unrelated person in the general population	0%	1%

Figure 1. Lifetime risks of developing schizophrenia are largely a function of how closely an individual is genetically related to a schizophrenic and not a function of how much their environment is shared. The observed risks, however, are much more compatible with a multifactorial polygenic theory of transmission than with a Mendelian model or one involving a single major locus, especially after allowance is made for some unsystematic environmental transmission. In the case of an individual with two schizophrenic parents, genetic relatedness cannot be expressed in terms of percentage, but the regression of the individual's "genetic value" on that of the parents is 1, the same as it is for identical twins. (Data from Gottesman and Shields 1982.)

ences that are implied by a genetic predisposition for schizophrenia? There are many studies of the biochemical parameters in schizophrenic patients, but as Kety (1980) has pointed out, differences between schizophrenics and controls often reflect not genetic differences but rather the patients' poor nutrition, chronic hospitalization, and history of medication.

Recent biochemical theorizing about schizophrenia has focused on dopamine, a catecholamine neurotransmitter. Evidence linking dopamine to schizophrenia may not be as strong as that for insulin and diabetes or cholesterol and cardiovascular disease, but it is compelling (Seeman 1980, in press). Antipsychotic drugs called neuroleptics block dopamine receptors, and the effectiveness of the drugs in treating schizophrenics is highly correlated with how well they block the receptors. Further, neuroleptics produce symptoms similar to those seen in Parkinson's disease, a disorder known to be due to insufficient dopamine in specific neurons. L-dopa, which increases stores of dopamine and which is used to treat Parkinson's disease, produces a psychosis with features resembling schizophrenia in some persons (e.g., Meltzer and Stahl 1976). Another clue from biochemical pharmacology is that amphetamine, which releases stored dopamine and thus stimulates the receptors, worsens psychotic symptoms in some schizophrenics and elicits the same symptoms in certain of their nonschizophrenic relatives. Amphetamine can also produce a schizophrenia-like syndrome in some apparently normal users.

These observations have led to the hypothesis that there is an excess of dopaminergic activity in schizophrenic patients. How does one go about testing such a theory? If genetic contributors to schizophrenia involve altered activity in dopamine pathways, one would expect to find a large variation in enzyme activities, concentrations of precursors, neurotransmitters, or metabolites, or functions of the receptors in the human dopamine system. We will give a brief overview of what is currently known about dopamine neurons before discussing studies of the biochemical genetics of this pathway.

Dopamine pathways

More is known about the enzymes involved in the synthesis and metabolism of dopamine than about the receptor proteins, in part because the enzymes either are soluble to begin with or can easily be made soluble in vitro, and thus they are more amenable to classical biochemical analysis. Dopamine is synthesized from tyrosine and dopa by the enzymes tyrosine hydroxylase and dopa decarboxylase (Fig. 2). In noradrenaline neurons, the enzyme dopamine-B-hydroxylase (DBH) converts dopamine to norepinephrine. However, DBH is absent from dopamine neurons, where dopamine is metabolized by the mitochondrial enzyme monoamine oxidase (MAO) and aldehyde oxidase to dihydroxyphenylacetic acid, or by the enzyme catechol-o-methyl transferase (COMT) to 3-methoxytyramine (Cooper et al. 1974).

Dopamine is stored in vesicles in the transmitting neuron and is released into the synaptic cleft when these vesicles merge with the presynaptic membrane. As is the case with certain other neurotransmitters, the rate of dopamine release is thought to be determined not only by the electrical activity in presynaptic neurons, but also by feedback from autoreceptors on such neurons (Meltzer 1980). After it is released into the synaptic cleft, dopamine can bind to receptors on postsynaptic membranes or to the autoreceptors on the presynaptic membrane, it can be metabolized by COMT, or it can be taken back up into the presynaptic neuron, where it is either metabolized by MAO or stored again in vesicles for later release.

Dopamine has only recently been shown to interact with at least two types of postsynaptic receptors: D_1 receptors that are linked to the enzyme dopamine-stimulated ade-

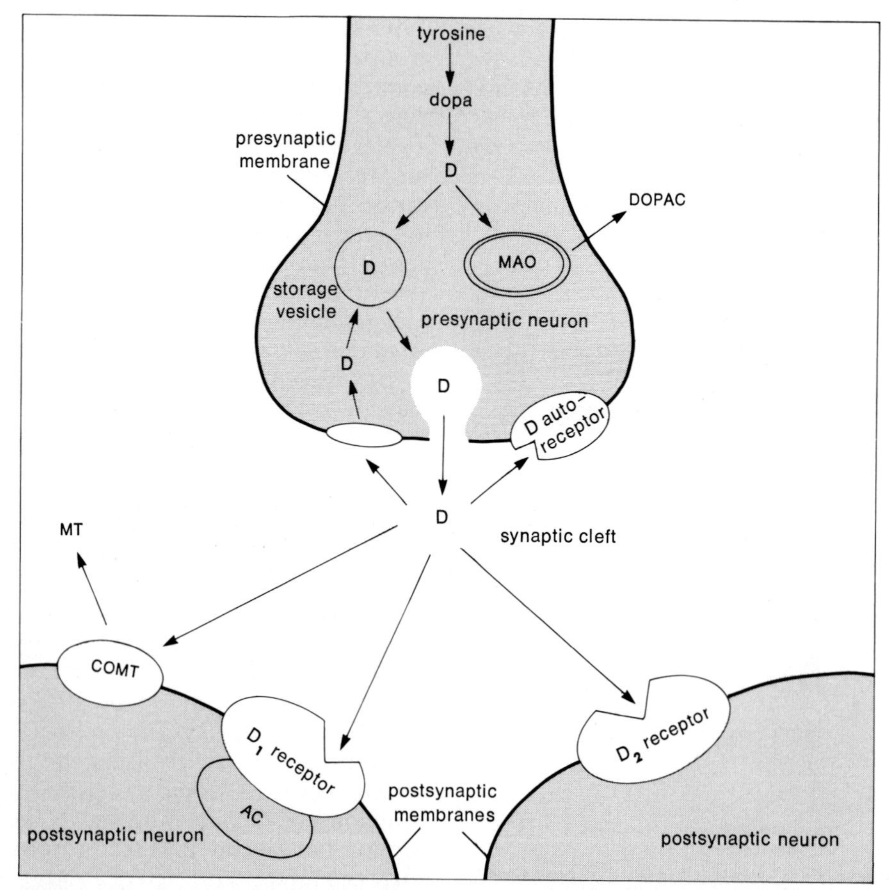

Figure 2. An excess of dopamine (D), which is synthesized from the amino acids tyrosine and dopa, may produce schizophrenic behavior. Dopamine can be metabolized to dihydroxyphenylacetic acid (DOPAC) by monoamine oxidase (MAO), or to 3-methoxytyramine (MT) by catechol-o-methyl transferase (COMT). After storage in the transmitting neuron, dopamine is released into the synaptic cleft, where it may bind to an autoreceptor and thus inhibit release of more dopamine. However, it may instead bind to a D_1 receptor associated with stimulating adenylate cyclase (AC), or to a D_2 receptor that inhibits AC; drugs that inhibit these receptors have proved effective in treating schizophrenia.

nylate cyclase (AC) (Kebabian and Calne 1979; Seeman 1980), and D_2 receptors that according to indirect evidence may be linked to dopamine-inhibited AC (Stoof and Kebabian 1981). In addition to differential association with AC, D_1 and D_2 sites have been distinguished by their relative affinity for dopamine agonists—agents that can imitate the actions of dopamine—and dopamine antagonists—agents that can block the effects of dopamine at receptor sites (Table 1). Seeman (1980) described two other dopamine sites identified by differing relative affinity for agonists and antagonists: the D_3 site and the D_4 site. It is known that at least a substantial proportion of D_1 and D_2 receptors are located on the postsynaptic neuron, but the location of D_3 and D_4 dopamine sites is less well established; there is some evidence that a proportion of D_3 receptors are found on the presynaptic neuron, perhaps as autoreceptors. Seeman distinguished dopamine sites from dopamine receptors based on the evidence that dopamine-related behaviors correlate best with activity of D_2 receptors. Other researchers do not make this distinction, referring to all sites where dopamine has the greatest affinity of any physiological compound as dopamine receptors.

Dopamine receptors were first identified biochemically in a peripheral tissue, the superior cervical ganglion (Greengard 1976). These receptors were shown to be associated with AC, as were the first dopamine receptors discovered in the brain. And it was the dopamine-sensitive AC in the brain that was first thought to be the site of action of the antipsychotic drugs (Clement-Cormier et al. 1974). Subsequent work demonstrated a good correlation between the clinical potency of the various phenothiazine antipsychotic drugs, such as chlorpromazine, and their inhibition of AC stimulated by dopamine, but the correlation between butyrophenone neuroleptics, such as haloperidol, and inhibition of this AC was not good (Fig. 3). The fact that the correlation between efficacy of both the phenothiazine and butyrophenone neuroleptics and their affinities for the D_2 receptor is excellent, as contrasted with the situation for the D_1 sites, definitively links the D_2 receptors to schizophrenia.

Dopamine and genetic liability

What changes in the dopamine pathway lead to schizophrenia? There are two ways this question can be asked with regard to neurons containing either dopamine or dopamine receptors: Of the many dopamine-containing neurons in the human body, which are relevant to schizophrenia? Of the many enzymes and receptors in these dopamine neurons, which may have pathological variations?

Although dopamine receptors exist both in the central nervous system and on its periphery, research on schizophrenia has focused primarily on the receptors in the brain. Investigations into the biochemical genetics of the disorder would be aided immeasurably if there were an easily sampled peripheral source of dopamine receptors, for instance on lymphocytes, platelets, or red blood cells. Impressive advances in the understanding of genetic and environmental influence on insulin receptor binding have been made with insulin receptors on blood cells (Jarett 1979). However, peripheral dopamine receptors are in areas that are not easily sampled, and they may be biochemically different from central dopamine receptors (Creese et al. 1981).

Beyond the issue of which dopamine neurons may play a role in the genetic predisposition to schizophrenia is the question of which of the components of Figure 2 may be involved. The predisposition may represent a number of protein variations resulting in different relations between the proteins' structure and function. Each of these protein variants may, under proper environmental conditions, produce a schizophrenic syndrome; alternatively, it may be the sum of variants at different loci that may place a person above the threshold of liability and lead to schizophrenia. We shall start the genetic overview with the synthetic and metabolic enzymes.

Table 1. Relative affinity of dopamine-sensitive sites and receptors for dopamine (agonist) and spiperone (antagonist)

Sites and receptors	*Affinity (nM)*	
	Dopamine	*Spiperone*
D_1 site (dopamine-sensitive adenylate cyclase)	~3,000	~2,000
D_2 receptor	~5,000	~0.3
D_3 site	~3	~1,500
D_4 site	~3	~1

SOURCE: Seeman 1980

Genetic regulation of the catecholamine-synthesizing enzymes has been demonstrated by Ciaranello and Boehme (1982), working with inbred mouse strains. Studies of human biochemical genetics have focused on the metabolic enzymes in catecholamine neurons, COMT and MAO. Weinshilboum and Raymond (1977) demonstrated that low erythrocyte COMT activity is inherited as an autosomal recessive mechanism. The relation between differences in COMT activity and vulnerability to disease has yet to be established. Work on MAO has generally proceeded in the opposite direction: instead of first establishing base rates for differences in enzyme activity in the general population and then trying to associate MAO activity with the presence of disease, the literature on MAO has focused on comparisons of schizophrenics and normal subjects, with conflicting results (*Schizophrenia Bull.* 1980). There is evidence from a number of studies that chronic schizophrenics may have less platelet MAO than control groups do, but the relation between platelet and brain MAO and the functional significance of a decrease in enzyme activity of the magnitude detected remain in question (Rice et al. 1982).

In a study of dopamine-stimulated AC in the human brain, Carenzi and his colleagues (1975) reported no significant differences in enzyme activity between the postmortem tissue of schizophrenics and that of controls, but the activity was shown, in rats, to be a function of how long after death the tissue was frozen: the interval for the human schizophrenics' tissue was significantly longer than for the normal human tissue. However, Nicol and her co-workers (1981; McSwigan et al. 1980) found that dopamine-stimulated AC in rat brain tissue was stable for at least ten hours after death when enzyme activity was measured in washed membrane preparations. This suggested the loss of a soluble activator rather than the degradation of the membrane enzyme itself during this period. Therefore, it is feasible to examine these enzyme activities in schizophrenia and other disorders, using human brain tissue obtained after death.

Several research groups have reported that postmortem brain tissue from schizophrenics contains more D_2 receptors and that the receptors have less affinity for butyrophenone neuroleptics than tissue from control subjects (e.g., Reisine et al. 1980; Mackay et al. 1982). However, Mackay and his colleagues (1982) have shown that the decrease in affinity of D_2 receptors for neuroleptics is reversed after the membranes are washed, which suggests that the decrease is due to the neuroleptic drugs the patients took before death. Whether the increase in the number of D_2 receptors is also an effect of the administration of neuroleptics before death remains controversial. Burt and his co-workers (1977) have established that long-term neuroleptic treatment does produce an increase in D_2 receptors in animals.

Lee and Seeman (1980) and Owen and his colleagues (1978) reported increases in D_2 receptors in tissue from a small number of schizophrenic patients who had been free of medication for at least a month before death. However, Mackay and his co-workers (1982) did not find any such increases in tissue from patients who had never taken neuroleptics, and concluded that the increases observed in medicated patients may be iatrogenic. There have been no reports of increases in the binding found at D_1 or D_3 receptors in postmortem brain tissue from schizophrenics, although these receptors have been less extensively studied (Seeman 1980).

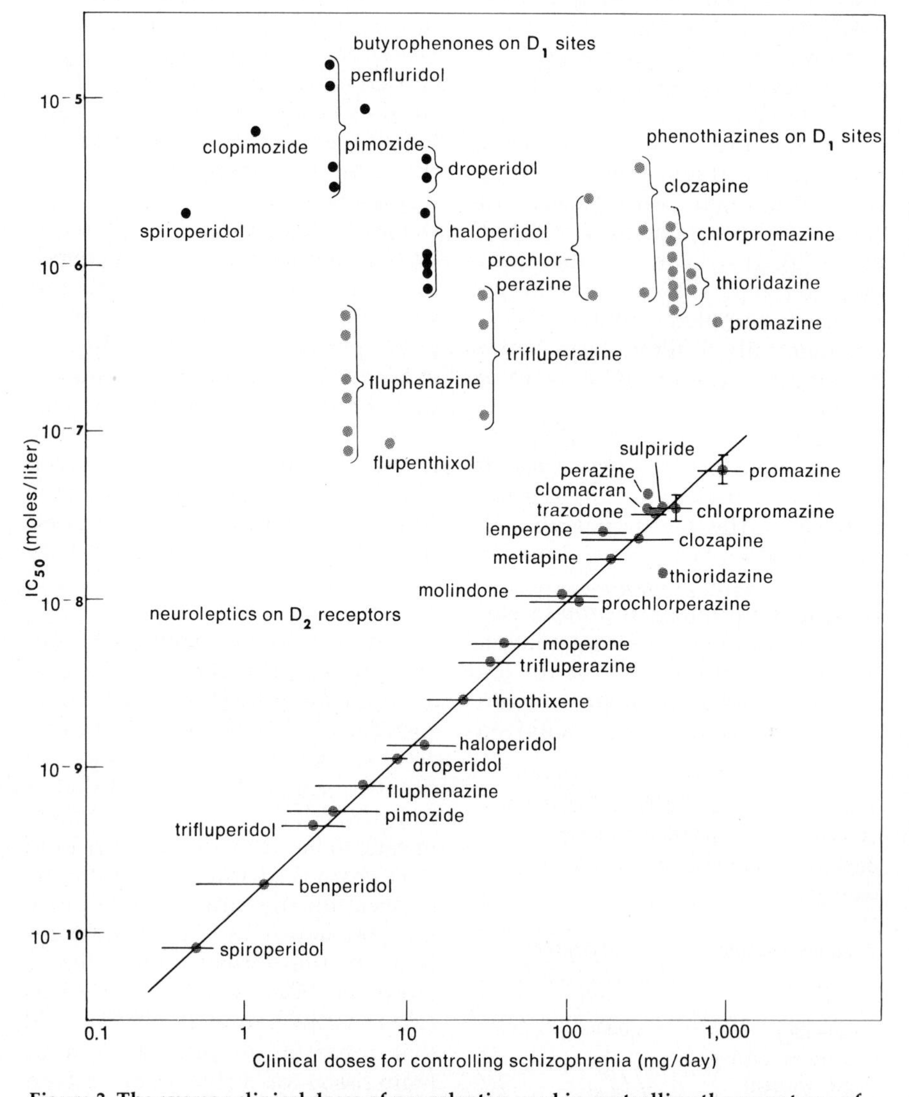

Figure 3. The average clinical doses of neuroleptics used in controlling the symptoms of schizophrenia correlate with the biochemical index (IC_{50}) of potency at D_2 dopamine receptors but not at D_1 sites. IC_{50} is a measure of the relative potency of the various antipsychotic drugs in blocking the dopamine sites. A range (*lines*) of clinical doses is depicted because of the variation reported for controlling the symptoms of schizophrenia in different individuals; the average doses are represented by dots. (After Seeman 1980.)

A number of investigators have measured the levels of dopamine and its metabolite, homovanillic acid (HVA), in schizophrenic patients and have reported conflicting results (see Meltzer and Stahl 1976 for dopamine metabolites in cerebrospinal fluid, and Haracz 1982 for these substances in postmortem tissue). The work by Spokes (1979), Bird and his colleagues (1979), and Mackay and his co-workers (1982) on increased dopamine in the nucleus accumbens and caudate nucleus in the brain deserves special mention. Beginning with the paper examining factors influencing the measurement of dopamine and other neurotransmitters and enzymes in postmortem tissue (Spokes 1979), these investigators reported increased dopamine in specific subcortical structures. In contrast to the changes in the concentration of D_2 receptors, increased levels of dopamine do not appear to be related to any history of neuroleptic treatment.

Although there was no correlation between dopamine concentration and the age at which the controls died, Mackay and his colleagues reported that schizophrenics who developed the disease early in life had the highest concentrations of dopamine after death. In their sample, an earlier age at death was also correlated with higher levels of dopamine in schizophrenics, and thus the relative importance of the age at onset of illness and the age at death is difficult to determine. However, the subjects who developed schizophrenia at an early age also had other biochemical abnormalities, such as decreased activity of the angiotensin-converting enzyme and low concentrations of gamma-aminobutyric acid (e.g., Arregui et al. 1980; Spokes et al. 1980).

These studies demonstrate both the existence of promising clues about the role of dopamine and the care with which such clues must be investigated. Many factors have to be controlled and analyzed. It may be that biochemical differences present in one group of schizophrenics, such as those who developed the disease at a young age, are not present in other groups. Alternatively, biochemical differences present at certain stages of the illness may not be present at other stages.

Crow (1980a, b) proposed two syndromes in schizophrenia. Type I, or acute schizophrenia characterized by "positive symptoms" (hallucinations, delusions, and disordered thought), is hypothesized to involve abnormalities in the dopamine pathway and to be responsive to neuroleptics. Type II, or chronic schizophrenia characterized by "negative symptoms" (reduced emotional response, poverty of speech, and loss of drive), is thought to involve structural changes in the brain and to be relatively resistant to neuroleptic medication. Crow emphasized, however, that both types often occur together, and that they are not separate disorders. Reports of structural abnormalities, such as enlarged cerebral ventricles, and neuropathological abnormalities in a proportion of schizophrenic patients give support to Crow's hypothesis of two syndromes (e.g., Johnstone et al. 1976; Weinberger and Wyatt 1980; Stevens 1982). Knowledge of structural abnormalities may be important both for treatment of schizophrenic patients and for research design.

Although a number of biochemical and neuroendocrinological parameters have shown promise in predicting the response to treatment of depressed and manic patients, few such predictors are now known for schizophrenia (see Uytdenhoef et al. 1982 for review). Assessment of cerebral ventrical abnormalities by CT scan and of cerebral metabolic functioning (e.g., Jaffe 1982) should result in more homogeneous groups for study.

This review has focused primarily on the pathophysiology of dopamine, but other mechanisms, including cellular loss and structural abnormalities, have also been proposed as the underlying causes of some schizophrenic symptoms or of these symptoms in some schizophrenic patients. Researchers continue to explore other mechanisms. Rapid advances in our knowledge of endorphins has spurred research on these compounds in schizophrenics. Most promising are the reports of the neuroleptic-like activity of a peptide related to endorphin (de Wied et al. 1980), and the discovery that a subgroup of schizophrenic patients responded to treatment by these gamma-type endorphins (Van Praag et al. 1982).

Since there are hundreds of neurotransmitters and modulators, researchers could spend years examining the biochemical genetics of each of these systems in relation to schizophrenia and other disorders of the central nervous system with known genetic components. One way to short-cut the process of locating suspected gene products is with linkage analysis—i.e., the method of proving that a gene connected with a Mendelian trait is located on a chromosome near a genetic marker. With the advent of techniques using restriction endonucleases (e.g., Botstein et al. 1980), the number of known polymorphic loci available for linkage analyses should increase dramatically in the next few years. (For a discussion of the problems and the promise of such an approach to genetic analysis, see, for example, Gershon et al. 1981.)

Although we cannot yet describe the biochemical genetics of schizophrenia, many leads are being explored. It is important to keep in mind the information accumulated from work on clinical psychiatric genetics as the search continues for complementary, rational neurobiological data that will allow us to solve the puzzle of schizophrenia.

References

Arregui, A., A. V. P. Mackay, E. G. Spokes, and L. L. Iversen. 1980. Reduced activity of angiotensin converting enzyme in basal ganglia in early onset schizophrenia. *Psychol. Med.* 10:307–13.

Bird, E. D., E. G. S. Spokes, and L. L. Iversen. 1979. Increased dopamine concentration in limbic areas of brain from patients dying with schizophrenia. *Brain* 102:347–60.

Bleuler, E. 1978. *The Schizophrenic Disorders: Long-Term Patient and Family Studies.* Trans. S. M. Clemens. Yale Univ. Press.

Botstein, D., R. L. White, M. Skolnick, and R. W. Davis. 1980. Construction of a genetic linkage map in man using restriction fragment length polymorphisms. *Am. J. Hum. Genet.* 32:314–31.

Burt, D. R., I. Creese, and S. H. Snyder. 1977. Antischizophrenic drugs: Chronic treatment elevates dopamine receptor binding in brain. *Science* 196:326–28.

Carenzi, A., et al. 1975. Dopamine-sensitive adenylyl cyclase in human caudate nucleus. *Arch. Gen. Psychiat.* 32:1056–59.

Ciaranello, R. D., and R. E. Boehme. 1982. Genetic regulation of neurotransmitter enzymes and receptors: Relationship to the inheritance of psychiatric disorders. *Beh. Genet.* 12:11–35.

Clement-Cormier, Y. C., J. W. Kebabian, G. L. Petzold, and P. Greengard. 1974. Dopamine-sensitive adenylate cyclase in mammalian brain: A possible site of action of antipsychotic drugs. *PNAS* 71:113–17.

Cooper, J. R., F. E. Bloom, and R. H. Roth. 1974. *The Biochemical Basis of Neuropharmacology.* Oxford Univ. Press.

Creese, I., D. R. Sibley, S. Leff, and M. Hamblin. 1981. Dopamine receptors: Subtypes, localization and regulation. *Fed. Proc.* 40: 147–52.

Crow, T. J. 1980a. Positive and negative schizophrenic symptoms and the role of dopamine. *Brit. J. Psychiat.* 137:383–86.

———. 1980b. Molecular pathology of schizophrenia: More than one disease process. *Brit. Med. J.*, 12 January, 66–68.

de Wied, D., J. M. van Ree, and H. M. Greven. 1980. Neuroleptic-like activity of peptides related to [DES-TYR[1]] γ-endorphin: Structure activity studies. *Life Sci.* 26:1575–79.

Fraser, F. C. 1981. The genetics of common birth defects and diseases. In *Genetic Issues in Pediatric and Obstetric Practice*, ed. M. Kaback, pp. 45–54. Chicago: Year Book Medical Publishers.

Gershon, E. S., S. Matthysse, X. O. Breakefield, and R. D. Ciaranello. 1981. *Genetic Research Strategies for Psychobiology and Psychiatry.* Pacific Grove, CA: Boxwood Press.

Goldberg, E. M., and S. L. Morrison. 1963. Schizophrenia and social class. *Brit. J. Psychiat.* 109:785–802.

Goldstein, J. L., and M. S. Brown. 1982. The LDL receptor defect in familial hypercholesterolemia: Implications for pathogenesis and therapy. *Med. Clinics of North Am.* 66: 335–62.

Gottesman, I. I., and J. Shields. 1982. *Schizophrenia: The Epigenetic Puzzle.* Cambridge Univ. Press.

Greengard, P. 1976. Possible role for cyclic nucleotides and phosphorylated membrane proteins in postsynaptic actions of neurotransmitters. *Nature* 260:101–08.

Haracz, J. L. 1982. The dopamine hypothesis: An overview of studies with schizophrenic patients. *Schiz. Bull.* 8:438–69.

Heston, L. L. 1966. Psychiatric disorders in foster home reared children of schizophrenic mothers. *Brit. J. Psychiat.* 112:819–25.

Jaffe, C. C. 1982. Medical imaging. *Am. Sci.* 70:576–85.

Jarett, L. 1979. Pathophysiology of the insulin receptor. *Human Pathol.* 10:301–11.

Johnstone, E. C., T. J. Crow, C. D. Frith, J. Husband, and L. Kreel. 1976. Cerebral ventricular size and cognitive impairment in chronic schizophrenia. *Lancet*, 30 October, 924–26.

Kebabian, J. W., and D. B. Calne. 1979. Multiple receptors for dopamine. *Nature* 277: 93–96.

Kety, S. S. 1980. The syndrome of schizophrenia: Unresolved questions and opportunities for research. *Brit. J. Psychiat.* 136: 421–36.

Kety, S. S., D. Rosenthal, P. H. Wender, and F. Schulsinger. 1976. Studies based on a total sample of adopted individuals and their relatives: Why they were necessary, what they demonstrated and failed to demonstrate. *Schiz. Bull.* 2:413–28.

Lee, T., and P. Seeman. 1980. Elevation of brain neuroleptics/dopamine receptors in schizophrenia. *Am. J. Psychiat.* 137:191–97.

McGue, M., I. I. Gottesman, and D. C. Rao. In press. The transmission of schizophrenia under a multifactorial threshold model. *Am. J. Hum. Genet.*

Mackay, A. V. P., et al. 1982. Increased brain dopamine and dopamine receptors in schizophrenia. *Arch. Gen. Psychiat.* 39: 991–97.

McKusick, V. 1983. *Mendelian Inheritance in Man.* 6th ed. Johns Hopkins Univ. Press.

McSwigan, J. D., S. E. Nicol, I. I. Gottesman, V. B. Tuason, and W. H. Frey II. 1980. Effect of dopamine on activation of rat striatal adenylate cyclase by free Mg^{2+} and guanyl nucleotides. *J. Neurochem.* 34:594–601.

Meltzer, H. Y. 1980. Relevance of dopamine autoreceptors for psychiatry: Preclinical and clinical studies. *Schiz. Bull.* 6:456–75.

Meltzer, H. Y., and S. M. Stahl. 1976. The dopamine hypothesis of schizophrenia: A review. *Schiz. Bull.* 2:19–76.

Morgan, T. H. 1911. The origin of five mutations in eye color in drosophila and their modes of inheritance. *Science* 33:534–37.

Nicol, S. E., et al. 1981. Postmortem stability of dopamine-sensitive adenylate cyclase, guanylate cyclase, ATPase, and GTPase in rat striatum. *J. Neurochem.* 37:1535–39.

O'Rourke, D., I. I. Gottesman, B. K. Suarez, J. Rice, and T. Reich. 1982. Refutation of the general single locus model for the etiology of schizophrenia. *Am. J. Hum. Genet.* 34: 630–49.

Owen, F., et al. 1978. Increased dopamine-receptor sensitivity in schizophrenia. *Lancet*, 29 July, 223–25.

Reisine, T. D., M. Rossor, E. Spokes, L. L. Iversen, and H. I. Yamamura. 1980. Opiate and neuroleptic receptor alterations in human schizophrenic brain tissue. In *Receptors for Neurotransmitters and Peptide Hormones*, ed. G. Pepeu, M. J. Kuhar, and S. J. Enna, pp. 443–50. New York: Raven.

Rice, J., P. McGuffin, and E. G. Shaskan. 1982. A commingling analysis of platelet monoamine oxidase activity. *Psychiat. Res.* 7: 325–35.

Rosenthal, D., et al. 1968. Schizophrenics' offspring reared in adoptive homes. In *The Transmission of Schizophrenia*, ed. D. Rosenthal and S. S. Kety, pp. 377–91. Pergamon Press.

Schizophrenia Bulletin. 1980. Schizophrenia and platelet monoamine oxidase (conf. proc.). 6:199–384.

Seeman, P. 1980. Brain dopamine receptors. *Pharm. Rev.* 32:229–313.

———. In press. Schizophrenia and dopamine receptors. In *A Method of Psychiatry*, 2nd ed., ed. S. E. Greben et al. Philadelphia: Lea and Febiger.

Spokes, E. G. S. 1979. An analysis of factors influencing measurements of dopamine, noradrenaline, glutamate decarboxylase and choline acetylase in human post-mortem brain tissue. *Brain* 102:333–46.

Spokes, E. G. S., N. J. Garrett, M. N. Rossor, and L. L. Iversen. 1980. Distribution of GABA in post-mortem brain tissue from control, psychotic and Huntington's chorea subjects. *J. Neurol. Sci.* 48:303–13.

Stevens, J. R. 1982. Neuropathology of schizophrenia. *Arch. Gen. Psychiat.* 39:1131–39.

Stoof, J. C., and J. W. Kebabian. 1981. Opposing roles for D-1 and D-2 dopamine receptors in efflux of cyclic AMP from rat neostriatum. *Nature* 294:366–68.

Uytdenhoef, P., P. Linkowski, and J. Mendlewicz. 1982. Biological quantitative methods in the evaluation of psychiatric treatment: Some biochemical criteria. *Neuropsychobiology* 8:60–72.

Van Praag, H. M., W. M. A. Verhoeven, J. M. van Ree, and D. de Wied. 1982. The treatment of schizophrenic psychoses with γ-type endorphins. *Biol. Psychiat.* 17:83–98.

Weinberger, D. R., and R. J. Wyatt. 1980. Schizophrenia and cerebral atrophy. *Lancet*, 24 May, 1130.

Weinshilboum, R. M., and F. A. Raymond. 1977. Inheritance of low erythrocyte catechol-o-methyltransferase activity in man. *Am. J. Hum. Genet.* 29:125–35.

Wender, P. H., D. Rosenthal, S. S. Kety, F. Schulsinger, and J. Welner. 1974. Cross-fostering: A research strategy for clarifying the role of genetic and experiential factors in the etiology of schizophrenia. *Arch. Gen. Psychiat.* 30:121–28.

Mozart's Starling

Meredith J. West
Andrew P. King

On 27 May 1784, Wolfgang Amadeus Mozart purchased a starling. Three years later, he buried it with much ceremony. Heavily veiled mourners marched in a procession, sang hymns, and listened to a graveside recitation of a poem Mozart had composed for the occasion *(1)*. Mozart's performance has received mixed reviews. Although some see his gestures as those of a sincere animal lover, others have found it hard to believe that the object of Mozart's grief was a dead bird. Another event in the same week has been put forth as a more likely cause for Mozart's funereal gestures: the death of his father Leopold *(2)*.

The scholars who have reported and interpreted this historical incident knew much about Mozart but little, if anything, about starlings. To put the incident into better perspective, we will provide here a profile of the vocal capacities of captive starlings. Mozart's skills as a musician and composer would have rendered him especially susceptible to the starling's vocal charms, and thus we will also propose that the funeral and the poem are not the end of the story. Mozart may have left another memorial to his starling, an offbeat requiem for rebels.

Like echo-locating bats or dolphins, some birds may bounce sounds off the animate environment, using behavioral reverberations to perceive the consequences of their vocal efforts

Mozart's starling was a European starling, *Sturnus vulgaris*. The species was later introduced to North America on an artistic note. The birds were imported from England in the 1890s in an effort to represent the avian cast of Shakespeare's plays in this country *(3)*. Fewer than 200 birds were released in New York's Central Park. Population estimates in the 1980s hovered around 200,000,000 birds, a millionfold increase, making starlings one of the most successful road shows in history.

The vocal talents of starlings have been known since antiquity *(4)*. The species possesses a rich repertoire of calls and songs composed of whistles, clicks, rattles, snarls, and screeches. In addition, starlings copy the sounds of other birds and animals, weaving these mimicked themes into long soliloquies that, in captive birds, can contain fragments of human speech. Pliny reported individual birds, mimicking Greek and Latin, that "practiced diligently and spoke new phrases every day, in still longer sentences." Shakespeare knew enough about their abilities to have Hotspur propose teaching a starling to say the name "Mortimer," an earl distrusted by Henry IV, to disturb the king's sleep *(Henry IV, Part I,* act 1, scene 3). In the song cycle *Die schöne Müllerin,* Schubert set to music a poem in which a starling is given a romantic mission: "I'd teach a starling how to speak and sing, / Till every word and note with truth should ring, / With all the skill my lips and tongue impart, / With all the warmth and passion of my heart" *(5)*.

Despite this wealth of anecdotal information, few scientists have studied the vocal behavior of starlings under the conditions necessary to separate fact from fiction. The problem with starlings is that they vocalize too much, too often, and in too great numbers, sometimes in choruses numbering in the thousands (a flock of starlings is labeled a murmuration). Even the seemingly elementary step of creating an accurate catalogue of the vocal repertoires of wild starlings is an intimidating task because of the variety of their sounds. Other well-known avian mimics, such as the mockingbird *(Mimus polyglottos)*, have proved as challenging, leaving unanswered key questions about the development and functions of mimetic behavior.

Some of the problems involved in the study of nonmimetic songbirds arise with mimics as well. Researchers must be able to find and raise songbirds from a young age or ideally from the egg under conditions in which their exposure to social and acoustic stimulation can be controlled. The birds must be observed for many months or sometimes years to capture fully the processes of cultural evolution and transmission of vocal motifs from generation to generation. And for all species, researchers must acquire expertise in the acoustic analysis of sounds to overcome their inability to hear much of the fine detail in avian vocalizations.

Because of these difficulties, many "definitive" pieces of work have been based on small sample sizes, often fewer than ten individuals, sometimes fewer than five. Larger samples are possible only with avicultural favorites, such as canaries *(Serinus canaria)* or zebra finches *(Poephila guttata)*. Even with these subjects, research schedules must be accommodated to seasonal

Meredith J. West and Andrew P. King received their Ph.D.s from the Department of Psychology at Cornell University. Meredith West is a professor of psychology at Indiana University, and Andrew King is a research associate professor at Duke University. Their research interests include learning, development, and communication. Address: Laboratory of Avian Behavior, Route #2 Box 315, Mebane, NC 27302.

cycles. The kinds of vocalizations produced by a species can differ considerably throughout the year, with the most "interesting" sounds in the form of territorial or mating signals occurring for only a few months each year. In sum, songbirds are a handful.

Mimetic species add another layer of difficulty by including sounds made by other birds, other animals, and even machines. Thus, in addition to exploring how members of a mimetic species develop species-typical calls and songs—that is, vocalizations with many shared acoustic properties within a population—investigators routinely encounter individual idiosyncracies. Why does one starling mimic a goat and another a cat? Given the abundance of sounds in the world, what processes account for the selection of models?

Baylis *(6)* advocated studying just part of the mimic's repertoire as a first step, suggesting the example of mockingbirds frequently mimicking cardinals *(Cardinalis cardinalis)*. Although mockingbirds mimic many species, cardinals are a favorite. Why? What consequences accrue for mimic or model? By focusing on one model-mimic system, scientists might answer a number of questions surrounding the nature and function of mimicry. Further control of the model-mimic system can be gained by exposing birds to human speech, a vocal code with a more favorable "signal-to-noise" ratio. This heightens the probability that investigators can detect mimicry and makes it easier to identify the origin of mimicked sounds and the environmental conditions facilitating or inhibiting interspecific mimicry *(7)*. Here, the use of human language is not comparable to efforts with apes or dolphins aimed at uncovering possible analogues to human language. Rather, the use of speech sounds is more properly compared to the use of a radioactive isotope to trace physiological pathways. Thus, when a captive starling utters, "Does Hammacher Schlemmer have a toll-free number?" it is easier to trace the phrase's origin and how often it has been said than to trace the history of the bird's production of "breep, beezus, breep, beeten, beesix."

Over the past decade, we have studied nine starlings, each hand-reared from a few days of age *(8)*. We have also collected information on the behavior of five other starlings (Fig. 1), raised under similar conditions by individuals unaware of our work and unaware of starlings' mimicking abilities when their relationship with the birds began *(9)*. Although many questions remain about the species's vocal capacities, the findings shed light on Mozart's response to his starling's death.

The 14 starlings experienced different social relationships with humans. Eight birds lived individually in what is called interactive contact with the humans who hand-reared them. Their cages were placed in busy parts of the home, and the birds had considerable freedom to associate with their caregivers in diverse ways: feeding from hands; perching on fingers, shoulders, or heads; exploring caregivers' possessions; and inserting themselves into activities such as meal preparation, piano lessons, baths, showers, and telephone conversations (Fig. 2). The humans spontaneously talked to the birds, whistled to them, and gestured by kissing, snapping fingers, and waving good-bye.

Explicit procedures to teach human words using methods prescribed for other mimicking species were not used. Six of the eight caregivers did not know that such training would have an effect until the birds themselves demonstrated their mimicking ability, and two refrained because they were instructed by us to do so. The birds could obtain food and water (and avian companionship in five of eight cases) without interacting with humans.

Three other starlings lived under conditions of limited contact with humans. After 30 days of hand rearing by us, they were individually placed in new homes, along with a cowbird *(Molothrus ater)*. They lived in cages, rarely flew free, and were passively exposed to humans. They heard speech but were not "spoken to" because they did not engage in the kinds of social interactions described for the first group. The final three starlings lived together in auditory contact with humans. They were housed in an aviary on a screened porch of the caregivers raising one of the freely interacting birds. As a result, their auditory environment was loosely yoked to that of the other bird.

The information gathered on the starling's mimicry

Hier ruht ein lieber Narr,
Ein Vogel Staar.
Noch in den besten Jahren
Musst er erfahren
Des Todes bittern Schmerz.
Mir blut't das Herz,
Wenn ich daran gedenke.
O Leser! schenke
Auch du ein Thränchen ihm.
Er war nicht schlimm;
Nur war er etwas munter,
Doch auch mitunter
Ein lieber loser Schalk,
Und drum kein Dalk.
Ich wett', er ist schon oben,
Um mich zu loben
Für diesen Freundschaftsdienst
Ohne Gewinnst.
Denn wie er unvermuthet
Sich hat verblutet,
Dacht er nicht an den Mann,
Der so schön reimen kann.

Den 4ten Juni 1787.

A little fool lies here
Whom I held dear—
A starling in the prime
Of his brief time,
Whose doom it was to drain
Death's bitter pain.
Thinking of this, my heart
Is riven apart.
Oh reader! Shed a tear,
You also, here.
He was not naughty, quite,
But gay and bright,
And under all his brag
A foolish wag.
This no one can gainsay
And I will lay
That he is now on high,
And from the sky,
Praises me without pay
In his friendly way.
Yet unaware that death
Has choked his breath,
And thoughtless of the one
Whose rime is thus well done.

differed by setting and caregiver. Extensive audio taping was carried out for the nine subjects studied under our supervision. For three of the remaining birds involved in interactive contact, we used repertoires available in published works, supplemented by personal inquiries. For the last two we obtained verbal reports from caregivers.

Social transmission of the spoken word

The starlings' mimetic repertoires varied consistently by social context: only the birds in interactive contact mimicked sounds with a clearly human origin. None of the other subjects imitated such sounds, although all mimicked their cowbird companions, each other, wild birds, and mechanical noises. For the purposes of this article, we have elected to focus solely on the actions of the birds in interactive contact.

All of these birds mimicked human sounds—including clear words, sounds immediately recognizable as speech but largely unintelligible, and whistled versions of songs identified as originating from a human source—and mechanical sounds whose source could be identified within the households. For the three audiotaped birds, roughly two-thirds of their vocalizations were related to the words or actions of caregivers. The same categories applied to the remaining five birds, who mimicked speech, whistles, and human-derived or mechanical sounds (Table 1).

Many of the more impressive properties of the starlings' vocal capacities defy simple categorization. The most striking feature was their tendency to mimic connected discourse, imitating phrases rather than single words. Words most often mimicked alone included the birds' names and words associated with humans' arrivals and departures, such as "hi" or "good-bye." All phrases were frequently recombined, sometimes giving the illusion of a different meaning. One bird, for example, frequently repeated, "We'll see you later," and "I'll see you soon." The phrase was often shortened to "We'll see," sounding more like a parental ploy than an abbreviated farewell. Another bird often mimicked the phrase "basic research" but mixed it with other phrases, as in "Basic research, it's true, I guess that's right."

Fewer than 200 starlings were released in Central Park in the 1890s; population estimates in the 1980s hovered around 200,000,000 birds, a millionfold increase

The audiotapes and caregivers' reports made clear, however, that nonsensical combinations (from a human speaker's point of view) were as frequent as seemingly sensible ones: the only difference was that the latter were more memorable and more often repeated to the birds. Sometimes, the speech utterances occurred in highly incongruous settings: the bird mentioned above blasted his owners with "Basic research!" as he struggled frantically with his head caught in string; another screeched, "I have a question!" as she squirmed while being held to have her feet treated for an infection. The tendency for the birds to produce comical or endearing combinations did much to facilitate attention from humans. It was difficult to ignore a bird landing on your shoulder announcing, "Hello," "Give me a kiss," or "I think you're right."

The birds devoted most of their singing time to rambling tunes composed of songs originally sung or whistled to them intermingled with whistles of unknown origin and starling sounds. Rarely did they preserve a melody as it had been presented, even if caregivers repeatedly whistled the "correct" tune. The tendency to sing off-key and to fracture the phrasing of the music at unexpected points (from a human perspective) was reported for seven birds (no information on the eighth). Thus, one bird whistled the notes associated with the words "Way down upon the Swa-," never adding "-nee River," even after thousands of promptings. The phrase was often followed by a whistle of his own creation, then a fragment of "The Star-spangled Banner," with frequent interpositions of squeaking noises. Another bird whistled the first line of "I've Been Working on the Railroad" quite accurately but then placed unexpectedly large accents on the notes associated with the second line, as if shouting, "All the livelong day!" Yet another routinely linked the energetically paced *William Tell* Overture to "Rockaby Baby."

One category of whistles escaped improvisation. Seven of the eight caregivers used a so-called contact whistle to call the birds, typically a short theme (e.g., "da da da dum" from Beethoven's Fifth Symphony). This fragment of melody escaped acoustic improvisation in all cases, although the whistles were inserted into other melodies as well. One bird, however, often mimicked her contact whistle several times in succession, with each version louder than the preceding one (perhaps a quite accurate representation of the sound becoming louder as her caregiver approached her).

All the birds in interactive contact showed an interest in whistling and music when it was performed. They often assumed an "attentive" stance, as shown in Figures 1 and 2: they stood very quietly, arching their necks and moving their heads back and forth. The birds did not vocalize while in this orientation. Records for all eight subjects contained verbal or pictorial reports of the posture.

Clear mimicry of speech was relatively infrequent, due in large part to the birds' tendency to improvise on the sounds, making them less intelligible although definitely still speechlike. Other aspects of their speech imitations were also significant. First, the birds would mimic the same phrase, such as "see you soon" or "come here," but with different intonation patterns. At times, the mimetic version sounded like a human speaking in a pleasant tone of voice, and at other times in an irritated tone. Second, when the birds repeated speech sounds, they frequently mimicked the sounds that accompany speaking, including air being inhaled, lips smacking, and throats being cleared. One bird routinely preceded his rendition of "hi" with the sound of a human sniffing, a combination easily traced to his caregiver being allergic to birds. Finally, the quality of the mimicry of the human voice was surprisingly high. Many visitors who heard the mimicry "live" looked for an unseen human. Those listening to tapes asked which sounds were the starlings' and which the humans', when the only voices were the birds'.

The particular phrases that were mimicked varied, although a majority fell into the broad semantic category of socially expressive speech used by humans as greetings or farewells, compliments, or playful responses to children and pets (see Table 1). Several of the starlings used phrases of greeting or farewell when they heard the sound of keys or saw someone putting on a coat or approaching a door. Several mimicked household events such as doors opening and closing, keys rattling, and dishes clinking together. One bird acquired the word "mizu" (Japanese for water), which she routinely used after flying to the kitchen faucet. Another chanted "Defense!" when the television was on, a sound that she apparently had acquired as she observed humans responding to basketball games.

Figure 1. Kuro is a starling who was hand-reared in captivity. Living in daily close contact with the Iizuka family, she has spontaneously developed, like other starlings in similar circumstances studied by the authors, a rich repertoire of imitations of human speech, songs, and household sounds. Here Kuro listens to whistling. (Photo by Birgitte Nielsen; reprinted by permission of Nelson Canada from *Kuro the Starling,* by Keigo Iizuka and family.)

Caregivers reported that it took anywhere from a few days to a few months for new items to appear in the birds' repertoires. Acquisition time may have depended on the kind of material: one of the birds in limited contact, housed with a new cowbird, learned its companion's vocalization in three days, while one bird in interactive contact took 21 days to mimic his cowbird companion. The latter bird, however, repeated verbatim the question, "Does Hammacher Schlemmer have a toll-free number?" a day after hearing it said only once.

Starlings copy the sounds of other birds and animals, weaving these mimicked themes into long soliloquies that, in captive birds, can contain fragments of human speech

Some whistled renditions of human songs also appeared after intervals of only one or two days. An important variable in explaining rate of acquisition and amount of human mimicry may be the birds' differential exposure to other birds. The three birds without avian cage mates appeared to have more extensive repertoires, but they were also older than the other subjects.

The birds did not engage much in mutual vocal exchanges with their caregivers—that is, a vocalization directed to a bird did not bring about an immediate vocal response, although it often elicited bodily orientation and attention. Thus, the mimicry lacked the "conversational" qualities that have been sought after in work with other animals *(10).* As no systematic attempt had been made to elicit immediate responding by means of food or social rewards, reciprocal exchanges may nevertheless be possible. Ongoing human conversation not involving the starlings, however, was a potent stimulus for simultaneous vocalizing. The birds chattered frequently and excitedly while humans were talking to each other in person or on the telephone.

The starlings' lively interest and ability to participate in the activities of their caregivers created an atmosphere of mutual companionship, a condition that may be essential in motivating birds to mimic particular models, as indicated by the findings with the birds in limited and auditory contact. The capacity of starlings to learn the sounds of their neighbors fits with what is known about their learning of starling calls, especially whistles, in nature. They learn new whistles as adults by means of social interactions, an ability that is quite important when they move into new colonies or flocks *(11).* Analyses of social interactions between wild starling parents and their young also indicate the use, early in ontogeny, of vocal exchanges between parent and young and between siblings *(12).* Thus, the capacities identified in the mimicry of human speech and their dependence on social context seem relevant to the starling's ecology.

Other mimics and songsters

Studies of another mimic, the African gray parrot *(Psittacus erithacus),* also indicate linkages between mimicry and social interaction *(13).* This species mimics human speech when stimulated to do so by an "interactive modeling technique" in which a parrot must compete for the attention of two humans engaged in conversation. Extrinsic rewards such as food are avoided. The reinforcement is physical acquisition of the object being talked about and responses from human caregivers. Such procedures lead to articulate imitation and often highly appropriate use of speech sounds. Pepperberg reports that one bird's earliest "words" referred to objects he could use: "paper," "wood," "hide" (from rawhide chips), "peg wood," "corn," "nut," and "pasta" *(14)*. The parrot also employed these mimicked sounds during exchanges with caregivers in which he answered questions about the names of objects and used labels identifying shape and color in appropriate ways. The parrot's use of "no" and "want" also suggested the ability to form functional relationships between speech and context, a capacity perhaps facilitated by the trainer's explicit attempts to arrange training sessions meaningful for the student.

Explanations of mimicry of human sounds in this and other species originate in the idea that hand-reared birds perceive their human companions in terms of the social roles that naturally exist among wild birds. Lorenz and von Uexküll elaborated on the kinds of relationships between and among avian parents, offspring, siblings, mates, and rivals *(15)*. In the case of captive birds, humans become the companion for all seasons, with the nature of the relationship shifting with the changing developmental and hormonal cycles in a bird's life.

Mimics are not the only birds to show clear evidence of the effects of companions on vocal capacities. Two examples from nonmimetic species are relevant. In the white-crowned sparrow *(Zonotrichia leucophrys)*, the capacity to learn the songs of other males differs according to the tutoring procedure used. For example, young males learn songs from tape recordings until they are 50 days of age but not afterward. They do acquire songs well after 50 days from live avian tutors with whom they can interact, copying the song of another species, even if

Explanations of mimicry of human sounds originate in the idea that hand-reared birds perceive their human companions in terms of the social roles that naturally exist among wild birds

they can hear conspecifics in the background. The potency of social tutors has led to a comprehensive reinterpretation of the nature of vocal ontogeny in this species *(16)*. We tried tutoring nine of the starlings using tapes of the caregiver's voice singing songs and reciting prose. There was no evidence of mimicry, except that one bird learned the sound of tape hiss. And thus, if we had relied on tape tutoring, as has been done with many species to assess vocal capacity, we would have vastly underestimated the starlings' skills.

What are the characteristics of live tutors that make them so effective? The studies of white-crowned sparrows suggest that it is not the quality of the tutor's voice, but the opportunity for interaction. Indeed, we have studied a case where voice could not be a cue at all because the "tutor" could not sing. In cowbirds, as in many songbirds, only males sing. Females are frequently the recipients of songs and display a finely tuned perceptual sensitivity to conspecific songs *(17)*. We have documented that acoustically naive males produce distinct themes when housed with female cowbirds possessing different song preferences. We have also identified one important element in the interaction. When males sang certain themes, females responded with distinctive wing movements. The males responded in turn to such behavior by repeating the songs that elicited the females' wing movements. Such data show that singers attend to visual, as well as acoustic, cues and that tutors can be salient influences even when silent. In this species, the social, as distinct from the vocal, conduct of a male's audience is of consequence.

Figure 2. Kuro adopts a listening posture during a music lesson, with neck arched and head moving back and forth. (Photo by Birgitte Nielsen; reprinted by permission of Nelson Canada from *Kuro the Starling*, by Keigo Iizuka and family.)

Studies of another avian group, domestic fowl *(Gallus gallus),* also direct attention to the importance of a signaler's audience *(18).* In this species, male cockerels produce different calls in the presence of different social companions. Emitting a food call in the presence of food is not an obligatory response but one modulated by the signaler's observations of his audience. Similar findings with cockerel alarm calls indicate the need to consider the multiple determinants of vocal production. Taken as a whole, the findings reveal that, for many birds, acoustic communication is as much visual as vocal experience.

Mozart as birdcatcher

Mozart knew how to look at, as well as listen to, audiences, especially when one of his compositions was the object of their attention. After observing several audiences watching *The Magic Flute,* he wrote to his wife, "I have at this moment returned from the opera, which was as full as ever. . . . But what always gives me most pleasure is the *silent approval!* You can see how this opera is becoming more and more esteemed" *(19).* Mozart's enjoyment of the less obvious reactions of his audience suggests that, like a bird, he too was motivated not only by auditory but by visual stimuli. The German word he used can be translated "applause" as well as "approval," suggesting his search for rewards more meaningful than the expected clapping of hands. We now turn to the case of Mozart's starling and to the kinds of social and vocal rewards offered to him by his choice of an avian audience.

Mozart recorded the purchase of his starling in a diary of expenses, along with a transcription of a melody whistled by the bird and a compliment (Fig. 3). He had begun the diary at about the same time that he began a catalogue of his musical compositions. The latter effort was more successful, with entries from 1784 to 1791, the year of his death. His book of expenditures, however, lapsed within a year, with later entries devoted to practice writing in English *(20).* The theme whistled by the starling must have fascinated Mozart for several reasons. The tune was certainly familiar, as it closely resembles a theme that occurs in the final movement of the Piano Concerto in G Major, K. 453 (see Fig. 3). Mozart recorded the completion of this work in his catalogue on 12 April in the same year. As far as we know, just a few people had heard the concerto by 27 May, perhaps only the pupil for whom it was written, who performed it in public for the first time at a concert on 13 June. Mozart had expressed deep concern that the score of this and three other concertos might be stolen by unscrupulous copyists in Vienna. Thus, he sent the music to his father in Salzburg, emphasizing that the only way it could "fall into other hands is by that kind of cheating" *(21).* The letter to his father is dated 26 May 1784, one day before the entry in his diary about the starling.

Mozart's relationship with the starling thus begins on a tantalizing note. How did the bird acquire Mozart's music? Our research suggests that the melody was certainly within the bird's capabilities, but how had it been transmitted? Given our observation that whistled tunes are altered and incorporated into mixed themes, we assume that the melody was new to the bird because it was so close a copy of the original. Thus, we entertain the possibility that Mozart, like other animal lovers, had already visited the shop and interacted with the starling before 27 May. Mozart was known to hum and whistle a good deal. Why should he refrain in the presence of a bird that seems to elicit such behavior so easily?

A starling in May would be either quite young, given typical spring hatching times, or at most a year old, still young enough to acquire new material but already an accomplished whistler. Because it seems unlikely to us that a very young bird could imitate a melody so precisely, we envision the older bird. The theme in question from K. 453 has often been likened to a German folk tune and may have been similar to other popular tunes already known to the starling, analogous to the highly familiar tunes our caregivers used. But to be whistled to by Mozart! Surely the bird would have adopted its listening posture, thereby rewarding the potential buyer with "silent applause."

Given that whistles were learned quite rapidly by the starlings we studied, it is not implausible that the Vienna starling could have performed the melody shortly after hearing it for the first time. Of course, we cannot rule out a role for a shopkeeper, who could have repeated Mozart's tune from its creator or from the starling. In any case, we imagine that Mozart returned to the shop and purchased the bird, recording the expense

Table 1. Sounds mimicked by starlings

Greetings and farewells		
hi	hey there	I'll (we'll) *see you* soon
good morning	*c'mon, c'mere*	breakfast
hello	go to your cage	it's time
hey buddy	night night	
Attributions		
you're a crazy bird	nutty bird	you're gorgeous
good girl	rascal	see you soon baboon
pretty bird	you're kidding	baby
silly bird		
Conversational fragments		
it's true	OK	have the kids called
I suggest	I have a question	*whatcha doing*
that's right	defense	what's going on
basic research	thank you	all right you guys
because	*right*	this is Mrs. Suthers calling
I guess	who is coming	
Human sounds		
sighing	*sniffing*	kissing
coughing	*lip smacking*	wolf whistle
throat clearing	*laughing*	
Household sounds		
door squeaking	alarm clock	dishes clinking
cat meowing	telephone beep	gun shots
dog barking	keys rattling	

Categories refer to social contexts in which humans produced the sounds, not necessarily the ones in which starlings repeated them. Italicized entries were imitated by four or more birds.

Figure 3. Wolfgang Amadeus Mozart was also the delighted owner of a pet starling. He recorded the purchase of the starling in an expense book, noting the date, price, and a musical fragment the bird was whistling. The pleasure he expressed at hearing the starling's song—"Das war schön!" (that was beautiful!)—is all the more understandable when one compares the beginning of the last movement of his Piano Concerto in G Major, K. 453, which was written about the same time. Somehow the bird had learned the theme from Mozart's concerto. It did however sing G sharp where Mozart had written G natural, giving its rendition a characteristically off-key sound.

out of appreciation for the bird's mimicry. Some biographers suggest an opposite course of transmission—from the starling to Mozart to the concerto—but the completion date of K. 453 on 12 April makes this an unlikely, although not impossible, sequence of events.

Given the sociable nature of the captive starlings we studied, we can imagine that some of the experiences that followed Mozart's purchase must have been quite agreeable. Mozart had at least one canary as a child and another after the death of the starling, suggesting that it would not be hard for him to become attached to so inventive a housemate. Moreover, he shared several behavioral characteristics with captive starlings. He was fond of mocking the music of others, often in quite irreverent ways. He also kept late hours, composing well into the night *(22)*. The caregivers of the starlings we studied uniformly reported—and sometimes complained about—the tendency of their birds to indulge in more than a little night music.

The mimicry of vocal acts such as lip noises, sniffs, and throat clearing brought to the attention of caregivers routine dimensions of their own behavior that they rarely took notice of

The text of Mozart's poem on the bird's death suggests other perceptions shared with the caregivers. Mozart dubbed his pet a "fool"—the German word could also be translated as "clown" or "jester"—an attribution in keeping with the modern starlings' vocal productions of "crazy bird," "rascal," "silly bird," and "nutty bird" and the even more frequent use of such terms in the written description of life with starlings. Mozart gets to the heart of the starling's character when he states that the bird was "not naughty quite, / But gay and bright, / And under all his brag, / A foolish wag." And thus, when we contemplate Mozart's emotions at the bird's death, we see no reason to invoke attributions of displaced grief. We regard Mozart's sense of loss as genuine, his epitaph as an apt gesture.

No other written records of Mozart's relationship with his pet are known. He may have said more, given his prolific letter writing, but much of his correspondence during this period has been lost. The lack of other accounts, however, cannot be considered to indicate a lack of interest in his starling. We are inclined to believe that other observations by Mozart on the starling do exist but have not been recognized as such. Our case rests in part on recent technical analyses of the original (autograph) scores of Mozart's compositions, investigations describing changes in handwriting, inks, and paper. Employing new techniques to date paper by analyzing the watermarks pressed into it at the time of its manufacture, Tyson *(23)* has established that the dates and places assigned to some of Mozart's compositions can be questioned, reaching the general conclusion that many pieces were written over an extended period of time and not recorded in his catalogue until the time of completion. The establishment of an accurate chronology of Mozart's compositions is obviously essential to those attempting to understand the development of his musical genius. It also serves our purposes in reconstructing events after the starling's funeral.

One composition examined by Tyson is a score entered in Mozart's catalogue on 12 June 1787, the first to appear after the deaths of his father and the starling. The piece is entitled *A Musical Joke* (K. 522). Consider the following description of it from a record jacket: "In the first movement we hear the awkward, unproportioned, and illogical piecing together of uninspired material . . . [later] the andante cantabile contains a grotesque cadenza which goes on far too long and pretentiously and ends with a comical deep pizzicato note . . . and by the concluding presto, our 'amateur composer' has lost all control of his incongruous mixture" *(24)*. Is the piece a musical joke? Perhaps. Does it bear the vocal autograph of a starling? To our ears, yes. The "illogical piecing together" is in keeping with the starlings' intertwining of whistled tunes. The "awkwardness" could be due to the starlings' tendencies to whistle off-key or to fracture musical phrases at unexpected points. The presence of drawn-out, wandering phrases of uncertain structure also is characteristic of starling soliloquies. Finally, the abrupt end, as if the instruments had simply ceased to work, has the signature of starlings written all over it.

Tyson's analysis of the original score of K. 522

indicates that it was not written during June 1787, but composed in fragments between 1784 and 1787, including an excerpt from K. 453. This period coincides with Mozart's relationship with the starling. A common interpretation is that *A Musical Joke* was meant to caricature the kinds of music popular in Mozart's day. Writing such music, a course of action urged on him by his father, might have earned Mozart more money. And thus, the composition has also been interpreted in regard to the father/son relationship *(25)*. Tyson disputes this view on the basis of the physical nature of the autograph score, as much of it was written before Leopold's death, and the lack of solid evidence that Mozart's relationship with his father was bitter enough to cause him to commemorate his first and foremost teacher with a parody.

Although we do not presume to explain all the layers of compositional complexity contained in K. 522, we propose that some of its starling-like qualities are pertinent to understanding Mozart's intentions in writing it. Given the propensities of the starlings we studied and the character and habits of Mozart, it is hard to avoid the conclusion that some of the fragments of K. 522 originated in Mozart's interactions with the starling during its three-year tenure. The completion of the work eight days after the bird's death might then have been motivated by Mozart's desire to fashion an appropriate musical farewell, a requiem of sorts for his avian friend.

Last words

We have offered these observations on starlings and on Mozart for two reasons. First, to give music scholars new insights with which to evaluate one of the world's most studied composers. The analyses of the autograph scores and recent reinterpretations of Mozart's illnesses and death demonstrate the power of present-day knowledge to inform our understanding of the past. We have provided the profile of captive starlings as another way to gain perspective on Mozart's genius.

Second, we hope to spark further interest in the analysis of the social stimulation of vocal learning. Although the role of social companions in motivating avian vocal learning is now well established, the mechanisms by which social influence exerts its effects have only begun to be articulated *(26)*. Part of the problem is defining the nature of social contexts. To say birds interact is to say something quite vague. Interact how? By fighting? By feeding? By flocking? By sitting next to one another? Measuring sound waves is easy compared to calibrating degrees of social influence. Moreover, social signals are multi-modal. The species described here make much use of visual, as well as vocal, stimulation. By what means do they link sights and sounds? Why are only certain linkages made? Answering these questions is the next challenge for students of communication.

One of the founders of the study of bird song, W. H. Thorpe, speculated that birds' imitation of sounds represents a quite simple cognitive process: "The essence of the point may be summed up by saying that while it is very difficult for a human being (and perhaps impossible for an animal) to see himself as others see him, it is much less difficult for him to hear himself as others hear him" *(27)*. Although we recognize the law of parsimony in Thorpe's remark, we are led by the evidence to seek a phylogenetic middle ground between self-awareness and vocal matching. We propose that some birds use acoustic probes to test the contingent properties of their environment, an interpretation largely in keeping with concepts of communication as processes of social negotiation and manipulation *(28)*. An analogy with the capacities of echo-locating animals may be appropriate. Like bats or dolphins emitting sounds to estimate distance, some birds may bounce sounds off the animate environment, using behavioral reverberations to gauge the effects of their vocal efforts. They are not using Thorpe's behavioral mirror, necessary for self-reflection, but instead a social sounding board with which to shape functional repertoires.

In the case of our starlings, we also conclude that social sonar works two ways: human caregivers cast many sounds in the direction of their starlings and were often educated by the messages returned. The mimicry of vocal acts such as lip noises, sniffs, and throat clearing brought to the attention of caregivers routine dimensions of their own behavior that they rarely took notice of. The birds' echoing of greetings, farewells, and words of affection conveyed a sense of shared environment with another species, a sensation hard to forget (Fig. 4). The caregivers' sadness in response to the illnesses, absence, or death of their avian companions also suggests that they had been beguiled by the chance to glimpse a bird's-eye view of the world. Most found themselves at a loss for words. And thus we turn to Mozart for fitting emotional expressions—his poem, his *Musical Joke*, and his appropriately grand burial for a "starling bird."

Figure 4. Relationships between starlings and human beings appear to reflect the behavior of birds in the wild. Hand-reared starlings interact with their human companions in terms of the social roles of wild birds. In particular, they learn by observing vocal and other responses to their own expressive efforts. (Photos by Birgitte Nielsen.)

References

1. G. Nottebohm. 1880. *Mozartiana*. Breitkopf and Härtel.
   O. E. Deutsch. 1965. *Mozart: A Documentary Biography*. Stanford Univ. Press.
2. O. Jahn. 1970. *Life of Mozart*, trans. P. D. Townsend. Cooper Square.
   B. Brophy. 1971. In *W. A. Mozart. Die Zauberflöte*. Universe Opera Guides.
   W. Hildesheimer. 1983. *Mozart*, trans. M. Faber. Vintage.
   P. J. Davies. 1989. *Mozart in Person: His Character and Health*. Greenwood.
3. F. M. Chapman. 1934. *Handbook of Birds of Eastern North America*. Appleton.
   E. W. Teale. 1948. *Days without Time*. Dodd, Mead.
4. E. A. Armstrong. 1963. *A Study of Bird Song*. Oxford Univ. Press.
   C. Feare. 1984. *The Starling*. Oxford Univ. Press.
5. R. Dyer-Bennet, trans. 1967. Impatience. In *The Lovely Milleress (Die schöne Müllerin)*. Schirmer.
6. J. R. Baylis. 1982. Avian vocal mimicry: Its function and evolution. In *Acoustic Communication in Birds*, vol. 2, ed. D. E. Kroodsma and E. H. Miller, pp. 51–84. Academic Press.
7. D. Todt. 1975. Social learning of vocal patterns and models of their application in grey parrots. *Zeitschrift für Tierpsychologie* 39:178–88.
   I. M. Pepperberg. 1981. Functional vocalizations by an African Grey Parrot *(Psittacus erithacus)*. *Zeitschrift für Tierpsychologie* 55:139–60.
8. M. J. West, A. N. Stroud, and A. P. King. 1983. Mimicry of the human voice by European starlings: The role of social interactions. *Wilson Bull*. 95:635–40.
9. H. B. Suthers. 1982. Starling mimics human speech. *Birdwatcher's Digest* 2:37–39.
   M. S. Corbo and D. M. Barras. 1983. *Arnie the Darling Starling*. Houghton Mifflin.
   K. Iizuka. 1988. *Kuro the Starling*. Nelson.
   M. S. Corbo and D. M. Barras. 1989. *Arnie and a House Full of Company*. Fawcett Crest.
   A. DeMotos, pers. com.
   W. R. Fox, unpubl. data.
   A. Peterson and T. Peterson, pers. com.
10. I. M. Pepperberg. 1986. Acquisition of anomalous communicatory systems: Implication for studies on interspecies communication. In *Dolphin Behavior and Cognition: Comparative and Ethological Aspects*, ed. R. J. T. Schusterman and F. Wood, pp. 289–302. Erlbaum.
11. M. Adret-Hausberger. 1982. Temporal dynamics of dialects in the whistled songs of sedentary starlings. *Ethology* 71:140–52.
    ———. 1986. Species specificity and dialects in starlings' whistles. In *Acta 19th Congr. Intl. Ornithol.*, vol. 2, pp. 1585–97.
12. M. Chaiken. 1986. Vocal communication among starlings at the nest: Function, individual distinctiveness, and development of calls. Ph.D. diss., Rutgers Univ.
13. I. M. Pepperberg. 1988. An interactive modeling technique for acquisition of communication skills: Separation of "labeling" and "requesting" in a psittacine subject. *App. Psycholing*. 9:59–76.
14. Pepperberg. Ref. *7*.
15. K. Lorenz. 1957. Companionship in bird life. In *Instinctive Behavior: The Development of a Modern Concept*, ed. C. H. Schiller, pp. 83–128. International Universities Press.
    J. von Uexküll. 1957. A stroll through the world of animals and men. In *Instinctive Behavior: The Development of a Modern Concept*, ed. C. H. Schiller, pp. 5–82. International Universities Press.
16. L. F. Baptista and L. Petrinovich. 1984. Social interaction, sensitive periods, and the song template hypothesis in the white-crowned sparrow. *Animal Behav*. 36:1753–64.
    L. Petrinovich. 1989. Avian song development: Methodological and conceptual issues. In *Contemporary Issues in Comparative Psychology*, ed. D. A. Dewsbury, pp. 340–59. Sinauer.
17. A. P. King and M. J. West. 1988. Searching for the functional origins of song in eastern brown-headed cowbirds, *Molthrus ater ater*. *Animal Behav*. 36:1575–88.
    M. J. West and A. P. King. 1988. Female visual displays affect the development of male song in the cowbird. *Nature* 334:244–46.
18. P. Marler, A. Dufty, and R. Pickert. 1986. Vocal communication in the domestic chicken. II. Is a sender sensitive to the presence of a receiver? *Animal Behav*. 34:194–98.
    S. J. Karakashian, M. Gyger, and P. Marler. 1988. Audience effects on alarm calling in chickens *(Gallus gallus)*. *J. Comp. Psychol*. 102:129–35.
19. E. Anderson, ed. 1989. *The Letters of Mozart and His Family*, p. 907. Norton.
20. Jahn. Ref. *2*.
21. Anderson. Ref. *19*, p. 877.
22. F. Niemtschek. 1956. *Life of Mozart*, trans. H. Mautner. Leonard Hyman.
    Jahn. Ref. *2*.
    Davies. Ref. *2*.
23. A. Tyson. 1987. *Mozart: Studies of the Autograph Scores*. Harvard Univ. Press.
24. W. A. Mozart. *A Musical Joke*. Liner notes by P. Cohen. Deutsche Grammophon. 400 065–2.
25. Ref. *2*.
26. Ref. *16*.
27. W. H. Thorpe. 1961. *Bird-Song*, p. 79. Cambridge Univ. Press.
28. D. W. Owings and D. F. Hennessy. 1984. The importance of variation in sciurid visual and vocal communication. In *The Biology of Ground-dwelling Squirrels: Annual Cycles, Behavioral Ecology, and Sociality*, ed. J. O. Murie and G. R. Michener, pp. 167–200. Univ. Nebraska Press.

Aerial Defense Tactics of Flying Insects

Preyed upon by echolocating bats, some night-flying insects have developed acrobatic countermeasures to evade capture

Mike May

Walking home late one summer night, I glimpsed a small mass slip through the air, past the halo of a street lamp, and into the darkness. Although I was fatigued by a long day, my curiosity was piqued; I crouched down in the darkness and waited for another sign of movement. Within minutes the elusive flyer returned, swerving momentarily in the light, and then shooting back into the night. It was a bat—apparently foraging for its nightly meal of insects. Soon there were others, darting and weaving by the lamp as they attempted to scoop up the insects attracted to the light. As I watched the aerial display, I was impressed by the remarkable speed at which a bat could change its flight path. I tossed a few pebbles into the air and watched as the bats easily pursued the decoys, but turned away when the deception became apparent. Surely, I thought, there was little hope for an insect once a bat had homed in on it.

I gave the matter no more thought until several years later, when I began my doctoral research—perhaps not coincidentally concerned with the flying abilities of insects. As a graduate student I learned there is a considerable history to the study of the aerial encounters between bat and insect. It proves to be a story with a number of surprising turns, and it begins almost 200 years ago with the discovery that bats use their ears, and not their eyes, to navigate.

Lazaro Spallanzani, an 18th-century pioneer of experimental biology, showed that blinded bats are not only able to avoid obstacles in their flight path—such as fine silk threads—but are also able to snag insects in midflight. After hearing of Spallanzani's research, Charles Jurine, a surgeon and entomologist, demonstrated that when the bats' ears are plugged, the animals collide with even relatively large objects in their path, and they are incapable of catching insects.

For over a century the observations of Spallanzani and Jurine were not widely accepted, primarily because no one could imagine how it was that a bat could hear the precise location of such small, essentially silent objects. No advance was made in understanding "Spallanzani's bat problem" until 1920, when the English physiologist H. Hartridge suggested that bats might somehow use sounds of very high frequency to detect the objects. Perhaps the frequencies might even extend beyond the upper limit of human hearing—about 20 kilohertz—to the part of the spectrum called ultrasound.

The mystery of bat navigation was ultimately solved by a Harvard undergraduate, Donald Griffin, in collaboration with the Harvard physicist G. W. Pierce—who invented a device that could detect ultrasound—and the Harvard physiologist Robert Galambos. In 1938 Pierce and Griffin pointed a "sonic detector" at bats flying in a room and found that the animals were, in fact, emitting signals at ultrasonic frequencies. Griffin and Galambos later showed that bats emit ultrasonic cries from their mouths and use their ears to detect the echoes of the sounds reflected from objects in their flight paths. Griffin called this process of navigation *echolocation.*

Echolocation turns out to be an extremely precise and effective method by which bats navigate and identify objects in the dark. In the early 1980s Hans-Ulrich Schnitzler and his colleagues at the Institute for Biology in Tübingen, and Nobuo Suga of Washington University, found that bats are able to analyze the ultrasonic echoes reflected from the bodies and wings of flying insects in such a way as to determine not only the location but also the speed and, perhaps, the type of insect that produces the echoes. All the evidence suggests that the echolocating bat is a very sophisticated hunter; not only is it an adept flyer, but it is equipped with a sensitive auditory system designed to locate and identify potential targets.

However, the bat's ability to find and capture a flying insect is just one side of the story. Some flying insects are able to detect the ultrasonic cries of a bat and take evasive action. Flying insects pursued by a bat do not follow simple ballistic trajectories; they are not such easy targets. To the contrary, the encounter between a bat and an insect is one that might rival the tactics of modern air-to-air combat, involving an efficient early-warning system, some clever aerodynamic engineering and

Mike May is a free-lance science writer. He acquired a taste for the breadth of biology as an undergraduate at Earlham College in Richmond, Indiana. While completing an M.S. in biological engineering at the University of Connecticut at Storrs, he discovered some electronic answers to biological questions. After pursuing bicycle mechanics for a year, he returned to biology and earned a Ph.D. as a biomechanic at Cornell University. Address: P.O. Box 141, Etna, New York 13062.

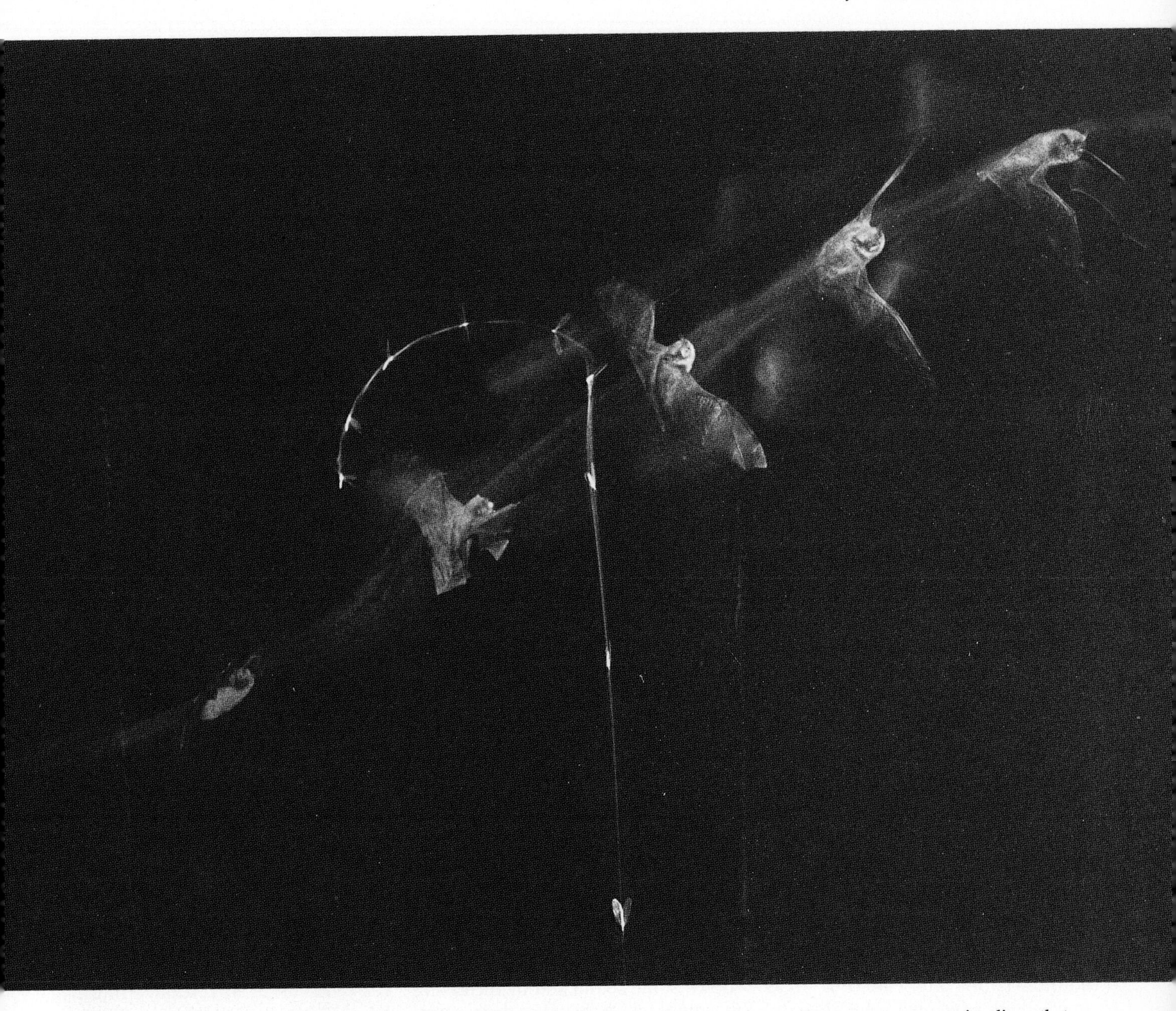

Figure 1. Aerial encounter between an insect-eating bat and a green lacewing reveals one of the evasive maneuvers—a passive dive—that an insect will use to escape a bat. Hunting bats locate their prey by emitting high-frequency (ultrasonic) cries and detecting the echoes of the sounds reflected from the insect's wings and body—a system of navigation called echolocation. Some insects are able to hear the bat's high-frequency sounds and can respond with rapid changes in their flight trajectories. In this stop-motion photograph, a stroboscopic flash reveals the relative positions of the insect and the bat—at intervals of less than one-tenth of a second—as they move from left to right in the scene. (Photograph courtesy of Lee Miller, Odense University, Denmark.)

the simple economics of making do with what is available.

Dodging a Speeding Bat

Almost 70 years before Griffin and Galambos demonstrated that bats can locate objects with ultrasound, F. Buchanan White of Perth, Scotland, proposed that moths can detect bats through the sense of hearing. Although White had no evidence for this conjecture, his idea was ultimately confirmed by behavioral studies in the 1950s, and especially by the work of Kenneth Roeder of Tufts University in the early 1960s. Roeder made hundreds of long-exposure photographs of free-flying moths and recorded their aerial maneuvers in response to a stationary source of artificial ultrasound. He found that if the moths were more than 10 feet from the source of the ultrasound, they simply turned away. But if they were closer to the sound, the moths performed a variety of acrobatic maneuvers, including rapid turns, power dives, looping dives and spirals. For a more natural touch, Roeder photographed wild bats attacking the flying moths; clearly visible in the photographs is the track of the bat zipping across the scene and the evasive path of the moth as it escapes the attack—sometimes.

Since Roeder's studies, a number of nocturnal flying insects have been found to perform evasive aerial maneuvers in response to the ultrasonic cries of bats. In 1979 Lee Miller and Jens Olesen of Odense University found that hunting bats or artificial ultrasonic pulses induce erratic flight in free-flying green lacewings. More recently, David Yager of Cornell University, Brock Fenton of York University in Toronto, and I have shown that some species of praying mantis will perform different types of escape procedures depending on the loudness of the ultrasound emission—not unlike Roeder's moths. In these experiments we

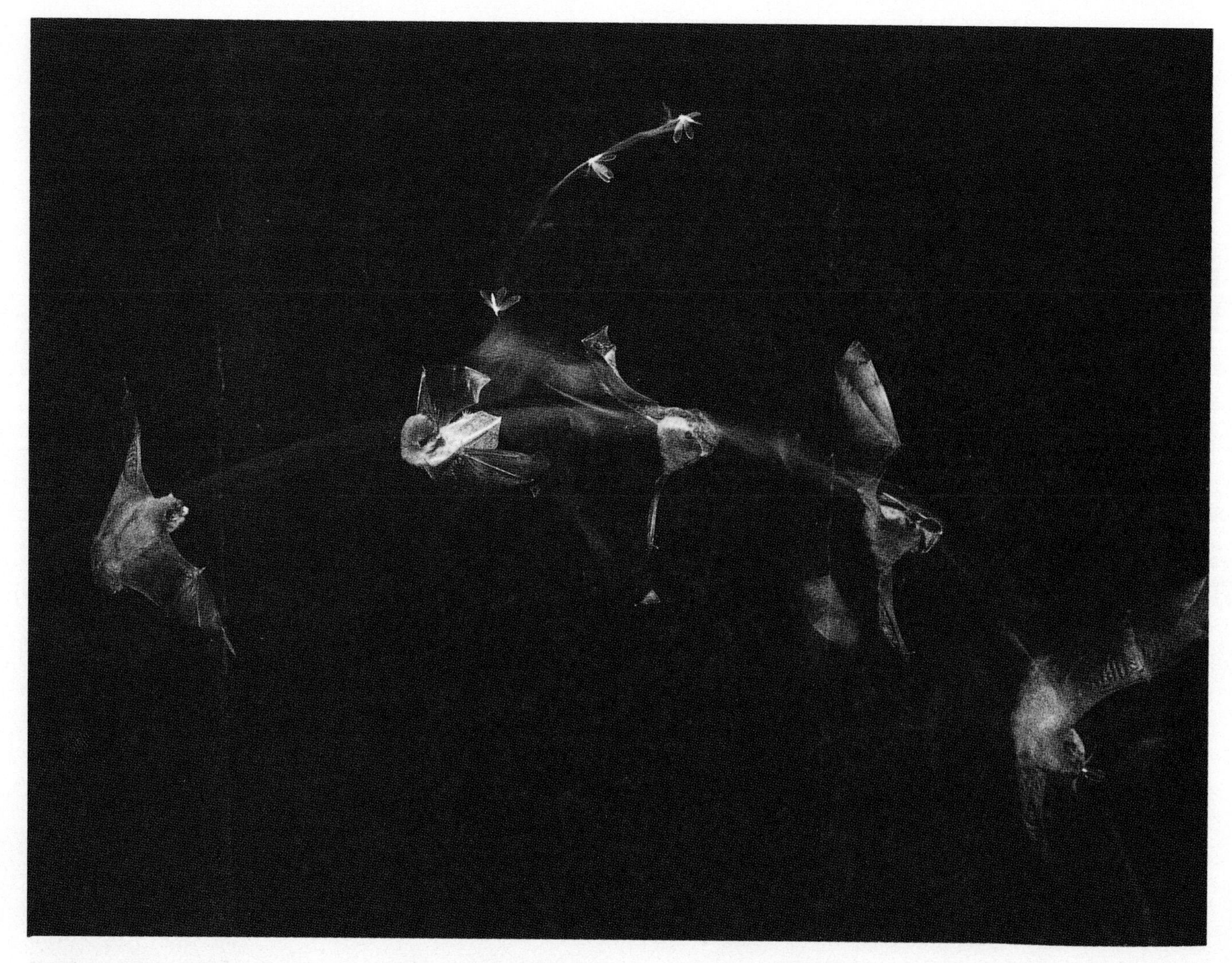

Figure 2. Failure to escape results in death by devourment for a green lacewing performing a diving arc into the embracing wings and tail membrane of an approaching bat. The wings of the captured insect can be seen in the mouth of the bat as it descends to the right. (Photograph courtesy of Lee Miller, Odense University, Denmark.)

used an artificial source of ultrasound—sportively called a "batgun"—with which we "shot" free-flying mantises. At distances greater than 10 meters, most of the mantises did not respond to the ultrasound. Within seven to nine meters of the batgun, however, the mantises would make a slight turn or a shallow dive. At still closer range—within five meters—the mantises would perform steep dives ranging from 45 degrees to nearly a vertical drop, occasionally even in a spiral. Just as Roeder had found in his studies of the moths, we found that the praying mantis will make its most drastic evasive maneuvers when the ultrasound is loudest.

Artificial ultrasound has also been shown to induce changes in the flying patterns of other insects. Daniel Robert, now at Cornell University, found that flying locusts respond to ultrasonic pulses by steering away from the source of the sound and by increasing the rate at which they beat their wings. Similarly, Hayward Spangler of the Carl Hayden Bee Research Center in Tucson showed that, immediately after hearing artificial ultrasonic pulses, tiger beetles fly toward the ground and land. Frederic Libersat, now at the Hebrew University in Jerusalem, and Ronald Hoy of Cornell University discovered that a tethered, flying katydid will stop flying immediately after hearing an ultrasonic stimulus, suggesting that it would perform a dive. Although no one has reported interactions between these insects and bats, it seems likely that such bat-avoidance responses will be found in many night-flying insects that hear ultrasound.

Certainly the value of a rapid escape mechanism for the survival of a flying insect is no longer in doubt. Roeder's studies showed that moths that dive in response to an ultrasonic stimulus were 40 percent less likely to be captured by a bat. The green lacewings, studied by Miller and Olesen, were even more successful at escaping bats—being captured only 30 percent of the time. On the other hand, deafened green lacewings were captured about 90 percent of the time.

More recently, Yager and members of Fenton's research group performed a series of field experiments in which they exposed two species of praying mantis to wild, hunting bats. One species, *Parasphendale agrionina*, makes rapid changes in its flight path in response to artificial pulses of ultrasound or in response to hunting bats. The other species, *Miomantis paykullii*, is an excellent flyer, but does not change its flying pattern in response to artificial ultrasound or in response to hunting bats. Of five attacks on *P. agrionina* in which the mantises performed evasive maneuvers the insects successfully escaped the hunting bats in every case. In contrast, during three attacks on *M. paykullii* and three attacks on *P. agrionina* in which neither species performed evasive maneuvers the insects were captured in five of the six cases. These experiments provide strong evidence that ultrasonic hearing and rapid changes in trajectory can help a flying insect evade an attack by a predatory bat.

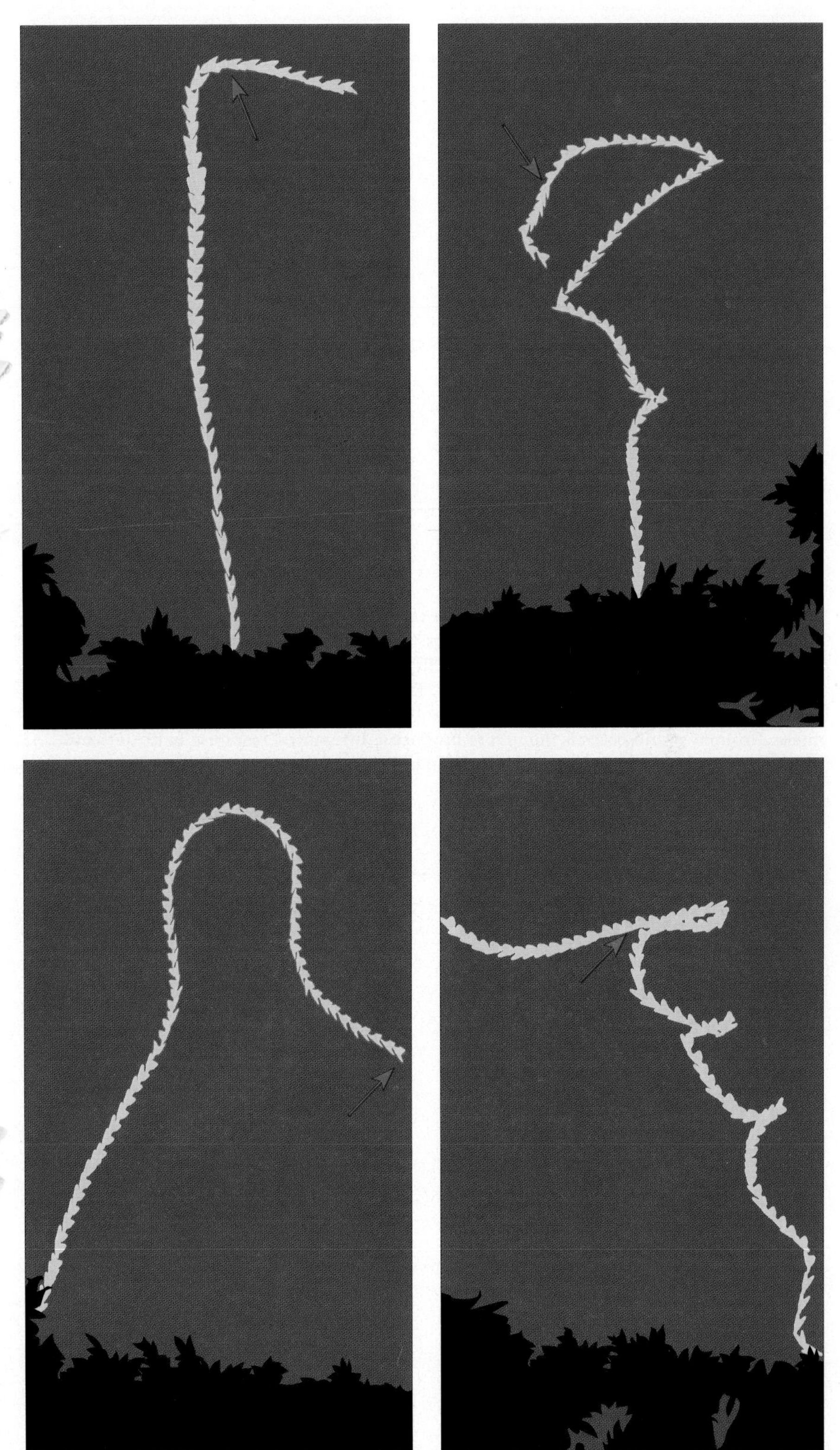

Figure 3. Flight paths of various insects in response to the onset of artificial pulses of ultrasound *(red arrows)* demonstrate some of the tactics used to escape a hunting bat. A single species of insect will often have several different evasive maneuvers in its behavioral repertoire—preventing bats from anticipating any single response. A passive dive *(upper left)*, resulting from the absence of any wing motion, is the simplest type of response. Erratic flight movements *(upper right)*, consisting of a looping turn and ending with a passive dive, is one of the evasive maneuvers performed by a small geometer moth. A powered dive (assisted by wingbeats) may be preceded by a rapid ascent *(lower left)*. A series of tight turns may also make the insect's descent to the ground somewhat more gradual *(lower right)*. (Adapted from Roeder 1962.)

Mating, Death and Phonotaxis

It seems clear that some insects are able to detect the ultrasonic pulses emitted by bats, and then use this information as an early warning system—much the same way a combat pilot in a fighter plane might detect the radar of an enemy plane. And like the combat pilot, the targeted insect must perform evasive maneuvers or suffer the consequences of being captured. In the case of the fighter pilot, however, we know the physics and the engineering behind the detection of radar, and the aerodynamics of flight maneuvers is the stuff of textbooks in flight school. But how does an insect do it? How does it detect the ultrasound, convert this signal into a message that says "take evasive action," and then perform its spectacular acrobatic maneuvers? We don't as yet know all the answers, but bits and pieces of the story are coming to light.

Part of the answer lies in the behavior known as phonotaxis, the movement of an animal in a direction determined by the location of a sound source. Phonotactic behavior of insects has been especially well studied in certain species of crickets. Phonotaxis takes two forms in these animals, based on the direction the cricket moves with respect to the sound source. When a female cricket moves toward the source of the calling song of a courting male cricket—the familiar chirp we hear on summer nights—the locomotory behavior of the female is described as positive phonotaxis. On the other hand, the same female will respond to another sound, of a higher frequency, by flying away from the source—a display of negative phonotaxis. Positive and negative phonotaxis in these instances suggest that the cricket is able to discern at least two distinct aspects of the sound: its fre-

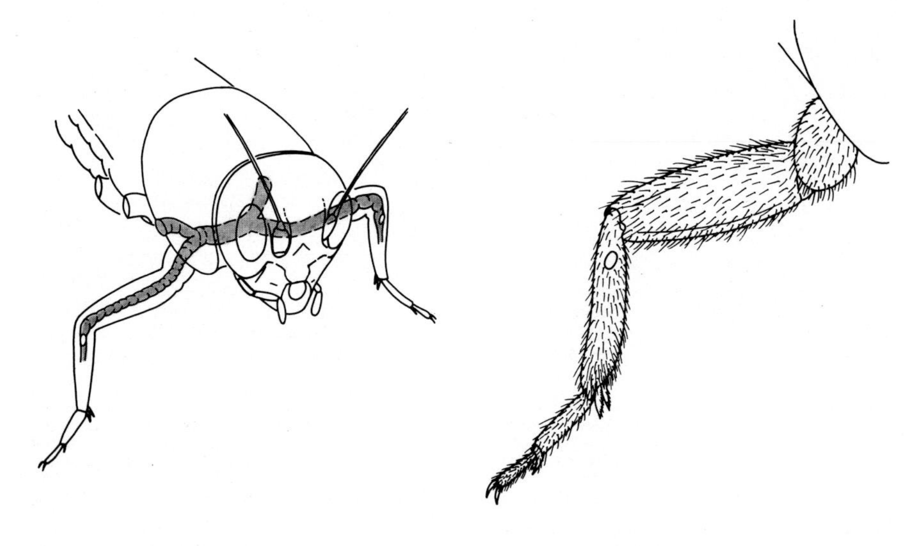

Figure 4. Crickets detect sounds—such as the cries of hunting bats—with a pair of ears located on the forelegs, just below the "knees." The ears are connected via air-filled tubes that meet at the insect's midline. Each of the tubes also has a branch leading to an opening, called a spiracle, behind each of the forelegs. The connections between the ears and the spiracles suggest that sound may reach an ear through separate channels. The presence of these different sound paths is thought to produce a differential response in the left and right ears that varies according to the location of the sound source. (Left illustration adapted from Hill and Boyan 1976.)

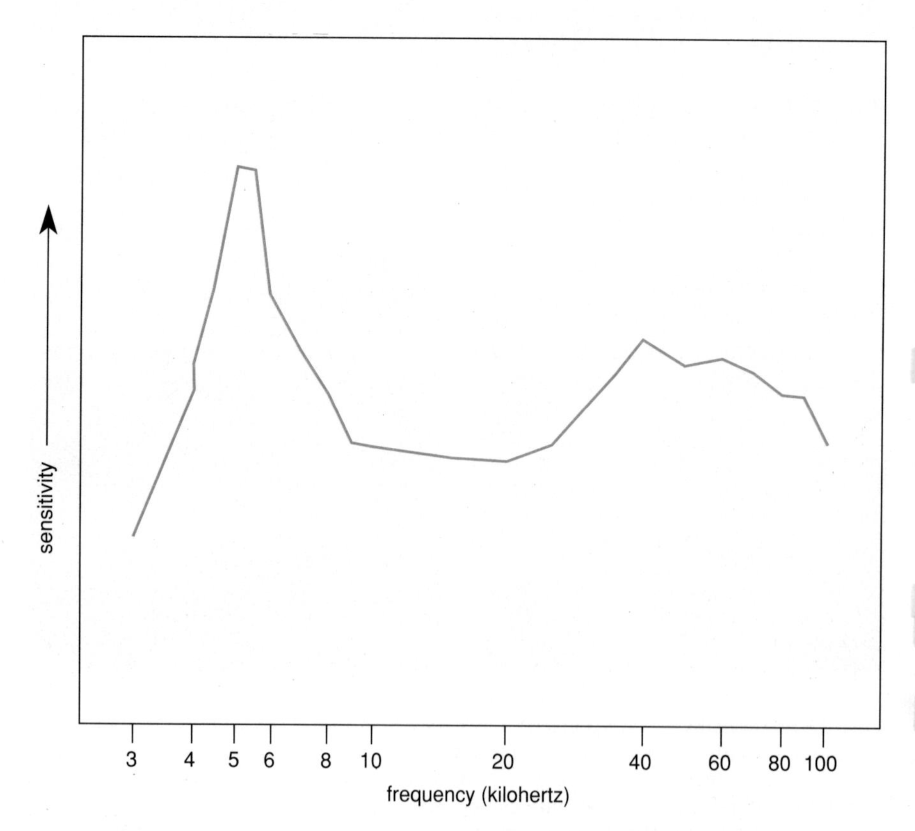

Figure 5. Frequency-sensitivity curve of a cricket's hearing exhibits two distinct peaks. A female cricket is most sensitive to sounds of about 5 kilohertz, which corresponds to the calling song of a courting male cricket. A second, broadly tuned peak—between 20 kilohertz and 100 kilohertz—lies within the frequency range of the ultrasonic cries emitted by hunting bats. The curve represents the average sensitivity of several animals as determined by observations of steering behavior toward or away from the sound. (Adapted from Moiseff, Pollack and Hoy 1978.)

quency and the location of its source.

An elegant demonstration of the frequency dependence of phonotactic behavior was devised in 1978 by Andrew Moiseff and his colleagues at Cornell. They investigated the behavior of flying female crickets (of the species *Teleogryllus oceanicus*) in response to electronically synthesized sounds ranging in frequency from 3 kilohertz to 100 kilohertz. The crickets were attached to a tether so that they were able to fly in place while suspended in an airstream—the aerial equivalent of a treadmill. Moiseff and his colleagues noticed that the sound stimulation caused the crickets to move their abdomens to one side or the other, an indication that they were attempting to steer in a particular direction. The female crickets seemed particularly responsive to two frequency ranges. They were sensitive to sounds with a frequency of about 5 kilohertz, but they also had a second sensitivity band that ranged from about 30 kilohertz to 90 kilohertz. The females steered toward the source of the 5-kilohertz sound—which corresponds to the frequency of the natural calling song of the male cricket—but steered away from the high-frequency sounds—which lie within the frequency range of the ultrasonic cries of echolocating bats. The phonotactic responses of the female cricket suggest that the auditory system of these animals is specialized not only for communication with other members of the same species but also for the detection and avoidance of the predatory bat.

The structure of the cricket's auditory system may help us to understand how it is that the cricket is able to determine the location of a sound source. The cricket's ears—consisting essentially of membranous eardrums—are not located on either side of its head but just below the "knee" on its foremost pair of legs. The ears on the left and right legs are connected via air-filled tubes that meet at the animal's midline. Each of the tubes also has a branch leading to an external opening, called a spiracle, on the cricket's body behind each of the forelegs.

Because the cricket's ears are connected to each other and to the spiracles, sound may reach the ear through any of three channels. First, of course, is the direct path, in which sound pressure waves strike the outside of the eardrum. But there are also two indirect paths, through which sound waves may strike the inside of the eardrum: through the air tube from the opposite ear and through the tube from the spiracles. When the sound originates from a source on one side of the cricket, the pressure wave takes a little longer to reach the eardrum on that side via the indirect routes than it does directly. The delay between the direct and the indirect routes is such that the sound pressure is at a maximum on the external part of the ear at the same time that it is at a minimum on the inside of the ear. Thus, when a sound wave coming from the cricket's left strikes the left ear, the eardrum is maximally excited. In contrast, a sound wave coming from the right strikes the left eardrum on the outside at about the same time that it reaches the inside of the left eardrum via the indirect route. Consequently, the internal and external pressures are the same, and there is little net movement of the eardrum. Such differential responses of the left and right ears to sound sources at various points in the cricket's acoustic space may allow the animal to determine the location of the sound source.

Having the means to discern the location of the sound source, the cricket must now translate this information into a movement toward or away from the sound. In 1983 Moiseff and Hoy investigated one part of the cricket's nervous system that may mediate this phonotactic response. They inserted a glass microelectrode into a cricket's prothoracic ganglion—a cluster of nerve cells that sends and receives information from the forelegs—while delivering sounds of various frequencies (from 3 kilohertz to 100 kilohertz) to the cricket's ear. They recorded from a nerve cell, which they called interneuron-1, that was excited by sounds over a wide range of frequencies (from 8 kilohertz to 100 kilohertz)—a range that covers the frequencies at which bats search during echolocation. Interneuron-1 was strongly excited when the sound stimuli mimicked not only the frequency but also the temporal patterns of the search stimulus used by the echolocating bat. Furthermore, sounds with frequencies of about 5 kilohertz—the frequency of the cricket's calling song—maximally inhibited the response of interneuron-1. Since interneuron-1 receives input from the auditory nerve that connects

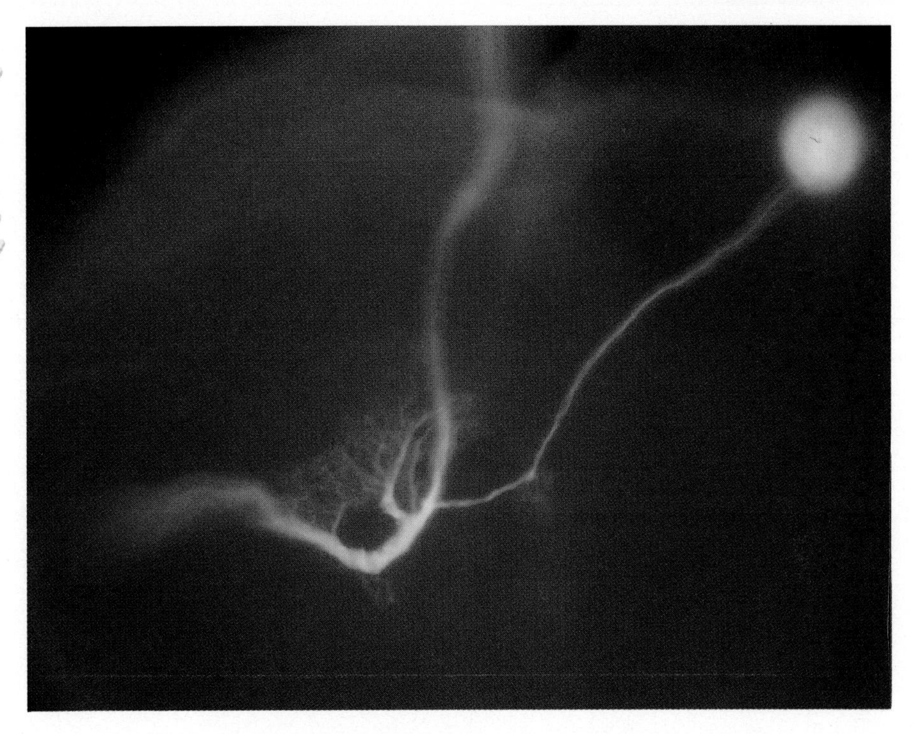

Figure 6. High-frequency-sensitive auditory neuron in a cricket—here labeled with a fluorescent dye called Lucifer Yellow—serves as an alarm that signals the presence of an echolocating bat. The neuron, called interneuron-1, is part of a neural circuit that elicits the insect's aerial escape maneuvers; excitation of the cell is both necessary and sufficient to elicit a bat-avoidance response. Fine branches called dendrites *(center)* receive signals from the auditory nerve *(not visible)*; a single axon *(exiting at top)* relays the information to the brain. The spherical cell body is at the upper right. (Photomicrograph courtesy of Andrew Moiseff, University of Connecticut at Storrs.)

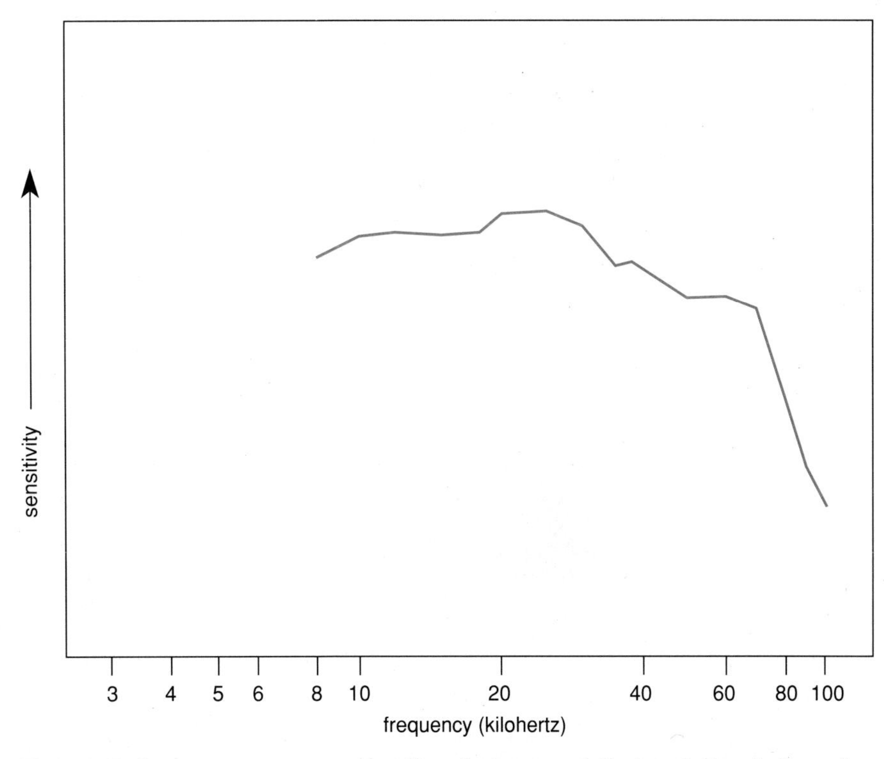

Figure 7. Excitatory response curve of auditory interneuron-1 displays the band of sound frequencies—ranging from 8 kilohertz to 100 kilohertz—that activate the nerve cell. The range of frequencies includes those sounds corresponding to the ultrasonic cries of hunting bats. The curve represents the average sensitivity of the neuron as determined by electrophysiological recordings in several animals. (Adapted from Moiseff and Hoy 1983.)

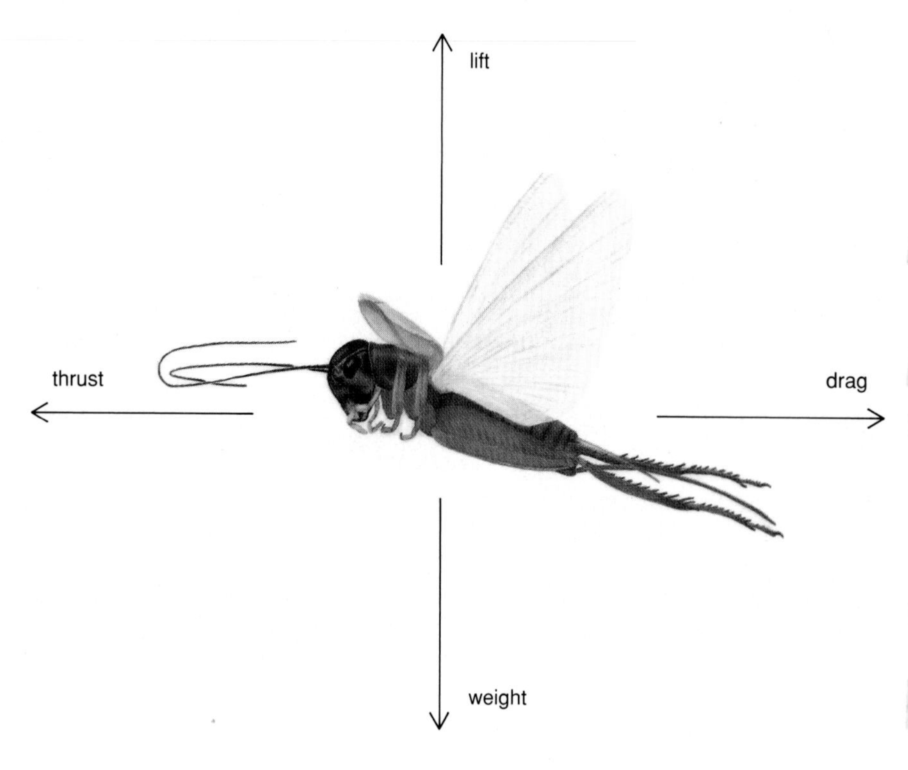

Figure 8. Balance of forces acting on a cricket in flight determines the insect's speed and altitude. Thrust propels the cricket forward and overcomes the drag imposed by the resistance of the air, whereas lift raises the cricket upward by overcoming the cricket's weight. A cricket modifies the balance of these forces by altering several factors: the frequency of its wingbeats, the extent of its wing strokes, the angle of its wings and the relative position of its body and its legs.

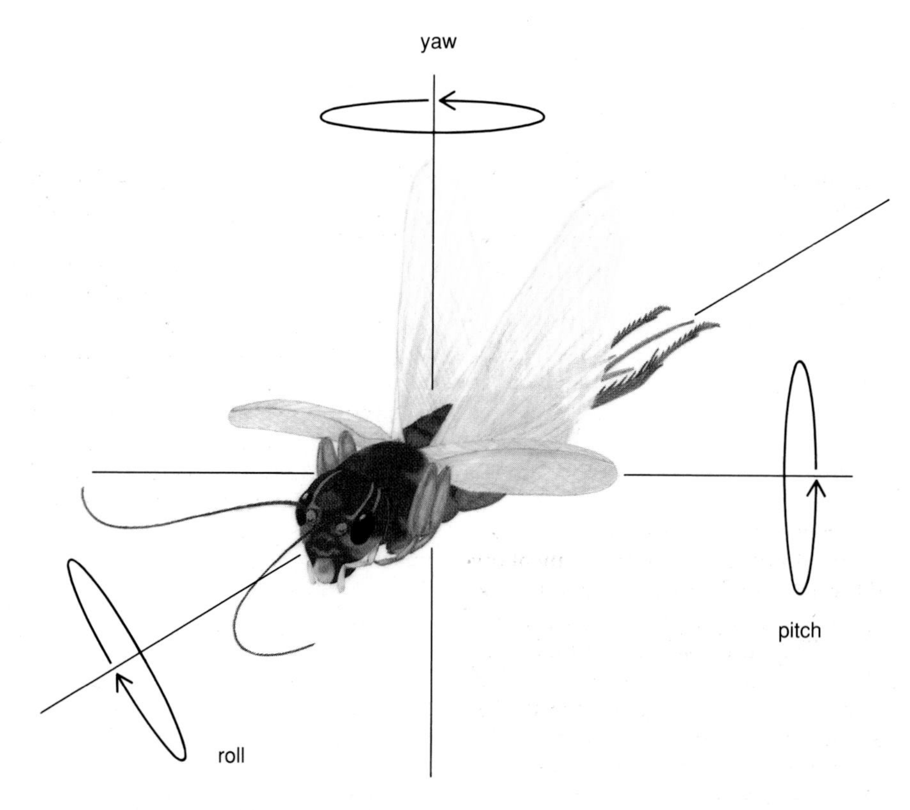

Figure 9. Body rotations in the three dimensions of space are employed by the cricket during evasive flight maneuvers. In response to ultrasonic stimulation from one side, a cricket will pitch downward, while yawing and rolling away from the sound source.

to the ear, and sends its signal to the brain, Moiseff and Hoy suggested that it might serve as an alarm that signals the presence of an echolocating bat.

Tom Nolen, now at the University of Miami, and Hoy later showed that interneuron-1 is both necessary and sufficient for eliciting negative phonotaxis in crickets. In other words, if interneuron-1 is inactivated, no level of ultrasound will induce a cricket to steer away. And, even without ultrasound, the activation of interneuron-1 by electrical stimulation will cause the cricket to steer away from the apparent source of the sound. This work seems to suggest a simple neuronal system: a single neuron that controls negative phonotaxis, perhaps by activating other neurons in the brain. Indeed, Peter Brodfuehrer, now at Bryn Mawr College, and Hoy showed that this bat-avoidance system diverges in the brain. At least 20 of the cricket's brain cells respond to ultrasound; they, in turn, must activate other neurons to produce the phonotactic response. At this time the complete circuit is still unknown.

In Two Strokes of a Cricket's Wing

My foray into the study of the cricket's response to ultrasound began in the fall of 1985, when I started my doctoral research in Hoy's laboratory at Cornell. The question I came to ask concerned not so much how a cricket knew to steer away from a predatory bat, but the aerodynamics behind the cricket's flying stunts. There were a number of ways the cricket might veer away from a bat; assessing the possibilities did not demand cleverness so much as a willingness to compromise. Although the creation of a natural environment plays a critical role—allowing the cricket to fly freely across the sky, for example—such a scenario allows few opportunities to measure subtle changes in the beating of the wings or the posture of the body. So, like the investigators before me, I accepted the now classical constraint of placing the insect on a tether.

The tether, a small wire attached to the cricket's back, holds the little acrobat in a flowing airstream where it can fly in place. In this situation the cricket beats its hindwings through large arcs, covering nearly 120 degrees, while the forewings essentially vibrate up and down. Both pairs of wings move rapidly, completing 32 wingbeats per second. In this situation a short pulse

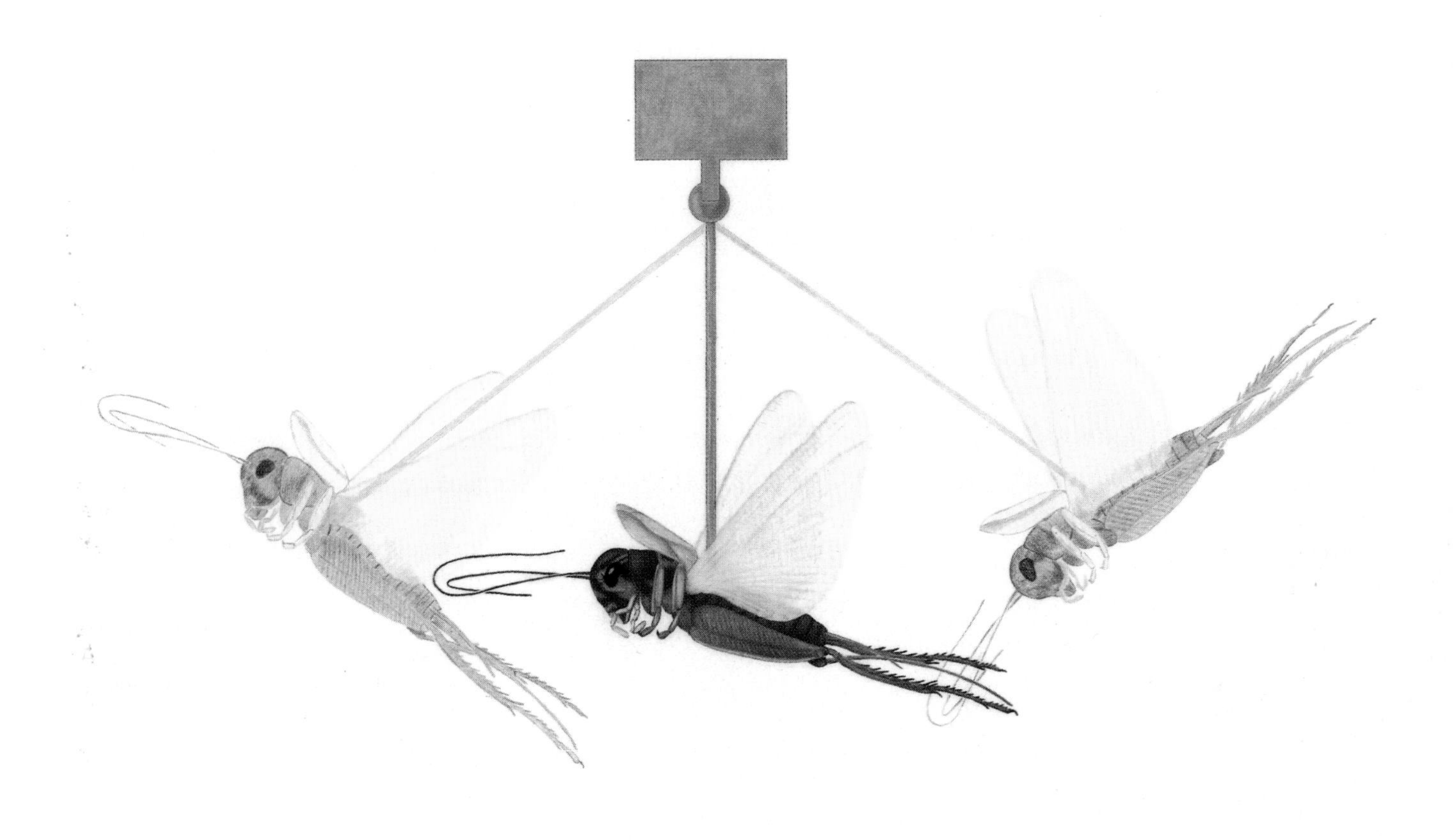

Figure 10. Crickets suspended on a pendulum tether may fly for over an hour, providing a useful model system in which to examine the aerodynamics of insect flight. The cricket swings forward by generating thrust with its wingbeats; it swings backward when the drag created by the airstream overpowers the cricket. When the thrust and the drag are equal, the cricket hangs directly below the pendulum, flying at a speed equal to that of the airstream. This technique reveals that crickets—and many other insects—fly at a speed of about two meters per second. Some hunting bats, on the other hand, fly at about nine meters per second.

of ultrasound (lasting 10 milliseconds), causes the cricket to respond immediately. The response is so strong that even a naive observer will interpret the cricket's movements as a turn, though the rigid tether prevents it.

One day in the fall of 1986, Brodfuehrer was watching the movements of a tethered, flying cricket. He noticed that whenever he jiggled a ring of keys—a makeshift source of ultrasound—the cricket tilted its forewings away from the sound. (This was the singular observation that launched me into the study of cricket aerodynamics.) Brodfuehrer and I made a few quick observations before I began a series of photographs to measure the angle of the forewing tilt. We noted that the cricket moved much like an airplane, banking its forewings into the turn. When I measured the angle of the tilt over a range of ultrasonic intensities, I found that the relationship was linear. That is, as the ultrasound gets louder, the cricket tilts its forewings farther away from the sound. This suggests that the degree of the turn is graded with the intensity of the ultrasonic stimulus.

Next, I decided to examine the frequency of the cricket's wingbeats. Others had found that a change in the frequency of the wingbeats accompanies steering in a number of insect species, including locusts, dragonflies, moths and several types of flies. Did the cricket also increase its pace in response to ultrasound? It did, by about three or four beats per second in response to a single pulse of ultrasound. Incredibly, the cricket will make this change of pace within the time it takes to beat its wings twice. This means that a cricket can detect the ultrasound, alter the rhythmic signals in its flight system and move its wings faster in as little as 60 milliseconds.

At this point I had a few valuable pieces of evidence. First, the linearity of the forewing tilt suggests that the cricket turns more sharply when the ultrasound is louder. That is important because an increase in the intensity of the ultrasound in the natural world corresponds to a closer bat; it behooves a cricket to turn a little more sharply. Moreover, by increasing the frequency of its wingbeats in response to ultrasound, the cricket probably flies faster and makes quicker changes in direction—useful attributes for any bat-avoidance system.

Flows Here, Forces There

I had learned a few things about the way a cricket flies, but I still had not answered the original question—how does a cricket evade ultrasound? I really wasn't sure that the forewing tilt made a cricket turn or that increasing the frequency of the wing stroke made it fly faster. The inference seemed reasonable, but I didn't want to be too hasty; more than any other endeavor I can think of, the field of aerodynamics punishes scientists for making too-quick assumptions. This happens largely because aerodynamics is extremely complicated; with flows here and forces there, how does one know which factors are going to be important?

To get an appreciation of the scientist's dilemma, consider an example from baseball: the curve ball. A well-thrown curve ball flies from the pitcher's hand, streaks toward the plate, and curves at the last instant. What does this say about cause and effect in aerodynamics? The cause (the manner in which the pitcher throws the ball) and the effect (the ball's curving trajectory) are substantially separated in time. It poses a challenge for a scientist intent on correlating cause and effect. Similar perplexities haunt all of aerodynamics.

Figure 11. Tethered praying mantis is seen in flight before *(top)* and after *(bottom)* the onset of an ultrasonic stimulus. During normal flight, the mantis has a streamlined posture—tucking its legs neatly into its straight body. On hearing the ultrasound, the mantis extends its forelegs, rolls its head back and bends its abdomen—all within about a tenth of a second. During free flight, this suite of behaviors results in a turning or spiralling dive that helps the insect to escape from a bat. (Photographs courtesy of David Yager, University of Maryland at College Park. From Yager and May 1990. Reproduced by permission of the Company of Biologists Limited.)

In an attempt to avoid such problems, I decided to start by looking at the effect, and then to search for the cause. I began by measuring the aerodynamic movements of the cricket—in pursuit of the forces involved.

A flying insect is balanced by two pairs of forces. First, in the vertical plane gravity produces a downward force on the cricket's mass, namely the force measured as weight; a flying insect offsets its weight with the upward force called lift. As long as the lift and the weight are equal, the insect maintains a constant altitude. Along the horizontal dimension, the force driving the insect forward is thrust; it is resisted by friction with the air, or drag, which pulls the insect backward. If the thrust and the drag are equal, then the insect moves at a constant speed. If a cricket is to increase its airspeed, there must be an imbalance: The thrust must exceed the drag on the cricket's body. But how does a cricket increase its airspeed? Does a change in the frequency of the wingbeats cause an imbalance between thrust and drag? It was a hypothesis waiting to be tested.

I used a tether designed much like a pendulum; it allowed the cricket to swing freely, forward and backward. When a cricket attached to such a pendulum starts to fly, thrust pushes the cricket forward, and so the cricket swings up, like a child in a swing. By turning on a fan, I created an airstream that pushes the cricket backward by increasing the drag. I adjusted the speed of the airstream until the pendulum hung straight down; this meant that the cricket's flight speed equaled the speed of the airstream—about two meters per second. As soon as I turned on the ultrasound, the flying cricket swung forward on the tether—because it flew faster. Now I could say that the frequency of the wingbeat and the speed of the cricket both increase in response to ultrasound.

The literature on insect aerodynamics also suggests that rotations of the animal's body are important for steering. Like an airplane, a flying insect can rotate about three axes, called the pitch, roll and yaw axes, that all pass through the insect's center of gravity. A pitching motion is a rotation around the transverse axis (parallel to an airplane's wings); roll is rotation around the longitudinal axis (parallel to the fuselage), and yaw describes rotation around the vertical axis.

I designed a tether that allowed a flying cricket the freedom to rotate about the pitch axis, and watched the insect's response to ultrasonic stimulation. When I delivered a pulse of ultrasound to the cricket, it pitched downward. As with the tilting of the forewings, the louder the ultrasonic stimulus, the greater the amount of downward pitch. Here was another linear response, one which probably translates into a dive.

But does the cricket use its forewings or its hindwings to adjust its pitch attitude? A few simple experiments—removing one or the other pair of wings—showed that crickets use both their hindwings and forewings for the control of pitch. But there was more. I set up an experiment to test the relative contributions of the hindwings and the forewings to the degree of pitch motion. I set the ultrasound to a constant intensity and then measured the angle of pitch in three circumstances: a cricket with both pairs of wings, a cricket with only its forewings and a cricket with only its hindwings. One might assume that the relative contributions of the hindwings and the forewings would be additive. That is, by adding the angle of pitch for a cricket that only had its forewings to the angle of pitch for a cricket that only had its hindwings, it might be expected that the sum would equal the angle of pitch for a cricket with both pairs of wings. But this isn't the case; the sum of the individual pitch angles is always smaller than the angle of pitch produced by a cricket with both pairs of wings. It was another pitfall in aerodynamics—the whole can be more than the sum of the parts.

Next, with a minor modification to the tether, I measured the angle of roll. Again, I found that the crickets tended to roll away from the source of the ultrasound. And, much like the pitch measurements, the crickets could roll with either their hindwings or their forewings. But in this case the sum of the parts (rolling with forewings or hindwings alone) did equal the whole (rolling with both pairs of wings). The aerodynamic corollary to be found here was that the whole *might*

Figure 12. Tethered cricket is seen in flight before *(top)* and after *(bottom)* the onset of an ultrasonic stimulus. During normal flight, the cricket tucks its forelegs and middle legs just behind its head—which it holds slightly above horizontal—while extending its hindlegs. After hearing the ultrasound, the cricket pitches downward, its body now nearly level. A hinge just above the cricket's back allows the cricket to pitch freely. (Photographs courtesy of the author.)

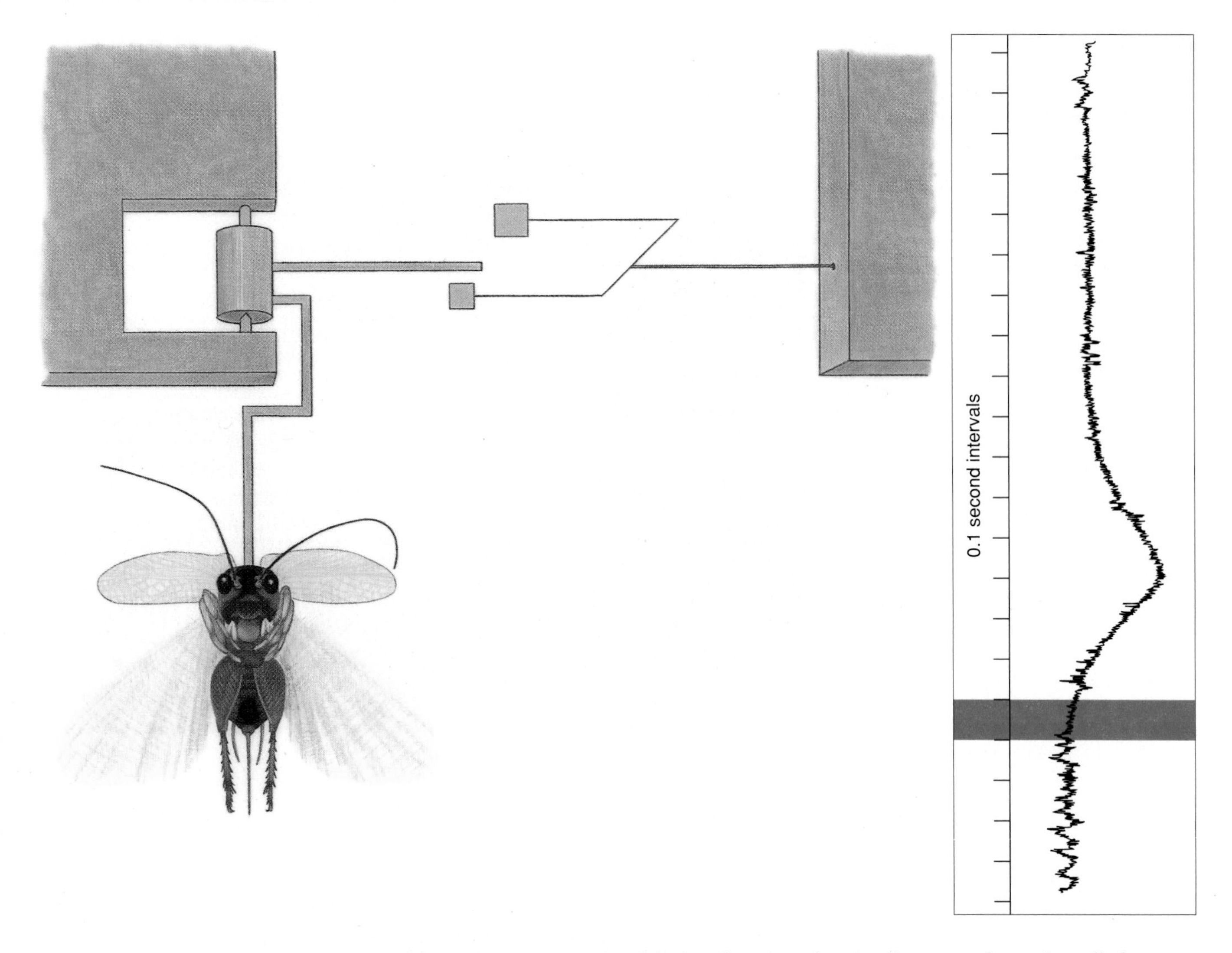

Figure 13. Yaw movements of a flying cricket are measured with a tether *(left)* that allows lateral motions by means of a rotating cylinder. Rotations of the cylinder swing a wand that is located between two sensors. The sensors record the size, the direction and the duration of the yaw. The recording *(right)* shows that a cricket will yaw in less than a tenth of a second after the onset of an ultrasonic stimulus *(red band)*.

equal the sum of the parts— sometimes.

One axis of rotation remained to be investigated, the yaw axis. When I had measured the pitch and the roll movements, my experimental apparatus determined only the change in the angle. For the measurement of yaw, I wanted both the angle itself and the change in the angle. This required the construction of a device that could keep track of the angle of yaw over time. The device also provided the latency of the response—the time from the onset of the stimulus to the onset of the yawing motion—and the angular velocity about the yaw axis.

I found that, like the pitching and rolling rotations, the magnitude of the yaw is linearly related to the intensity of the ultrasound. And both the forewings and the hindwings participate in the yaw movement. But I also found that the intensity of the ultrasound affects the time course of the yaw. As the ultrasound gets louder, the latency of the yaw decreases, whereas the angular velocity increases. At the highest intensities of ultrasound— around 90 decibels, which corresponds to a shrieking bat within a few meters of a cricket's ear—the latency is down to about a tenth of a second, and the angular velocity is about 100 degrees per second.

There was one surprise in my studies of the cricket's ability to yaw. Not only is the cricket able to yaw with either pair of wings, but it can also yaw without any wings at all. The movement is small, but significant. This "body yaw" is reminiscent of an abdominal swing that crickets are known to perform in response to ultrasound. Previous studies had suggested that the abdomen might act like a rudder; by swinging the abdomen to one side, the cricket increased the drag on that side. It seems reasonable to think that increasing the drag on one side would produce yaw. Might that be the cause of the yaw in a wingless cricket? It could be another aerodynamic trap between cause and effect. I put the hypothesis to the test by removing the cricket's wings and fixing the abdomen so that it could not swing. Surprisingly, this treatment did not stop the crickets; they were still able to yaw. In a second approach I used wingless crickets again, but this time I freed the abdomen and fixed the front portion of the cricket's body, the thorax. This time the crickets were unable to yaw. The swinging abdomen did not produce the yaw; it was the twisting thorax that did it. The rudder is in the front.

Some Leg Action

Not all useful observations come as a result of premeditated experimental design; sometimes it's useful just to play with the equipment. That is what happened one day while I was toying with the ultrasound stimulus and watching the responses of a tethered cricket. I was looking at the cricket's wings—which appear to be a transparent blur, beating about 30 times a second—when I noticed that the hindwing farther from the source of the ultrasound didn't go down as far. With every burst of ultrasound, the blurred wing seemed to stop short. Was this another mechanism by which the cricket was trying to turn? If the wing farther from the ultrasound did not make a full stroke, it should produce less thrust—inducing a turn toward that side.

On closer inspection, more than the hindwing was changing. For every pulse of ultrasound, one of the cricket's hindlegs would swing up, and appear to collide with the hindwing. Was the hindleg impeding the movement of the hindwing? Jeffrey Camhi of the Hebrew University in Jerusalem had noticed a similar phenomenon in tethered locusts, but had not shown that there were any aerodynamic consequences.

To establish that there was an interaction between the hindwing and the hindleg I needed photographic evidence. I posed the crickets in three situations: an intact cricket flying without ultrasound, an intact cricket flying with ultrasound, and a cricket that had a hindleg removed and that was flying with ultrasound. The photographs confirmed my suspicions. When there is no ultrasonic stimulus, the hindleg does not interfere with the hindwing, but with the ultrasound the hindleg farther from the sound does impede the hindwing's downstroke. In a cricket without the hindleg, the hindwing completes a full stroke even during an ultrasound stimulus.

Is there an aerodynamic effect to sticking a leg into a beating wing? To test this possibility I tethered some crickets in the device I had used to measure yaw motions. The ultrasonic stimulus was given from the left and the right sides of the cricket, both before and after removing the cricket's left hindleg. Removing the left hindleg means that all of the ultrasonic pulses from the left serve as controls (the hindleg only hits the hindwing when the ultrasound comes from the *opposite* side). And indeed, the yaw induced by an ultrasound stimulus from the left looked the same before and after removing the left hindleg. Removing the

Figure 14. Hypothetical flight path shows how a free-flying cricket might respond to the ultrasonic cries of an echolocating bat. Normal flight *(top)* is interrupted after the cricket detects the ultrasound coming from the insect's left. Within 40 milliseconds *(middle)* the cricket swings its right hindleg into its right wing, thereby reducing the thrust on that side. At about the same time, the cricket tilts its forewings, increases its speed, and pitches downward while rolling and yawing to its right. All factors should contribute to a powered dive away from the hunting bat *(bottom)*.

hindleg that was on the same side as the ultrasound stimulus did not change the flight behavior. But when the ultrasound came from the right, there was a large change in the cricket's responses. When the cricket had both hindlegs, ultrasound from the right would cause the cricket to yaw after 100 milliseconds; after removing the left hindleg, the yaw did not begin until after 140 milliseconds. The latency of the yaw response increased by 40 percent. The magnitude of the yaw also changed: it was 17 degrees before removing the leg and only 9 degrees after removing the leg. The hindleg does affect steering—it makes the yaw bigger and faster.

The hindleg might affect steering in two different ways. It might work alone to cause some drag, or its interaction with the hindwing could be the crucial factor. I repeated the yaw experiments with just one change—I removed the hindwings. This time there was no change in the yaw; it looked the same whether the ultrasound came from the left or the right, or whether the hindleg was present or absent. The result was clear: The hindleg steers the cricket by impeding the hindwing's downstroke.

Although the mechanism may appear to be crude—somewhat like stopping a bicycle by sticking a tire pump into the spokes of the rear wheel—it is functional and economical engineering design. To appreciate this, consider another example from baseball. During a fast-ball pitch, the pitcher's arm is pulled through a high-speed arc to throw the ball toward the plate. Attempting to stop this motion halfway through the pitch is humanly impossible. A cricket winging its way from an echolocating bat faces the same difficulty. Its wing muscles are not designed to halt the downstroke in the brief time needed to steer out of the bat's way. But the hindleg is able to block the downstroke of the hindwing in such a way as to produce a nearly instantaneous change in the length of the stroke—which causes the cricket to veer away. It is a clever yet simple solution.

All told, however, the swing of the hindleg is merely one small part of the cricket's acrobatic maneuvers. We now know that the cascade of responses consists of swinging legs, tilting forewings, twisting thoraxes and rapidly beating wings. It all adds up—in a fraction of a second—to an elegant ballet that whisks the cricket beyond the grasp of a hungry bat.

Bibliography

Brodfuehrer, P. D. and R. Hoy. 1989. Integration of ultrasound and flight inputs on descending neurons in the cricket brain. *Journal of Experimental Biology* 146:157-171.

Brodfuehrer, P. D. and R. Hoy. 1990. Ultrasound-sensitive neurons in the cricket brain. *Journal of Comparative Physiology, A* 166:651-662.

Camhi, J. M. 1970. Yaw-correcting postural changes in locusts. *Journal of Experimental Biology* 52:519-531.

Cooter, R. J. 1979. Visually induced yaw movements in the flying locust, *Schistocerca gregaria* (Forsk). *Journal of Comparative Physiology* 99:1-66.

Cranbrook, T. E. O., and H. G. Barrett. 1965. Observations of nocturnal bats (*Nyctalus noctula*) captured while feeding. *Proceedings of the Zoological Society of London* 144:1-24.

Easteria, D. A. and J. O. Whitaker, Jr. 1972. Food habits of some bats from Big Bend National Park, Texas. *Journal of Mammalogy* 53(4):887-890.

Griffin, D. R. and R. Galambos. 1941. The sensory basis of obstacle avoidance by flying bats. *Journal of Experimental Zoology* 86:481-506.

Griffin, Donald R. 1984. *Listening in the Dark.* Dover Publications.

Hill, K. G. and G. S. Boyan. 1976. Directional hearing in crickets. *Nature* 262:390-391.

May, M. L., Brodfuehrer, P. D., and R. R. Hoy. 1988, Kinematic and aerodynamic aspects of ultrasound-induced negative phonotaxis in flying Australian field crickets (*Teleogryllus oceanicus*). *Journal of Comparative Physiology* 164:243-249.

May, M. L. and R. R. Hoy. 1990a. Ultrasound-induced yaw movements in the flying Australian field cricket (*Teleogryllus oceanicus*). *Journal of Experimental Biology* 149:177-189.

May, M. L. and R. R. Hoy. 1990b. Leg-induced steering in flying crickets. *Journal of Experimental Biology* 151:485-488.

Miller, L. A. and J. Olesen. 1979. Avoidance behavior in green lacewings. I. Behavior of free flying green lacewings to hunting bats and ultrasound. *Journal of Comparative Physiology* 131:113-120.

Moiseff, A., Pollack, G. S. and R. R. Hoy. 1978. Steering responses of flying crickets to sound and ultrasound: mate attraction and predator avoidance. *Proceedings of the National Academy of Sciences of the U.S.A. Biological Sciences.* 75:4052-4056.

Moiseff, A. and R. R. Hoy. 1983. Sensitivity to ultrasound in an identified auditory interneuron in the cricket: a possible neural link to phonotactic behavior. *Journal of Comparative Physiology* 152:155-167.

Nachtigall, W. and D. M. Wilson. 1967. Neuromuscular control of dipteran flight. *Journal of Experimental Biology* 47:77-97.

Nolen, T. G. and R. Hoy. 1984. Initiation of behavior by single neurons: The role of behavioral context. *Science* 226:992-994.

Pollack, G. S. and N. Plourde. 1982. Phonotaxis in flying crickets: Neural correlates. *Journal of Insect Physiology* 146:207-215.

Popov, A. V. and V. F. Shuvalov. 1977. Phonotactic behavior of crickets. *Journal of Comparative Physiology* 119:111-126.

Robert, D. 1989. The auditory behavior of flying locusts. *Journal of Experimental Biology* 147:279-301.

Roeder, K. D. 1962. The behavior of free flying moths in the presence of artificial ultrasonic pulses. *Animal Behavior* 10:300-304.

Roeder, K. D. 1967. Turning tendency of moths exposed to ultrasound while in stationary flight. *Journal of Insect Physiology* 13:873-888.

Roeder, K. D. and A. E. Treat. 1961. The detection and evasion of bats by moths. *American Scientist* 49:135-148.

Rüppell, G. 1989. Kinematic analysis of symmetrical flight manoeuvres of Odonata. *Journal of Experimental Biology* 144:13-42.

Schnitzler, H.-U., D. Menne, R. Kober and K. Heblich. 1983. The acoustical image of fluttering insects in echolocating bats. pp. 235-250, in *Neuroethology and Behavioral Physiology*, F. Huber and H. Markl (eds.). Springer: Heidelberg.

Spangler, H. G. 1988. Hearing in tiger beetles (Cicindelidae). *Physiological Entomology* 13:447-452.

Suga, N. 1984. Neural mechanisms of complex-sound processing for echolocation. *Trends in Neurosciences* 7:20-27.

Whitaker, J. O., Jr. and H. Black. 1976. Food habits of cave bats from Zambia, Africa. *Journal of Mammalogy* 57(1):199-205.

Yager, D. D., May, M. L., and B. M. Fenton. 1990. Ultrasound-triggered, flight-gated evasive maneuvers in the flying praying mantis, *Parasphendale agrionina*. I: Free flight. *Journal of Experimental Biology* 152:17-39.

Yager, D. D. and M. L. May. 1990. Ultrasound-triggered, flight-gated evasive maneuvers in the flying praying mantis, *Parasphendale agrionina*. II: Tethered flight. *Journal of Experimental Biology* 152:41-58.

M. Brock Fenton
James H. Fullard

Moth Hearing and the Feeding Strategies of Bats

Variations in the hunting and echolocation behavior of bats may reflect a response to hearing-based defenses evolved by their insect prey

In the eighteenth century several natural historians suggested that the seemingly chaotic flight of moths was somehow related to attacks by hunting bats (see Kirby and Spence 1826), but the details of this relationship were clarified only after we understood something about moth hearing and about bat echolocation, or biosonar—the bat's use of echoes of the sounds it produces to detect objects from which these echoes rebound. Moths in several families have ears sensitive to ultrasonic sound located variously on their thoraxes, abdomens, or mouth parts (see Michelsen 1979), and these moths respond neurologically and behaviorally to the orientation cries of insectivorous bats. Moths that can hear such cries are 40% less likely to be caught than those that cannot (Roeder 1967), and there are similar data for some lacewings (Miller and Olesen 1979). The evasive response of moths is thus an example of a predator-specific defense (Edmunds 1974).

Two main lines of study have contributed to our understanding of the interaction between moths and bats: a detailed consideration of what moths hear and a close examination of the echolocation, prey selection, and feeding behavior of bats (see Fig. 1). This research suggests that moths may, by their ability to detect marauding bats, exert a selective pressure on echolocation strategies. Here we review the data on moth hearing and bat echolocation, analyze the implications of these data for detectability of bats by moths, and examine available information on the diets of insectivorous bats to find out if our predictions are correct. We also consider some of the intriguing questions raised by this research. If all bats are not equally detectable by all moths, do bats that are acoustically inconspicuous take advantage of this characteristic by specializing in moths? What is the function of the clicks produced by some tiger moths (Arctiidae) immediately before they are attacked by a bat?

M. Brock Fenton has been a member of the Department of Biology at Carleton University since 1969. Trained at the Royal Ontario Museum and at the University of Toronto, where he received his Ph.D. in 1969, he is currently studying the ecology and behavior of bats. James H. Fullard received his Ph.D. from Carleton University in 1979 for work on the role of audition in moth defensive behavior. He is now an Assistant Professor of Zoology at Erindale College, University of Toronto, and is continuing his studies of moth defensive behavior and auditory systems. Address: M. Brock Fenton, Department of Biology, Carleton University, Ottawa, Canada K1S 5B6.

Moth hearing

Roeder (1967) used the behavioral responses of flying moths—both individuals mounted in stationary flight on pins in the laboratory and free-flying moths in the field—to show that the response of a flying moth to an approaching bat varies according to the situation. Impulses, or spikes, from a pair of auditory neurons inform the moth of the direction and proximity of an echolocating bat. Weak stimulation, encoded as a few nerve spikes, implies the presence of a distant bat and results in negative phonotaxis: the moth turns and flies away from the bat. Strong stimulation, perceived as many nerve spikes from more than one neuron, causes the moth to fold its wings and dive to the ground.

Roeder and his coworkers have demonstrated this pattern of response through experiments in which they monitored activity in the auditory nerves, usually by attaching a stainless steel extracellular hook electrode, and have described the anatomy and neurology of various moth ears. The ear of a typical moth, such as an arctiid or a noctuid, includes three neurons: two A cells that transmit data about sound, and a B cell that may function mainly as a stretch receptor for monitoring changes in pressure on the ear. Some moths—notodontids, for example—have simpler wiring, with only one auditory neuron, while geometrids, at the other extreme, have four auditory neurons.

The relatively simple wiring of moth ears and the fact that they respond to any sound of appropriate intensity and frequency suggest that they react to ultrasonic stimuli rather than to specific bat sounds. However, echolocating bats are the major source of ultrasonic sound in the nocturnal environment, and they are moreover important predators of nocturnal insects. Significantly, the ears of moths are not equally sensitive to sounds at all frequencies, but seem to have their greatest sensitivity in the range of frequencies used by most of the bats that share their habitat (Roeder 1967; Fullard 1979a).

In a recent study, we further explored the ability of moth ears to detect sounds at various frequencies by monitoring activity in the auditory nerves of moths while presenting signals of known intensity and frequency (Fenton and Fullard 1979). The threshold for response was defined as the first detectable activity occurring in the auditory nerve cells in conjunction with presentation of the stimuli; some of the resulting data

Figure 1. The little brown bat (*Myotis lucifugus*), the subject of early research on bat echolocation and its use in locating insect prey, continues to offer insights into the workings of the predator-prey relationship between bat and moth. The high-intensity echolocation calls produced by this bat have been found to be readily detectable by moths sharing its habitat at distances of up to 40 m. However, the little brown bat appears to be capable of detecting the moths only at a range of 2 m or less. (Photo by C. Hill.)

are shown here as audiograms, or tuning curves (Fig. 2).

Moths from two different zoogeographic regions—North America and Africa—showed significant differences in their audiograms. Species from Ontario were generally similar in their responses, although there was some variation among individuals and species. By contrast, moths from Ivory Coast showed greater interspecific variation in their hearing ability, and differed from the Ontario species in other ways as well: their range of best frequencies—frequencies at which they were most sensitive—was broader, and they were more sensitive over the entire range of frequencies tested (5 to 110 kHz), particularly to sounds above 65 kHz (Fullard 1979a).

The neural audiograms appear to reflect the behavioral thresholds for response to sounds. Dogbane tiger moths (*Cycnia tenera*) show three gradations in their response to sound: the threshold for continuous neural response requires the weakest auditory signal, a stronger signal leads to the production of characteristic clicking sounds, and a signal 15–25 dB more intense results in cessation of flight (Fullard 1979b). Thus there is a continuum of behavioral responses from negative phonotaxis to cessation of flight and, in the case of some arctiids, a burst of sound just before the moth stops flying and the bat completes its attack.

The data presented in the audiograms can be used to predict the distances over which a moth could detect sound at given frequencies. The maximum distance of detection (Fig. 3) is calculated by taking into account the neural threshold of the moth's ear, the initial intensity of the sound, and the rate of attenuation of sound at different frequencies and air humidities (Griffin 1971). This distance is probably a crucial element in bat-moth interaction, since it represents the time available to predator and prey for appropriate subsequent action. It will be useful at this point to review some aspects of bat echolocation and to consider in detail the options available to an echolocating bat, given the maximum distances of detection for different moths and the effect of sound intensity and frequency on these values.

Bat echolocation

In the late 1700s Lazarro Spallanzani conducted a series of experiments in an effort to determine how bats orient themselves in total darkness. From his tests, which eliminated the use of smell, vision, and touch, he concluded that bats had a "sixth sense." In the 1930s Griffin repeated some of Spallanzani's experiments while monitoring bats with microphones sensitive to ultrasonic sound. He demonstrated that the little brown bat (*Myotis lucifugus*) produces pulses of ultrasonic sound and that it uses the echoes of these sounds to orient itself with respect to obstacles (see Griffin 1958). He also showed (Griffin et al. 1960) that bats use calls and their echoes to locate insect prey. Griffin provided a basic understanding of echolocation and some appreciation of the diversity involved in this form of orientation. Recent work (e.g. Busnel and Fish 1980) has embellished the picture considerably by providing data on bats hunting in the field and on how information is processed in their brains.

Several aspects of bat echolocation calls, including intensity, the pattern of frequency change over time, the presence of harmonics, and the rate at which calls are produced, are relevant to the interaction between moths and bats.

Differences in the intensities of bat echolocation calls have been recognized for some time, because not all bats proved equally easy to detect with microphones sensitive to ultrasonic sound (Griffin 1958). Intensities measured 10 cm in front of the bats' mouths or nostrils range from very high (≥110 dB) to low (65 dB) and include a variety of intermediate readings (75 to 100 dB). For comparison, consider that a typical smoke detector produces a signal of 113 dB at 10 cm. It is difficult to measure the intensities of bat calls accurately in the field or in the laboratory—among other problems, the calls tend to be short in duration—and many recent

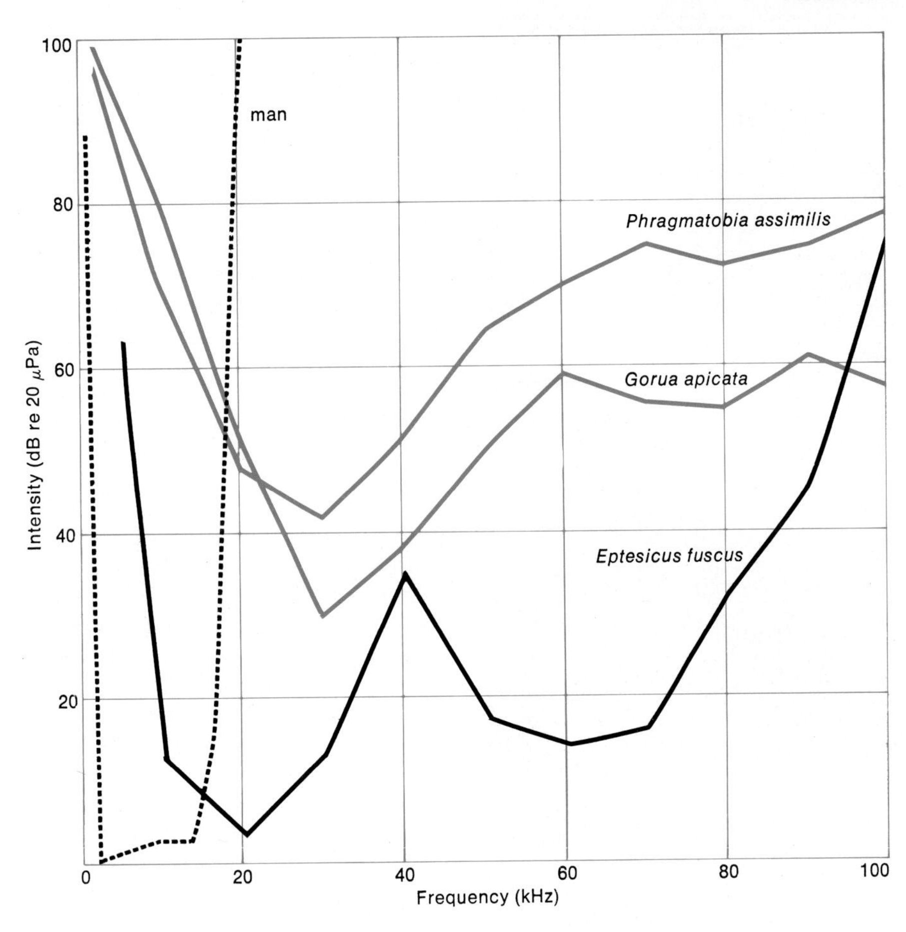

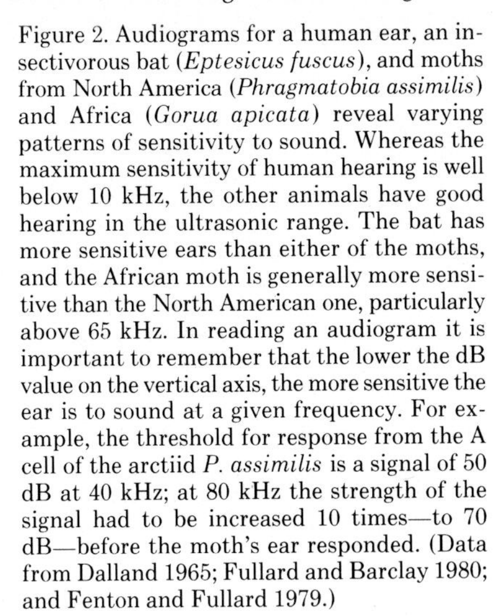
Figure 2. Audiograms for a human ear, an insectivorous bat (*Eptesicus fuscus*), and moths from North America (*Phragmatobia assimilis*) and Africa (*Gorua apicata*) reveal varying patterns of sensitivity to sound. Whereas the maximum sensitivity of human hearing is well below 10 kHz, the other animals have good hearing in the ultrasonic range. The bat has more sensitive ears than either of the moths, and the African moth is generally more sensitive than the North American one, particularly above 65 kHz. In reading an audiogram it is important to remember that the lower the dB value on the vertical axis, the more sensitive the ear is to sound at a given frequency. For example, the threshold for response from the A cell of the arctiid *P. assimilis* is a signal of 50 dB at 40 kHz; at 80 kHz the strength of the signal had to be increased 10 times—to 70 dB—before the moth's ear responded. (Data from Dalland 1965; Fullard and Barclay 1980; and Fenton and Fullard 1979.)

studies that deal in depth with echolocation (e.g. Gustafson and Schnitzler 1979) avoid the topic completely.

However, by using a broadband bat detector that is sensitive to sounds from 10 to over 200 kHz (Simmons et al. 1979a) it is possible to classify bats in a general way by the intensity of their orientation calls (Fig. 4). Species characterized by calls of high intensity (≥110 dB) are readily detectable by these microphones at distances of up to, and in some cases more than, 10 m when the bat is facing the microphone, while the faint "whispering bats" (65 dB) are usually detectable only at 0.5 m or less. Bats using calls of intermediate intensity (75 to 100 dB) are detectable at distances of 1.5 to 2.0 m. This arbitrary classification is convenient but must be treated with caution, since the intensity of these emissions for one species varies with the situation, making the continuum from very high to low intensity even more complex.

At one time it seemed appropriate to classify bats by the frequency-time patterns of their echolocation calls, distinguishing frequency-modulated (FM) from constant-frequency (CF) species or from bats using some combination of the two components, for example CF-FM (Simmons et al. 1975). However, examination of the calls emitted by different species under different conditions has shown that although calls or their components can be classified in this way, the designations are not necessarily correlated with the bats' use of the calls (Simmons et al. 1978; Gustafson and Schnitzler 1979).

Several basic patterns of frequency change over time provide the bat with various kinds of information about targets (Fig. 5). Steep FM calls rapidly sweep from higher to lower frequencies, while shallow FM calls are longer and cover a smaller range of frequencies (Simmons and Stein 1980). Steep FM components have been detected in some segments of

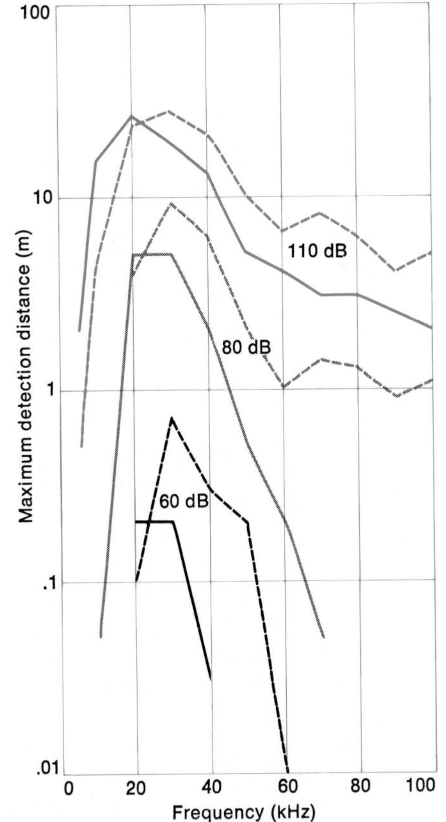

Figure 3. The maximum distance at which a moth can detect a sound of known intensity and frequency is an important strategic factor in the moth's confrontation with its predator. Using data provided by the audiograms shown in Fig. 2, maximum detection distances are calculated for *P. assimilis* (*solid line*) and *G. apicata* (*dashed line*) at three levels of sound corresponding to low, intermediate, and high-intensity echolocation calls. The dB values correspond to intensities measured 10 cm from the bats' mouths or nostrils.

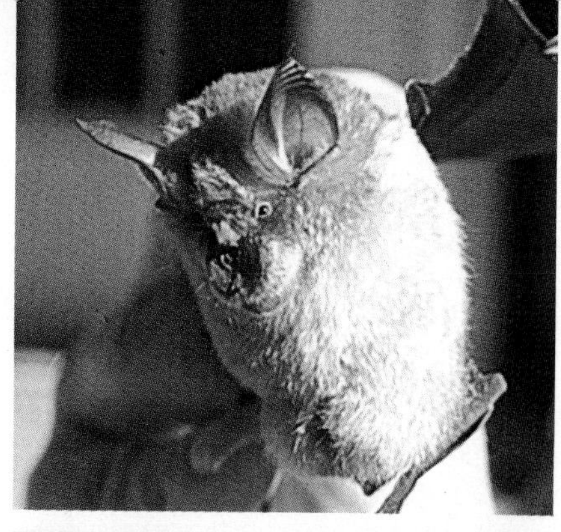

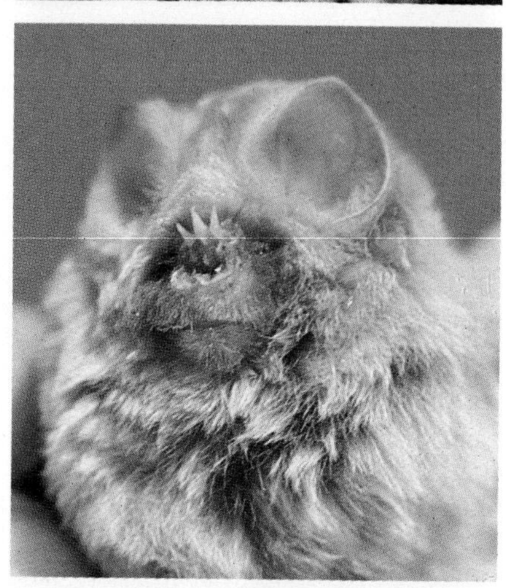

Figure 4. Bats exhibit a wide range of echolocation strategies, varying such crucial elements as the intensity of the call. From top to bottom are shown five species whose calls are representative of the variety of intensities used. Geoffroy's horseshoe bat (*Rhinolopus clivosus*) emits high-intensity pulses through its nostrils, whereas the Western big-eared bat (*Plecotus townsendii*) is commonly considered to produce a low-intensity call. The short-eared trident bat (*Cloeotis percivali*) uses an intermediate-intensity call, also emitted through the nostrils, while the Mexican long-eared bat (*Myotis auriculus*) emits pulses of intermediate intensity through its open mouth. The Mexican free-tailed bat (*Tadarida brasiliensis*) produces high-intensity calls, adjusting them according to the conditions under which it is hunting. (Photos by M. B. Fenton.)

the hunting sequences of all bats examined to date, although they may only occur immediately before the bat attacks its target. The difference in the information available to the bat is clear: the steeper FM components provide more precise data about both the location of the target and its surface details.

CF components are less common in the echolocation calls of bats, but provide an excellent means of detection in some species. An integral part of the Doppler-shift compensating system found in some bats, they serve as carrier or marker frequencies, permitting the monitoring of small changes in frequency associated with insect wing beats. This Doppler-shift compensation is accomplished by a zone of maximal frequency sensitivity in the inner ear known as an acoustic fovea (Schuller and Pollak 1979), and species capable of it appear to be well equipped to detect flying insects (Griffin and Simmons 1974; Goldman and Henson 1977; Schnitzler 1978; Simmons et al. 1979b).

Echolocating bats may alter the harmonic structure of their orientation calls. Some species use multiple harmonics of calls, apparently to increase precision of target resolution (Fig. 5). Others, for example the Mexican free-tailed bat (*Tadarida brasiliensis*), add harmonics when hunting in cluttered surroundings but omit them in more open terrain (Simmons et al. 1978).

The sequence of calls produced by an echolocating bat as it searches for, locates, and then closes with its prey (Fig. 6) demonstrates two important points: (1) the design of the call changes as this process takes place, and (2) so does the rate of pulse production. Changes in the design of the call include differences in duration and frequency that produce information required by the attacking bat. The characteristic increase in the rate of pulse production appears to apprise the bat of last-millisecond changes in the position of the target. Mexican long-eared *Myotis* (*Myotis auriculus*) did not increase their rate of pulse repetition when taking stationary moths from a wall, but did exhibit an increase when hunting flying prey (Fenton and Bell 1979). By using their wing or tail membranes to grab insect prey, bats can compensate for imprecise information about target location, perhaps the result of last-ditch defensive maneuvers by prey.

Although most biologists associate echolocation in bats with ultrasonic sound, many species use lower-frequency components in their orientation calls. In some instances the components of bat echolocation calls audible to human observers are the "Ticklauts" associated with the production of each pulse—the low-frequency transients that helped the Dutch biologist Dijkgraaf unravel the role of sound in bat orientation. In other cases the human ear hears only the lowest-frequency components in a call whose energy is mainly in the ultrasonic range. However, some bats in temperate and tropical areas—for example the spotted bat (*Euderma maculatum*) of western North America or Martienssen's free-tailed bat (*Otomops martiensseni*) of East Africa—use echolocation calls that are entirely within the range of human hearing (Fenton and Bell, in press).

Since most bats appear regularly to include an FM component in their echolocation calls, it is not possible to associate a particular frequency with a given species of bat. The frequency range covered by species studied to date is 8 to 215 kHz. It is possible, however, to use pulse repetition rate and the pattern of frequency change over time to identify bats in the field (Fenton and Bell, in press); the most convenient means of "seeing" these characteristics is through the oscil-

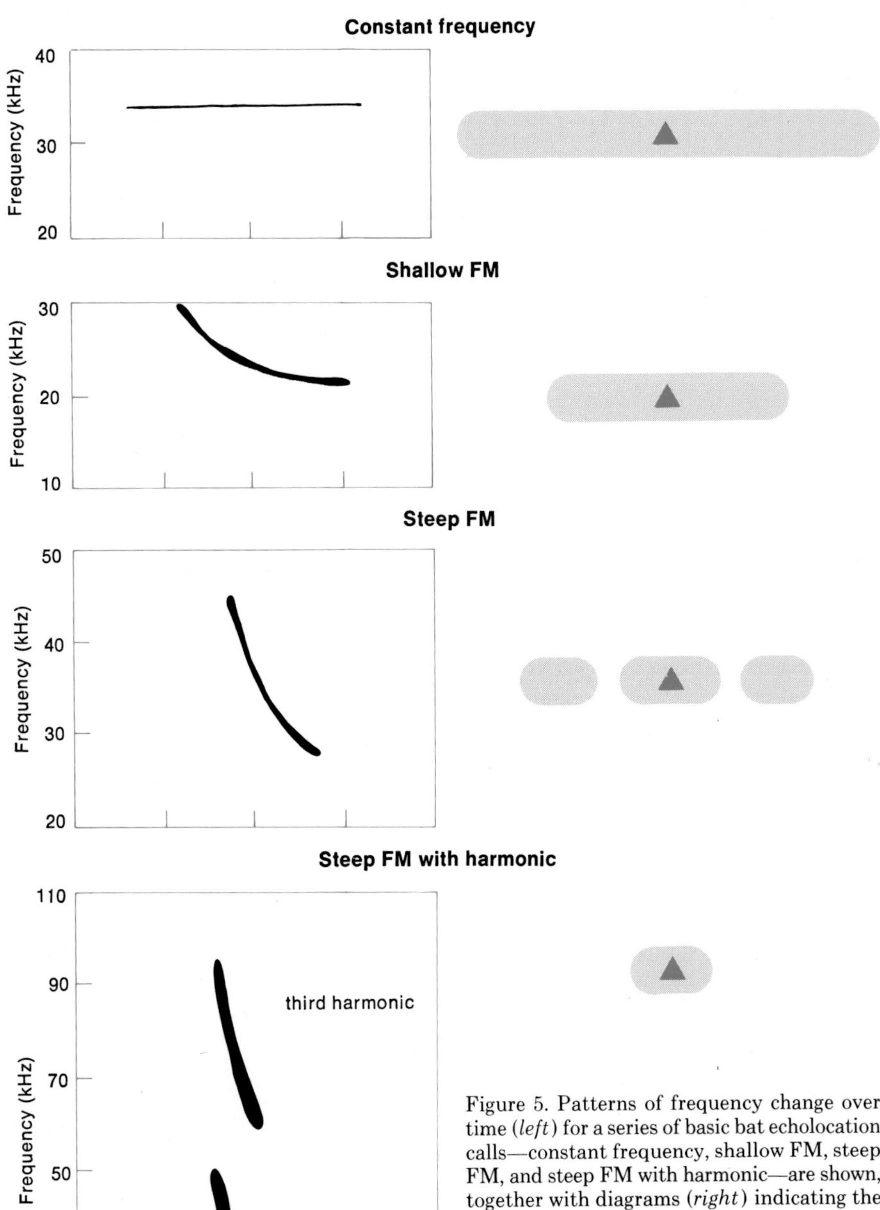

Figure 5. Patterns of frequency change over time (*left*) for a series of basic bat echolocation calls—constant frequency, shallow FM, steep FM, and steep FM with harmonic—are shown, together with diagrams (*right*) indicating the information about the relative position of a target made available to the bat by each pattern. The area shown in color indicates the target area as perceived by the bat; the triangular shape shows the position of the target itself. The addition of harmonics to a steep FM sweep results in a greatly heightened resolution of the target.

loscope display of the output of a zero-crossing period meter (Simmons et al. 1979a).

Echolocating bats, then, vary the intensity, rate of pulse repetition, frequency-time structure, and harmonic components of their echolocation calls according to the situation encountered. It is appropriate to consider the implications of this circumstance for the hearing-based defense of moths.

Detection of bats by moths

Roeder has demonstrated (1967) that several species of moths detect approaching bats at distances of up to 40 m. Indeed, evidence suggests that moths' ears are much more sensitive to bat calls than the most sensitive bat-detecting microphones now available (Griffin and Roeder, pers. comm.). The distances at which Roeder's moths detected approaching bats coincide with predictions based on the hearing sensitivity of the typical North American moth (Fig. 3), but the data on maximum distances of detection and echolocation suggest that not all bats would be equally conspicuous to all moths.

African insectivorous bats show greater variety in the design of their calls than do their North American counterparts (Fenton and Bell, in press). We therefore tested the ability of the African notodontid moth *Desmeocraera graminosa* to detect bats using echolocation calls of different intensities and frequencies by monitoring the activity in the moth's auditory nerve as various known species of bats flew in the laboratory (Fenton and Fullard 1979).

The effect of the design of the echolocation calls on the ability of the moth to detect a given bat was easily demonstrated. The banana bat (*Pipistrellus nanus*), whose high-intensity FM echolocation calls sweep from 120 to 75 kHz, was detectable anywhere in the small laboratory (up to a distance of 3 m from the moth), whereas Noack's African leaf-nosed bat (*Hipposideros ruber*), whose high-frequency (139 kHz in the CF portion of the call) calls are of intermediate intensity (75–80 dB), was detected only when it flew within 1.5 m of the moth. Large-eared slit-faced bats (*Nycteris macrotis*), which use echolocation calls of low intensity (60 dB) with frequency components from 120 to 50 kHz, were detected only when they flew within 0.2 m of the moth. In the field the ears of *D. graminosa* apparently detected echolocation calls in the 15 to 25 kHz range at distances exceeding 25 m, but there the identity of the bats was unknown, and precise figures on range of detection are lacking.

The results of such experiments generally support predictions of maximum distance of detection (Fig. 3). They suggest, moreover, that bats using sounds over 65 kHz would be less conspicuous to moths than those using sounds from 20 to 50 kHz, and that the differences in the hearing sensitivities of North American and African moths coincide with differences in the range of echolocation calls to which the two groups are exposed (Fenton and Fullard 1979). In North America most echolocating bats include FM components of between 25 and 60 kHz in their calls, whereas in Africa many species use components above or below this range (Fenton and Bell, in press).

The possible costs associated with the use of higher-frequency calls are greater atmospheric attenuation and heightened directionality of the sound. Greater directionality narrows the area from which the bat gathers information, increasing its chances of losing contact with a moving target. Either limitation could reduce the effective range of the bat's echolocation and affect its flight speed. Lower intensity would also reduce the effective range of echolocation; Griffin (1971) has estimated that the fawn horseshoe bat (*Hipposideros galeritus*), whose calls are of intermediate intensity, cannot detect targets beyond 2 m.

The distances at which echolocating bats respond to targets vary from about 1 m to at least 10 m (Griffin et al. 1960; Novick 1977; Fenton and Bell 1979; Fenton 1980). However, using behavioral responses to determine the range of echolocation is questionable, since the feature being measured is the reaction, not the detection distance. Our knowledge of the maximum distance at which a bat can detect a target is poor, but at present there is no evidence for detection of targets beyond 20 m, or even beyond 10 m for most bats (Fenton 1980).

Distance of detection is the essential point in the predator-prey confrontation between bats and moths. If a bat detects a moth at the same time or before the bat itself is detected by the moth, then the bat has an opportunity to lock onto its target and pursue it through subsequent evasive maneuvers. The maximum distance at which some moths detect bats greatly exceeds the range of echolocation in many bats, especially those using high-intensity calls in the area of maximum acoustical sensitivity of the moths.

In determining the optimal intensities and frequencies that a bat might use to locate prey, several factors must be taken into account. The first consideration is the incidence of prey equipped with bat detectors. If the existence of functional ears in some species implies a broader occurrence of this feature within families or su-

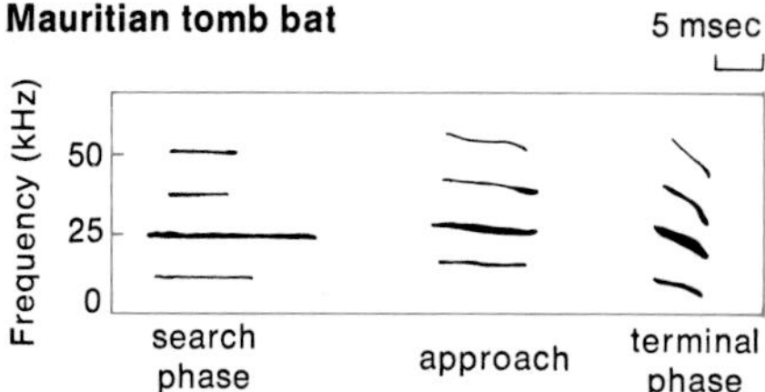

Figure 6. The sequence of echolocation calls produced by a big brown bat (*Eptesicus fuscus*) as it searches for, approaches, and attacks a June bug is compared with calls used by a Mauritian tomb bat (*Taphozous mauritianus*) as it closes with a handful of sand thrown up into the air. Notice how both bats alter the design of their calls as the attack unfolds, progressively shortening the calls, decreasing the CF or shallow FM components, and adjusting the presence of harmonics. Calls in the search phase tend to be longer and either shallow FM (*above*) or CF (*below*) in nature. In approaching its target the big brown bat relied more heavily on steep FM calls than did the tomb bat, but in the terminal phase both species used short calls with steep FM components, the big brown bat exclusively, the tomb bat more intermittently. The modulation of the design of the call from phase to phase is accompanied in both cases by an increased rate of pulse production.

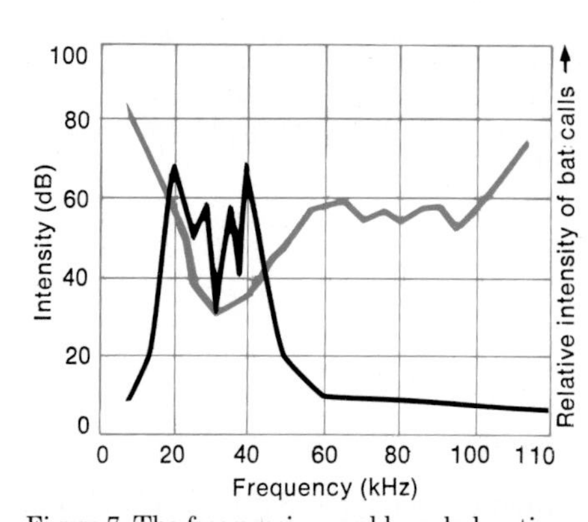

Figure 7. The frequencies used by echolocating bats at a site in Ivory Coast are those to which *G. apicata*, a noctuid moth that shares this habitat, is most sensitive. The bat calls (*black line*) show increasing power progressing up the vertical axis, while moth hearing sensitivity (*colored line*) increases down the vertical axis. Characteristically, the calls of this community of African bats extend both above and below the frequency range of 25 to 60 kHz used by comparable bat communities in Ontario.

perfamilies, then most moths have ears that are sensitive in some degree to the calls of echolocating bats. In Ontario over 95% of the species of macrolepidopteran moths possess functional ears, while in southern Africa we estimate that 85% of the species in families for which there are any data have bat detectors, although either of these figures could be influenced by variables such as infestation by ear mites (Treat 1975).

Outside the Lepidoptera many insects have ears (Michelsen 1979). There are ultrasonic components in the calls of many Orthoptera (Sales and Pye 1974), suggesting that these insects could hear bats, and there is some behavioral evidence that crickets avoid high-frequency sound (Moiseff et al. 1978). Some lacewings have bat detectors, but there are few data on the many other insects that are potential bat food. A number of beetles, notably some scarabs, possess sound-producing structures, although data on their ears are lacking. While we do not yet have enough information to evaluate the incidence of bat detectors among nocturnal insects precisely, it appears that a bat feeding heavily on moths equipped with bat detectors should have some means of foiling the insects' warning system.

Second, it is important to determine which bats make the greatest contribution to sounds in a given habitat from the insect's point of view. In monitoring bat activity with a broadband microphone we found (1979) that with one exception FM calls of molossids and vespertilionids with frequency components from 15 to 60 kHz accounted for most of the ultrasonic sounds in a variety of habitats. The exception was Gallery Forest in Ivory Coast, where bats using high-intensity calls with FM components were absent and species using intermediate-intensity calls with CF and FM components dominated. In both North America and West Africa the frequencies most commonly used by the bats were those to which the moths were most sensitive (Fig. 7).

The tuning of moth ears to the frequencies most commonly used by sympatric bats is of potential strategic importance because it leaves the way open for other bats to exploit this tuning curve by using calls outside the region of maximum sensitivity. In some tropical and subtropical areas, especially in the Ethiopian, Oriental, and Australasian tropics, about a third of the insectivorous species use calls that by virtue of their intensity and frequency might be outside the moths' early warning systems. These species could be more effective predators of moths because of their relative acoustic inconspicuousness.

A third important point to be considered is variation in the hearing ability of insects. The moths we studied showed considerable interspecific and somewhat less intraspecific variability in hearing sensitivity. Roeder (1975) has commented on the "evitability" of moth behavioral and neurological responses in general, pointing out that there was more variability in the former than in the latter in the moths he studied.

The predatory behavior of bats would ensure continuous selection for variability in the defensive response of prey. Echolocation clearly permits a bat to recognize a moth by the echo it creates (Simmons et al. 1979a), and as soon as a moth's response to an approaching bat can be predicted by the bat, the defense would be less effective. Miller (pers. comm.) found that Pipistrelles (*Pipistrellus pipistrellus*) feeding on green lacewings learned to exploit the defensive response of the prey and thus increased their rate of capture. Bats also learn to exploit predictable food sources such as the ultraviolet lights of resident entomologists. Unpredictability of response on the part of the individual prey could therefore make it more difficult for a hunting bat to circumvent or exploit a defensive response.

Diets of insectivorous bats

The information about moth hearing ability and bat echolocation presented here might lead one to suspect that some bats would specialize in moths, whereas others would not. But studies of the diets of insectivorous bats do not unequivocally support this suspicion. Such diets have been analyzed by examination of the contents of stomachs or feces, direct observation of the hunting behavior of individuals, or a combination of these techniques.

With each of these methods it is important to know where the bats have fed and the incidence of different insects there. Although we know of no study that has obtained this array of data for insectivorous bats, some broad outlines are emerging. Analyses of prey in stomach contents or feces have demonstrated a combination of variability (several prey taxa per sample) and selectivity (few prey taxa per sample), usually within a relatively narrow size range (see Buchler 1976; Anthony and Kunz 1977). However, in the case of the large slit-faced bat (*Nycteris grandis*) such analysis showed variability in the type, weight, and size of prey consumed, with prey ranging from 10-g sunbirds to 15-mm caddisflies (Fenton et al., in press). By contrast, evidence of specialization comes from a study that tabulated the presence of specific insect parts (beetle pieces or moth scales), rather than analyzing complete samples (Black 1974).

The diet of the little brown bat, a species for which our data are relatively complete (Belwood and Fenton 1976; Buchler 1976; Anthony and Kunz 1977), shows a combination of selectivity and variation. The species occasionally feeds selectively on particular types of insects, usually aquatic species, but at other times it consumes a range of prey. Variability may reflect differences in sex, reproductive condition, or age; lactating females include more moths in their diets than do males and nonparous females, and subadults show great variation in the prey they capture.

Such variability in diet also appears to be related to prey availability, and is moreover compatible with what we know of bats' energy budgets (Anthony and Kunz 1977; Kunz 1980). Insectivorous bats tend to be small and to employ an energy-intensive means of locomotion, and so incur a high cost in foraging. Given this situation, they minimize the intervals between captures by feeding in swarms of insects (Fenton and Morris 1976; Bell 1980) and exhibit masticatory specializations that permit rapid feeding (Kallen and Gans 1972).

An overall assessment of a wide range of feeding studies (Table 1) suggests four important points: (1) bats with high-intensity FM calls tend to show greater variation in their diets than species mainly using CF-FM calls; (2) some bats with low-intensity calls (e.g. *Nycteris* spp. and *Micronycteris* spp.) have extremely variable diets; (3) many species show a mosaic of extreme selectivity and great variation in prey taxa consumed per feeding period; and (4) there is a tendency among some rhinolophids and hipposiderids to feed heavily on moths.

The incomplete data we now have on the diets of some rhinolophids and hipposiderids support predictions, based on the intensities and frequencies of their echolocation calls, that these bats are well suited to thwart the auditory defenses of some moths. Dent's horseshoe bat (*Rhinolophus denti*), a rhinolophid, uses

Table 1. Results of studies of the incidence of Lepidoptera in the diets of insectivorous bats, arranged according to intensity and frequency of echolocation calls

			Incidence of Lepidoptera		
Echolocation strategy	*Number of species studied*	*Number of studies*	*Mean % by volume* ± *SD*[a]	$N_{100\%}$[b]	$N_{0\%}$[c]
High-intensity FM	35	45	32.4 ± 29	1	8
High-intensity CF	8	15	77.2 ± 21	4	1
Intermediate-intensity CF	3[d]	3	80.2 ± 15	1	0
Low intensity	3	11	25.8 ± 28	0	2

SOURCE: see Fenton and Fullard 1979
[a] SD = standard deviation
[b] Number of studies where Lepidoptera constituted 100% of diet
[c] Number of studies where Lepidoptera constituted 0% of diet
[d] *Hipposideros ruber*, *Cloeotis percivali*, and *Rhinolophus denti*

high-frequency calls (~ 100 kHz in the CF portion) of intermediate intensity, as does Noack's African leaf-nosed bat (CF at 139 kHz), a hipposiderid, and both bats feed heavily on moths.

Another hipposiderid, the small short-eared trident bat (*Cloeotis percivali*), also feeds heavily on moths (Whitaker and Black 1976), and from its size and diet we predicted that it too would use calls of high frequency and intermediate intensity—a prediction confirmed by recent field observations in Zimbabwe, which showed that the CF portion of its call was at 210 kHz (Fenton and Bell, in press). By contrast, the larger Hildebrandt's horseshoe bat (*R. hildebrandti*), a rhinolophid using more intense calls of lower frequency (CF at 50 kHz), commonly takes a variety of beetles, moths, and other insects.

It may be significant that rhinolophids and hipposiderids are probably capable of Doppler-shift compensation, which would permit them to exploit information about the wing beats of targets by means of an acoustic fovea (Schuller and Pollak 1979). The acoustic fovea of the greater horseshoe bat (*R. ferrumequinum*) may allow this species to distinguish between different species of insects by their wing-beat signatures. However, fluctuations in insect populations from night to night in many habitats may deny bats the opportunity to learn to recognize prey by wing-beat signatures even though they have the neural capacity for such identification. Present data on the diets of rhinolophids and hipposiderids, while indicating that they feed on moths, do not include any information about where or when these bats feed, what the insect populations of their habitats are, or what strategy they use to capture prey. Until there are field observations of the feeding behavior of the bats and the associated responses of the prey, definite conclusions are impossible.

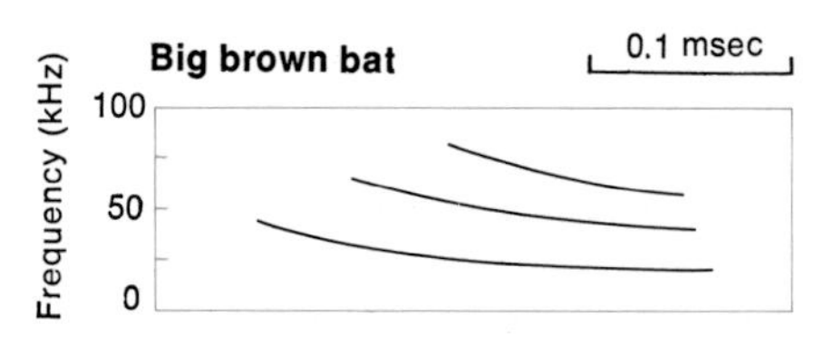

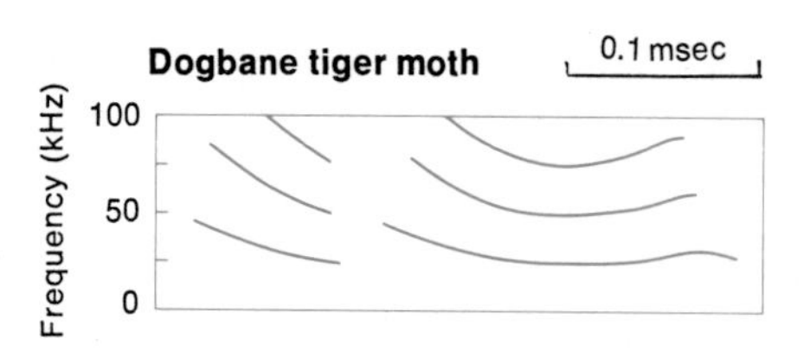

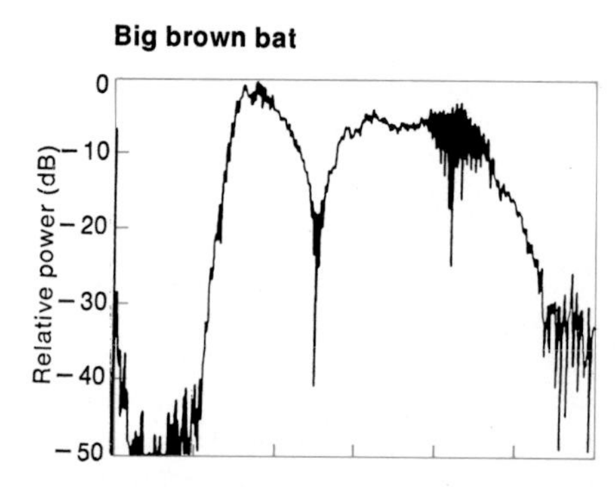

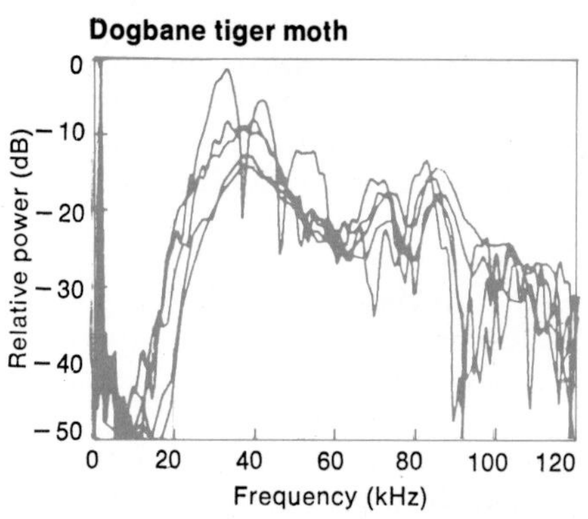

Figure 8. The power spectra of five clicks produced by the tymbal of the dogbane tiger moth (*Cycnia tenera*), an arctiid, appear to mimic closely the power spectra of the echolocation call of a big brown bat as it closes with a target (*above*). Moreover, both the intensity and the frequency-time structure of the clicks resemble comparable features of the echolocation call (*below*).

Jamming bats

Some arctiid moths produce trains of ultrasonic clicks when presented with high-intensity acoustic stimulation typical of a bat in the terminal phase of its attack. Not all arctiids click, however, and many silent species have functional ears. In southeastern Ontario, clicking arctiids are more common in summer than in spring, and in genera such as *Phragmatobia* the spring species have vestigial and the summer forms functional clickers (Fullard 1977). Clicking is typical of arctiids emerging during times of highest bat-predation pressure (Fullard and Barclay 1980).

Dunning and Roeder (1965) have demonstrated that little brown bats in the terminal stage of approach to their prey veer sharply away from their targets when presented with arctiid clicks. A variety of functions have been ascribed to these clicks, ranging from jamming the echolocation system of the bat, to startling the bat, to conveying a warning that arctiids are distasteful.

In the past, jamming has been rejected as an explanation for the response of the bat on the grounds that the moths' clicks are too low in intensity and the bats' echolocation systems too sophisticated. However, a recent reconsideration of the situation (Fullard et al. 1979) has led to renewed support for the jamming theory. The power spectra of the clicks of the dogbane tiger moth (*Cycnia tenera*) are surprisingly similar to those of calls produced by some bats during their feeding buzz (Fig. 8). Furthermore, the clicks are similar in intensity to echoes the bats would be receiving from a target at close range, and the frequency-time structures of the clicks resemble those of the bat calls (Fig. 8).

Taken together, these data suggest that the moth click could penetrate the information-processing system of the bat by passing through filters designed to remove extraneous information. The moth clicks could then disrupt the bat's processing of information and elicit the startle response reported by Dunning and Roeder (1965).

Roeder (pers. comm.) has suggested that moths could not mimic bats. He compares the situation to a cocktail party. Although it would be difficult for someone to convince you that you said something you did not say at such an event, chemical stimuli notwithstanding, if in the middle of a conversation you heard someone else talking about you, this would interfere with your concentration and your ability to process information. In this way moth clicks could interfere with bats' processing of information and effectively jam them.

Moth vs. bat

The interaction between moth and bat is an example of an eye-opening discovery about animal behavior. It involves much more than a simple response to predator-specific signals, and the theory has complications from either the moth's or the bat's point of view.

Nocturnal activity allows moths to avoid a range of diurnal predators, and possession of ears sensitive to the frequencies most commonly used by echolocating bats further reduces the odds against them. The average moth in North America or Africa detects the average bat long before the bat is aware of the moth, and negative phonotaxis may keep the alerted moth out of harm's way. Closer, more immediate threats, perceived by the moth as more intense stimuli, lead to complicated and unpredictable flight patterns, effective at close range against attacking bats.

There is variability in the responses of flying moths at both the individual and the faunal level, in part reflecting variation in hearing sensitivity. This variability enhances the defensive effect of bat detection by making it impossible for an attacking bat to predict the evasive response of its target.

Yet another dimension is added to the contest between moth and bat by the clicks of some tiger moths. These clicks are well suited to jam the echolocation of some bats by interfering with echo processing.

On the other side, bats appear to have several ways of countering the hearing-based defense of insects. Since the ears of moths are most effective at detecting the average bat, echolocation calls differing from the average may make an attacking bat less conspicuous acoustically. The use of calls either above or below the range of frequencies to which moths are most sensitive appreciably reduces the distance at which most moths can detect bats, as does the use of calls of lower intensity. Either alternative, however, imposes costs on the bat.

Another way to circumvent insect defenses is to stop echolocating and rely on other cues, a strategy adopted by some bats (Vaughan 1976; Fiedler 1979; Barclay et al., in press). Fringe-lipped bats (*Trachops cirrhosus*) may use the sounds produced by their prey in conjunction with echolocation to locate their targets, while Indian false vampire bats (*Megaderma lyra*) often use the sounds of their prey alone. We still do not know the extent to which animal-eating bats exploit other cues from prey, or how much they rely on their vision in finding food.

Observations of Mexican long-eared bats feeding on moths in Arizona (Fenton and Bell 1979) suggested that they might circumvent hearing-based defenses by taking moths resting on surfaces. However, more recent studies (Werner 1981) have shown that moths sitting or walking on surfaces also respond behaviorally to ultrasonic sound.

Although we know that bats use echolocation calls to detect prey and that some prey avoid attack at least some of the time by listening for bat calls, we lack several items of information needed to complete the picture. It appears that some bats—for example, rhinolophids and hipposiderids—may exploit the hearing sensitivities of moths to permit closer undetected approach, but more precise information about bat hunting, selection of prey, and the defensive behavior of the moths is required to confirm this possibility. It would also be interesting to find out whether the clicks of arctiids are more effective against some bats than against others; the properties of different echolocation calls suggest that this should be true.

Echolocation may have been a vital element in the origin and radiation of insectivorous bats, permitting them or their immediate ancestors to hunt flying insects regardless of the amount of light available. The appearance of hearing-based insect defenses may in turn have reduced bats' effectiveness against some prey. We are now making progress in exploring the third twist in this story: the possibility that the variety we find in bat hunting and echolocation behavior may be in part a response to these insect defenses.

References

Anthony, E. L. P., and T. H. Kunz. 1977. Feeding strategies of the little brown bat, *Myotis lucifugus*, in southern New Hampshire. *Ecology* 58:775–86.

Barclay, R. M. R., M. B. Fenton, M. D. Tuttle, and M. J. Ryan. In press. Echolocation calls produced by *Trachops cirrhosus* (Chiroptera: Phyllostomatidae) while hunting for frogs. *Can. J. Zool.*

Bell, G. P. 1980. Habitat use and response to patches of prey by desert insectivorous bats. *Can. J. Zool.* 58:1876–83.

Belwood, J. J., and M. B. Fenton. 1976. Variation in the diet of *Myotis lucifugus* (Chiroptera: Vespertilionidae). *Can. J. Zool.* 54: 1674–78.

Black, H. L. 1974. A north temperate bat community: Structure and prey populations. *J. Mamm.* 55:138–57.

Buchler, E. R. 1976. Prey selection by *Myotis lucifugus* (Chiroptera: Vespertilionidae). *Amer. Nat.* 110:619–28.

Busnel, R. -G., and J. F. Fish, eds. 1980. *Animal Sonar Systems.* NATO Advanced Study Institute Series A, vol. 28. Plenum Press.

Dalland, J. I. 1965. Hearing sensitivity in bats. *Science* 150:1185–86.

Dunning, D. C., and K. D. Roeder. 1965. Moth sounds and the insect-catching behavior of bats. *Science* 147:173–74.

Edmunds, M. 1974. *Defence in Animals.* Longman.

Fenton, M. B. 1980. Adaptiveness and ecology of echolocation in terrestrial (aerial) systems. In *Animal Sonar Systems*, ed. R. -G. Busnel and J. F. Fish, pp. 427–46. NATO Advanced Study Institute Series A, vol. 28. Plenum Press.

Fenton, M. B., and G. P. Bell. 1979. Echolocation and feeding behaviour in four species of *Myotis* (Chiroptera). *Can. J. Zool.* 57: 1271–77.

———. In press. Recognition of species of insectivorous bats by their echolocation calls. *J. Mamm.*

Fenton, M. B., and J. H. Fullard. 1979. The influence of moth hearing on bat echolocation strategies. *J. Comp. Physiol.* 132:77–86.

Fenton, M. B., and G. K. Morris. 1976. Opportunistic feeding by desert bats (*Myotis* spp.). *Can. J. Zool.* 54:526–30.

Fenton, M. B., G. P. Bell, and D. W. Thomas. 1980. Echolocation and feeding behaviour of *Taphozous mauritianus* (Chiroptera: Emballonuridae). *Can. J. Zool.* 58:1774–77.

Fenton, M. B., D. W. Thomas, and R. Sasseen. In press. *Nycteris grandis* (Nycteridae): An African carnivorous bat. *J. Zool. London.*

Fiedler, J. 1979. Prey catching with and without echolocation in the Indian false vampire (*Megaderma lyra*). *Behav. Ecol. Sociobiol.* 6:155–60.

Fullard, J. H. 1977. Phenology of sound-producing arctiid moths and the activity of insectivorous bats. *Nature* 267:42–43.

———. 1979a. Auditory components in the defensive behaviour of certain tympanate moths. Ph.D. Thesis, Carleton University.

———. 1979b. Behavioral analyses of auditory sensitivity in *Cycnia tenera* Hübner (Lepidoptera: Arctiidae). *J. Comp. Physiol.* 129:79–83.

Fullard, J. H., and R. M. R. Barclay. 1980. Audition in spring species of arctiid moths as a possible response to differential levels of insectivorous bat predation. *Can. J. Zool.* 58:1745–50.

Fullard, J. H., M. B. Fenton, and J. A. Simmons. 1979. Jamming bat echolocation: The clicks of arctiid moths. *Can. J. Zool.* 57: 647–49.

Goldman, L. J., and O. W. Henson, Jr. 1977. Prey recognition and selection by the constant frequency bat, *Pteronotus parnellii parnellii*. *Behav. Ecol. Sociobiol.* 2:411–20.

Griffin, D. R. 1958. *Listening in the Dark.* Yale Univ. Press.

———. 1971. The importance of atmospheric

attenuation for the echolocation of bats (Chiroptera). *Anim. Behav.* 19:55–61.

Griffin, D. R., and J. A. Simmons. 1974. Echolocation of insects by horseshoe bats. *Nature* 250:731–32.

Griffin, D. R., F. A. Webster, and C. R. Michael. 1960. The echolocation of flying insects by bats. *Anim. Behav.* 8:141–54.

Gustafson, Y., and H. -U. Schnitzler. 1979. Echolocation and obstacle avoidance in the hipposiderid bat *Asellia tridens. J. Comp. Physiol.* 131:161–67.

Kallen, F. C., and C. Gans. 1972. Mastication in the little brown bat, *Myotis lucifugus. J. Morph.* 136:385–420.

Kirby, W., and W. Spence. 1826. *An Introduction to Entomology.* London.

Kunz, T. H. 1980. *Daily Energy Budgets of Free-Living Bats.* Proc. Fifth Int. Bat Res. Conf. Lubbock: Texas Tech Press.

Michelsen, A. 1979. Insect ears as mechanical systems. *Amer. Sci.* 67:696–706.

Miller, L. A., and J. Olesen. 1979. Avoidance behavior in green lacewings. I. Behavior of free flying green lacewings to hunting bats and to ultrasound. *J. Comp. Physiol.* 131: 113–20.

Moiseff, A., G. S. Pollack, and D. R. Hoy. 1978. Steering responses of flying crickets to sound and ultrasound: Mate attraction and predator avoidance. *Proc. NAS:* 75:4052–56.

Novick, A. 1977. Acoustic orientation. In *Biology of Bats*, ed. W. A. Wimsatt, vol. 3, pp. 74–289. Academic Press.

Roeder, K. D. 1967. *Nerve Cells and Insect Behavior.* Rev. ed. Harvard Univ. Press.

———. 1975. Neural factors and evitability in insect behavior. *J. Exp. Zool.* 194:75–88.

Roeder, K. D., and M. B. Fenton. 1973. Acoustic responsiveness of *Scoliopteryx libatrix* (Lepidoptera: Noctuidae), a moth that shares hibernacula with some insectivorous bats. *Can. J. Zool.* 51:291–99.

Sales, G., and J. D. Pye. 1974. Ultrasonic communication by animals. London: Chapman and Hall.

Schnitzler, H. -U. 1978. Die detektion von Bewegungen durch Echoortung bei Fledermäusen. *Verh. Deut. Zool. Ges.* 1978:16–33.

Schuller, G., and G. Pollack. 1979. Disproportionate frequency representation in the inferior colliculus of Doppler-compensating greater horseshoe bats: Evidence for an acoustic fovea. *J. Comp. Physiol.* 132:47–57.

Simmons, J. A., and R. A. Stein. 1980. Acoustic imaging in bat sonar: Echolocation signals and the evolution of echolocation. *J. Comp. Physiol.* 135:61–84.

Simmons, J. A., M. B. Fenton, W. R. Ferguson, M. Jutting, and J. Palin. 1979a. *Apparatus for Research on Animal Ultrasonic Signals.* Life Sci. Misc. Pub. Royal Ontario Museum.

Simmons, J. A., M. B. Fenton, and M. J. O'Farrell. 1979b. Echolocation and pursuit of prey by bats. *Science* 203:16–21.

Simmons, J. A., D. J. Howell, and N. Suga. 1975. Information content of bat sonar echoes. *Amer. Sci.* 63:204–215.

Simmons, J. A., W. A. Lavender, B. A. Lavender, J. E. Childs, K. Hulebak, M. R. Rigden, J. Sherman, B. Woolman, and M. J. O'Farrell. 1978. Echolocation by free-tailed bats (*Tadarida*). *J. Comp. Physiol.* 125:291–99.

Treat, A. E. 1975. *Mites of Moths and Butterflies.* Cornell Univ. Press.

Vaughan, T. A. 1976. Nocturnal behavior of the African false vampire bat (*Cardioderma cor*). *J. Mamm.* 57:227–48.

Werner, T. K. 1981. Responses of non-flying moths to ultrasound: The threat of gleaning bats. *Can. J. Zool.* 59:525–29.

Whitaker, J. O., Jr., and H. L. Black. 1976. Food habits of bats from Zambia, Africa. *J. Mamm.* 57:199–204.

Carl Hirschie Johnson
J. Woodland Hastings

The Elusive Mechanism of the Circadian Clock

The quest for the chemical basis of the biological clock is beginning to yield tantalizing clues

"The early bird catches the worm." A worm that delays its return to the burrow at dawn, however, exposes itself to the whole flock. Indeed, eukaryotic organisms, unlike bacteria and other prokaryotes, have evolved a circadian system, a biological clock that serves to schedule biological events at ecologically appropriate times during the day. In addition, the circadian system provides the chronometer necessary for birds and bees to orient themselves by the sun and stars and thus may function in navigation and migration. Moreover, the clock provides the timing mechanism that monitors annual changes in the length of the day so that seasonal processes like flowering or animal reproduction are stimulated (Pittendrigh 1981).

Beyond such interactions of the organism with the external environment, biological clocks are also responsible for the appropriate timing of many different physiological processes within the organism and even within the cell. In humans, disruption of this internal temporal program has clinical consequences. Our goal here is to describe briefly the basic properties of circadian systems and then to discuss some of the experimental strategies that are currently being applied to elucidate the biochemical mechanism of these timekeepers.

Although we presume that the daily fluctuations of light intensity and temperature were important in the origin and evolution of the biological clock (and still serve to regulate phase), the clock itself is endogenous and can function independently of environmental cycles. The intrinsic nature of daily rhythmicity is deduced from the fact that under conditions of constant illumination and temperature the free-running period of rhythms is close to but usually different from 24 hours, generally being in the range of 22 to 26 hours—hence the name *circadian*, which means literally about (*circa*) a day (*diem*) (Halberg et al. 1959). Furthermore, the periods of circadian oscillators, which would be precisely 24 hours if they depended on a geophysical cycle for their timing, may be altered by mutation, indicating that the clock mechanism is somehow encoded in the genes (Feldman 1982).

These observations have discredited alternative models of timekeeping, which had hypothesized that subtle geophysical cues, such as daily fluctuations in cosmic irradiation or the electromagnetic fields of the earth, control the timing of biological rhythms (Brown 1972). Recent experiments in the NASA space shuttle have underscored the endogenous basis of the clock by showing that circadian rhythms of fungi persist in earth orbit under conditions presumably devoid of such external daily signals (Sulzman et al. 1984).

To be an accurate reference for local time, a clock must be synchronized with the solar cycle of light and dark, and it must have a stable period that is isolated from unpredictable environmental fluctuations. The synchronization, or *entrainment*, of the clock to the solar day is necessary in order for clock time to match local time and also to compensate for the fact that the circadian period is often not 24 hours. Even when the circadian period differs only slightly from 24 hours, the biological clock would eventually drift out of phase with local time were it not for the daily resynchronizing. The predominant signal that accomplishes this entrainment is the daily cycle of sunlight and darkness (Pittendrigh 1976). In the absence of light, other periodic signals may synchronize biological clocks; in humans, for example, periodic social factors may play an important role in phasing the biological clock (Wever 1979; Winfree 1982). Nevertheless, the cycle of light and dark appears to be the most important daily synchronizing cue of human clocks (Czeisler et al. 1981).

As Pittendrigh (1954) first recognized, a reliable endogenous timekeeper must also have a period that is not greatly affected by environmental fluctuations in temperature. Indeed, circadian clocks compensate for temperature; that is, if the temperature changes, the rate of these circadian pacemakers changes only slightly, if at all, even in organisms that cannot regulate their body temperature. Surprisingly, in some organisms, such as the dinoflagellate alga *Gonyaulax*, which we use as a research organism, the clock actually runs more slowly at warmer temperatures. If one considers that the rate of most biochemical processes increases significantly with temperature, the mechanism of this temperature compensation—and sometimes overcompensation—poses an interesting problem, still unsolved.

Carl Johnson, an NIH postdoctoral fellow at Harvard University, has investigated biological clocks in many organisms, ranging from breadmold to cockroaches and earthworms to vertebrates. He is a graduate of the University of Texas and Stanford University (Ph.D. 1982), where he studied with Colin Pittendrigh and David Epel. J. Woodland Hastings is Professor of Biology at Harvard University, where he joined the faculty in 1966. His research on the biochemical nature of circadian rhythmicity has focused on the unicellular alga Gonyaulax. *The research reported here was supported by grants from the NIH. Address: The Biological Laboratories, Harvard University, 16 Divinity Avenue, Cambridge, MA 02138.*

The nature of clock systems

Although the cellular and biochemical nature of the circadian pacemaker is not known, it is clear that it controls more than one observable process. This is the case in the unicellular alga *Gonyaulax*, for which four rhythms have been characterized: the capacity to fix carbon by photosynthesis, the timing of cell division, and two modes of bioluminescent emission, a spontaneous glow and a capacity for flashing in response to stimulation or disturbance (Fig. 1). Although the phases at which these rhythms peak are different, it is generally believed that all are controlled by a single pacemaker. Experimental conditions have not yet been discovered that allow the expression of either different frequencies or different phase relationships among the different rhythms, as would be expected if each rhythm were controlled by a separate pacemaker (McMurry and Hastings 1972). Therefore, even though the possibility of tightly coupled but distinct pacemakers controlling the different overt rhythms is not excluded in *Gonyaulax*, the various phase positions of the rhythms shown in Figure 1 are believed to be controlled by a single pacemaker to which these rhythms lock on in different phase relationships.

These observed rhythmic processes seem to have no feedback to the pacemaker mechanism and thus appear to be merely driven systems. For example, photosynthesis can be completely inhibited with a specific herbicide, but the bioluminescence rhythm continues and its phase is not altered (Hastings 1960). Likewise, the expression of the bioluminescence rhythm can be blocked at the same time that the underlying pacemaker can be shown to be still operating (Hobohm et al. 1984).

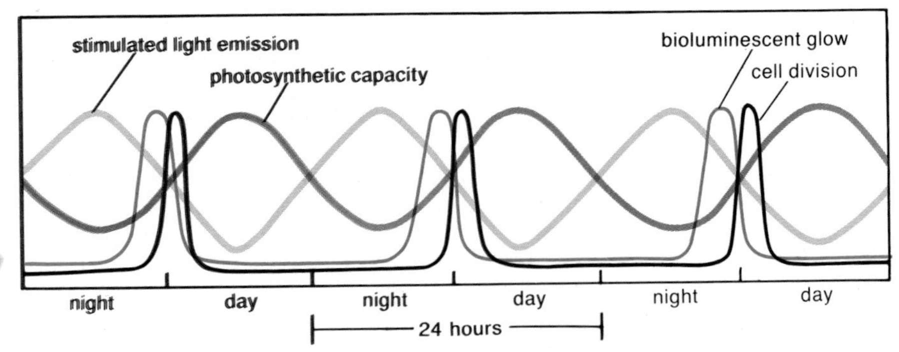

Figure 1. Circadian rhythmicity has been identified in at least four distinct biological processes of the unicellular alga *Gonyaulax*. Although the four rhythms peak at different phases of the 24-hour day, as represented here diagrammatically, they are all synchronized with the circadian clock. Perturbations that reset the clock will phase-shift all four rhythms similarly. This observation plus the fact that it has not been possible to dissociate the four rhythms from each other into different frequencies suggest that these overt rhythms are driven by a single pacemaker, as diagrammed in Figure 2.

Such observations in *Gonyaulax* and other organisms suggest models of circadian organization like that diagrammed in Figure 2. The central pacemaker is phased to the solar light cycle by means of a photoreceptor and an entraining pathway, and it controls the expression of the different rhythms, such as cell division, bioluminescence, and photosynthesis. These rhythms thus provide a monitor of the pacemaker's phase and period but are not themselves considered to be a part of the pacemaking mechanism. The observed rhythms are like the hands of a mechanical clock, which indicate the clock's position without being a gear or cog of the timing mechanism. Nevertheless, the characterization of the pathways by which these overt rhythms are controlled can help to discover the clock mechanism.

Since single cells like *Gonyaulax* have circadian pacemakers, should one think of multicellular organisms as composed of many separate pacemakers coupled together (Pittendrigh 1960)—what Winfree (1980) calls the "clockshop"? One example supporting this concept is the "internal" desynchronization of rhythms that is sometimes observed in humans isolated from time cues (Wever 1979); unlike photosynthesis and bioluminescence in *Gonyaulax*, human cycles of body temperature and sleep may uncouple and exhibit distinctly different frequencies (Fig. 3). This type of observation has been taken as evidence that humans have at least two circadian pacemakers which are usually but not necessarily coupled (Wever 1979; Kronauer et al. 1982).

The clockshop seems to be organized as a hierarchy of oscillators. Studies of rhythmic behaviors in multicellular organisms have offered persuasive phenomenological evidence indicating that overt rhythms

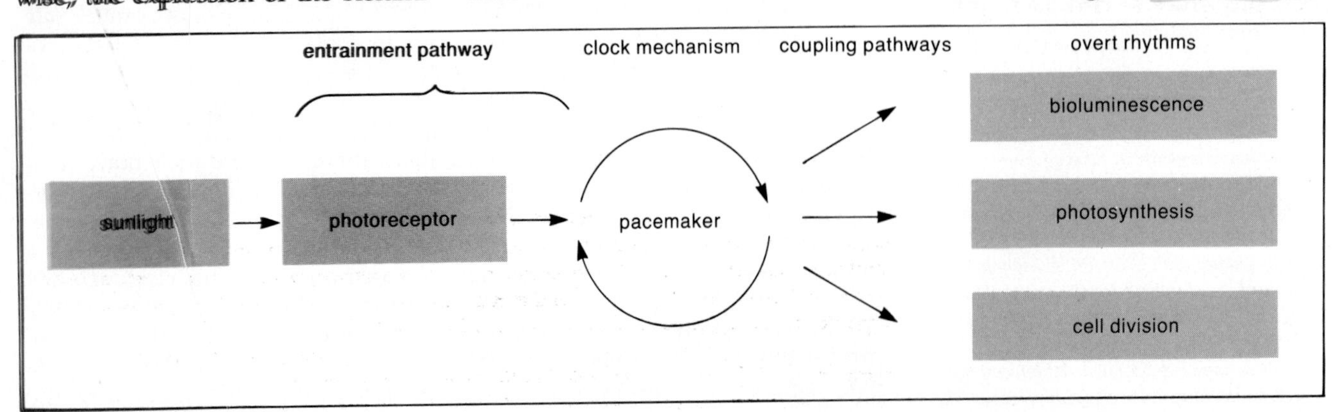

Figure 2. Although the biochemical basis for the circadian rhythms observed in living organisms is not yet understood, many researchers are **investigating the possibility** that a single endogenous clock within each cell can control a multiplicity of observed rhythms, as in this model **of circadian organization** in the alga *Gonyaulax*.

are directly controlled by individual "slave" oscillators, each of which is, in turn, coupled to a "master" clock (Pittendrigh 1981). This concept is supported by recent physiological discoveries of multiple oscillators in moths (Mizoguchi and Ishizaki 1982; Truman 1984) and of anatomically discrete master pacemakers localized in molluscan eyes, cockroach optic lobes, the pineal glands of lizards and birds, and the suprachiasmatic nuclei of mammals (Takahashi and Menaker 1984; Jacklet 1984).

The model of a master-slave oscillator strongly predicts that when the master clock is reset by a light signal, one slave oscillator may take longer to resynchronize to the shifted master clock than another slave having a different period, damping coefficient, or strength of coupling to the master oscillator (Pittendrigh 1981). Different lags in the resynchronization of various rhythmic functions have indeed been observed in humans subjected to phase changes such as those experienced after traveling across time zones (Wever 1979). A consequence of this circadian structure is that resetting the master clock may result in a transient disturbance of the relative phase relationships among the slave oscillators; the organism would then be in considerable temporal disarray, an experience familiar to humans as jet lag.

Clinical implications

The temporal disruption inherent in jet lag may also have important implications for mental health. Cycles of human body temperature and REM sleep in depressed patients are often phased earlier in the day than in healthy subjects. Wehr and Goodwin (1983), reasoning that the depressive symptoms of manic-depressive patients might be due in part to just such an abnormal phase relationship, sought to ameliorate the symptoms by changing the phase of patients in order to reestablish temporarily a normal phase relationship between the temperature-sleep cycles and the cycle of light and dark. In some cases, the induced jet lag seemed to be an effective, albeit temporary, antidepressant; the antidepressant effect lasts only about as long as the disruption of circadian organization induced by jet lag in healthy subjects. This and other intriguing studies suggest that mental health is influenced by circadian biology and may be partially dependent on internal temporal harmony.

Principles of circadian biology are also relevant to healthy humans. An important example is that of shift work. By many criteria, human performance is clearly impaired during jet-lag disruption, yet many shift workers are exposed to repeated jet lag in the form of schedules that change weekly (Winfree 1982; Moore-Ede et al. 1983). The efficiency of these workers could probably be improved by applying our current knowledge of circadian biology, such as by rotating schedules less frequently and, when a schedule is rotated, by making the shift a delay rather than an advance (Czeisler et al. 1982; Reinberg and Halberg 1971). The issue is not simply one of efficiency or personal comfort, particularly in the case of workers such as medical personnel on night duty, air traffic controllers, airline pilots, and operators of nuclear energy plants, whose tasks involve the health of others. Especially frightening is the practice on US Navy submarines of assigning crews responsible for nuclear missiles to a very disruptive schedule of 6 hours on duty, 12 hours off duty; this 18-hour cycle forces sailors to phase-advance by six hours in every cycle and has the alarming consequence of placing our security in the hands of personnel who continually suffer jet lag (Moore-Ede et al. 1983).

There are also important pharmacological implications of circadian rhythmicity. The efficacy or the toxicity of many drugs varies strikingly with the time of day at which the drug is administered; thus, it is clear that, whenever it is feasible to do so, drug intake should be timed to maximize pharmacological effectiveness (Reinberg and Halberg 1971). Furthermore, the phase of maximal efficacy of some drugs does not necessarily coincide with the phase of maximal toxicity. Cancer chemotherapy, for example, might be timed to minimize toxicity to normal tissues while maximizing destruction of tumors, a tactic that has proved effective in treating leukemic mice (Scheving et al. 1977).

Several kinds of drugs are known to perturb the clock. Basic investigations of the pharmacological sensitivity of the circadian clock may benefit clinical therapy by identifying drugs useful for the treatment of diseases related to circadian organi-

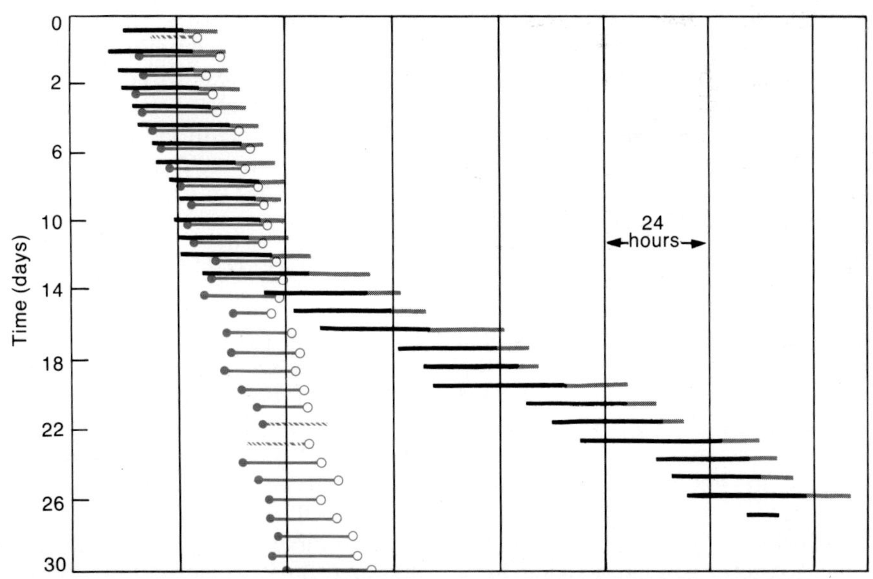

Figure 3. Experiments in which subjects lived for many days in isolation from any time cues suggest that humans have at least two distinct circadian pacemakers, one involving the rhythm of waking (*black bars*) and sleeping (*gray*), the other involving the observed cycle of maximum temperature (*color dots*) and minimum (*color circles*). The two cycles normally coincide and have periods that typically exceed 24 hours, as is the case with this subject, whose cycles averaged 25.7 hr in the first 14 days. However, after fourteen days the two cycles uncoupled; whereas the period of the temperature cycle remained about the same, averaging 25.1 hr, that of the sleeping-waking cycle averaged 33.4 hr. Thus, while this subject experienced the passage of 25 "days" of waking and sleeping, 31 days had actually elapsed (each vertical line represents one additional day "lost"). (After Wever 1979.)

zation. For example, a drug that quickly resets the clock could be used to reduce the duration of jet lag or to help a shift worker adjust to a new schedule. Or a drug that changes the clock's period or the phase relationship among various overt rhythms might be useful for treating manic-depression or sleep disorders. In fact, drugs commonly used in manic-depressive therapy, such as lithium, tricyclics, and monoamine oxidase inhibitors, are known to change the circadian period in humans and in other mammals (Engelmann 1973; Wehr and Goodwin 1983; Moore-Ede et al. 1983). Therefore, discovering drugs that can manipulate the biological clock may lead to impressive therapeutic tools for clock-related disorders, and understanding the clock's fundamental biochemical mechanism would obviously contribute to this end.

Strategies for research

What is the mechanism of this biological clock? Does it use or depend upon a known biochemical pathway, perhaps in a heretofore unsuspected way, such as through a circadian feedback loop? Or is the circadian pacemaker driven by a totally unknown system? Although we don't yet know, many models of the clock's biochemistry have been proposed that variously involve sequential gene transcription, membrane properties, ion transport, mitochondrial oxidative phosphorylation, cyclic nucleotide levels, messenger RNA production, or translation on 80S ribosomes (Sweeney 1983; Jacklet 1984). Some investigators have also suggested that circadian and cell division cycles are inextricably linked (Edmunds and Adams 1981). But none of the proposed models has yet received substantial experimental support, and many do not even address all the key features of circadian systems.

Not only is the search for the clock's mechanism like looking for a needle in a haystack, it is also complicated by the fact that the haystack has so many counterfeit needles. That is, many cellular processes are rhythmic without being a part of the pacemaking mechanism, analogous to a mechanical clock having many hands. A necessary but not sufficient criterion to decide whether a rhythmic process is a component in the clock mechanism is to test whether changes in the process have a predictable impact upon the pacemaker's phase, period, or both. The difficulties in applying this test to a plethora of cellular processes means that the attempt to discover the clock mechanism by mere cataloging of rhythmic processes is likely to be a futile quest.

Concrete experimental approaches are needed to establish how biochemical entities that control or are controlled by the clock are linked with its mechanism. Knowledge concerning the mechanism might thereby be inferred. Eskin (1979) has pointed out three such approaches based on the circadian organization depicted in Figure 2. One is to perturb the phase or period of the clock by drug treatment or mutation and then determine the biochemical site of action of the drug or the nature of the mutation. Another would be to identify the photoreceptor for the clock and characterize the entrainment pathway that leads to the clockwork. The third approach is to study an output pathway by choosing an overt, observed rhythm and working out step-by-step the nature of the coupling of this rhythm upstream to the pacemaking mechanism. Each of these three strategies has recently yielded insights into circadian organization at the biochemical level, and each will be considered here in turn.

Perturbing the clock

Certain pharmacological agents alter the phase or period of the clock and might thereby help to identify its mechanism. For example, a brief exposure (minutes or a few hours) to an inhibitor of protein synthesis on 80S ribosomes can reset the phase of circadian oscillators by many hours in a large number of organisms, including unicellular algae, fungi, higher plants, and molluscs (Sweeney 1983; Jacklet 1984). Figure 4 illustrates an example of this resetting. Such studies suggest that the phase of the biological clock is dependent on the synthesis of one or more specific proteins. A substantial number of other agents have been found to cause phase shifts or period changes. Because their specificities are not so clear, it is not known whether few or many different biochemical pathways are critically involved in the circadian mechanism.

Mutants exhibiting altered frequencies can also be helpful in the analysis of the clock; such mutants have been isolated in the fruit fly *Drosophila,* the fungus *Neurospora,* and the alga *Chlamydomonas* (Feldman 1982). Mutations at a single gene locus (the *per* gene) of *Drosophila* were identified that alter the circadian period (Konopka and Benzer 1971). Almost a decade later, it was discovered that these same mutations affect a *Drosophila* rhythm having a very different frequency, the 1-minute cycle of the courtship song (Kyriacou and Hall 1980). The mutations modify both rhythms similarly—that is, both are shortened or lengthened, or both are abolished, becoming arrhythmic. Thus, a single gene locus in *Drosophila* can control the period of two rhythms of very different frequencies.

More recent results using the molecular-genetics approach also indicate its promise for discovering the clock mechanism. The *per* gene has now been cloned (Bargiello and Young 1984; Reddy et al. 1984); when cloned DNA segments from the *per* locus are introduced into arrhythmic mutant flies, rhythmicity is partially or fully restored (Zehring et al. 1984; Bargiello et al. 1984). In addition, a particular messenger RNA transcribed from the gene locus undergoes a 15-fold oscillation in its amount during the circadian cycle (Reddy et al. 1984). Perhaps this oscillating messenger RNA is involved in coupling genetic information to overt rhythmicity. Another recent study has shown that the *per* gene contains an unusual DNA sequence that is homologous to DNA of chickens, mice, and humans (Shin et al. 1985). It is not yet known whether this DNA also has a clock function in these vertebrates.

Despite this hopeful beginning, however, results obtained by the genetic approach must be interpreted cautiously since the genetic lesions might be acting indirectly, altering not a component of the pacemaking mechanism but some other cellular process that impinges on it. For example, one model of the temperature-compensation mechanism suggests that it is independent of any pacemaking reaction but has inhibitory feedback into the pacemaker (Hastings and Sweeney 1957). If this is true, then mutations or drugs could easily affect the temperature-compensation properties of the clock—thereby modifying its period—without directly interfering with any pacemaking reaction.

This possibility is supported by a study of Woolum and Strumwasser (1983) demonstrating that an increase in the intracellular level of free calcium ions caused by different drugs also causes changes in the periods of both a high-frequency neural pacemaker, with a period of approximately one minute, and of the 24-hour circadian pacemaker of the isolated eye of the mollusc *Aplysia*. They interpret these results to mean that the constant frequency of this circadian oscillator is not directly regulated for changes in the level of free calcium ions but depends on a constant intracellular environment. This simultaneous impact upon a short- and long-period oscillation is similar to the effects of the *per* gene on the song and circadian cycles of *Drosophila*. On the other hand, the properties of the *per* locus in *Drosophila* and the *frq* locus in *Neurospora* suggest that the products of these genes play a key role in the clock (Feldman 1982). Nevertheless, it is important to remember that agents or mutations which appear to act directly on the pacemaker may not be doing so (Woolum and Strumwasser 1983).

Characterizing the entrainment pathway

A second strategy for discovering the circadian mechanism is to delineate the pathway of entrainment by light—or, to put it another way, to follow the photon to the pacemaker. For example, a clock photoreceptor might first be identified by its action spectrum—the spectral response of the clock—and then the sequence of steps by which the light signal is transformed into a phase shift of the clock might be characterized. Action spectra for circadian phase-shifting have been determined for *Gonyaulax*, *Neurospora*, *Drosophila*, the moth *Pectinophora*, and the hamster *Mesocricetus*, and have shown that the photoreceptor pigments involved are not the same among these organisms.

Not only is the identification of the pigment by its action spectrum the first step in discovering the entrainment pathway, but studying the basic photobiology of phase-shifting by light might yield other information. For example, a particularly intriguing study of the clock photoreceptor in hamster retina showed that this photoreceptor, for which the photopigment appears to be rhodopsin, has a very high threshold, about one million times higher than rod vision, but the reciprocity between intensity and duration holds for light pulses as long as 45 minutes (Takahashi et al. 1984). This unusual photoreceptive system therefore acts very much like a photon integrator. It remains to be seen whether other photoreceptors of circadian pacemakers will behave similarly.

The entrainment pathway can be dissected pharmacologically by using drugs to block the resetting action of light pulses (Hastings 1960). Drugs preventing the effect of light that do not themselves perturb the clock, and that act at some specific biochemical-physiological site, can be used to infer the entrainment pathway (Eskin 1979). This approach has been useful in characterizing the physiology of the light entrainment pathway in the isolated *Aplysia* eye (Eskin 1977). The circadian oscillator in the *Aplysia* eye is also reset by pulses of the neurotransmitter serotonin. Although this neurotransmitter pathway is distinct from the light entrainnment pathway, its phase-shifting pathway can be similarly dissected pharmacologically. Eskin and his co-workers have found that this entraining agent acts by increasing the intracellular concentration of the ubiquitous regulatory molecule, cyclic AMP (cAMP), and they are now trying to determine the mechanism by which cAMP resets the clock (Eskin et al. 1982). Cyclic AMP might

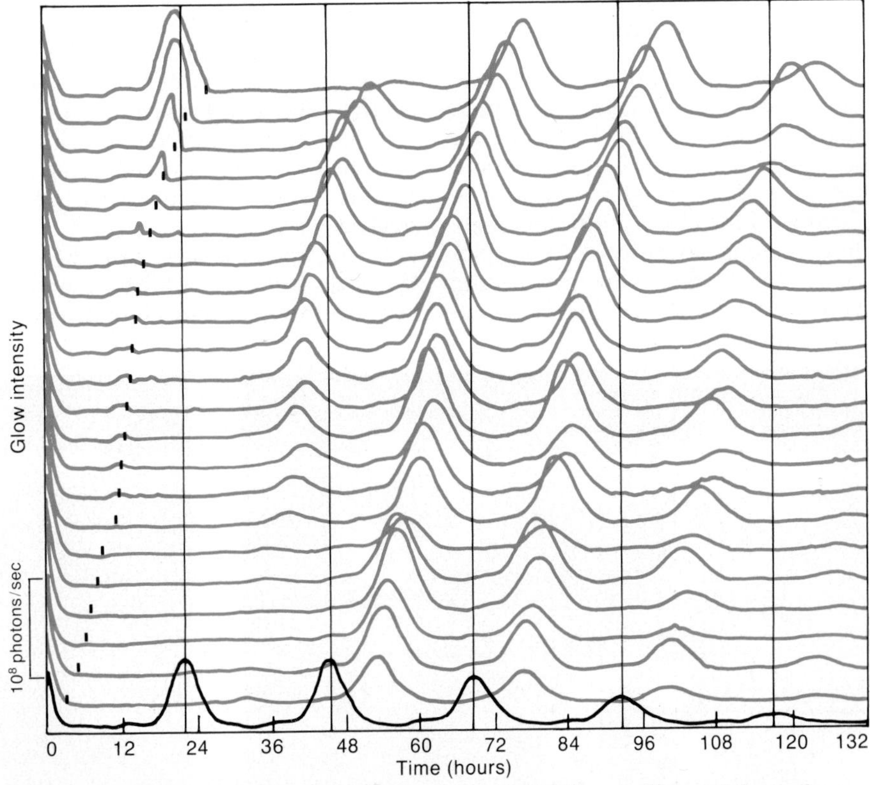

Figure 4. The circadian rhythms of bioluminescence in 22 different cultures of *Gonyaulax* represented here were altered by exposure to an inhibitor of protein synthesis, anisomycin. The cultures, which were initially in phase with the control culture (*black curve*) and which remained in constant light conditions throughout the experiment, were exposed to the drug for one hour at the times indicated by the black bars. The new phases generally parallel the times at which the drug was applied, indicating that the clock is reset by the drug and therefore that the clock depends on the synthesis of one or more proteins. (After Taylor et al. 1982.)

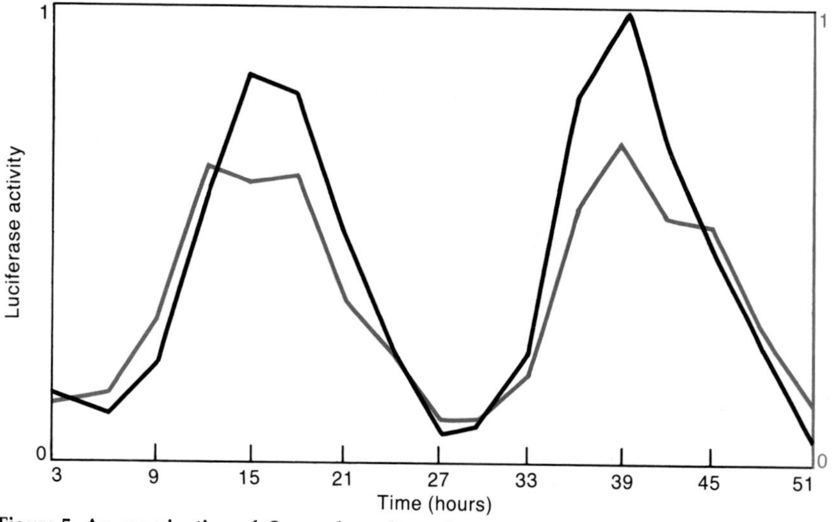

Figure 5. An examination of *Gonyaulax* cultures in constant light shows that the level of activity of luciferase, the enzyme that catalyzes the circadian bioluminescence in this alga, is itself rhythmic. The cyclic activity of the enzyme can be attributed to a corresponding rhythm in its amounts. Hence the synthesis and degradation of this enzyme is, at least in part, the method by which the endogenous clock controls the daily rhythm of bioluminescence. (After Johnson et al. 1984.)

even be a component of *Aplysia*'s clock mechanism itself.

Characterizing the coupling pathway

The organization of circadian systems illustrated in Figure 2 implies that another productive strategy for disclosing their mechanisms would be to select an overt rhythm, a "hand" of the clock, and to elucidate the coupling pathway and the biochemistry involved by proceeding backward step-by-step to the pacemaker. Quite apart from its usefulness in unveiling the central mechanism, the question of how the clock controls its overt rhythms is interesting in its own right. As a strategy to discover the mechanism of the clock, however, it has rarely been used, because the overt rhythms are so complex in many systems that no obvious biochemical correlate of the rhythm can be measured and tracked.

However, *Gonyaulax* has been a favorable subject for this strategy, because there are biochemical correlates of its overt rhythms. For example, knowledge of how the clock controls photosynthesis in *Gonyaulax* has recently implicated one of the two light reaction systems in photosynthesis (photosystem II) as the site of regulation (Samuelsson et al. 1983). We have concentrated on the coupling pathway of the *Gonyaulax* clock to bioluminescence, a relatively simple reaction consisting of two key components: an enzyme, luciferase, and a substrate, luciferin; their reaction with oxygen results in light emission:

$$\text{luciferin} + O_2 \xrightarrow{\text{luciferase}} \text{light}$$

The first experiments demonstrated that the activity of the enzyme is rhythmic (Hastings and Bode 1962). The rhythmic activity might have been controlled by any of a number of different mechanisms, examples of which include the modification of the enzyme molecule, such as by phosphorylation or methylation, and the activation or inhibition by some other molecule; but we have found rather that the intracellular concentration of the enzyme oscillates (Dunlap and Hastings 1981; Johnson et al. 1984). Therefore, more enzyme activity is measured at nighttime simply because more enzyme is present (Fig. 5). This clock-coupled enzyme thus undergoes daily synthesis and degradation.

In addition, the amount of the substrate molecule, luciferin, also un-

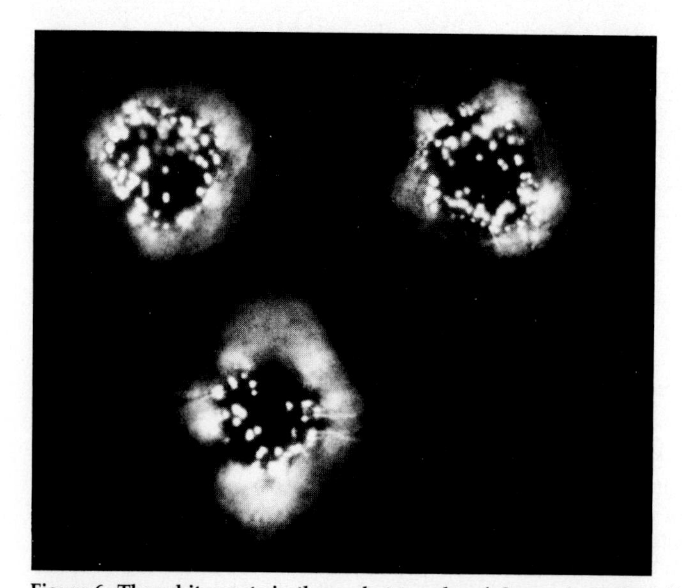

Figure 6. The white spots in these photographs of *Gonyaulax* cells indicate the fluorescence and thus the localization of luciferin, the substrate that reacts with the enzyme luciferase and oxygen to produce light. The intensity of luciferin fluorescence is greater in the night phase (*left*) of the circadian cycle of bioluminescence than in the day phase (*right*). Thus, like the amount of luciferase (Fig. 5), the intracellular amount of luciferin oscillates during the daily cycle. (From Johnson et al. 1985).

dergoes a daily cycle, as measured by bioluminescence in vitro (Hastings and Bode 1962). This observation was recently confirmed by measuring in vivo the intracellular concentration of luciferin with an assay based on its endogenous fluorescence, as illustrated in Figure 6 (Johnson et al. 1985). Therefore, the clock rhythmically modulates the intracellular concentrations of both bioluminescence reactants so that the capability of the cell to emit light upon stimulation is also rhythmic.

Now that the intracellular concentrations of luciferin and luciferase are known to be rhythmic, the next step is to determine how luciferase synthesis is turned on and off. In particular, the question is whether the control is transcriptional, in which case luciferase mRNA concentration would be rhythmic, or translational, with luciferase mRNA concentration being constant but translated rhythmically. An answer may be provided by cloning luciferase cDNA and using it as a probe to measure luciferase mRNA at different circadian times.

Studies of clock-controlled cellular compounds have been productive in other organisms also. Goto (1984) has found that in the duckweed *Lemna* the intracellular concentration of the coenzyme NADP oscillates. This cycle of a coenzyme level appears to be due to a rhythm of activity of the enzyme NAD kinase, which phosphorylates the coenzyme. This is an interesting enzyme because its activity is exquisitely sensitive to changes in calcium ions and calmodulin. Goto proposes that the rhythm of NADP is controlled by rhythmic changes in calcium ions, calmodulin, or both. If true, many metabolic pathways in duckweed, and perhaps other organisms as well, might be controlled by a coupling pathway involving calmodulin and calcium ions.

Another clock-controlled enzyme that has been intensively studied is N-acetyltransferase (NAT), the enzyme that limits the overall rate in the synthetic pathway of the hormone melatonin. Melatonin appears to be involved in hormonal coupling within the circadian system of higher organisms, especially birds and lizards, and is probably also important in mediating photoperiodic responses of the reproductive system in these vertebrates as well as in mammals (Takahashi and Menaker 1984). Rhythmic melatonin synthesis seems to be a result of rhythmic NAT activity (Binkley 1983). Rhythmic activity of NAT has been observed in eyes of the toad *Xenopus* that have been isolated in vitro (Besharse and Iuvone 1983). But the best-characterized tissue exhibiting NAT rhythmicity is the vertebrate pineal gland; even when isolated in vitro, the pineal gland of some vertebrates still displays circadian rhythms of NAT activity and of melatonin secretion that are entrainable by light and temperature-compensated (Takahashi and Menaker 1984). Some evidence suggests that the rhythm of NAT activity, like that of *Gonyaulax* luciferase, is accomplished by daily synthesis and degradation of the enzyme (Binkley 1983). And in the next step backward in this coupling pathway, there are indications that the rhythm of NAT activity seems to be itself regulated by a rhythm of intracellular CAMP concentration in pineal cells (Takahashi and Menaker 1984).

Although synthesis and degradation of enzymes might seem wasteful, two factors in *Gonyaulax* make turnover of the enzyme level an appropriate mode of control: luciferase accounts for only 0.1 to 1.0% of the total soluble protein pool; and as a photosynthetic marine alga, *Gonyaulax* is probably limited by nitrogen availability but is not energy-limited. These facts suggest that *Gonyaulax* may conserve nitrogen by degrading enzymes whose function is temporarily unnecessary and by recycling their amino acids while expending abundant photosynthetic energy.

However, the control of luciferase by way of turnover may be an example of a more general strategy whereby the clock regulates enzymes. In modulating metabolic pathways, the clock regulates the activity of enzymes that are rate-limiting for their respective pathways (e.g., *Gonyaulax* luciferase, pineal NAT, liver tyrosine aminotransferase, and liver HMG CoA reductase). Rate-limiting enzymes are usually degraded rapidly, as with tyrosine aminotransferase and HMG CoA reductase, which have a half-life of 2 to 3 hours. If the clock regulates rate-limiting enzymes whose half-lives are only a small fraction of the circadian period, enzyme turnover may be a common mode of circadian control over biochemical pathways (Johnson et al. 1984).

The case of vertebrate NAT also illustrates the advantages of developing a tissue or cell culture that exhibits a circadian oscillation in vitro. Not only have in vitro systems served as the ultimate test of self-sustained circadian oscillation within the suprachiasmatic nuclei of mammals, the pineal gland of birds and lizards, and the eyes of amphibians and molluscs, but they will undoubtedly advance our understanding of the clock's biochemistry in higher organisms (Jacklet 1984; Takahashi and Menaker 1984).

Realistic modeling of the circadian clock requires new experimental data, a requirement that may well be satisfied by following the three strategies outlined here—perturbing the clock directly or characterizing the entrainment or coupling pathways. These approaches have yielded new insights into the biochemical nature of circadian systems, but all have limitations as well. Thus, the wisest plan for future research is to focus on organisms that allow more than one approach.

In addition, we must not lose sight of the possibility that the clock mechanism might not be the same in different organisms. The amalgamation of experimental results from unicellular plants, molluscan eyes, and vertebrate pinealocytes may have pitfalls. As Pittendrigh (1976) points out, the assumption that the pacemaking mechanism is conserved through evolution is largely based on its phenomenological properties such as entrainability and temperature compensation. Any functional clock, however, must have these properties, and the similarities of clocks in different phyla could therefore be a result of convergent evolution. Furthermore, future models should also take into account the possibility that many different cellular processes may be critically involved, and thus that several existing models may need to be combined (Schweiger and Schweiger 1977).

It is an exciting time to be investigating circadian clocks. Molecular genetics promises new insight into the mechanism of circadian rhythmicity, and the development of new tissue-culture systems should permit a cellular analysis of the vertebrate clock. Moreover, other aspects of circadian biology, such as the implementation of the lessons of circadian organization into public health policy and mental health therapy, are being pursued with new vigor (Moore-Ede et al. 1983). Clock research is ticking.

References

Bargiello, T. A., and M. W. Young. 1984. Molecular genetics of a biological clock in *Drosophila*. PNAS 81:2142–46.

Bargiello, T. A., F. R. Jackson, and M. W. Young. 1984. Restoration of circadian behavioral rhythms by gene transfer in *Drosophila*. *Nature* 312:752–54.

Besharse, J. C., and P. M. Iuvone. 1983. Circadian clock in *Xenopus* eye controlling retinal serotonin N-acetyltransferase. *Nature* 305: 133–35.

Binkley, S. 1983. Rhythms in ocular and pineal N-acetyltransferase: A portrait of an enzyme clock. *Comp. Biochem. Physiol.* 75A:123–29.

Brown, F. A. 1972. The "clocks" timing biological rhythms. *Am. Sci.* 60:756–66.

Czeisler, C. A., G. S. Richardson, J. C. Zimmerman, M. C. Moore-Ede, and E. D. Weitzman. 1981. Entrainment of human circadian rhythms by light-dark cycles: A reassessment. *Photochem. Photobiol.* 34:239–47.

Czeisler, C. A., M. C. Moore-Ede, and R. M. Coleman. 1982. Rotating shift work schedules that disrupt sleep are improved by applying circadian principles. *Science* 217:460–63.

Dunlap, J. C., and J. W. Hastings. 1981. The biological clock in *Gonyaulax* controls luciferase activity by regulating turnover. *J. Biol. Chem.* 256:10509–18.

Edmunds, L. N., and K. J. Adams. Clocked cell cycle clocks. *Science* 211:1002–13.

Engelmann, W. 1973. A slowing down of circadian rhythms by lithium ions. *Zeitschrift für Naturforschung* 28:733–36.

Eskin, A. 1977. Neurophysiological mechanisms involved in photo-entrainment of the circadian rhythm from the *Aplysia* eye. *J. Neurobiol.* 8:273–99.

———. 1979. Identification and physiology of circadian pacemakers. Circadian system of the *Aplysia* eye: Properties of the pacemaker and mechanisms of its entrainment. *Fed. Proc.* 38:2570–79.

Eskin, A., G. Corrent, C. -Y. Lin, and D. J. McAdoo. 1982. Mechanism for shifting the phase of a circadian rhythm by serotonin: Involvement of cAMP. PNAS 79:660–64.

Feldman, J. F. 1982. Genetic approaches to circadian clocks. *Ann. Rev. Plant Physiol.* 33:583–608.

Goto, K. 1984. Causal relationships among metabolic circadian rhythms in *Lemna*. *Zeitschrift für Naturforschung* 39c:73–84.

Halberg, F., E. Halberg, C. P. Barnum, and J. J. Bittner. 1959. Physiologic 24-hour periodicity in human beings and mice, the lighting regimen and daily routine. In *Photoperiodism*, ed. R. D. Withrow, pp. 803–78. AAAS.

Hastings, J. W. 1960. Biochemical aspects of rhythms: Phase shifting by chemicals. *Cold Spring Harbor Symp. Quant. Biol.* 25:131–43.

Hastings, J. W., and V. C. Bode. 1962. Biochemistry of rhythmic systems. *Ann. N. Y. Acad. Sci.* 98:876–89.

Hastings, J. W., and B. M. Sweeney. 1957. On the mechanism of temperature independence in a biological clock. PNAS 43:804–11.

Hobohm, V., G. Cornelius, W. Taylor, and L. Rensing. 1984. Is the circadian clock of *Gonyaulax* held stationary after a strong pulse of anisomycin? *Comp. Biochem. Physiol.* 79A:371–78.

Jacklet, J. W. 1984. Neural organization and cellular mechanisms of circadian pacemakers. *Int. Rev. Cytol.* 89:251–94.

Johnson, C. H., J. F. Roeber, and J. W. Hastings. 1984. Circadian changes in enzyme concentration account for rhythm of enzyme activity in *Gonyaulax*. *Science* 223:1428–30.

Johnson, C. H., S. Inoué, A. Flint, and J. W. Hastings. 1985. Compartmentalization of algal bioluminescence: Autofluorescence of bioluminescent particles in the dinoflagellate *Gonyaulax* as studied with image-intensified video microscopy and flow cytometry. *J. Cell Biol.* 100:1435–46.

Konopka, R. J., and S. Benzer. 1971. Clock mutants of *Drosophila melanogaster*. PNAS 68:2112–16.

Kronauer, R. E., C. A. Czeisler, S. F. Pilato, M. C. Moore-Ede, and E. D. Weitzman. 1982. Mathematical model of the human circadian system with two interacting oscillators. *Am. J. Physiol.* 242:R3–R17.

Kyriacou, C. P., and J. C. Hall. 1980. Circadian rhythm mutations in *Drosophila melanogaster* affect short-term fluctuations in the male's courtship song. PNAS 77:6729–33.

McMurry, L., and J. W. Hastings. 1972. No desynchronization among four circadian rhythms in the unicellular alga *Gonyaulax polyedra*. *Science* 175:1137–39.

Mizoguchi, A., and H. Ishizaki. 1982. Prothoracic glands of the saturniid moth *Samia cynthia ricini* possess a circadian clock controlling gut purge timing. PNAS 79:2726–30.

Moore-Ede, M. C., C. A. Czeisler, and G. S. Richardson. 1983. Circadian timekeeping in health and disease. *New Engl. J. Med.* 309:469–76, 530–36.

Pittendrigh, C. S. 1954. On temperature independence in the clock-system controlling emergence in *Drosophila*. PNAS 40:1018–29.

———. 1960. Circadian rhythms and the circadian organization of living systems. *Cold Spring Harbor Symp. Quant. Biol.* 25:159–84.

———. 1976. Circadian clocks: What are they? In *Dahlem Workshop on the Molecular Basis of Circadian Rhythms*, pp. 11–48. Berlin: Dahlem Konferenzen.

———. 1981. Circadian systems: General perspective. In *Handbook of Behavioral Neurobiology*, ed. J. Aschoff, vol. 4, pp. 57–80. Plenum.

Reddy, P., et al. 1984. Molecular analysis of the period locus in *Drosophila melanogaster* and identification of a transcript involved in biological rhythms. *Cell* 38:701–10.

Reinberg, A., and F. Halberg. 1971. Circadian chronopharmacology. *Ann. Rev. Pharmacol.* 11:455–92.

Samuelsson, G., B. M. Sweeney, H. A. Matlick, and B. B. Prezelin. 1983. Changes in photosystem II account for the circadian rhythm in photosynthesis in *Gonyaulax polyedra*. *Plant Physiol.* 73:329–31.

Scheving, L. E., E. R. Burns, J. E. Pauly, F. Halberg, and E. Haus. 1977. Survival and cure of leukemic mice after circadian optimization of treatment with cyclophosphamide and 1-B-D-arabinofuranosylcytosome. *Cancer Res.* 37:3648–55.

Schweiger, H.-G., and M. Schweiger. 1977. Circadian rhythms in unicellular organisms: An endeavor to explain the molecular mechanism. *Int. Rev. Cytol.* 51:315–42.

Shin, H. -S., T. A. Bargiello, B. T. Clark, F. R. Jackson, and M. W. Young. 1985. An unusual coding sequence from a *Drosophila* clock gene is conserved in vertebrates. *Nature* 317:445–48.

Sulzman, F. M., D. Ellman, C. A. Fuller, M. C. Moore-Ede, and G. Wassmer. 1984. *Neurospora* circadian rhythms in space: A reexamination of the endogenous-exogenous question. *Science* 225:232–34.

Sweeney, B. M. 1983. Circadian time-keeping in eukaryotic cells, models and hypotheses. *Prog. Phycol. Res.* 2:189–225.

Takahashi, J. S., P. J. DeCoursey, L. Baumann, and M. Menaker. 1984. Spectral sensitivity of a novel photoreceptive system mediating entrainment of mammalian circadian rhythms. *Nature* 308:186–88.

Takahashi, J. S., and M. Menaker. 1984. Circadian rhythmicity: Regulation in the time domain. In *Biological Regulation and Development*, ed. R. Goldberger and K. Yamamoto, vol. 3B, pp. 285–303. Plenum.

Taylor, W., R. Krasnow, J. C. Dunlap, H. Broda, and J. W. Hastings. 1982. Critical pulses of anisomycin drive the circadian oscillator in *Gonyaulax* towards its singularity. *J. Comp. Physiol.* 148:11–25.

Truman, J. W. 1984. Physiological aspects of the two oscillators that regulate the timing of eclosion in moths. In *Photoperiodic Regulation of Insect and Molluscan Hormones*, Ciba Foundation Symposium 104, pp. 221–32. London: Pitman.

Wehr, T. A., and F. K. Goodwin. 1983. Introduction; Biological rhythms in manic-depressive illness. In *Circadian Rhythms in Psychiatry*, ed. T. A. Wehr and F. K. Goodwin, pp. 1–15; 129–84. Boxwood Press.

Wever, R. A. 1979. *The Circadian System of Man*. Springer-Verlag.

Winfree, A. T. 1980. *The Geometry of Biological Time*. Springer-Verlag.

———. 1982. Human body clocks and the timing of sleep. *Nature* 297:23–27.

Woolum, J. C., and F. Strumwasser. 1983. Is the period of the circadian oscillator in the eye of *Aplysia* directly homeostatically regulated? *J. Comp. Physiol.* 151:253–59.

Zehring, W. A., et al. 1984. P-element transformation with *period* locus DNA restores rhythmicity to mutant, arrhythmic *Drosophila melanogaster*. *Cell* 39:369–76.

Part III
Evolutionary History and Behavioral Ecology

Behavioral ecology is the study of how behavior is influenced by natural selection in relation to ecological conditions. The six articles in this section introduce readers to the twin goals of behavioral ecology: reconstruction of historical pathways and determination of adaptive value. These are complementary, not mutually exclusive, ways to study behavior because they involve different levels of analysis (as defined in the article in Part I by Holekamp and Sherman). Researchers interested in the history of a behavior or in its adaptive value often employ the comparative method in their explorations of these issues.

For example, in the first article, Bert Hölldobler and Edward Wilson compare some remarkable ants that build silken nests in trees, rather like our familiar tent caterpillars. Among the four living weaver-ant genera there are many behavioral similarities but also some important differences. The authors examine these behavioral characteristics and try to figure out what changes occurred over evolutionary history that led to nest weaving. Detailed comparisons of adult and larval behaviors enable the authors to array the genera along a "phylogenetic grade" from least to most complex. For example, in the "advanced" genus *Oecophylla,* middle-aged larvae produce silk and workers use them as living shuttles, carrying the larvae to and fro to bind leaves together into a nest; by contrast, in the "primitive" genus *Dendromyrmex,* larvae produce silk only when they are about to pupate, and workers simply plop the larvae down and let them create small patches of silk by thrashing about.

Hölldobler and Wilson suggest that the primitive weaver ants may be on an ascending evolutionary trajectory that will carry them through an orderly series of steps, including stages exemplified by the genera that are intermediate in complexity (i.e., *Polyrhachis, Camponotus*). What are the merits of arraying species according to phylogenetic grade? Is evolution unidirectional, and if so why? Does natural selection tend to promote complexity, or is the idea that more complexity indicates further advancement perhaps an artifact of our admiration for technological complexity?

In any case, are "advanced" species really better adapted? If so, why are there so many "intermediate" and "primitive" species that appear to possess "imperfect" adaptations? Finally, why do you suppose only a handful of the estimated 20,000 species of ants have taken to building silken nests in the treetops? To answer this question, it would be useful to develop hypotheses about the possible adaptive significance of communal nest weaving in ants (and tent caterpillars).

Complex nest construction by weaver ants, like any phenotypic attribute, can be analyzed both in terms of its past history and its current fitness consequences. Hölldobler and Wilson treat these two types of evolutionary analyses as complementary. But in the next article, Stephen Jay Gould suggests that understanding a trait's history may make it unnecessary to discuss its current adaptive value. He scrutinizes both levels of analysis in detail and discusses how biologists can rigorously test evolutionary hypotheses. Although it is often thought that the "scientific method" depends entirely on carefully controlled experiments, Gould argues that Darwin had it right when he developed the comparative method, a nonexperimental approach to evolutionary hypothesis testing. This method takes advantage of the thousands of "natural experiments" that have occurred over evolutionary time: the formation of new species and the development of their special characteristics.

According to Gould it is possible to establish the evolutionary relationships among species by identifying shared traits with distinctive similarities that were probably inherited from the common ancestor of those species. For example, one can see in the flower parts of many beautiful and bizarre orchids features demonstrating that these parts derive from petals and sepals possessed by less spectacular plants. To Gould, this suggests that history may supercede current adaptive value as the evolutionary explanation for many traits. What do you think of this claim? For example, does the hypothesis that many human traits were inherited from Pleistocene ancestors render insignificant any attempt to analyze the current fitness consequences of those traits?

Having digested Gould's article, readers may wish to review Hölldobler and Wilson's use of the comparative method to reconstruct the history of nest weaving by ants. If Gould is correct that evolution is more like a bush than a ladder, are Hölldobler and Wilson mistaken to infer "progress" toward more complex forms of behavior? Is there any middle ground between these approaches?

In reconstructing the history of a trait, researchers typically find evidence implying that the current trait is "jury-rigged" from traits present in ancestral species. As a result, traits give the appearance of being designed by a committee that sought consensus rather than by an engineer who knew what the most appropriate design was. This being so, is Gould correct to say that evolved traits are generally less than optimally functional, and less than fully adaptive?

Finally, how can complicated structures, like an orchid flower or the vertebrate eye, evolve in stages when it would seem that anything less sophisticated than the current structures would not work at all? Don't orchid flowers and human eyes suggest that evolution, in fact, has a purpose or goal? To solve this paradox, try applying the comparative method to either of these structures.

A colony of honey bees is another outstanding example of functional complexity. Here thousands of workers must coordinate their efforts precisely and perform dozens of activities in the proper sequences if the colony is to survive and reproduce. Mark Winston and Keith Slessor spent a decade decoding the complex chemical "language" that controls the inner workings of a honey bee nest.

Queen bees are not physically imposing or aggressive, but they possess something very special—a pheromonal scent that suppresses worker reproduction and enforces loyalty to the queen. Winston and Slessor used traditional experimental procedures to establish precisely what chemicals—and in what proportions—are necessary to control the workers. Their painstaking chemical sleuthing paid off: The "essence of royalty" was finally isolated and identified as a complex mixture of three decenoic acids and two aromatic compounds.

When reading about Winston and Slessor's procedures for studying chemical communication, contrast them with those used to study vocal or visual communication. Why do you suppose honey bees rely so heavily on chemicals (instead of sound or visual cues) to coordinate their nesting activities? And why is the queen substance such a complicated chemical blend? Was this an historical accident, as Gould might argue, or could pheromone complexity be the adaptive product of worker–queen conflict?

In the nest, worker bees actively seek access to the queen and to her pheromones. This is somewhat curious, because the queen substance blocks worker reproduction. How would you explain the evolution of special behaviors associated with procuring the pheromone when the outcome of those behaviors is reproductive suppression? Winston and Slessor note that workers eat the queen substance, and imply that they do so to remove it from circulation. Can you think of any alternative hypotheses to explain "internalization" by the workers (and by the queen herself)?

An interesting property of queen substance is that it causes workers to aggregate, for example during swarming. Winston and Slessor realized the practical value of being able to attract workers at will. By spraying synthetic queen pheromone on cultivated plants, they were able to improve pollination and increase crop yields. Clearly behavioral ecologists are not all "ivory tower" types!

Chemical sleuthing is taken a step further in the article by Aileen Morse. Consider, as she did, the plight of a baby abalone the size of a speck of dust that is trying to find a home on the vast, heterogeneous bottom of the Pacific Ocean. For seven days the larva is whirled and tossed by currents and tides. Then, suddenly, it drops out of the water column and comes to rest smack dab on a patch of coralline red algae, the preferred larval habitat. Was this an accident? A lucky chance perhaps? Apparently not.

Morse and her colleagues have discovered that larval abalone possess complex and sophisticated chemosensory systems. In particular, they are highly sensitive to a chemical that emanates from coralline algae, and they will not settle unless it is present. The chemical triggers a cascade of biochemical changes in the larva that result in a metamorphosis: the larva stops swimming and begins the transformation into the adult form—a bottom-dwelling gastropod.

The author's studies have exploded the myth that invertebrate larvae generally settle randomly. After reading Morse's article, you may wish to consider why habitat selection is so critical for larval life stages. And what do you suppose it is about coralline red algae that makes it the preferred substrate for so many types of larvae—abalones, honeycomb worms, and lettuce corals, among others?

Morse points out interesting parallels between the signal–receptor systems of larval abalones and those that exist in the vertebrate nervous system. Are these similarities analogies or homologies, in Gould's terminology? What, if anything, can studies of the biochemistry of settling by invertebrate larvae tell us about the mechanisms and evolution of cell–cell communication in our own bodies?

Habitat and foraging selectivity also come into play in the next article, but in a quite different way. Kim Hill and Magdalena Hurtado spent several years living with a tribe of hunter–gatherers, the Ache, in the rainforests of Paraguay. By following hunting parties the authors discovered that individuals did not forage randomly. Indeed, both men and women were highly selective, seeking animals and plants that yielded the highest caloric returns per unit of time devoted to the chase or harvest, while ignoring many species that are edible but hard to obtain.

Ache foraging patterns were thus predictable from simple economic models based on the assumption that individuals attempt to maximize their net rate of food acquisition. Just prior to the authors' arrival in Paraguay, a few of the Ache had obtained shotguns. Interestingly, although the firearms made it easier to bring down certain elusive prey, access to guns did not change the general foraging "rules," based on energy maximization.

The adaptability of the Ache to firearms did not surprise Hill and Hurtado, but something else did. When successful hunters returned to the group's encampment, they butchered their kill, divided it up, and then *gave it all away!* Hunters generally didn't eat what they had worked so hard to obtain. You may wish to consider why Ache men are apparently so selfless. Is there any conceivable way that successful hunters could benefit by giving away their prizes? What is there to prevent a man from "sluffing off" and allowing other tribesmen to provide for his family's needs?

Hill and Hurtado believe that studies of hunter–gatherers offer a window on our own evolutionary past. The reasoning is somewhat similar to that employed by Höll-dobler and Wilson in comparing "primitive" and "advanced" nest construction by weaver ants. What do you think about applying a sort of "sociological grade" to our own species? Are hunter–gatherer societies really less complex than our own? If so, then why did our massive brains evolve under such "simple" societal conditions?

Hill and Hurtado emphasize that different hunter–gatherer societies have different social and demographic characteristics. For example, the Ache are well fed, the birth rate is high, and childhood mortality is low, whereas among the !Kung of eastern Africa all these trends are reversed. What hypotheses would you propose to account for such differences among hunter–gatherers? Could you test your hypotheses by making predictions about what traits should occur in different societies subjected to similar selective pressures? Despite many intriguing differences among the Ache, the Hiwi, and the !Kung, in all three groups men hunt for meat and women gather or cultivate plant foods. Is there an adaptive explanation for this sexual division of labor? How could you make comparisons among these and still other cultures to test your hypothesis?

In considering these questions you have been using the comparative method—not to reconstruct the history of human behaviors, but rather to test ideas about their adaptive value. Likewise, Thomas Seeley uses the comparative method to help understand differences in the antipredator behaviors of various species of honey bees. Seeley points out that although honey bees are well known as sources of honey the world over, how they live in nature has only recently been studied. In our North American species (*Apis mellifera*), foragers search an area of some 25 km^2 to obtain the pollen and nectar that will sustain the colony through the cold winter months. Colonies maintain a high "body" (nest) temperature, allowing the bees to start reproducing in January. This, in turn, enables colonies to get large enough to divide (swarm) in May and June, giving the young colonies ample time to collect and store honey before the next winter sets in. The bees' "dance language" and chemical communication systems (recall the Winston and Slessor article), respectively, focus the foragers' efforts on flower patches that yield the highest net energy returns and coordinate the efforts of thousands of workers.

In addition to their social skill in foraging, honey bees also work together to deal with predators. Seeley makes detailed comparisons of predator avoidance behavior among three tropical species that are closely related to *A. mellifera*. The three Asian bees differ considerably in body size, number of workers per nest, and in nest conspicuousness and dispersion. In general, large-bodied, aggressive bees build huge, long-lasting exposed-comb nests in conspicuous places, whereas diminutive, timid species hide their small nests in the foliage and move them frequently.

North American honey bees have behavioral characteristics that combine those of several tropical bees (e.g., *A. mellifera* hide their nests in tree holes but their workers aggressively defend the nest). How would our bees fit into Seeley's comparative picture? The Africanized honey bees (so-called "killer bees") that have recently invaded the United States are extremely aggressive around their nest. Does knowing this enable you to predict how they live in their native habitat (i.e., open vs. enclosed nests, high vs. low nests, large vs. small colonies) and what types of predators attack them? Could such knowledge help you devise ways to avoid being harmed or enable you to control the bees without widespread use of pesticides?

Although Seeley uses the comparative method, he avoids applying the idea of phylogenetic grade to the nesting biology of tropical honey bees, in part because he is not trying to explain the history of behavioral changes in the genus *Apis*. But why not array antipredator adaptations on a continuum from "primitive" to "advanced," like the nest weaving traits of ants? Is there something different about predator avoidance responses, or is there something inadequate about the phylogenetic grade idea?

Twenty-five years ago an evolutionary biologist named Michael Ghiselin published *The Triumph of the Darwinian Method*, a book championing the comparative method as a way to answer questions about adaptation. The six articles in this section testify to the range and power of the approach Darwin pioneered as they successfully apply the comparative method to creatures as different as weaver ants and abalones, honey bees and humans. Indeed, because gaining knowledge about complex behaviors (e.g., social behaviors, mating systems) from the fossil record is extremely difficult, the comparative method is virtually the only way to reconstruct historical pathways, and it can be a useful way to infer adaptive significance as well.

Bert Hölldobler
Edward O. Wilson

The Evolution of Communal Nest-Weaving in Ants

Steps that may have led to a complicated form of cooperation in weaver ants can be inferred from less advanced behavior in other species

One of the most remarkable social phenomena among animals is the use of larval silk by weaver ants of the genus *Oecophylla* to construct nests. The ants are relatively large, with bodies ranging up to 8 mm in length, and exclusively arboreal. The workers create natural enclosures for their nests by first pulling leaves together (Fig. 1) and then binding them into place with thousands of strands of larval silk woven into sheets. In order for this unusual procedure to succeed, the larvae must cooperate by surrendering their silk on cue, instead of saving it for the construction of their own cocoons. The workers bring nearly mature larvae to the building sites and employ them as living shuttles, moving them back and forth as they expel threads of silk from their labial glands.

The construction of communal silk nests has clearly contributed to the success of the *Oecophylla* weaver ants. It permits colonies to attain populations of a half million or more, in spite of the large size of the workers, because the ants are freed from the spatial limitations imposed on species that must live in beetles' burrows, leaf axils (the area between the stems of leaves and the parent branch), and other preformed vegetative cavities. This advance, along with the complex recruitment system that permits each colony to dominate up to several trees at the same time, has helped the weaver ants to become among the most abundant and successful social insects of the Old World tropics (Hölldobler and Wilson 1977a, b, 1978; Hölldobler 1979). A single species, *O. longinoda*, occurs across most of the forested portions of tropical Africa, while a second, closely related species, *O. smaragdina*, ranges from India to Queensland, Australia, and the Solomon Islands. The genus is ancient even by venerable insect standards: two species are known from Baltic amber of Oligocene age, about 30 million years old (Wheeler 1914). *O. leakeyi*, described from a fossil colony of Miocene age (approximately 15 million years old) found in Kenya, possessed a physical caste system very similar to that of the two living forms (Wilson and Taylor 1964).

Our recent studies, building on those of other authors, have revealed an unexpectedly precise and stereotyped relation between the adult workers and the larvae. The larvae contribute all their silk to meet the colony's needs instead of their own. They produce large quantities of the material from enlarged silk glands early in the final instar rather than at its end, thus differing from cocoon-spinning ant species, and they never attempt to construct cocoons of their own (Wilson and Hölldobler 1980). The workers have taken over almost all the spinning movements from the larvae, turning them into passive dispensers of silk.

It would seem that close attention to the exceptional properties of *Oecophylla* nest-weaving could shed new light on how cooperation and altruism operate in ant colonies, and especially on how larvae can function as an auxiliary caste. In addition, a second, equally interesting question is presented by the *Oecophylla* case: How could such extreme behavior have evolved in the first place? As is the case with the insect wing, the vertebrate eye, and other biological prodigies, it is hard to conceive how something so complicated and efficient in performance might be built from preexisting structures and processes. Fortunately, other phyletic lines of ants have evolved communal nest-weaving independently and to variably lesser degrees than *Oecophylla*, raising the prospect of reconstructing the intermediate steps leading to the extreme behavior of weaver ants. These lines are all within the Formicinae, the subfamily to which *Oecophylla* belongs. They include all the members of the small Neotropical genus *Dendromyrmex*, the two Neotropical species *Camponotus* (*Myrmobrachys*) *senex* and *C.* (*M.*) *formiciformis*, which are aberrant members of a large cosmopolitan genus, and various members of the large and diverse Old World tropical genus *Polyrhachis*.

Two additional but doubtful cases have been reported outside the Formicinae. According to Baroni Urbani (1978), silk is used in the earthen nests of some Cuban species of *Leptothorax*, a genus of the subfamily Myrmicinae. However, the author was uncertain whether the material is obtained from larvae or from an extraneous source such as spider webs. Since no other myrmicine is known to produce silk under any circumstances, the latter alternative seems the more probable.

The authors take pleasure in dedicating this article to Caryl P. Haskins, fellow myrmecologist and distinguished scientist and administrator, on the occasion of his seventy-fifth birthday and his retirement from the chairmanship of the Board of Editors of American Scientist.

Bert Hölldobler is Alexander Agassiz Professor of Zoology at Harvard University. After completing his Dr. rer. nat. at the University of Würzburg in 1965 and his Dr. habil. at the University of Frankfurt in 1969, he served at the latter institution as privatdocent and professor until 1973, when he joined the Harvard faculty. Edward O. Wilson is Frank B. Baird Jr. Professor of Science and Curator in Entomology of the Museum of Comparative Zoology, Harvard University. He received his Ph.D. from Harvard in 1955, held a Junior Fellowship in the Society of Fellows during 1953–56, and has served on the faculty continuously since that time. Address: Museum of Comparative Zoology, Harvard University, Cambridge, MA 02138.

Similarly, the use of silk to build nests was postulated for the Javan ant *Technomyrmex bicolor textor*, a member of the subfamily Dolichoderinae, in an early paper by Jacobson and Forel (1909). But again, the evidence is from casual field observations only, and the conclusion is rendered unlikely by the fact that no other dolichoderines are known to produce silk.

During the past ten years we have studied the behavior of both living species of *Oecophylla* in much greater detail than earlier entomologists, and have extended our investigations to two of the other, poorly known nest-weaving genera, *Dendromyrmex* and *Polyrhachis*. This article brings together the new information that resulted from this research and some parallel findings of other authors, in a preliminary characterization of the stages through which the separate evolving lines appear to have passed.

In piecing together our data, we have utilized a now-standard concept in organismic and evolutionary biology, the phylogenetic grade. The four genera of formicine ants we have considered are sufficiently distinct from each other on anatomical evidence as to make it almost certain that the communal nest-weaving displayed was in each case independently evolved. Thus it is proper to speak of the varying degrees of cooperative behavior and larval involvement not as the actual steps that led to the behavior of *Oecophylla* but as grades, or successively more advanced combinations of traits, through which autonomous evolving lines are likely to pass. Other combinations are possible, even though not now found in living species, and they might be the ones that were actually traversed by extreme forms such as *Oecophylla*. However, by examining the behavior of as many species and phyletic lines as possible, biologists are sometimes able to expose consistent trends and patterns that lend convincing weight to particular evolutionary reconstructions. This technique is especially promising in the case of insects, with several million living species to sample. Within this vast array there are more than 10,000 species of ants, most of which have never been studied, making patterns of ant behavior exceptionally susceptible to the kind of analysis we have undertaken and are continuing to pursue on communal nest-building.

The highest grade of cooperation

The studies conducted on *Oecophylla* prior to our own were reviewed by Wilson (1971) and Hemmingsen (1973). In essence, nest-weaving with larval silk was discovered in *O. smaragdina* independently by H. N. Ridley in India and W. Saville-Kent in Australia, and was subsequently described at greater length in a famous paper by Doflein (1905). Increasingly detailed accounts of the behavior of *O. longinoda*, essentially similar to that of *O. smaragdina*, were provided by Ledoux (1950), Chauvin (1952), Sudd (1963), and Hölldobler and Wilson (1977a).

The sequence of behaviors by which the nests are constructed can be summarized as follows. Individual workers explore promising sites within the colony's territory, pulling at the edges and tips of leaves. When

Figure 1. To make a nest out of leaves and larval silk, worker ants of the species *Oecophylla smaragdina*, the Australian green tree ant, first choose a pliable leaf. They then form a row and pull in unison, as shown in the photograph, until they force two leaves to touch or one leaf to curl up on itself. (All photographs are by Bert Hölldobler unless otherwise indicated.)

Figure 2. If a single ant cannot bridge the gap between two leaves to be used in a nest, *O. smaragdina* workers arrange themselves in chains and pull together to close the gap. The photograph at the left indicates how many individuals can be involved in this stage of the work; at the right, several parallel chains of ants are shown in more detail.

a worker succeeds in turning a portion of a leaf back on itself, or in drawing one leaf edge toward another, other workers in the vicinity join the effort. They line up in a row and pull together (Fig. 1), or, in cases where a gap longer than an ant's body remains to be closed, they form a living chain by seizing one another's petiole (or "waist") and pulling as a single unit. Often rows of chains are aligned so as to exert a powerful combined force (Fig. 2). The formation of such chains of ants to move objects requires intricate maneuvering and a high degree of coordination. So far as is known, it is unique to *Oecophylla* among the social insects.

When the leaves have been maneuvered into a tentlike configuration, workers carry larvae out from the interior of the existing nests and use them as sources of silk to bind the leaves together (Figs. 3, 4). Our previous studies (Wilson and Hölldobler 1980) showed that the *O. longinoda* larvae recruited for this purpose are all in the final of at least three instars, and have heads in excess of 0.5 mm wide. However, their bodies (exclusive of the rigid head capsule) are smaller than those of the larvae at the very end of the final instar, which are almost ready to turn into prepupae and commence adult development. Thus the larvae used in nest-weaving are well along in development and possess large silk glands, but they have not yet reached full size and hence are more easily carried and manipulated by the workers.

species

In *O. longinoda*, all the workers we observed with spinning larvae have been majors, the larger adults that possess heads between 1.3 and 1.8 mm in width. Hemmingsen (1973) reported that majors of *O. smaragdina* perform the weaving toward the exterior, while minor workers—those with heads 1–1.2 mm wide—weave on the inner surfaces of the leaf cavities. We have observed only major workers performing the task in *O. longinoda*, but admittedly our studies of interior activity have been limited. Hemmingsen also recorded that exterior weaving is rare during the daytime but increases sharply at night, at least in the case of *O. smaragdina* working outdoors in Thailand. We have seen frequent exterior weaving by *O. lon-*

ginoda during the day in a well-lit laboratory, as well as by *O. smaragdina* outdoors in Queensland.

In recent studies reported here for the first time, we followed the spinning process of *O. longinoda* through a frame-by-frame analysis of 16-mm motion pictures taken at 25 frames per second. The most distinctive feature of the larval behavior, other than the release of the silk itself, is the rigidity with which the larva holds its body. There is no sign of the elaborate bending and stretching of the body or of the upward thrusting and side-to-side movements of the head that characterize cocoon-spinning in other formicine ant larvae, particularly in *Formica* (Wallis 1960; Schmidt and Gürsch 1971). Rather, the larva keeps its body stiff, forming a straight line when viewed from above but a slightly curved, **S**-shaped line when seen from the side, with its head pointing obliquely downward as shown in Figure 4. Occasionally the larva extends its head for a very short distance when it is brought near the leaf surface, giving the impression that it is orienting itself more precisely at the instant before it releases the silk. The worker holds the larva in its mandibles between one-fourth and one-third of the way down the larva's body from the head, so that the head projects well out in front of the worker's mandibles.

The antennae of the adult workers are of an unusual conformation that facilitates tactile orientation along the edges of leaves and other vegetational surfaces. The last four segments are shorter relative to the eight segments closest to the body than in other ants we have examined, including even communal silk-spinning formicines such as *Camponotus senex* and *Polyrhachis acuta*. They are also unusually flexible and can be actively moved in various directions in a fashion seen in many solitary wasps.

As the worker approaches the edge of a leaf with a larva in its mandibles, the tips of the antennae are brought down to converge on the surface in front of the ant. For 0.2 ± 0.1 sec ($\bar{x} \pm$ SD, n = 26, involving a total of 4 workers), the antennae play along the surface, much in the manner of a blindfolded person feeling the edge of a table with his hands. Then the larva's head is touched to the surface and held in contact with the leaf for 1 sec (0.9 ± 0.2 sec, n = 26). During this time, the tips of the worker's antennae are vibrated around the larva's head, stroking the leaf surface and touching the larva's head about 10 times (9.2 ± 3.6, n = 26). At some point the larva releases a minute quantity of silk, which attaches to the leaf surface.

About 0.2 sec before the larva is lifted up again, the worker spreads and raises its antennae. Then it carries the larva directly to the edge of the other leaf, causing the silk to be drawn out as a thread. While moving between leaves, the worker holds its antennae well away from the head of the larva. When it reaches the other leaf, it repeats the entire procedure exactly, except that the larva's head is held to the surface for only 0.5 sec (0.4 ± 0.01 sec, n = 26); during this phase the worker's antennae touch the larva about 5 times (5.2 ± 2.4, n = 26). In other words, the workers alternate between a longer time spent at one leaf surface and a shorter time at the opposing surface.

To summarize, the weaving behavior of the *Oecophylla* worker is even more complicated, precise, and distinctive than realized by earlier investigators. The movements are rigidly stereotyped in form and sequence. The antennal tips are used for exact tactile orientation, a "topotaxis" somewhat similar to that employed by honeybee workers to assess the thickness of the waxen walls of the cells in the comb (Lindauer and Martin 1969). The worker ant also appears to use its flexible antennal tips to communicate with the larva, presumably to induce it to release the silk at the right moment. Although we have no direct experimental proof of this effect, we can report an incidental observation consistent with it. One worker we filmed held the larva upside down, so that the front of the larva's head and its silk-gland openings could not touch the surface or be stroked by the antennal tips. The worker went through the entire sequence correctly, but the larva did not release any silk.

For its part, the larva has evolved distinctive traits and behaviors that serve communal weaving. It releases some signal, probably chemical, that identifies it as being in the correct phase of the final instar. When a worker picks it up, it assumes an unusual **S**-shaped posture. And when it is held against the surface of a leaf and touched by a worker's antennae, it releases silk, in a context and under circumstances quite o... the ordinary for most immat... sects.

Figure 3. The nest of the African *Oecophylla longinoda*, the most sophisticated in design of weaver ants' nests, is formed basically of living leaves and stems bound together with larval silk. Some of the walls and galleries are constructed entirely of the silk.

Figure 4. Once leaves have been pulled together to form a nest, the workers hold them in place with larval silk. A simple form of weaving is practiced by the workers of an Australian *Polyrhachis* species similar to *doddi*; at the top, one worker holds a larva above the surface, allowing it to perform most of the weaving movements. The most sophisticated type of weaving has been developed by *O. smaragdina*; in the bottom photograph, *O. smaragdina* workers perform almost all the movements while the larvae serve principally as passive shuttles.

Intermediate steps

The existence of communal nest-weaving in *Polyrhachis* was discovered in the Asiatic species *Polyrhachis* (*Myrmhopla*) *dives* by Jacobson (Jacobson and Wasmann 1905). However, few details of the behavior of these ants have been available until a recent study by Hölldobler, reported here for the first time.

A species of *Polyrhachis* (*Cyrtomyrma*), tentatively classified near *doddi*, was observed in the vicinity of Port Douglas, Queensland, where its colonies are relatively abundant. The ants construct nests among the leaves and twigs of a wide variety of bushes and trees (Fig. 5). Most of the units are built between two opposing leaves, but often only one leaf serves as a base or else the unit is entirely constructed of silk and is well apart from the nearest leaves.

Polyrhachis ants have never been observed to make chains of their own bodies or to line up in rows in the manner routine for *Oecophylla*. Occasionally a single *Polyrhachis* worker pulls and slightly bends the tip or edge of a leaf, but ordinarily the leaves are left in their natural position and walls of silk and debris are built between them.

The weaving of *Polyrhachis* also differs markedly from that of *Oecophylla*. The spinning larvae are considerably larger and appear to be at or near the end of the terminal instar (Fig. 4). The workers hold them gently from above, somewhere along the forward half of their body, and allow the larvae to perform all of the spinning movements. In laying silk on the nest wall, the larvae use a version of the cocoon-spinning movements previously observed in the larvae of *Formica* and other formicine ants. Like these more "typical" species, which do not engage in communal nest-building, *Polyrhachis* larvae begin by protruding and retracting the head relative to the body segments while bending the forward part of the body downward. Approximately this much movement is also seen in *Oecophylla* larvae prior to their being touched to the surface of a leaf.

The *Polyrhachis* larvae are much more active, however, executing most of the spinning cycle in a sequence very similar to that displayed by cocoon-spinning formicines. Each larva begins with a period of bending and stretching, then returns to its original position through a series of arcs directed alternately to the left and right; in sum, its head traces a rough figure eight. Because the larvae are held by the workers, the movements of their bodies are restricted. They cannot complete the "looping-the-loop" and axial rotary movements described by Wallis (1960), by which larvae of other formicine ants move around inside the cocoon to complete its construction. In fact, the *Polyrhachis* larvae do not build cocoons. They pupate in the naked state, having contributed all their expelled silk to the communal nest. In this regard they fall closer to the advanced *Oecophylla* grade than to the primitive *Dendromyrmex* one, discussed below.

Polyrhachis ants are also intermediate between *Oecophylla* and *Dendromyrmex* in another important respect. The *Polyrhachis* workers do not move the larvae constantly like living shuttles as in *Oecophylla*, nor do they hold the larvae in one position for long periods of time or leave them to spin on their own as in *Dendromyrmex*. Rather, each spinning larva is held by a worker in one spot or moved slowly forward or to the side for a variable period of time (range 1–26 sec, mean 8 sec, SD 7.1 sec, n = 29). After each such brief episode the larva is lifted up and carried to another spot inside the nest, where it is permitted to repeat the stereotyped spinning movements. While the larva is engaged in spinning, the worker touches the substrate, the silk, and the front half of the larva's body with its antennae. However, these antennal movements are less stereotyped than in *Oecophylla*.

The product of this coordinated activity is an irregular, wide-meshed network of silk extending throughout the nest. The construction usually begins with the attachment of the silk to the edge of a leaf or stem. As the spinning proceeds, some workers bring up small particles of soil and bark, wood chips, or dried leaf material that the ants have gathered on the ground below. They attach the detritus to the silk, often pushing particles into place with the front of their heads, and then make the larvae spin additional silk around the particles to secure them more tightly to the wall of the nest. In this way a sturdy outside shell is built, consisting in the end of several layers of silk reinforced by solid particles sealed into the fabric. The ants also weave an inside layer of pure silk, which covers the inner face of the outer wall and the surfaces of the supporting leaves and twigs. Reminiscent of wallpaper, this sheath is thin, very finely meshed, and tightly applied so as to follow the contours of the supporting surface closely. When viewed from inside, the nest of the *Polyrhachis* ant resembles a large communal cocoon (Fig. 5).

A very brief description of the weaving behavior of *Polyrhachis* (*Myrmhopla*) *simplex* by Ofer (1970) suggests that this Israeli species constructs nests in a manner similar to that observed in the Queensland species. The genus *Polyrhachis* is very diverse and widespread, ranging from Africa to tropical Asia and the Solomon Islands. Many of the species spin communal nests, apparently of differing degrees of complexity, and further study of their behavior should prove very rewarding.

A second intermediate grade is represented by *Camponotus* (*Myrmobrachys*) *senex*, which occurs in moist forested areas of South and Central

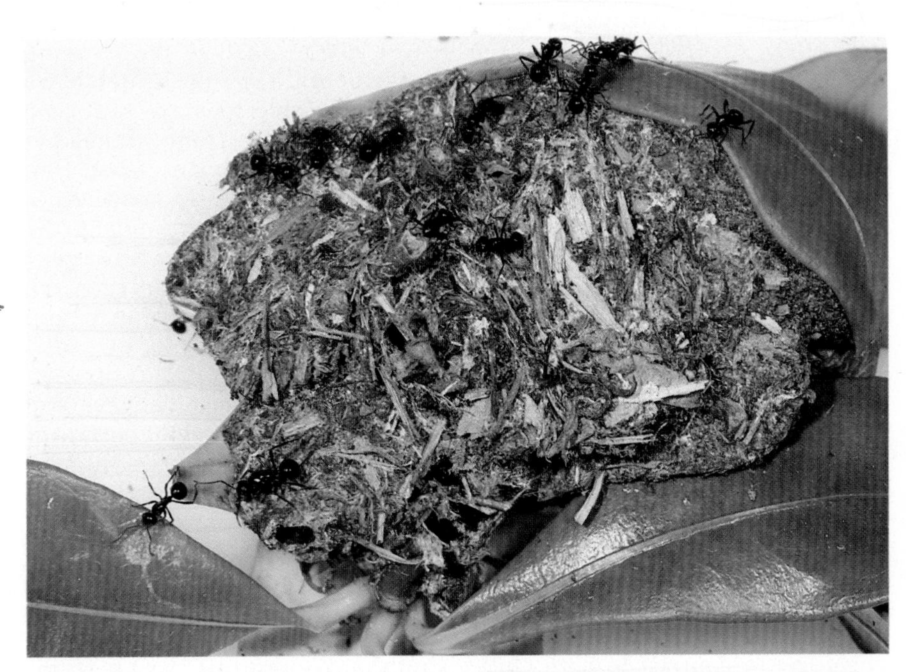

Figure 5. The nest of the Australian *Polyrhachis* species (*top*) is at an intermediate level of complexity, consisting of sheets of silk woven between leaves and twigs and reinforced by soil and dead vegetable particles. The interior of this type of nest (*bottom*) has a layer of silk tightly molded to the supporting leaf surfaces.

America. It is one of only two representatives of the very large and cosmopolitan genus *Camponotus* known to incorporate larval silk in nest construction (although admittedly very little information is available about most species of this genus), and in this respect must be regarded as an evolutionarily advanced form. The most complete account of the biology of *C. senex* to date is that of Schremmer (1972, 1979a, b).

Unlike the other weaver ants, *C. senex* constructs its nest almost entirely of larval silk. The interior of the nest is a complex three-dimensional maze of many small chambers and connecting passageways. Leaves are often covered by the silken sheets, but they then die and shrivel, and thereafter serve as no more than internal supports. Like the Australian *Polyrhachis*, *C. senex* workers add small fragments of dead wood and dried leaves to the sheets of silk along the outer surface. The detritus is especially thick on the roof, where it serves to protect the nest from direct sunlight and rain.

As Schremmer stressed, chains of worker ants and other cooperative maneuvers among workers of the kind that characterize *Oecophylla* do not occur in *C. senex*. The larvae employed in spinning are relatively large and most likely are near the end of the final instar. Although they contribute substantial amounts of silk collectively, they still spin individual cocoons—in contrast to both *Oecophylla* and the Australian *Polyrhachis*. Workers carrying spinning larvae can be most readily seen on the lower surfaces of the nest, where walls are thin and nest-building unusually active. During Schremmer's observations they were limited to the interior surface of the wall and consequently could be viewed only through the nascent sheets of silk. Although numerous workers were deployed on the outer surface of the same area at the same time, and were more or less evenly distributed and walked slowly about, they did not carry larvae and had no visible effect on the workers inside. Their function remains a mystery. They could in fact be serving simply as guards.

Although Schremmer himself chose not to analyze the weaving behavior of *C. senex* in any depth, we have been able to make out some important details from a frame-by-frame analysis of his excellent film (Schremmer 1972). In essence, *C. senex* appears to be very similar to the Australian *Polyrhachis* in this aspect of their behavior. Workers carry the larvae about slowly, pausing to hold them at strategic spots for extended periods. They do not contribute much to the contact between the heads of the larvae and the surface of the nest. Instead, again as in *Polyrhachis*, the larvae perform strong stretching and bending movements, with some lateral turning as well. When held over a promising bit of substrate, larvae appear to bring the head down repeatedly while expelling silk. We saw one larva perform six "figure eight" movements in succession, each time touching its head to the same spot in what appeared to be typical weaving movements. The duration of the contact between its head and the substrate was measured in five of these cycles; the range was 0.4–1.5 sec and averaged 0.8 sec. During the spinning movements the workers play their antennae widely over the front part of the body of the larva and the adjacent substrate.

The nest-weaving of *C. senex*, then, is the same as that of the Australian *Polyrhachis*. The only relevant difference between the two is that *C. senex* larvae construct individual cocoons and *Polyrhachis* larvae do not.

The simplest type of weaving

A recent study of the tree ants *Dendromyrmex chartifex* and *D. fabricii* has revealed a form of communal silk-weaving that is the most elementary conceivable (Wilson 1981). The seven species of *Dendromyrmex* are concentrated in Brazil, but at least two species (*chartifex* and *fabricii*) range into Central America. The small colonies of these ants build oblong carton nests on the leaves of a variety of tree species in the rain forest (Weber 1944).

The structure of the nests is reinforced with continuous sheets of larval silk (Fig. 6). When the nest's walls are deliberately torn to test their strength, it can be seen that the silk helps hold the carton together securely. Unlike *Oecophylla* larvae, those of *Dendromyrmex* contribute silk only at the end of the final instar, when they are fully grown and ready to pupate. Moreover, only part of the silk is used to make the nest. Although a few larvae become naked pupae, most enclose their own bodies with cocoons of variable thickness. Workers holding spinning larvae remain still while the larvae perform the weaving movements; in *Oecophylla*, the larvae are still and the workers move. Often the larvae add silk to the nest when lying on the surface unattended by workers. Overall, their nest-building movements differ from those of cocoon-spinning only by a relatively small change in orientation. And, not surprisingly, this facultative communal spinning results in a smaller contribution to the structure of the nest than is the case in *Oecophylla* and other advanced weaver ants.

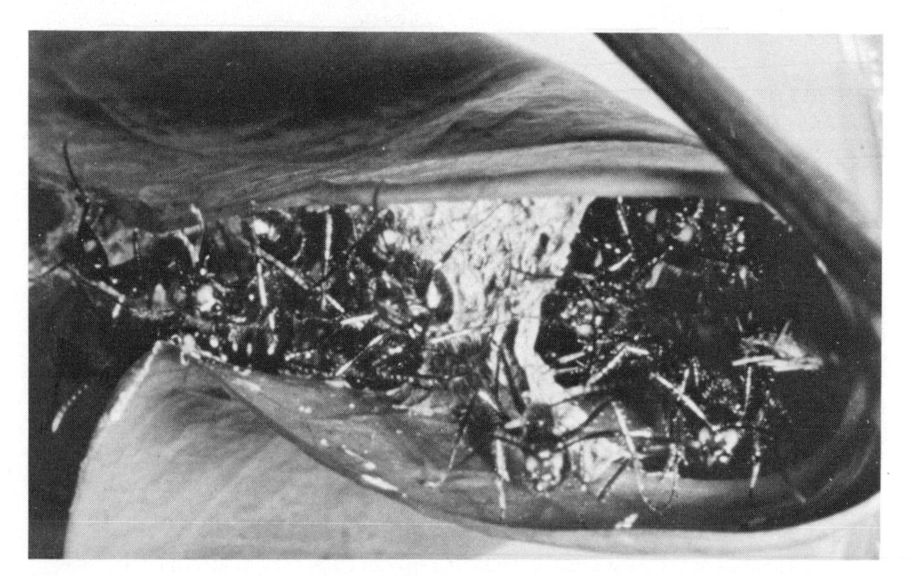

Figure 6. *Dendromyrmex chartifex*, of Central and South America, makes the simplest type of woven nest, a carton-like structure of chewed vegetable fibers reinforced with larval silk. (Photograph from Wilson 1981.)

Anatomical changes

The behavior of communally spinning ant larvae is clearly cooperative and altruistic in nature. If general notions about the process of evolution are correct, we should expect to find some anatomical changes correlated with the behavioral modifications that produce this cooperation. Also, the degree of change in the two kinds of traits should be correlated to some extent. And finally, the alterations should be most marked in the labial glands, which produce the silk, and in the external spinning apparatus of the larva.

These predictions have generally been confirmed. *Oecophylla*, which has the most advanced cooperative behavior, also has the most modified external spinning apparatus. The labial glands of the spinning larvae of *Oecophylla* and *Polyrhachis* are in fact much larger in proportion to the size of the larva's body than is the case in other formicine ant species whose larvae spin only individual cocoons (Karawajew 1929; Wilson and Hölldobler 1980). On the other hand, *C. senex* larvae do not have larger labial glands than those of other *Camponotus* larvae. Schrem-

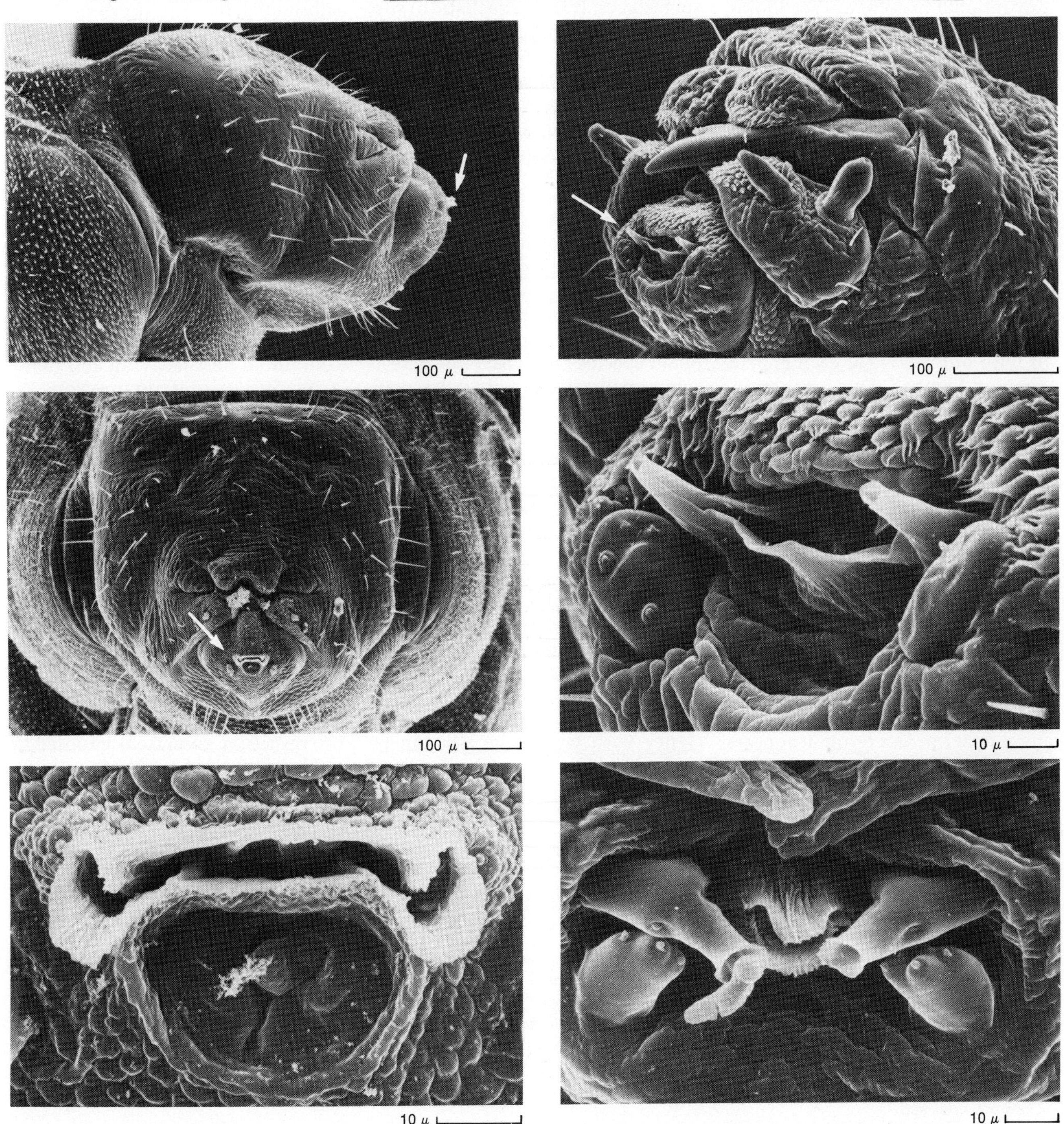

Figure 7. Scanning electron micrographs reveal adaptations in the spinning apparatus of ant larvae in *Oecophylla*. At the left, the head of an *O. longinoda* larva is shown from the side (*top*) and front (*middle*); the arrows indicate the slit-shaped opening of the silk glands, which is modified substantially from the more primitive forms at the right. The reduced lateral nozzles in *O. longinoda* and the larger central nozzle are clearly visible at the bottom left. In *Nothomyrmecia macrops*, a living Australian ant thought to be similar to the earliest formicines, there is no central nozzle and the lateral nozzles are much more prominent; the arrow at the top right points to the area enlarged at the middle right. The silk-gland opening of the Australian weaver ant, a species of *Polyrhachis* (*bottom right*), is similar in structure to that in *Nothomyrmecia*. (Micrographs by Ed Seling.)

mer (1979a) tried to fit this surprising result into the expected pattern by suggesting that the *C. senex* larvae produce silk for longer periods of time than other species that weave nests communally, and therefore do not need larger glands. This hypothesis has not yet been tested.

Until recently, little was known about the basic structure of the spinning apparatus of formicines. Using conventional histological sectioning of larvae in the ant genera *Formica* and *Lasius*, Schmidt and Gürsch (1970) concluded that the silk glands open to the outside by three tube-like projections, or nozzles. They were indeed able to pull three separate silken strands away from the heads of larvae with forceps. However, our studies, which combine histology and the use of the scanning electron microscope, have led us to draw a somewhat different picture.

In general, the labial gland opens to the outside through a small slit with one nozzle at each end, as shown in Figure 7. This is the structure found in the Australian *Nothomyrmecia macrops*, which is considered to be the living species closest to the ancestors of the Formicinae, as well as in a diversity of formicines themselves. Among the formicines examined, including those engaged in communal nest-weaving, only *Oecophylla* has a distinctly different external spinning apparatus. The labial-gland slit of these extremely advanced weaver ants is enlarged into a single nozzle, incorporating and largely obliterating the lateral nozzles. As a result, it appears that each *Oecophylla* larva is capable of expelling a broad thread of silk—the kind of thread needed to create the powerful webs binding an aboreal nest together.

The uncertain climb toward cooperation

In order to summarize existing information on the evolution of communal spinning, the grades in Table 1 are defined according to the presence or absence of particular traits associated with communal nest-weaving. We believe that it is both realistic and useful to recognize three such stages. It is also realistic to suppose that the most advanced weaver ants, those of the genus *Oecophylla*, are derived from lines that passed through lower grades similar to, if not identical with, those exemplified by *Dendromyrmex*, *Polyrhachis*, and *Camponotus senex*.

On the other hand, we find it surprising that communal nest-weaving has arisen only four or so times during the one hundred million years of ant evolution. Even if new cases of this behavior are discovered in the future, the percentage of ant species that weave their nests communally will remain very small. It is equally puzzling that the most advanced grade was attained only once. The separate traits of *Oecophylla* nest-weaving provide seemingly clear advantages that should predispose arboricole ants to evolve them. The remarkable cooperative maneuvers of the workers allow the colony to arrange the substrate in the best positions for the addition of the silk bonds and sheets. By taking over control of the spinning movements from the larvae, the workers enormously increase the speed and efficiency with which the silk can be applied to critical sites. For their part the larvae have benefited the colony by moving the time when they produce silk forward in the final instar, thus surrendering once and for all the ability to construct personal cocoons but allowing workers to carry and maneuver them more effectively because of their smaller size.

The case of *Dendromyrmex* is especially helpful in envisioning the first steps of the evolution in behavior that culminated in the communal nest-weaving of *Oecophylla*. Although the contribution of the larvae to the structure of the nest is quite substantial, the only apparent change in their behavior is a relatively slight addition to their normal spinning cycle, so that the larva releases some silk onto the floor of the nest while weaving its individual cocoon. It is easy to imagine such a change occurring with the alteration of a single gene affecting the weaving program. Thus, starting the evolution of a population toward communal weaving does not require a giant or otherwise improbable step.

There is another line of evidence indicating the general advantage of communal nest-weaving and hence a relative ease of progression. We discovered that both male and female larvae contribute silk to the nest in the case of *Oecophylla* (Wilson and Hölldobler 1980) and *Dendromyrmex* (Wilson 1981); male contribution has not yet been investigated in *Polyrhachis* and *Camponotus*. Because cooperation and altruism on the part of male ants is rare, it is always worthy of close examination. Bartz (1982) has recently shown that in social Hymenoptera, natural selection will favor the evolution of either male workers or female workers, but not both, and the restrictive conditions imposed by the haplodiploid mode of sex determination—used by all Hymenoptera—favor all-female worker castes. In fact, the sterile workers of hymenopterous societies are always female

Table 1. Grades of communal nest-weaving

	Larvae contribute silk to nest	Workers always hold spinning larvae	Larvae no longer make individual cocoons	Workers repeatedly move larvae	Workers cooperate in adjusting substrate	Workers perform most spinning movements	Silk is produced before end of final instar
Grade 1							
Dendromyrmex spp.	+	–	–	–	–	–	–
Grade 2							
Polyrhachis ?doddi	+	+	+	+	–	–	–
Camponotus senex	+	+	–	+	–	–	–
Grade 3							
Oecophylla spp.	+	+	+	+	+	+	+

(Oster and Wilson 1978). In boreal carpenter ants of the genus *Camponotus*, where the males do contribute some labor to the colony, it is in the form of food-sharing, an apparent adaptation to the lengthy developmental cycle of *Camponotus*. The males are kept in the colonies from late summer or fall to the following spring, and it benefits both the colony and the individual males to exchange liquid food (Hölldobler 1966).

The contribution of silk by male weaver-ant larvae is a comparable case. When the queens of *Oecophylla* and *Dendromyrmex* die, some of the workers lay eggs, which produce males exclusively (Hölldobler and Wilson 1983). Such queenless colonies can last for many months, until the last of the workers have died. During this period it is clearly advantageous for male larvae to add silk to the nest, for their own survival as well as that of the colony as a whole.

In summary, then, weaver ants exemplify very well an important problem of evolutionary theory: why so many intermediate species possess what appear to be "imperfect" or at least mechanically less efficient adaptations. Two hypotheses can be posed to explain the phenomenon that are fully consistent with the manifest operation of natural selection in such cases. The first is that some species remain in the lower grades because countervailing pressures of selection come to balance the pressures that favor the further evolution of the trait. In particular, the tendency for larvae to collaborate in the construction of nests could be halted or even reversed in evolution if surrendering the ability to make cocoons reduces the larvae's chance of survival. In other words, the lower grade might represent the optimum compromise between different pressures.

The second, quite different hypothesis is that the communal weavers are continuing to evolve—and will eventually attain or even surpass the level of *Oecophylla*—but species become extinct at a sufficiently high rate that most such evolutionary trends are curtailed before they are consummated. Even a moderate frequency of extinction can result in a constant number of species dispersed across the various evolutionary grades.

At present we see no means of choosing between these two hypotheses or of originating still other, less conventional evolutionary explanations. The greatest importance of phenomena such as communal nest-weaving may lie in the prospects they offer for a deeper understanding of arrested evolution, the reasons why not all social creatures have attained what from our peculiar human viewpoint we have chosen to regard as the pinnacles of altruistic cooperation.

References

Baroni Urbani, C. 1978. Materiali per una revisione dei *Leptothorax* neotropicali appartenenti al sottogenere *Macromischa* Roger, n. comb. (Hymenoptera: Formicidae). *Entomol. Basil.* 3:395–618.

Bartz, S. H. 1982. On the evolution of male workers in the Hymenoptera. *Behav. Ecol. Sociobiol.* 11:223–28.

Chauvin, R. 1952. Sur la reconstruction du nid chez les fourmis Oecophylles (*Oecophylla longinoda* L.). *Behaviour* 4:190–201.

Doflein, F. 1905. Beobachtungen an den Weberameisen (*Oecophylla smaragdina*). *Biol. Centralbl.* 25:497–507.

Hemmingsen, A. M. 1973. Nocturnal weaving on nest surface and division of labour in weaver ants (*Oecophylla smaragdina* Fabricius, 1775). *Vidensk. Meddr. Dansk Naturh. Foren.* 136:49–56.

Hölldobler, B. 1966. Futterverteilung durch Männchen im Ameisenstaat. *Z. Vergl. Physiol.* 52:430–55.

———. 1979. Territories of the African weaver ant (*Oecophylla longinoda* [Latreille]): A field study. *Z. Tierpsychol.* 51:201–13.

Hölldobler, B., and E. O. Wilson. 1977a. Weaver ants. *Sci. Am.* 237:146–54.

———. 1977b. Weaver ants: Social establishment and maintenance of territory. *Science* 195:900–02.

———. 1978. The multiple recruitment systems of the African weaver ant *Oecophylla longinoda* (Latreille) (Hymenoptera: Formicidae). *Behav. Ecol. Sociobiol.* 3:19–60.

———. 1983. Queen control in colonies of weaver ants (Hymenoptera: Formicidae). *Ann. Entomol. Soc. Am.* 76:235–38.

Jacobson, E., and A. Forel. 1909. Ameisen aus Java und Krakatau beobachtet und gesammelt von Herrn Edward Jacobson, bestimmt und beschrieben von Dr. A. Forel. *Notes Leyden Mus.* 31:221–53.

Jacobson, E., and E. Wasmann. 1905. Beobachtungen ueber *Polyrhachis dives* auf Java, die ihre Larven zum Spinnen der Nester benutzt. *Notes Leyden Mus.* 25:133–40.

Karawajew, W. 1929. Die Spinndrüsen der Weberameisen (Hym. Formicidae). *Zool. Anz.* (Wasmann Festband) 1929:247–56.

Ledoux, A. 1950. Recherche sur la biologie de la fourmi fileuse (*Oecophylla longinoda* Latr.). *Ann. Sci. Nat.* (Paris), ser. 11, 12:313–416.

Lindauer, M., and H. Martin. 1969. Special sensory performances in the orientation of the honey bee. In *Theoretical Physics and Biology*, ed. M. Marois, pp. 332–38. North-Holland.

Ofer, J. 1970. *Polyrhachis simplex:* The weaver ant of Israel. *Ins. Soc.* 17:49–81.

Oster, G. F., and E. O. Wilson. 1978. *Caste and Ecology in the Social Insects.* Princeton Univ. Press.

Schmidt, G. H., and E. Gürsch. 1970. Zur Struktur des Spinnorgans einiger Ameisenlarven (Hymenoptera, Formicidae). *Z. Morph. Tiere* 67:172–82.

———. 1971. Analyse der Spinnbewegungen der Larve von *Formica pratensis* Retz. (Form. Hym. Ins.). *Z. Tierpsychol.* 28:19–32.

Schremmer, F. (prod.). 1972. *Die südamerikanische Weberameise* Camponotus senex (*Freilandaufnahmen*). 16 mm film. Distributed by Inst. Wissenschaft, Göttingen. Film no. W1161.

———. 1979a. Die nahezu unbekannte neotropische Weberameise *Camponotus* (*Myrmobrachys*) *senex* (Hymenoptera: Formicidae). *Ent. Gen.* 5:363–78.

———. 1979b. Das Nest der neotropischen Weberameise *Camponotus* (*Myrmobrachys*) *senex* Smith (Hymenoptera, Formicidae). *Zool. Anz.* 203:273–82.

Sudd, J. H. 1963. How insects work in groups. *Discovery* (London), June, 15–19.

Wallis, D. I. 1960. Spinning movements in the larvae of the ant, *Formica fusca. Ins. Soc.* 7:187–99.

Weber, N. A. 1944. The tree-ants (*Dendromyrmex*) of South and Central America. *Ecology* 25:117–20.

Wheeler, W. M. 1914. The ants of the Baltic amber. *Schrift. Phys.-Ökon. Ges. Königsberg* 55:1–142.

Wilson, E. O. 1971. *The Insect Societies.* Harvard Univ. Press.

———. 1981. Communal silk-spinning by larvae of *Dendromyrmex* tree-ants (Hymenoptera: Formicidae). *Ins. Soc.* 28:182–90.

Wilson, E. O., and B. Hölldobler. 1980. Sex differences in cooperative silk-spinning by weaver ant larvae. *PNAS* 77:2343–47.

Wilson, E. O., and R. W. Taylor. 1964. A fossil ant colony: New evidence of social antiquity. *Psyche* (Cambridge) 71:93–103.

Evolution and the Triumph of Homology, or Why History Matters

Stephen Jay Gould

In 1912, when the nation both needed and still had a good five cent cigar, Sigma Xi spent three dollars to rent a hall for its annual banquet. Receipts for 1912 totaled $646.42 against expenses of $160.22 (including that three bucks), leaving a balance of $486.20, a fine improvement from the 1911 surplus of $295.67. Our society then included 8,200 members, 2,176 listed as active. In that year, Sigma Xi also decided, for the first time, to publish a journal, the *Sigma Xi Quarterly* (renamed *American Scientist* in 1942). In his "Salutatory" to the very first issue, president (and paleontologist) S. W. Williston wrote on page 1, volume 1, number 1: "Since its beginning Sigma Xi has stood for the encouragement of investigation, of research, rather than for the mere acquisition of knowledge."

How Darwin's "long argument" has changed the path of scientific thought during the past 100 years

In 1886 the founders of Sigma Xi had chosen for their motto "Companions in Zealous Research"—a phrase that we have happily retained despite its archaic ring. The original zealots were an uncompromising lot. Some roamed public places with hidden daggers to strike down supporters of Rome; others committed mass suicide at Masada. They were, above all, men of action—the *doers* of their generation. Our founders chose their words well. Science is doing, not just clever thinking. As Williston noted, our society stands for action expressed as research.

I have been assigned the impossible task of encapsulating the intellectual impact of evolution, both on other sciences and upon society in general, during the past 100 years. I have chosen this fundamental definition of Sigma Xi as prologue because I want to argue that Darwin's most enduring impact has generally been underestimated (or underesteemed). I will hold that his theory is, first and foremost, a guide to action in research—the first *workable* program ever presented for evolution. Darwin was, above all, a *historical methodologist*. His theory taught us the importance of history, expressed in doing as the triumph of homology over other causes of order. History is science of a different kind—pursued, when done well, with all the power and rigor ascribed to more traditional styles of science. Darwin taught us why history matters and established the methodology for an entire second style of science.

Stephen Jay Gould is Alexander Agassiz Professor of Zoology at Harvard University.

Darwin as a historical methodologist

Introducing the final chapter of the *Origin of Species* (1859), Darwin writes: "this whole volume is one long argument" (p. 459). Since Darwin was not a conscious or explicit philosopher, and since he crammed the *Origin* so full of particulars collected during twenty long years of preparation, readers often miss the unity of intellectual design. Indeed, Huxley commented that readers often misinterpret the *Origin* as a "sort of intellectual pemmican—a mass of facts crushed and pounded into shape, rather than held together by the ordinary medium of an obvious logical bond" (1893, p. 25).

What, then, is Darwin's "long argument," so deftly hidden amidst his particulars? It is not merely the specific defense of "natural selection," for most of the *Origin* is a basic argument for descent, not for any particular mechanism governing the process. But neither is it the most general marshaling of support for evolution—for transmutation was among the commonest of nineteenth-century heresies, and Darwin had something more special and personal to say.

The "long argument," as I read it, is the claim that *history* stands as the coordinating reason for relationships among organisms. Darwin's argument possesses a simple and beautiful elegance. Before the *Origin*, scientists had sought intrinsic purpose and meaning in taxonomic order. Darwin replied that the ordering reflects a historical pathway, pure and simple. (As just one example, numerological systems of taxonomy flourished in the decades before Darwin. These attempts to arrange all creatures in groups neatly ordered and numbered according to simple mathematical formulae—see Oken 1809–11 or Swainson 1835, for example—make sense if a rational intelligence created all organisms in an ordered scheme, but devolve to absurdity if taxonomy must classify the results of a complex and contingent history.) The eminent historian Edward H. Carr writes: "The real importance of the Darwinian revolution was that Darwin, completing what Lyell had already begun in geology, brought history into science" (1961, p. 71).

So much so good. But the simple statement that Darwin made history matter contains a dilemma, especially if we wish to defend the cardinal premise of Sigma Xi: that science is productive doing, not just clever thinking. History is the domain of narrative—unique, unrepeatable, unobservable, large-scale, singular events. One of the oldest saws of freshman philosophy classes asks: Can history be science? Many professors answer "no" because science seeks immanence by experiment and prediction, while the narrative quality of history seems to preclude just these defining features.

How then can we marry these two apparently contradictory statements—the claim that Darwin's "long argument" made history matter and the usual impression that Darwin was a great scientist? The problem vanishes when we locate Darwin's singular greatness in his extended campaign to establish a scientific methodology for history—to make history doable for the zealous researchers of science. Darwin was, more than anything else, a historical methodologist.

Michael Ghiselin's landmark book (1969) was the first to analyze all of Darwin's writing (not just the central trilogy of the *Origin*, the *Descent of Man*, and the *Expression of the Emotions*). He was also the first to suggest with proper documentation that Darwin's greatness as a scientist lay in the middle ground between his most basic elucidation of evolution as a fact, and his most general development of the radical implications (randomness, materialism, nonprogressionism) that so upset Western culture (but produced less immediate impact upon the day-to-day practice of science). This middle ground embodies Darwin's arguments for a methodology of research, for the actual *doing* itself. For Ghiselin, Darwin succeeded because he consistently used the hypothetico-deductive method so celebrated by recent philosophers of science—even though, as a loyal Victorian, he usually misportrayed himself (primarily in his *Autobiography*) as a patient and rigorous Baconian inductivist. (Ruse 1979 supports Ghiselin in the major work on Darwin's methodology written since.)

While I applaud Ghiselin's insight that Darwin must be viewed primarily as a methodologist—as someone who taught scientists how to proceed—I disagree that the central theme and sustaining power of Darwin's methodology lies in its hypothetico-deductive format. Philosopher of science Philip Kitcher has recently written that "if Darwin was a scientist practicing by the canons favored by Ghiselin and Ruse, then he was a poor practitioner"; in Kitcher's account, Darwin "answers to rather different methodological ideals" (1985, footnote 11). Kitcher views Darwin's theories as sets of "problem-solving patterns, aimed at answering families of questions about organisms, by describing the histories of these organisms" (p. 135)—the very aspect of life that had no relevance in the pre-Darwinian world of created permanence. But how can a naturalist do history in a scientific way, especially given the poor reputation of history as a ground for testable hypotheses? How can history be incorporated into science?

Darwin's long argument is not a simple brief for evolution; it is, above all, a claim for *knowability*—a set of methods that subjected evolution, for the first time, to zealous research. Previous briefs for evolution—and many had appeared to much comment (Lamarck 1809; Chambers 1844)—had presented speculative systems suggesting little in the way of doable research. Lyell's strong distaste for Lamarck (an opinion shared by Darwin) centered upon the methodological vacuousness of his system: "There were no examples to be found. . . . When Lamarck talks of 'the efforts of internal sentiment,' 'the influence of subtle fluids,' and 'acts of organization,' as causes . . . he substitutes names for things; and, with a disregard to the strict rules of induction, resorts to fictions, as ideal as the 'plastic virtue,' and other phantoms, of the geologists of the middle ages" (1842, pp. 10–11).

Darwin's claim for knowability centers upon two themes: first, the uniformitarian argument that one should work by extrapolating from small-scale phenomena that can be seen and investigated; second, the establishment of a graded set of methods for inferring history when only large-scale results are available for study.

The uniformitarian argument

In a famous letter for once not overly immodest, Darwin stated that half his work came out of Lyell's brain. Lyell's insistent argument through three volumes of the *Principles of Geology*—that a historical scientist must work with observable, gradual, small-scale changes and extrapolate their effects through immense time to encompass the grand phenomena of history—won Darwin's allegiance, with a central commitment for its transfer, in toto, to biological realms. But, as a fateful event in the history of nineteenth-century science (Gould 1965; Rudwick 1972; Hooykaas 1963), Lyell advanced uniformity as more than a methodological postulate—work with small-scale events when you can, because they are all you have. It became a strong substantive claim as well—the world really works that way, all the time.

In Lamarck's system, small-scale adaptations (the giraffe's neck, the long legs and webbed feet of wading birds) are tangential—literally orthogonal—to a different, virtually unobservable, and more essential process that moves organisms up the ladder of life toward ever-greater complexity. Savor the paradox for a scientist committed by definition to doing: what you can know and manipulate is unimportant or irrelevant; what is essential cannot be directly observed.

Darwin broke through this disabling paradox by proclaiming that the tangible small-scale evidences of change—artificial selection as practiced by breeders and farmers, tiny differences in geographic variation among races of a species, for example— are, by smooth extrapolation, *the* stuff of all evolution. Darwin, for the first time, made evolution a workable research program, not just an absorbing subject for speculation. This methodological breakthrough was his finest achievement.

But Darwin, like Lyell, then ventured beyond the methodological issue, thereby setting the pathways of evolutionary debate ever since (including all the hubbub of the last decade). He argued that all change, at whatever apparent level, really did arise as the extrapolated result of accumulated selection within populations. The distinctive features of strict Darwinism—particularly its location of causality in struggles among organisms (denial of hierarchy), and its argument for continuity in rate, style, and effect from the smallest observable to the largest inferred events of change (see Gingerich 1983 for a modern defense; Gould 1984 for a rebuttal)—emerge from this substantive extension of the uniformitarian argument.

The questioning of this extension unifies the apparently diverse critiques prominently discussed during the past decade. Critics are denying the reductionistic causal premise (struggles among organisms) by outlining a hierarchical theory of selection, independent at several levels of genes, organisms, demes, and species, but with complex interactions across all levels (Gould 1982a; Vrba and Eldredge 1984; for a philosopher's defense of hierarchy as logically sound see Sober 1984). They are also denying causal continuity from competition in a crowded world (Darwin's "wedging") to relays and replacements of faunas in mass extinctions (Raup and Sepkoski 1984), thereby defending more randomness and discontinuity in *change* (rather than merely for raw material) than Darwin's vision allowed. (Punctuated equilibrium does not challenge the continuationist claim per se, since paleontological punctuations proceed tolerably slowly in ecological time, but rather the reductionistic argument about the primacy of selection on organisms, since trends mediated by differential speciation offer such scope for true species selection—see Eldredge and Gould 1972; Gould 1982b. Thus punctuated equilibrium leads to hierarchy, not saltationism.)

In short, Darwin made evolution doable for the first time, but by holding so strongly to the substantive side of uniformitarianism, ultimately offered a restrictive version that hierarchies of causal levels and tiers of time (Gould 1985a) must extend. This discussion has raged for ten years and continues unabated, but I shall pursue it no farther here because I want to concentrate on the character and meaning of Darwin's second great contribution to a science of history.

Inferring history from its results

The uniformitarian argument constructs history from an observable, small-scale present. But how can scientists proceed when they have only results before them? Past processes are, in principle, unobservable, yet science traffics in process. How, then, can we make history doable if our data feature only its results?

I have come to view Darwin's sequence of books as proceeding at two levels—an explicit and conscious treatment of diverse subjects (from coral reefs, to orchids, to insectivorous plants, to climbing plants, to worms, to evolution); and a covert, perhaps unconscious extended treatise on historical methodology, with each book featuring a different principle of historical reconstruction. We may arrange these principles—three in number—in terms of decreasing information for making inferences.

First, the large-scale results may lie before us, and we can also measure the rate and effect of the process that presumably produces them. In such cases of maximal information, we can use the uniformitarian method in its purest form: make rigorous measurements of the modern process and extrapolate into available time to render the full result. *The Formation of Vegetable Mould, Through the Action of Worms* (1881) is Darwin's finest example of this method. This book, Darwin's last, may also be his most misunderstood. Often seen as a pleasant trifle of old age, Darwin's worm book is a consciously chosen exemplar of historical reasoning at its most complete. What better choice of object than the humble, insignificant worm, working unnoticed literally beneath our feet? Could something so small really be responsible both for England's characteristic topography and for the upper layer of its soil?

***Punch*'s commentary on Darwin's last work on worms (*Punch* 22 October 1881, p. 190)**

Darwin's argument is pure uniformitarianism, carefully extended in stages. He counts worms to see if the soil contains enough for the work needed. He collects castings to measure the rate of churning (about 1/10 inch per year). He then extends the time scale to decades, via natural experiments on layers, once at the surface, that now lie coherently below material brought up by worms (burned coals, rubble from demolition, flints on his own ploughed field)—and then even farther by measuring the rate of foundering for historical objects (the "Druidical" rocks of Stonehenge, for example).

Second, we may have insufficient data about modern rates and processes simply to extrapolate their effects, but we can document several kinds or categories of results and seek relationships among them. Here we face a problem of taxonomy. Darwin argues that we may still proceed in the absence of direct data for uniformitarian extrapolation. We must formulate a historical hypothesis and then arrange the observed results as stages of its operation. The historical process, in other words, becomes the thread that ties all results together causally. This method succeeds because the process works on so many sequences simultaneously, but beginning at different times and proceeding at different rates in its various manifestations; therefore, all stages exist somewhere in the world at any one time (just as we may infer the course of a star's life by finding different stars in various stages of a general process, even though we trace no actual history of any individual star).

Darwin's first theoretical book, *The Structure and Distribution of Coral Reefs* (1842), illustrates this powerful guide to history. Its argument rests upon a classification of reefs into three basic categories of fringing, barrier, and atoll. Darwin proposed a common theme—subsidence of islands—to portray all three as sequential stages of a single historical process. Since corals build up and out from the edges of oceanic platforms, reefs begin by fringing their islands, become barriers as their islands subside, and finally atolls when their platforms submerge completely. But the taxonomy itself guarantees no history. During the nineteenth century, Darwin's opponents developed a series of counterproposals that may be called, collectively, "antecedent platform" theories. They argued that since corals build at the edges of platforms, these fringing reefs, barriers, and atolls only record the extent of previous planation, not a historical sequence in coral growth—platforms eroded to a small notch develop fringing reefs; those planed flat by waves become substrates for atolls. (Lyell, before Darwin convinced him otherwise, had advanced an even simpler ahistorical theory: that atolls develop on the circular rims of volcanoes.) The two theories can be distinguished by a crucial test not available in Darwin's time: no correlation between vertical extent of the reef and its form for antecedent platforms, progressive thickening from fringing reef to atoll in Darwin's subsidence. Twentieth-century drilling into Pacific atolls has affirmed Darwin's view.

Third, we must sometimes infer history from single objects; we have neither data for extrapolation from modern processes, nor even a series of stages to arrange in historical sequence. But how can a scientist infer history from single objects? This most common of historical dilemmas has a somewhat paradoxical solution. Darwin answers that we must look for imperfections and oddities, because any perfection in organic design or ecology obliterates the paths of history and might have been created as we find it. This principle of imperfection became Darwin's most common guide (if only because the fragmentary evidence of history often fails to provide better data in the preceding categories). I like to call it the "panda principle" in honor of my favorite example—the highly inefficient, but serviceable, false thumb of the panda, fashioned from the wrist's radial sesamoid bone because the true anatomical first digit had irrevocably evolved, in carnivorous ancestors of the herbivorous panda, to limited motility in running and clawing. (The herbivorous panda uses its sesamoid "thumb" for stripping leaves off bamboo shoots.) I titled one of my books *The Panda's Thumb* (1980) to honor this principle of historical reasoning.

The panda principle is a basic method of all historical science, linguistics, and history itself, for example, not just a principle for evolutionary reasoning. In *The Various Contrivances by which Orchids Are Fertilized by Insects* (1862), the book that followed the *Origin of Species*, Darwin shows that the wondrously complex adaptations of orchids, so intricately fashioned to aid fertilization by insects, are all jury-rigged from the ordinary parts of flowers, not built to the optimum specifications of an engineer's blueprint. "The use of each trifling detail of structure," Darwin writes, "is far from a barren search to those who believe in natural selection" (1888 ed., p. 286).

Throughout all these arguments, Darwin also showed his keen appreciation for the other great principle of historical science—the importance of proper taxonomies. In a profession more observational and comparative than experimental, the ordering of diverse objects into sensible categories becomes a sine qua non of causal interpretation. A taxonomy is not a mindless allocation of objective entities into self-evident pigeonholes, but a theory of causal ordering. Proper taxonomies require two separate insights: the identification and segregation of the basic phenomenon itself, and the division of its diverse manifestations into subcategories that reflect process and cause. Consider, for example, Steno's *Prodromus* (1669). This founding document of geology is, fundamentally, a new taxonomy (see Gould 1983). Steno identifies solid objects enclosed in other solids as the basic phenomenon (a stunningly original and peculiar choice in the light of ordering principles generally accepted in his time); he then makes a fundamental division into objects hard before surrounded (fossils) and those introduced without initial solidity into a rigid matrix (crystals in geodes, for example). Using these divisions, Steno could identify the organic origins of fossils and the temporal nature of strata—the cornerstones of historical geology.

We must look for imperfections and oddities, because any perfection in organic design or ecology obliterates the paths of history and might have been created as we find it

Darwin's *Different Forms of Flowers on Plants of the Same Species* (1877) is a fine illustration of how taxonomy informs history. The basic recognition of the phenomenon itself as a unitary puzzle poses a historical question. The work then becomes a long argument about subdivisions by function (heterostyly to assure cross-fertilization, cleistogamy to allow some advantageous selfing while retaining other forms of flowers for occasional crossing, for example), and about ancestral states and the paths of potential transformation.

The *Origin of Species* achieves its conceptual power by using all these forms of historical argument: uniformitarianism in extrapolating the observed results of artificial selection by breeders and farmers; inference of history from temporal ordering of coexisting phenomena (in constructing, for example, a sequence leading from variation within a population, to small-scale geographic differentiation of races, to separate species, to the origin of major groups and key innovations in morphology); and, most often and to such diverse effect, the panda principle of imperfection (vestigial organs, odd biogeographic distributions made sensible only as products of history, adaptations as contrivances jury-rigged from parts available).

I do not know whether Darwin operated by conscious design to construct his multivolumed treatise on historical method. Since great thinkers so often work by what our vernacular calls intuition (though the process involves no intrinsic mystery, as logical reconstruction by later intellectual biographers can attest), conscious intent is no criterion of outcome. Still, I like to think that the last paragraph of Darwin's last book records his own perception of connection—for he closes his last treatise on worms by remembering his first on corals, thereby linking both his humble subjects and his criteria of history:

It may be doubted whether there are many other animals which have played so important a part in the history of the world, as have these lowly organized creatures [worms]. Some other animals, however, still more lowly organized, namely corals, have done far more conspicuous work in having constructed innumerable reefs and islands in the great oceans.

Using the panda principle

Historical science is still widely misunderstood, underappreciated, or denigrated. Most children first meet science in their formal education by learning about a powerful mode of reasoning called "*the* scientific method." Beyond a few platitudes about objectivity and willingness to change one's mind, students learn a restricted stereotype about observation, simplification to tease apart controlling variables, crucial experiment, and prediction with repetition as a test. These classic "billiard ball" models of simple physical systems grant no uniqueness to time and object—indeed, they remove any special character as a confusing variable—lest repeatability under common conditions be compromised. Thus, when students later confront history, where complex events occur but once in detailed glory, they can only conclude that such a subject must be less than science. And when they approach taxonomic diversity, or phylogenetic history, or biogeography—where experiment and repetition have limited application to systems in toto—they can only conclude that something beneath science, something merely "descriptive," lies before them.

These historical subjects, placed into a curriculum of science, therefore become degraded by their failure to match a supposedly universal ideal. They become, in our metaphors, the "soft" (as opposed to "hard") sciences, the "merely descriptive" (as opposed to "rigorously experimental"). Every year Nobel prizes are announced to front-page fanfare, and no one who works with the complex, unrepeated phenomena of history can win—for the prizes only recognize science as designated by the stereotype. (I'm not bitching—since it makes for a much more pleasant profession—only making a social comment.) Plate tectonics revolutionized our view of the earth, but its authors remain anonymous to the public eye; molecular phylogeny finally begins to unravel the complexities of genealogy, and its accomplishments rank as mere narrative. Harvard organizes its Core Curriculum and breaks conceptual ground by dividing sciences into the two major styles of experimental-predictive and historical, rather than, traditionally, by discipline. But guess which domain becomes "Science A" and which "Science B"?

In a perverse way, the best illustration of this failure to understand the special character of history can be found in writings by opponents of science, who use clever rhetoric to argue against evolution because it doesn't work like their simplistic view of physics. In his book *Algeny* (1983), for example (see Gould 1985b for a general critique), Jeremy Rifkin dismisses Darwin because evolution can't be turned into a controlled laboratory experiment: "To qualify as a science, Darwin's theory should be provable by means of the scientific method. In other words, its hypotheses should be capable of being tested experimentally" (p. 117). Rifkin then cites Dobzhansky's statement about history as though it represents a fatal confession: "Dobzhansky laments the fact that 'evolutionary happenings are unique, unrepeatable, and irreversible. It is as impossible to turn a vertebrate into a fish as it is to effect the reverse transformation.' Dobzhansky is chagrined" (p. 118). But Dobzhansky in this passage is neither sheepish nor troubled; he is simply commenting upon the nature of history. Rifkin concludes nonetheless: "Embarrassing, to say the least. Here is a body of thought, incapable of being scientifically tested. . . . If not based on scientific observation, then evolution must be a matter of personal faith" (pp. 118–19).

When creationist lawyer Wendell Bird took my deposition in pretrial hearings on the constitutionality of Louisiana's "creation-science" law (mercifully tossed out without trial by a federal judge), he spent an inordinate amount of time (and verbal trickery) trying to make me admit that a suspension of natural law—a miracle—might fall within the purview of science. At the close of this lamentable episode, I was astounded when he asked "Are you familiar with the term singularistic?" and then tried to argue that complex historical events (singularistic in that sense), as unrepeatable, are somehow akin to miracles (singularities) and therefore make such historical sciences as evolution either as good or as bad as the Genesis-literalism of so-called creation-science!

These arguments about history are red herrings and we would do well to suppress them by acknowledging the *different* strengths of historical science, lest the simplistic stereotype be turned, as in these cases, against all of science.

The "lesser" status of historical science may be rejected on two grounds. First, it is not true that standard techniques of controlled experimentation, prediction, and repeatability cannot be applied to complex histories. Uniqueness exists in toto, but "nomothetic undertones" (as I like to call them) can always be factored out. Each mass extinction has its endlessly fascinating particularities (trilobites died in one, dinosaurs in another), but a common theme of extraterrestrial impact (Alvarez et al. 1980) may trigger a set of such events, even perhaps on a regular cycle (Raup and Sepkoski 1984). Nature, moreover, presents us with experiments aplenty, imperfectly controlled compared with the best laboratory standards, but having other virtues (temporal extent, for example) not attainable with human designs.

Second, as argued earlier, Darwin labored for a lifetime to meet history head on, and to establish rigorous methods for inference about its singularities. History, by Darwin's methods, is knowable in principle (though not fully recoverable in every case, given the limits of evidence), testable, and different. We do not attempt to predict the future (I could already retire in comfort if someone paid me a dollar for each rendition of my "paleontologists don't predict the future" homily following the inevitable question from the floor at all presentations to nonscientists—"Well, where is human evolution going anyway?"). But we can postdict about the past—and do so all the time in historical science's most common use of repeatability (every new iridium anomaly at the Cretaceous-Tertiary boundary is a repeated postdicted affirmation of Alvarez's conjecture about impact, based upon just three sites in the original article).

Finally, history's richness drives us to different methods of testing, but testability (via postdiction) is our method as well. Huxley and Darwin maintained interestingly different attitudes toward testing in history. Huxley, beguiled by the stereotype, always sought a crucial observation or experiment (the destruction of theory by a "nasty, ugly, little fact" of his famous aphorism). Darwin, so keenly aware of both the strengths and limits of history, argued that iterated pattern, based on types of evidence so numerous and so diverse that no other coordinating interpretation could stand—even though any item, taken separately, could not provide conclusive proof—must be the criterion for evolutionary inference. (The great philosopher of science William Whewell had called this historical method "consilience of inductions.") Huxley sought the elusive crucial experiment; Darwin strove for attainable consilience. Di Gregorio's recent treatise on Huxley's scientific style contains an interesting discussion of this difference (1984). (Ironically, Whewell, a conservative churchman, later banned Darwin's *Origin* from the library of Trinity College, Cambridge, where he was master. What greater blow than the proper use of one's own arguments in an alien context.)

In an essay of 1860, for example, Huxley wrote: "but there is no positive evidence, at present, that any group of animals has, by variation and selective breeding, given rise to another group which was, even in the least degree, infertile with the first. Mr. Darwin is perfectly aware of this weak point, and brings forward a multitude of ingenious and important arguments to diminish the force of the objection" (quoted by di Gregorio, p. 61). Note particularly Huxley's subtle misunderstanding of Darwin's methodology. What Huxley views as a set of indirect arguments, presented faute de mieux in the absence of experimental proof, *is* Darwin's consilience, positively developed as the proper method of historical inference. Darwin complained of just this misunderstanding in a letter to Hooker in 1861: "change of species cannot be directly proved . . . the doctrine must sink or swim according as it groups and explains phenomena. It is really curious how few judge it in this way, which is clearly the right way" (di Gregorio, p. 62). And, more specifically, in his *Variation of Animals and Plants Under Domestication* (1868):

> Now this hypothesis [natural selection] may be tested—and this seems to me the only fair and legitimate manner of considering the whole question—by trying whether it explains several large and independent classes of facts; such as the geological succession of organic beings, their distribution in past and present times, and their mutual affinities and homologies. If the principle of natural selection does explain these and other large bodies of facts, it ought to be received. [I, 657]

Historical scientists try to import bodily an oversimplified caricature of "hard" science

Despite the ready availability of these powerful, yet different, modes of inference, historical scientists have often been beguiled by the stereotype of direct experimental proof, and have wallowed in a curious kind of self-hate in trying to ape, where not appropriate, supposedly universal procedures of *the* scientific method. Many of the persistent debates within evolutionary biology are best viewed as a series of attempts to divest evolution of history under the delusion that scientific rigor gains thereby—with responses by defenders that history cannot be factored away, and that good science can be done just splendidly with it.

In extreme versions, for example, the welcome and powerful movement of "equilibrium" biogeography and ecology, which developed in the 1960s and peaked in the 1970s, not only denied history, but viewed its singularities as impediments to real science. Equilibrium models avoid history by explaining current situations as active balances maintained between competing forces now operating. Such models apply a reverse panda principle by identifying nonequilibria as signs of history—situations not yet balanced and therefore in a relevant state of history. Equilibria, when reached, are timeless, changing only when the measurable inputs alter, and not by any historically bound "evolution" of the system.

Ironically, the founding document of this movement was written by two fine historical scientists who understood proper limits, and who used their models to identify nomothetic undertones of a valued history (MacArthur and Wilson 1967). They also explicitly discussed the interactions of history and equilibrium, and the long-term evolutionary adjustments that continue, albeit at slower rates, within systems at equilibrium: "The equilibrium model has the virtues of making testable predictions which were not immediately obvious and of making the individual vagaries of island history seem somewhat less important in understanding the diversity of the island's species. Of course the history of islands remains crucial to the understanding of the taxonomic composition of species" (1967, p. 64). But in the hands of singleminded and less thoughtful supporters, equilibrium ecology moved from suggestive simplification (or search for repeated undertones) to a hard substantive claim spearheading a crusade for bringing "real" science into an antiquated domain of descriptive natural history. The campaign quickly stalled, however; nature fights back effectively.

The most common denial of history made by self-styled Darwinian evolutionists resides in claims for optimality—conventionally for the mechanics of morphology, more recently for behavior and ecology. Again, optimality theory has its place and uses (primarily in designating ideals for assessing natural departures). Committed votaries think that they are celebrating evolution by showing how inexorably and fine the mills of natural selection can grind; in fact, they are attempting to abrogate Darwin's most important criterion of history—the panda principle of imperfection and oddity as signs of previous lifestyles and affinities genealogically preserved. Under certain conditions of minimal constraint, we may legitimately seek optimality (animal color patterns, often less subject than morphology to developmental covariance, represent one promising domain—see Cott 1940 for the classic statement). Usually, history and complexity must assert themselves prominently.

Evolution is a bush, not a ladder

The sad tale goes on and on. Historical scientists, who should take legitimate pride in their different ways, try to import bodily an oversimplified caricature of "hard" science, or simply bow to pronouncements of professions with higher status. Some accepted Kelvin's last and most restrictive dates for a young earth, though fossils and strata spoke differently; many more foreswore their own data when physicists proclaimed that continents cannot move laterally. Charles Spearman misused factor analysis to identify intelligence as a measurable physical thing in the head, and then said of psychology that "this Cinderella among the sciences has made a bold bid for the level of triumphant physics itself" (see Gould 1981, chap. 6).

But the great historical scientists have always treasured both their rigorous, testable methods and their singular data. D'Arcy Thompson, whose own vision of optimal form, impressed directly upon organisms by physical forces, must rank among the most ahistorical of approaches to evolution (see Gould 1971), nonetheless knew that a retrievable history pervaded all objects—and that the panda principle can recover it. He wrote in his incomparable prose (1942):

> Immediate use and old inheritance are blended in Nature's handiwork as in our own. In the marble columns and architraves of a Greek temple we still trace the timbers of its wooden prototype, and see beyond these the tree-trunks of a primeval sacred grove; roof and eaves of a pagoda recall the sagging mats which roofed an earlier edifice; Anglo-Saxon land-tenure influences the planning of our streets and the cliff-dwelling and cave-dwelling linger on in the construction of our homes! So we see enduring traces of the past in the living organism—landmarks which have lasted on through altered function and altered needs. [pp. 1020–21]

The triumph of homology

Louis Agassiz chose an enigmatic name for his institution—and for good cause. He called it the Museum of Comparative Zoology (I am sitting in its oldest section as I write this article) in order to emphasize that the sciences of organic diversity do not usually seek identity in repeated experiment, but work by comparing the similarities among objects of nature as given. Kind, extent, and amount of similarity provide the primary data of historical science.

As a problem, recognized since Aristotle, natural similarities come in two basic, largely contradictory styles. We cannot simply measure and tabulate; we must factor and divide. Similarities may be homologies, shared by simple reason of descent and history, or analogies, actively developed (independently, but to similar form and effect) as evolutionary responses to common situations.

Systematics (the science of classifying organisms) is the analysis of similarity in order to exclude analogy and recognize homology. Such an epigram may sacrifice a bit of subtlety for crisp epitome, but it does capture the first goal of historical science. Homologous similarity is the product of history; analogy, as independent tuning to current circumstance, obscures the paths of history.

The major brouhaha about cladistics (see Hennig 1966; Eldredge and Cracraft 1980; any issue of *Systematic Zoology* for the past decade, or, for self-serving misuse by yet another opponent of science, Bethell 1985) has unfolded in needless acrimony because cladistics has not been properly recognized, even by some of its strongest champions, as a "pure" method for defining historical order and rigidly excluding all other causes of similarity. Cladistics allies objects in branching hierarchies defined only by relative times of genealogical connection. Closest, or "sister-group," pairs share a unique historical connection (a common ancestor yielding them as its only descendants). The system then connects sister-group pairs into ever-more inclusive groups sharing the same genealogical uniqueness (common ancestry that includes absolutely all descendant branches and absolutely no other groups). Cladistics is the science of ordering by genealogical connection, *and nothing else*. As such, it is the quintessential expression of history's primacy above any other cause or expression of similarity—and, on this basis alone, should be received with pleasure by evolutionists.

Nonetheless, several of the most forceful cladists never grasped this central point clearly; they buried their subject in frightful terminology and such exaggerated or extended claims that they antagonized many key systematists and never won the general approbation they deserved. In an almost perverse interpretation (literal, not ethical), some supporters, the self-styled "transformed" or "pattern" cladists, have actually negated the central strength of their method as a science of history by claiming—based on a curiously simplistic reading of Karl Popper—that science should eschew all talk of "process" (or cause) and work only with recoverable "pattern" (or the branching order of cladograms). Pattern cladists are not anti-evolution (as misportrayed by Bethell 1985); rather, as a result of narrow commitment to an extreme

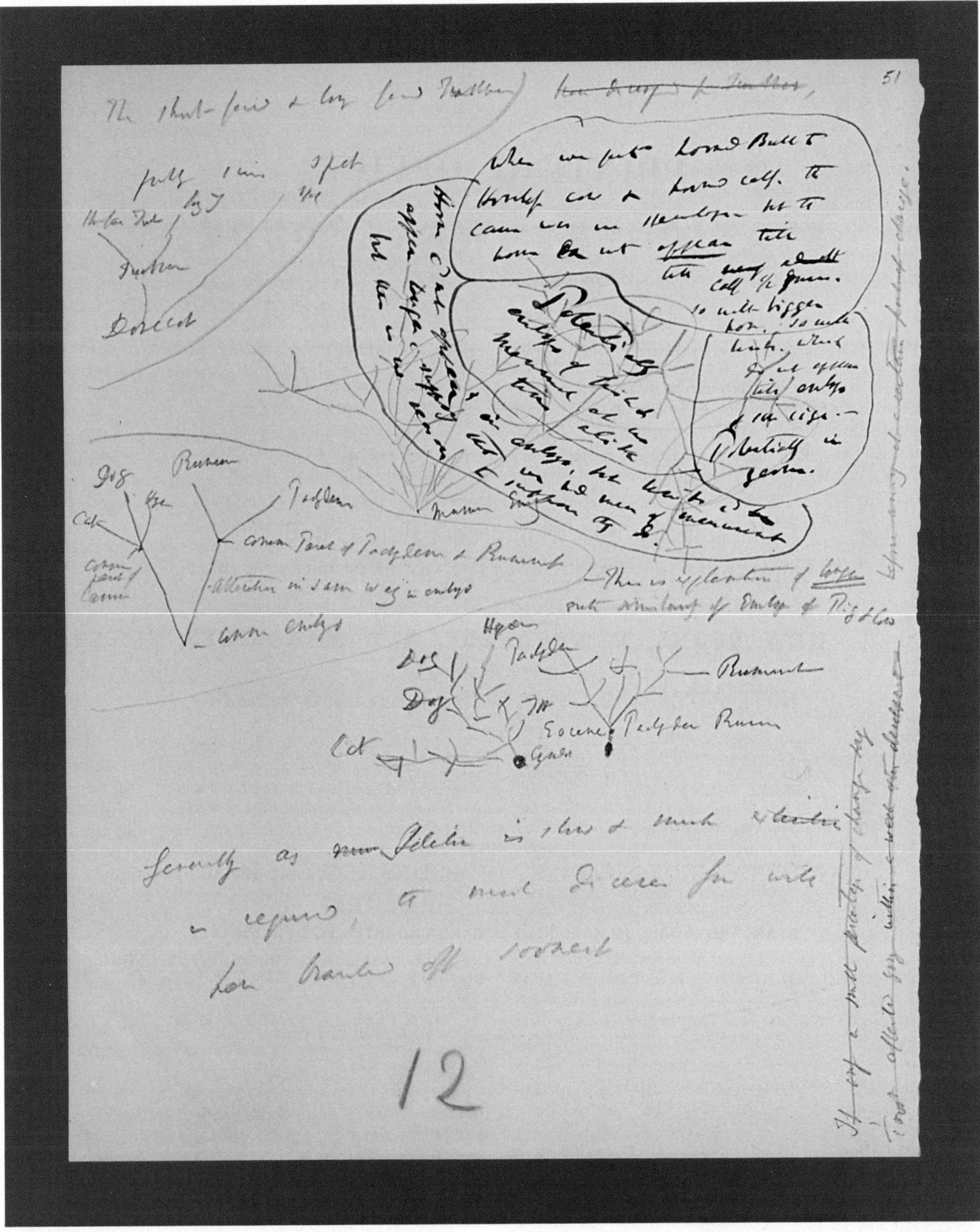

This unpublished page from Darwin's file "Embryology" shows Darwin constructing congruent trees of relationship, based in one case on developmental similarities, and in the other on time of divergence. The page, which dates from the mid-1850s, though by no means his first use of trees, is one of his most sophisticated. In the upper left, Darwin shows the descent of two breeds of tumbler pigeons from the common pigeon. Below, he arranges carnivores on one great branch and ungulates (divided into pachyderms and ruminants) on the other, which unite at a base labeled "common embryo." The central tree relates the same groups, but grows through geological time, rather than developmental space; here Darwin writes "Eocene Pachyderm Ruminant." On the right, he discusses the inheritance, expression, timing, and alteration of characters in development. (Summary by S. Rachootin; photograph of DAR 205.6:51 reproduced by permission of the Syndics of the Cambridge University Library.)

view of empiricism and falsification, they choose to ignore evolution as unprovable talk about process, and to concentrate on recovery of branching pattern alone. All well and good; it can be done with cladistic methods. But unless we wish to abandon a basic commitment to cause and natural law, branching order must arise for a reason (by a process, if you will). And that process is history, however history be made. In other words, the clearest method ever developed for discerning history has been twisted by some supporters to a caricatured empiricism that denies its subject. What an irony. We need to recover cladistics—a pure method for specifying those histories that develop by divergent and irreversible branching—from some of its own most vociferous champions.

Cladistics, as a methodology, is conceptually sound. In biological systematics, it has often proved inconclusive in practice (leading to a false suspicion that the method itself might be logically flawed) because taxonomists have worked, generally faute de mieux, or by simple weight of tradition, with the inappropriate data of morphology. Homology can often be recovered from morphology, but the forms of organisms often include an inextricable mixture of similarities retained by history (homology) and independently evolved in the light of common function (analogy). Morphology is not the best source of data for unraveling history. Thoughtful systematists have known this ever since Darwin, but have proceeded as best they could in the absence of anything better.

In principle, the recovery of homology only requires a source of information with two properties: sufficiently numerous and sufficiently independent items to preclude, on grounds of mathematical probability alone, any independent origin in two separate lineages. The "items" of morphology are too few and too bound in complex webs of developmental correlation to yield this required independence. Yet the discoveries and techniques of molecular biology have now provided an appropriate source for recovering homology—a lovely example of science at its unified best, as a profession firmly in the camp of repetition and experiment provides singular data for history. Molecular phylogenies work not because DNA is "better," more real, or more basic than morphology, but simply because the items of a DNA program are sufficiently numerous and independent to ensure that degrees of simple matching accurately measure homology. The most successful technique of molecular phylogeny has not relied upon sophisticated sequencing, but rather the "crudest" brute-force matching of all single-copy DNA—for such a method uses a maximum number of independent items (Sibley and Ahlquist 1981, 1984, and 1985).

Consider just one elegant example of the triumph of homology (as determined by molecular methods)—a case that should stand as sufficient illustration of the primacy of history as both a basic cause of order and a starting point for all further analyses. The songbirds of Australia have posed a classic dilemma for systematists because they seem to present a biogeographic and taxonomic pattern so different from the mammals. The marsupials of Australia form a coherent group bound by homology and well matched to the geographic isolation of their home. But the songbirds include a large set of creatures apparently related to several Eurasian and American groups: warblers, thrushes, nuthatches, and flycatchers, for example. This classification has forced the improbable proposal that songbirds migrated to Australia in several distinct waves. Even though birds fly and mammals do not, the discordance of pattern between the two groups has been troubling, since most birds are not nearly so mobile as their metaphorical status suggests. But Sibley and Ahlquist (1985) have now completed the molecular phylogeny (true homology) of Australian songbirds—and they form a coherent group of homologues, just like the marsupials. The similarities to Eurasian and American taxa are all convergences—independently evolved analogies—not homologies. Marsupials, of course, have also evolved some astounding convergences upon placentals (marsupial "wolves," "mice," and "moles"), but systematists have not been confounded because all marsupials retain signatures of homology in their pouches and epipubic bones. We now realize that the apparent and deeply troubling discordance between Australian birds and mammals arose only because the birds did not retain such morphological signatures of homology, and systematists were therefore fooled by the striking analogies.

Once we map homologies properly, we can finally begin to ask interesting biological questions about function and development—that is, we can use morphology for its intrinsic sources of enlightenment, and not as an inherently flawed measure of genealogical relationships. The historical flow of protoplasm through branching systems of genealogy *is* the material reality that structures biology. But analogies have a different (and vital) meaning that the resolution of history finally permits us to appreciate. They are functional themes that stand, almost like a set of Platonic forms, in a domain for biomechanics, functional morphology, and an entire set of nonhistorical disciplines. "Wrenness," "mouseness," and "wormness" may not have the generality of forms so basic that they have rarely been confused with homologies (bilateral symmetry, for example), but they still represent iterated themes, standards of design external to history. The material reality of history—phylogeny—flows through them again and again, forming a set of contrasts that define the fascination of biology: homology and analogy, history and immanence, movement and stability.

If the primacy of history is evolution's lesson for other sciences, then we should explore the consequences of valuing history as a source of law and similarity, rather than dismissing it as narrative unworthy of the name science. I argued in the first section that Darwin cut through a tangle of unnecessary complexity by proposing "just history" as a disconcertingly simple answer in domains where science had sought a deeper, rational immanence (as for the ordering of taxonomy). I wonder how far we might extend this insight. Consider our relentless search for human universals and our excitement at the prospect that we may thereby unlock something at the core of our being. So Jung proposes archetypes of the human psyche in assessing the similarities of mythic systems across cultures, while others invoke brain structure and natural selection as a source of uncannily complex repetitions among human groups long separate. Yet evolution is a bush, not a ladder. *Homo sapiens* had a discrete and recent origin, presumably as an isolated local population, crowned with inordinate later success. Many of these similarities may therefore be simple homologies of a contingent history, not deep immanences of the soul. Such an offbeat idea might provide an astonishingly simple solution to some of the oldest dilemmas born of the Socratic injunction—know thyself.

Finally, history seems to be extending its influence to ever-widening domains. Soon we may no longer be able even to maintain the basic division of two scientific styles discussed in this paper (which only advocates equal treatment for the equally scientific, but different, historical sciences). The latest researches in cosmology are suggesting that the laws of nature themselves, those supposed exemplars of timeless immanence, may also be contingent results of history! Had the universe passed through a different (and possible) history during its first few moments after the big bang, nature's laws might have developed differently. Thus everything, ultimately, may be a product of history—and we will need to understand, appreciate, and use the principles of historical science throughout our entire domain of zealous research.

This has been an unconventional discharge of an appointed duty—to write a centennial essay on the impact of evolution. I have not written a traditional review. I have not chronicled the major advances and discoveries of evolution. I have tried, instead, to suggest that evolution's essential impact upon the practice of science has been methodological—validating the historical style as equally worthy and developing for it a rigorous methodology, outlined by Darwin himself in his most distinctive (but largely untouted) contribution, and continually refined to the kind of ultimate triumph for homology that molecular phylogeny can provide.

I have presented nothing really new, only a plea for appreciating something so basic that we often fail to sense its value. With a bow to that overquoted line from T. S. Eliot, I only ask you to return to a place well known and see it for the first time.

References

Alvarez, L. W., W. Alvarez, F. Asaro, and H. V. Michel. 1980. Extraterrestrial cause for the Cretaceous-Tertiary extinction. *Science* 208:1095–1108.

Bethell, T. 1985. Agnostic evolutionists: The taxonomic case against Darwin. *Harper's*. Feb., pp. 49–61.

Carr, E. H. 1961. *What Is History?* Vintage Books.

Chambers, R. 1844. *Vestiges of the Natural History of Creation*. London: John Churchill. (Published anonymously.)

Cott, H. B. 1940. *Adaptive Coloration in Animals*. Methuen.

Darwin, C. 1842. *The Structure and Distribution of Coral Reefs*. London: Smith, Elder.

———. 1859. *On the Origin of Species*. London: John Murray.

———. 1862. *The Various Contrivances by which Orchids Are Fertilized by Insects*. London: John Murray.

———. 1868. *Variation of Animals and Plants Under Domestication*. London: John Murray.

———. 1877. *Different Forms of Flowers on Plants of the Same Species*. London: John Murray.

———. 1881. *The Formation of Vegetable Mould, Through the Action of Worms*. London: John Murray.

di Gregorio, M. A. 1984. *T. H. Huxley's Place in Natural Science*. Yale Univ. Press.

Eldredge, N., and J. Cracraft. 1980. *Phylogenetic Patterns and the Evolutionary Process*. Columbia Univ. Press.

Eldredge, N., and S. J. Gould. 1972. Punctuated equilibria: An alternative to phyletic gradualism. In *Models in Paleobiology*, ed. T. J. M. Schopf, pp. 82–115. Freeman, Cooper, and Co.

Ghiselin, M. 1969. *The Triumph of the Darwinian Method*. Univ. of California Press.

Gingerich, P. D. 1983. Rates of evolution: Effects of time and temporal scaling. *Science* 222:159.

Gould, S. J. 1965. Is uniformitarianism necessary. *Am. J. Sci*. 263:223–28.

———. 1971. D'Arcy Thompson and the science of form. *New Lit. Hist*. 2 (2):229–58.

———. 1980. *The Panda's Thumb*. W. W. Norton.

———. 1981. *The Mismeasure of Man*. W. W. Norton.

———. 1982a. Darwinism and the expansion of evolutionary theory. *Science* 216:380–87.

———. 1982b. The meaning of punctuated equilibrium and its role in validating a hierarchical approach to macroevolution. In *Perspectives on Evolution*, ed. R. Milkman, pp. 83–104. Sunderland, Mass.: Sinauer Assoc.

———. 1983. *Hen's Teeth and Horse's Toes*. W. W. Norton.

———. 1984. Smooth curve of evolutionary rate: A psychological and mathematical artifact. *Science* 226:994–95.

———. 1985a. On the origin of specious critics. *Discover*. Jan., pp. 34–42.

———. 1985b. The paradox of the first tier: An agenda for paleobiology. *Paleobiology* 11:2–12.

Hennig, W. 1966. *Phylogenetic Systematics*. Univ. of Illinois Press.

Hooykaas, R. 1963. *The Principle of Uniformity in Geology, Biology, and Theology*. Leiden: E. J. Brill.

Huxley, T. H. 1893. The origin of species [1860]. In *Collected Essays*, vol. 2: *Darwiniana*, pp. 22–79. Appleton.

Kitcher, P. 1985. Darwin's achievement. In *Reason and Rationality in Science*, ed. N. Rescher, pp. 123–85. Pittsburgh: Stud. Phil. Sci.

Lamarck, J. B. 1809. *Philosophie zoologique*. Trans. 1984 by H. Elliot as *Zoological Philosophy*. Univ. of Chicago Press.

Lyell, C. 1842. *Principles of Geology*. 6th ed., 3 vols. Boston: Hilliard, Gray, and Co.

MacArthur, R. H., and E. O. Wilson. 1967. *The Theory of Island Biogeography*. Princeton Univ. Press.

Oken, L. 1805–11. *Lehrbuch der Naturphilosophie*. Jena: F. Frommand.

Raup, D. M., and J. J. Sepkoski, Jr. 1984. Periodicity of extinctions in the geologic past. *PNAS* 81:801–05.

Rifkin, J. 1983. *Algeny*. Viking Press.

Rudwick, M. J. S. 1972. *The Meaning of Fossils*. London: MacDonald.

Ruse, M. 1979. *The Darwinian Revolution: Science Red in Tooth and Claw*. Univ. of Chicago Press.

Sibley, C. G., and J. E. Ahlquist. 1981. The phylogeny and relationships of the ratite birds as indicated by DNA-DNA hybridization. In *Evolution Today*, ed. G. G. E. Scudder and J. L. Reveal, pp. 301–35. Proc. Second Intl. Cong. Syst. Evol. Biol.

———. 1984. The phylogeny of the hominoid primates, as indicated by DNA-DNA hybridization. *J. Mol. Evol*. 20:2–15.

———. 1985. The phylogeny and classification of the Australo-Papuan passerine birds. *The Emu* 85:1–14.

Sober, E. 1984. *The Nature of Selection*. MIT Press.

Steno, N. 1669. *De solido intra solidum naturaliter contento dissertationis prodromus*. Trans. 1916 by J. G. Winter as *The Prodromus of Nicolaus Steno's Dissertation*. Macmillan.

Swainson, W. 1835. *Classification of Quadrupeds*. London: Dr. Lardner's Cabinet Cyclopedia.

Thompson, D. W. 1942. *Growth and Form*. Macmillan.

Vrba, E. S., and N. Eldredge. 1984. Individuals, hierarchies and processes: Towards a more complete evolutionary theory. *Paleobiology* 10:146–71.

The Essence of Royalty: Honey Bee Queen Pheromone

A queen bee induces thousands of worker bees to submit to her hegemony by secreting a potent chemical blend, which is spread by physical contact throughout the colony

Mark L. Winston and Keith N. Slessor

Beekeeping, like most skills, has developed its own shorthand language, by which a word or simple phrase can evoke complex concepts that quickly are understood by those who practice the craft. One such word is "queenright," meaning that a colony's queen is present and thousands of her worker offspring are collaborating diligently on communal tasks. Hidden in this beekeeper jargon, however, lies the most fundamental mystery of social insects: the mechanisms of coordination and integration that mediate the tasks of thousands of individuals into the smoothly functioning unit of the colony. In this article, we shall describe some new findings concerning one aspect of these integrative mechanisms, the honey bee queen's pheromonal influence over worker bees. The discovery of the nature and function of this "essence of royalty" has provided profound insights into the functioning of social-insect colonies, and also has led to some significant commercial applications for crop pollination and beekeeping.

Mark L. Winston, who earned B.A. and M.Sc. degrees at Boston University and his Ph.D. at the University of Kansas, is the author of two recent books about honey bees, The Biology of the Honey Bee *(1987) and* Killer Bees: The Africanized Honey Bee in the Americas *(1992), both published by Harvard University Press. Keith N. Slessor, who received his Ph.D. at the University of British Columbia, is a native British Columbian who finds organic chemistry an excellent medium for both teaching and exploration, and regards it as a challenge to decipher the communication skills of economically important insects. Winston and Slessor are professors at Simon Fraser University, Burnaby, B.C. V5A 1S6, Canada, in the departments of Biological Sciences and Chemistry/Biochemistry, respectively.*

The study of social-insect pheromones, or sociochemistry, is not a new discipline. Many pheromones are known throughout the social insects that are used in myriad tasks such as alarm behavior, orientation, trail marking, nest recognition, mate attraction and others (Bradshaw and Howse 1984; Duffield, Wheeler and Eickwort 1984; Free 1987; Hölldobler and Wilson 1990; Howse 1984; Wilson 1971; Winston 1987). The honey bees, for example, are estimated to produce at least 36 pheromones, which together constitute a chemical language of some intricacy. However, virtually all of the known chemicals are secretions that "release," or elicit, a specific behavior. The queen sociochemicals belong to another class, called primer pheromones, that exercise a more fundamental level of control by mediating worker and colony reproduction and influencing broad aspects of foraging and other behaviors. Further, they can be both stimulatory and inhibitory, depending on the specific function. The existence of primer sociochemicals is well known, but their identification and synthesis have proved difficult.

Indeed, the only primer pheromone that has been identified comes from the queen honey bee (*Apis mellifera* L.); it is secreted by the mandibular glands, a paired set of glands on either side of the queen's head (Figure 2). Here we discuss the research that led to the pheromone's identification, new findings concerning its effects on worker bees and colony functions, transmission mechanisms within colonies, and some applications for crop pollination and beekeeping.

Figure 1. Retinue response is a stereotyped honey bee behavior by which various pheromones are transferred from the queen to the worker bees. In the photograph on the opposite page workers turn to face a queen, touching her with their antennae and licking her. By this means they pick up the pheromone that she exudes on her body. For the next half hour, the workers act as dispersers of pheromone, travelling throughout the nest and contacting other workers more frequently than normal. It is the pheromone itself—including the mixture of substances secreted by the queen's mandibular glands—that induces the retinue response. The effectiveness of the pheromone is demonstrated by experiments in which worker bees form a retinue around a glass lure coated with mandibular-gland extract, as in the photograph above. The response has formed the basis of a bioassay for synthetic pheromones. (Photograph opposite courtesy of Kenneth Lorenzen; photograph above by Keith N. Slessor.)

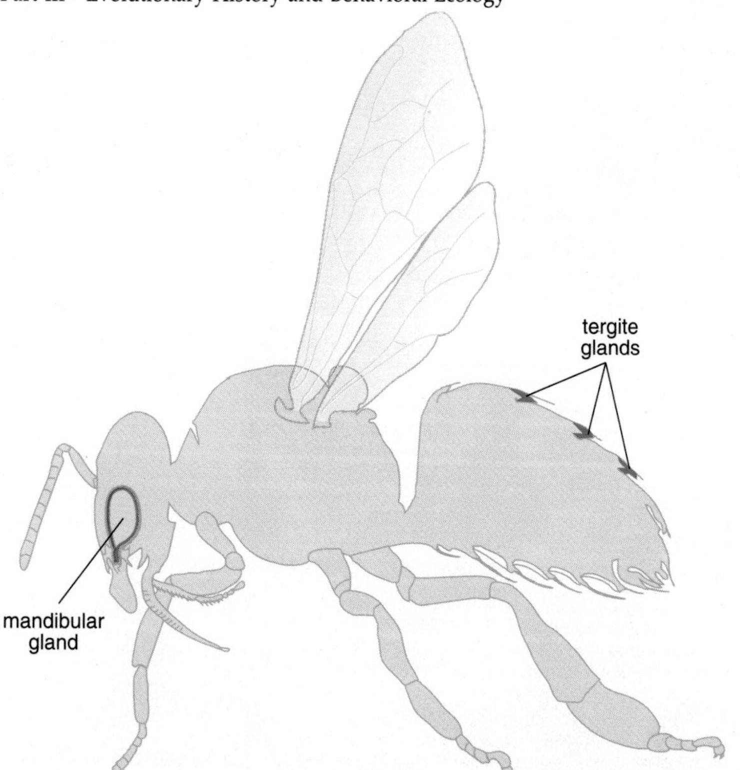

Figure 2. Two sets of glands in the queen honey bee are thought to secrete primer pheromones, the class of pheromones that regulate reproductive behavior in the colony. The best-studied primer pheromones are those produced in the mandibular glands on either side of the head. Each gland is connected to the mandible by a duct; a valve allows the bee to regulate the discharge of secretions. The queen may also synthesize primer pheromone in the tergite glands, which are large subepidermal complexes of glandular cells near the rear of some of the tergites, or dorsal plates. The functions of the secretions from the tergite glands have not been definitively established.

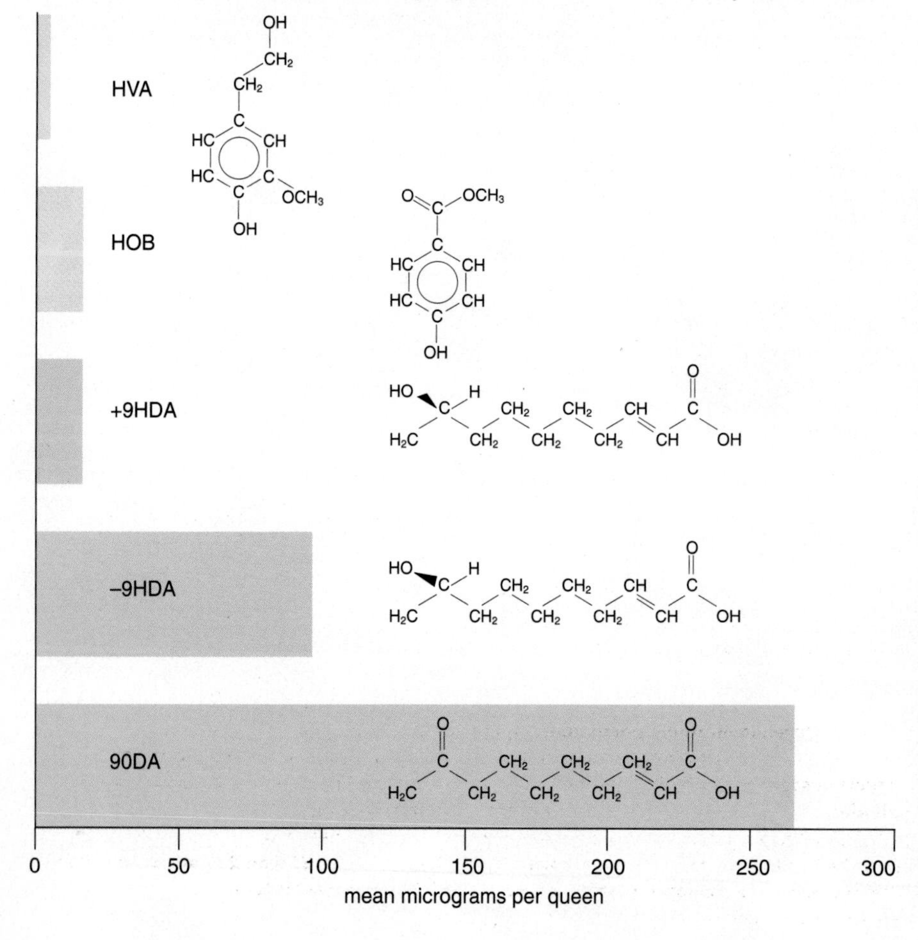

Figure 3. Queen mandibular pheromone is a complex of five substances. Shown here is the average composition of one queen equivalent (Qeq) of mandibular pheromone obtained by analyzing mandibular-gland extracts from 55 queens. (The composition of the pheromone changes over the queen's life span and varies considerably even among mature, mated queens.) A queen equivalent is the average amount of pheromone in a queen's glands at a given time. The most abundant ingredient is (E)-9-keto-2-decenoic acid (9ODA). Another decenoic acid has two optical isomers, both of which must be present to achieve full activity. The isomers are designated (R,E)-(–)-9-hydroxy-2-decenoic acid (–9 HDA) and (S,E)-(+)-9-hydroxy-2-decenoic acid (+9 HDA). The remaining two ingredients are aromatic compounds: methyl *p*-hydroxybenzoate (HOB) and 4-hydroxy-3-methoxyphenylethanol (HVA).

Early Research

For more than 30 years it has been recognized that queen pheromones mediate many colony activities. The pheromones were known to inhibit the rearing of new queens and the development of ovaries in worker bees; they were also observed to attract workers to swarm clusters, stimulate foraging, attract workers to a queen and allow workers to recognize their own queen.

Surprisingly, only two active substances had been identified. These two most abundant compounds in queen mandibular pheromone are the organic acids (E)-9-keto-2-decenoic acid, or 9ODA, which was identified in 1960 (Callow and Johnston 1960, Barbier and Lederer 1960) and (E)-9-hydroxy-2-decenoic acid, or 9HDA, which was identified in 1964 (Callow, Chapman and Paton 1964; Butler and Fairey 1964).

Experiments with the synthetic acids, however, showed that they did not fully duplicate the effects of mandibular extracts or of the queen's presence (reviewed by Free 1987; Winston 1987). Although analysis of crushed bee heads produced laundry lists of chemicals, none was shown on closer examination to be active, and progress stalled.

One reason it took so long to identify queen mandibular pheromone is that bioassays for primer pheromones are difficult and time-consuming. The potency of candidate releaser pheromones can be evaluated by relatively simple means. Orientation pheromones, for example, can be evaluated by counting the number of bees entering a hive after pheromone is deposited at the hive opening. Demonstrating that a synthetic blend or a natural extract is fully equivalent to a primer pheromone, however, may require a battery of bioassays, some of which may take months to complete. To further complicate matters, there is often disagreement over a primer's functions.

The chemical composition of queen mandibular pheromone and its biological effects were not the only controversial topics. Even the means by which the pheromone reaches worker bees was unknown. Is the pheromone transmitted when the workers exchange food? Does it volatilize and spread through the air? Or is it spread by body contact? Here the main obstacle was the minute amount involved: A queen produces less than four ten-thousandths of a gram per day, and a worker bee is able to sense a ten-millionth of the queen's daily production.

Discovery

One of the stereotyped behaviors thought to be elicited by queen pheromones is the retinue response. Worker bees turn toward the queen, forming a dynamic retinue around her. About 10 workers at a time contact her with their antennae, forelegs or mouthparts for periods of minutes. Then the workers leave the queen, groom themselves and move through the nest, making frequent reciprocated contacts with other workers for the next half-hour (Allen 1960; Seeley 1979).

Bioassays involving retinue behavior had given particularly confusing results. Mandibular extracts seemed to be important to worker recognition of the queen and attraction to her. However, the most abundant component of the glands, 9ODA, did not by itself elicit retinue behavior. To add to the confusion, bees had been shown to exhibit the retinue response to queens whose mandibular glands had been surgically removed.

This is how matters stood in 1985 when, quite by accident, we put a glass lure coated with queen mandibular extract next to some stray workers on a laboratory bench. To our surprise, the workers formed a retinue around the lure. Clearly the extract had properties that its major known constituent did not. This was exciting because the retinue response is easily recognizable and takes place immediately. For both reasons it might serve as the basis for a practical bioassay for mandibular pheromone.

Indeed we were able to develop a bioassay based on the retinue response that was a particularly sensitive indica-

Scott Camazine (Photo Researchers, Inc.)

Figure 4. Queen rearing is one of the behaviors inhibited by queen mandibular pheromone. Queens are reared in large cells that hang from the comb. Workers, in contrast, are reared in cells that are capped at the level of the comb. The first sign of queen rearing is the construction of new cups suitable for rearing queens. Most eggs in these cups are laid there by the queen, but workers sometimes move fertilized eggs or very young larvae from worker cells into queen cups. Once the eggs hatch, the larvae are fed a specialized diet that induces them to develop into queens. As the larval queens grow, the cells are elongated downward. They are finally sealed at the end of the larval feeding period.

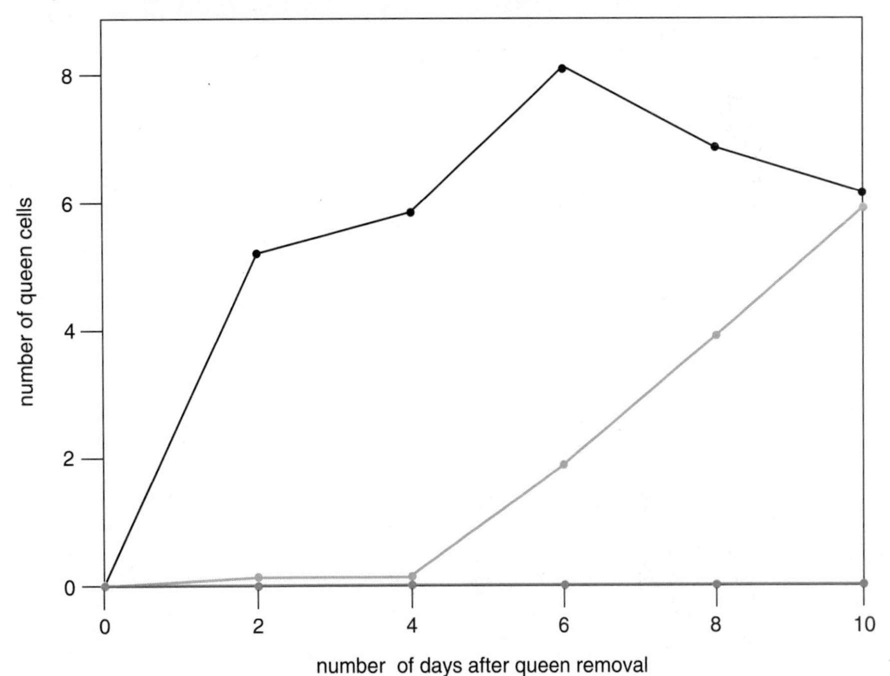

Figure 5. Queen mandibular pheromone suppresses queen rearing. This effect was demonstrated by comparing the number of queen cells in queenless colonies receiving various dosages of synthetic mandibular pheromone with the number of cells in queenright colonies. Shown here are three cases: a queenless colony receiving a glass slide coated only with the solvent used in preparing pheromone *(black)*; a queenless colony receiving one Qeq of synthetic pheromone *(gold)*; and a queenright colony *(red)*. Brood reared more than six days after the queen was removed would have developed into queens of inferior quality or into workers rather than queens.

tor of pheromonal activity. In a series of experiments we used this bioassay to evaluate complete or fractionated mandibular extracts. Interestingly, fractions that contained a single substance were relatively inactive. We began to see retinue behavior around the lure only when we combined certain fractions. Eventually, we identified the components of the active fractions and devised a synthetic blend of five components that duplicated the activity of the natural mandibular secretion (Figure 3) (Kaminski et al. 1990, Slessor et al. 1988).

The ingredients include two forms of 9HDA. This substance is chiral: It has a left-handed and a right-handed form. Earlier, we had demonstrated that one form, or enantiomer, is significantly more effective in evoking some behavioral responses than the other (Winston et al. 1982). We had also determined that one enantiomer predominates in the natural pheromone (Slessor et al. 1985). But it turned out that both enantiomers also have to be present to evoke the full retinue response.

The remaining missing ingredients were two small, aromatic compounds. One of the aromatic compounds, methyl *p*-hydroxybenzoate, or HOB, had been identified earlier (Pain, Hugel and Barbier 1960) but had not been considered to be active as a pheromone. The other aromatic molecule, 4-hydroxy-3-methoxyphenylethanol, or HVA, was first identified in the course of our experiments.

The pheromonal activity of the aromatic molecules was quite unexpected. Most insect pheromones are aliphatic compounds, built up out of straight or branched chains of carbon atoms; they are derived from fatty acids or terpenes. The decenoic acids in mandibular gland pheromone fit this description, for example. The aromatics, which feature closed rings of carbon atoms, are a different class of chemical entirely. Indeed we probably found them only because we reversed the customary experimental procedure. Instead of identifying substances in gland extracts and then screening them for pheromonal activity, we screened for activity and then identified the compounds we had isolated.

Our research into the mandibular gland's composition and activity cleared up several points of confusion. First, the complete ensemble of molecules is necessary for full efficacy. Removing any of the components reduced the pheromone's effect by as much as 50 percent, and individual components are inactive when they are tested alone. This is why the pheromone's major component by itself had failed to elicit retinue behavior.

Second, although we managed to duplicate the effects of mandibular secretions, we should emphasize that we did not duplicate the queen's complete arsenal of pheromonal effects. The mandibular glands are not the only site of production for queen pheromones; they are also probably secreted by the tergite glands, which are located in some of the membranes between the queen's abdominal segments (Renner and Baumann 1964, Velthuis 1970). Moreover, the mandibular and other pheromones may have closely linked or overlapping effects, which may explain why demandibulated queens still evoked the retinue response.

Third, the pheromonal blend a queen produces varies with her age. A newly emerged virgin queen has virtually no pheromone in her glands, but when she begins to mate some six days later, she is secreting the decenoic acids, and her glands contain about half the amount of acid a mated queen's contain. The full blend, includ-

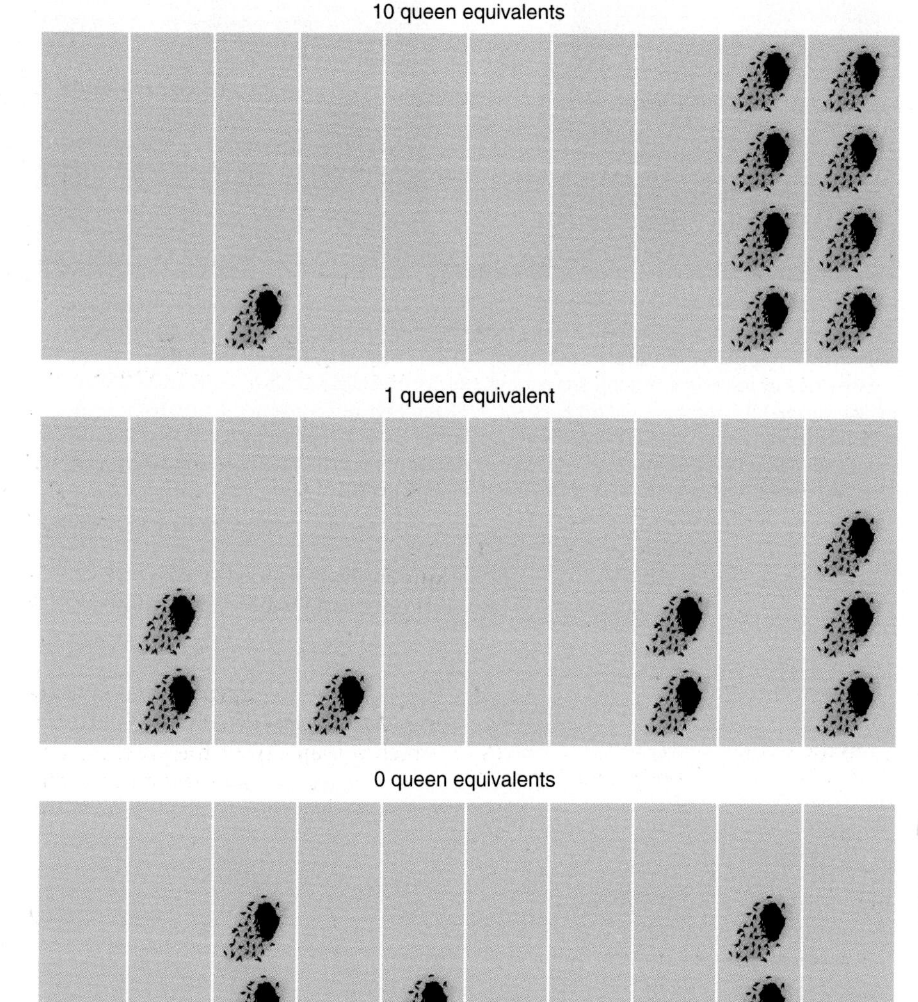

Figure 6. Queen mandibular pheromone plays a role in the timing of swarming. Experiments in which supplemental pheromone was administered on lures to queenright colonies suggest that hives swarm when congestion causes the levels of pheromone reaching workers to fall below a threshold. In the experiments summarized here, 10 Qeq of supplemental pheromone significantly delayed swarming, but one Qeq of supplemental pheromone did not. Other experiments suggest that a lower dose is effective if the pheromone is well dispersed.

Figure 7. Attracting flying workers to swarm clusters is another function of queen mandibular pheromone. The fence posts in this photograph were baited with lures containing one Qeq of synthetic pheromone. The synthetic pheromone was almost as effective as a queen in attracting workers. (Photograph courtesy of K. Naumann.)

ing the aromatic compounds, is secreted only after a queen has finished mating and begins to lay eggs (Slessor et al. 1990). This variation in the composition of the pheromone with the queen's age also may help to explain differing reports of effectiveness.

Finally, the mandibular pheromone is effective over a wide range of dosages. We defined the average amount of pheromone found in the glands of a mated queen to be one queen equivalent (Qeq). Worker bees respond to dosages as low as 10^{-7} Qeq, or a ten-millionth of her daily production (Kaminski et al. 1990, Slessor et al. 1988). This finding was particularly interesting in relation to hypotheses about pheromone transmission.

Functions

Once we were satisfied that we had completely identified queen mandibular pheromone, we began to test the effects of the pheromone on colony functions. One of the first pheromone-mediated behaviors we examined was the inhibition of queen rearing. This function can be easily demonstrated by removing a queen from her colony; the workers become agitated after half an hour and begin rearing new queens within 24 hours. Beekeepers call this "emergency queen rearing," because failure to promptly rear a new queen results in the death of the colony.

To rear queens, workers elongate the cells around a few newly hatched larvae (Figure 4). These larvae are fed a highly enriched food called royal jelly that adult workers produce in their brood-food glands. The specialized diet directs larval development away from the worker pathway and onto the queen pathway. Brood older than six days (three days as eggs and three as larvae) lose their ability to develop into queens. Since no new eggs are laid after the queen is lost, workers have only six days to begin queen rearing before the colony loses its ability to produce a new, functional queen.

We examined the role of mandibular pheromone in emergency queen rearing by comparing queen rearing in queenright colonies, queenless colonies and queenless colonies receiving daily doses of synthetic mandibular pheromone (Winston et al. 1989, 1990). The results were dramatic. When queenless colonies received one Qeq or more of pheromone per day, they made almost no attempt to rear new queens for four days after the queen was removed. Even six days after queen loss, there was no significant difference between the number of queen cells in queenless colonies treated with pheromone and the number in queenright control colonies (Figure 5). We concluded that the queen's mandibular pheromone is largely responsible for the inhibition of queen rearing in queenright colonies.

Colonies also rear new queens in preparation for reproductive swarming, the process of colony division in which a majority of the workers and the old queen leave the colony and search for a new nest. Left behind in the old colony are developing queens, some of which may issue with additional swarms once they emerge, and one of which will reign over the old nest once swarming is completed.

Swarming poses something of a puzzle because it does not occur unless new queens are being reared, but the old queen is present during the initial stages of queen rearing and is presumably still secreting her inhibitory chemicals (Butler 1959, 1960, 1961; Seeley and Fell 1981; Simpson 1958). One hypothesis is that the transmission of queen pheromones is slowed as colonies grow and become more congested prior to swarming. As the amounts of pheromone reaching workers diminish, the workers begin to escape the queen's control.

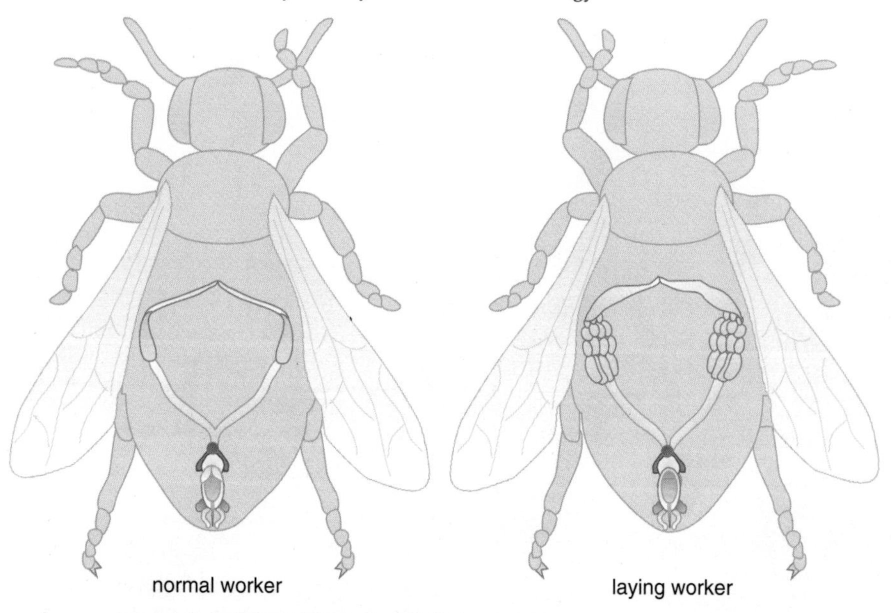

Figure 8. Suppression of worker ovary development, which had been thought to be a function of queen mandibular pheromone, is probably controlled by other primer pheromones, including substances produced by larval and pupal bees. Egg-laying is pre-eminently the queen's function but, under certain conditions, workers' ovaries also develop, and they can then lay eggs. In queenright colonies, a few workers have developed ovaries, but the workers lay few eggs unless the colony becomes queenless. If the queen is lost and the colony fails to rear a new queen, the workers with enlarged ovaries begin to lay eggs. Because workers lack the genital structures that allow the queen to mate and store sperm, any eggs they lay are unfertilized and usually develop into drones.

We tested the role of queen pheromone in reproductive swarming by supplementing the queen's normal secretion with additional, synthetic pheromone. Again, the pheromone had a significant effect on queen rearing. Colonies receiving 10 Qeq per day swarmed an average of 25 days later than control colonies (Figure 6). The pheromone-treated colonies became extraordinarily crowded before they would swarm, indicating just how powerful a suppressant the queen pheromone can be.

For these experiments, the pheromone was presented on stationary lures, but queen bees move about in the nest. In another set of experiments, we tried to better mimic natural conditions by administering the pheromone as a spray. We found that a dose of one Qeq per day of supplemental pheromone was sufficient to suppress swarming. The fact that the better-dispersed spray was active at a much lower dose than the stationary lure supports the hypothesis that the slowing of pheromonal dispersal in congested colonies triggers swarming (Winston et al. 1991).

The queen mandibular pheromone plays two other roles in swarming as well: It attracts flying workers to the swarm cluster, and it stabilizes the cluster, preventing workers from becoming restless and leaving before the swarm reaches a new nest site. We found that the synthetic mandibular pheromone is almost as effective as the queen in attracting workers to the cluster and in keeping them together, at least for the first few hours following cluster formation (Figure 7).

Next we examined the effect of mandibular pheromone on worker ovary development and egg laying. Both are known to be suppressed by queen pheromones. Indeed pheromonal suppression of worker reproduction is one of the hallmarks of advanced insect societies. Primitive social insects suppress nestmate reproduction primarily through dominance interactions. Many wasp and bumblebee queens, for example, seem to maintain dominance by biting and harassing the other females in the colony.

Honey bee workers have the potential to lay unfertilized eggs that can develop into males, or drones. However, some combination of pheromones secreted by the queen and the brood (the developing larvae or pupae) inhibits the maturation of the workers' ovaries, which usually remain small and nonfunctional (Figure 8) (Jay 1970, 1972). Nevertheless, even in queenright colonies a few workers manage to lay eggs (Ratnieks and Visscher 1989, Visscher 1989) and many workers begin egg-laying two or three weeks after the queen and brood are removed (Winston 1987).

The ovary-suppressing substance produced by the queen was thought to be mandibular pheromone, particularly 9ODA (see reviews by Free 1987; Willis, Winston and Slessor 1990; Winston 1987). To test this hypothesis, we removed queens from groups of colonies and applied various doses of mandibular pheromone to the colonies for the next 43 days. Some colonies received daily doses as high as 10 Qeq. To our surprise, workers in queenless, pheromone-treated colonies developed ovaries at the same rate as workers in queenless, pheromone-free colonies (Willis, Winston and Slessor 1990). Apparently queen mandibular pheromone is not involved in the suppression of worker ovary development. Another primer pheromone must be responsible for this effect, possibly the unidentified queen tergite pheromone in combination with brood pheromone.

A final set of experiments demonstrated that queen mandibular pheromone is not just an inhibitor. In addition to suppressing reproductive activity, it can stimulate foraging and brood rearing. To establish this, we applied pheromone to newly founded, queenright colonies and counted the number of foragers, the size of their nectar and pollen loads and the extent of brood rearing (Figure 9). The pheromone did not increase overall foraging activity, but colonies that received one Qeq of supplemental pheromone daily had more pollen foragers, and those foragers carried heavier pollen loads. Pheromone applications increased the amount of pollen entering the nest by 80 percent, and pheromone-treated colonies reared 18 percent more brood, possibly due to the increased amounts of protein-rich pollen being brought into those colonies (Higo et al. 1992).

Transmission

The set of experiments we have just described demonstrated that queen mandibular pheromone has a wide range of functions, but we still did not know how much is produced and secreted by the queen daily or how these important compounds are transmitted to workers. The chemical components of the pheromone are not particularly volatile, and so they probably cannot spread throughout the colony by diffusion alone. The number of workers in an established colony and the rapidity with which the pheromone is lost make it unlikely that the queen is the sole disseminator of the pheromone. Some experiments had suggested that retinue workers act as messengers, transmitting the queen's pheromones to other workers when they touch antennae or mouthparts (Juška, Seeley and Velthuis 1981; Seeley 1979). Pheromone could not be found on worker bees, however, possibly because it was present in concentrations below the limit of detection of the instruments then in use.

We were able to follow pheromone secretion and transmission using the more sensitive chromatographic and spectroscopic equipment now available, as well as radiolabelled pheromone provided by Glenn Prestwich and Francis Webster of the State University of New York at Stony Brook and Syracuse (Webster and Prestwich 1988). Quantitative measurements of rates of production, transfer and loss allowed us to construct a mathematical model of pheromone dispersal. Our results essentially confirm the messenger-bee hypothesis, but they raise many interesting questions as well (Figure 10) (Naumann et al. 1991).

We first determined that a queen typically has five micrograms, or 0.001 Qeq, of pheromone on her body at any one time. To determine how much a queen secretes daily, we removed queens from colonies and measured the pheromone that built up on their cuticles. It turned out that a queen secretes between 0.2 and 2 Qeq of pheromone per day. These results fit well with our within-colony function studies, where pheromonal effects typically appeared at a dosage of about 1 Qeq.

We then quantified the processes by which pheromone is transferred or lost. For example, we allowed workers to contact the queen for different periods and then measured the amount of pheromone they had picked up. We found that each process could be approximated by a first-order rate equation. In other words, the amount of pheromone transferred during any short interval is proportional to the quantity present at the source during that interval; the constant of proportionality is called the rate constant. The first-order equation yields an exponential loss of material at the source; the "half life" is inversely proportional to the rate constant. We calculated the rate constants for all of the transmission pathways from our experimental data and then used the value we had obtained for the queen's daily secretion and the rate equations to determine the amount of pheromone that would reach worker bees each day.

Our confidence in the model was bolstered when it predicted rates of pheromone transfer that are consistent with bee behavior. For example, the model predicts a flux of about one Qeq of pheromone through the nest per day; in our function experiments, a similar dose was typically the most active. The model also predicts that the amount of pheromone passed between workers will drop below the detectable level (10^{-7} Qeq) about 30 minutes after it is picked up from the queen, which is about how long it takes for workers to become restless after the queen is removed. Finally, the model predicts that pheromone deposited by the queen in the wax comb will become undetectable in a few hours, which is about how long it takes the bees to begin rearing new queens.

The retinue workers do indeed remove the greatest fraction of the queen's pheromonal production. Not all bees in the retinue pick up and transfer the same amount of pheromone, however. There are two types of retinue bees, which we call licking and antennating messengers (Figure 11). The lickers touch the queen with their tongues, forelegs and mouthparts. The antennators brush the tips of their antennae lightly and quickly over her body. Only about 10 percent of the bees are lickers, but they pick up over half the queen's pheromonal secretion. Each antennating worker picks up much less pheromone and passes on much less in subsequent contacts. Why some bees lick and others antennate is not known, nor do we understand the implications of this dual transmission system for the colony's social structure.

The wax comb also plays a role in the transfer of pheromone, although the queen deposits only 1 percent of her production there. Workers pick up a little of the pheromone the queen deposits. The remaining pheromone is probably released slowly over the next few hours. The slow release may explain why empty comb that has had brood in it is attractive to adult workers. The queen spends most of her time in the brood area, and her pheromonal footprints would be concentrated there.

Figure 9. Pollen foraging is yet another behavior influenced by queen mandibular pheromone, but in this case the pheromone stimulates rather than suppresses the behavior. A bee transports pollen in corbiculae, or pollen baskets, on the outer surfaces of the hind legs. The photograph shows a bee with heavily laden pollen baskets. Under some circumstances, queen mandibular pheromone stimulates workers to collect more pollen. (Photograph courtesy of Kenneth Lorenzen.)

One of our more surprising findings is that pheromone is lost primarily through internalization. The queen herself internalizes some 36 percent of the pheromone she produces. She swallows some of it, some is adsorbed on or bound to her cuticle, and some moves through the cuticle into the blood (hemolymph) system. What purpose internalization serves is unclear. Even if the internalized pheromone retains the same chemical identity, it is no longer available to the colony. It seems odd that such inefficient use is made of the queen's pheromone secretion. It is possible that the model somewhat overesti-

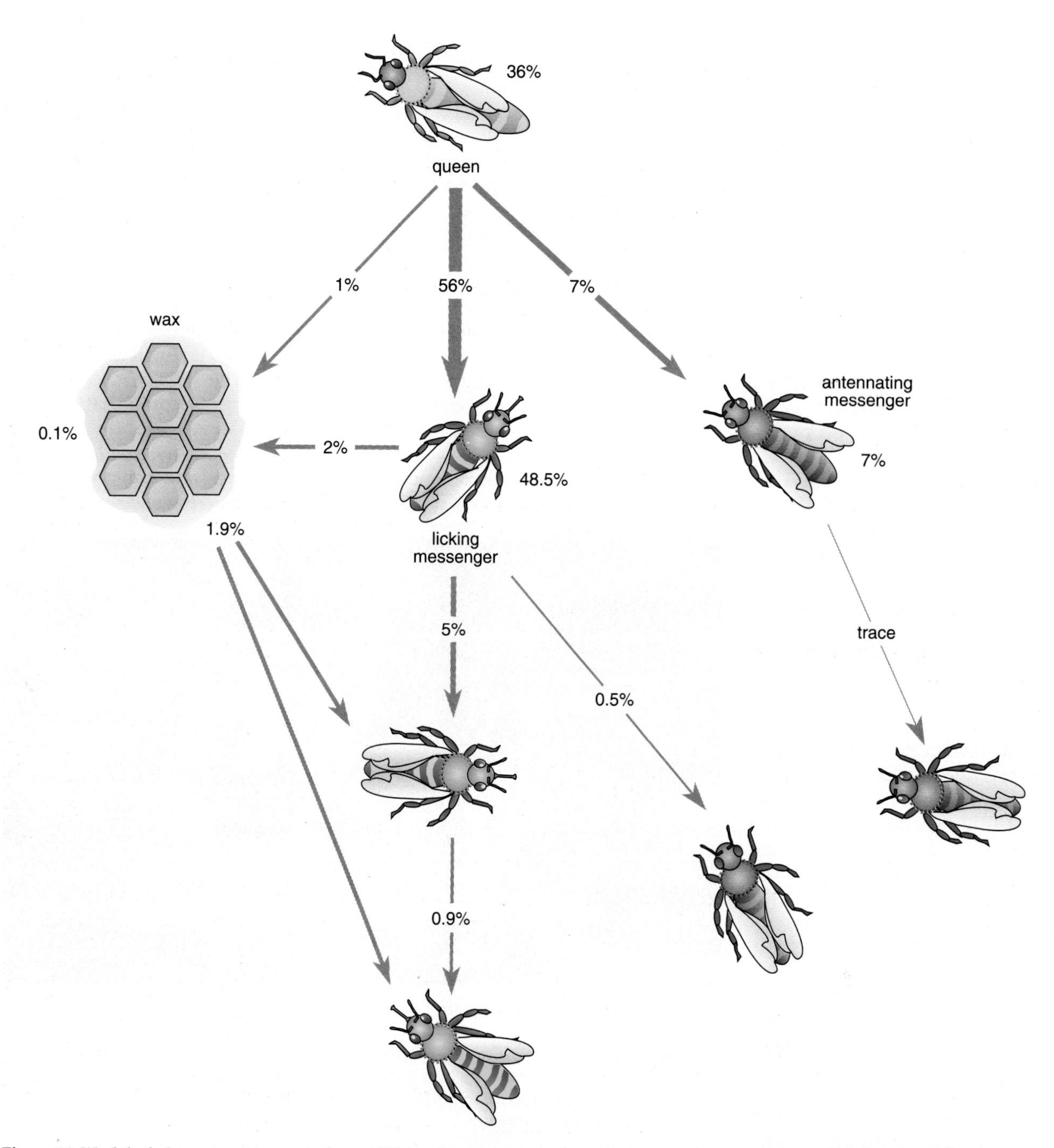

Figure 10. Model of pheromonal transmission within a colony was constructed from data obtained in a series of experiments with radiolabelled 9ODA as part of the synthetic pheromone blend. The pheromone is transmitted primarily by messenger bees, which gather it from the queen either by licking her body or by stroking with the antennae. Some pheromone is also deposited by the queen on the wax comb and picked up by worker bees. Only one bee in 10 is a licker, but the lickers gather much more pheromone than antennating bees and transfer more in subsequent contacts. For unknown reasons, large amounts of pheromone are internalized both by the queen herself and by the messenger bees. The numbers give the percentage of the queen's daily production that is transmitted in the indicated directions.

mates the queen's internalization of pheromone, since this was the only process that did not seem to conform to a first-order rate equation.

On the other hand, the messenger bees also internalize pheromone at a surprising rate (Figure 12). The licker bees swallow 40 percent of the pheromone they pick up. Most of the rest is quickly transferred from the worker's head to her abdomen by a combination of passive transport and active grooming (Naumann 1991). There the pheromone passes into and through the cuticle. Some of it ends up in the blood system within 30 to 60 minutes, and some remains bound to or adsorbed on the cuticle. In short, between the queen and the workers, nearly all of the pheromone is eventually internalized.

The mechanisms by which worker bees perceive the pheromone are not well understood. Sensory structures on a bee's antennae allow it to detect some odors at a concentration equal to a tenth or a hundredth of the concentration people can detect. The arrival of odorant at receptor cells inside the antennae causes associated nerve cells to fire, and the firing can initiate further behavioral and physiological changes. In most insects pheromones act by similar mechanisms.

But what about the large amounts of pheromone that are internalized? One hypothesis is that once the pheromone is translocated across the cuticle, its constituents or their breakdown products become active as hormones rather than as pheromones. We are only beginning to explore this intriguing hypothesis.

It also is unclear why the queen produces such a large amount of mandibular pheromone. The answer may be rooted in queen-worker conflict and the evolution of queen dominance in highly social insects. As we have mentioned, there are almost no dominance interactions in a normally functioning honey bee colony. Instead the queen controls the workers entirely through her sophisticated arsenal of pheromones. Perhaps over evolutionary history the workers and queen engaged in a kind of chemical arms race, where the workers began to break down the queen's compounds more rapidly and the queen responded by secreting more pheromone. Thus the swallowing and absorbing of pheromone by workers that we find so puzzling may be an attempt to catabolize the queen's pheromonal dominance and escape her control. Although this idea is speculative, it is certainly plausible. Indeed, this concept is reminiscent of most families and societies, which exhibit a complex blend of cooperation and conflict, with some objectives in common but with each individual also having its own goals.

Commercial Applications

We have been investigating commercial applications of queen pheromone for beekeeping and crop pollination. Several beekeeping applications for queen mandibular pheromone are already close to commercial realization. For example, we have shown that the pheromone allows packages of worker bees to be shipped without queens (Naumann et al. 1990). Beekeepers often establish a colony by buying a kilogram of workers and a queen in a wire package, but queens are typically produced by beekeepers who specialize in this art, and so they may come from a different source. The pheromone's ability to delay swarming also might be useful in bee management, since frequent swarming reduces honey production and can threaten colony survival. The pheromone's ability to stimulate pollen foraging and brood rearing might have commercial importance as well, since substantial increases in colony growth rates can be achieved by these means. Finally, we are investigating the use of mandibular pheromone as an attractant for swarms. Such an attractant would be useful for the monitoring and control of honey bee diseases and the Africanized bee (the so-called "killer bee").

The most significant application of queen mandibular pheromone, however, may be in assisting crop pollination. Managed crop pollination is economically the most important function of bees. Colonies of bees are moved to blooming crops in order to provide enough bees to ensure good pollination and seed set. Beekeepers receive up to $45 per colony for this service, which is essential for the production of many fruits and vegetables. Not only does pollination affect yield, it also is often closely associated with the quality of fruits, vegetables and seeds. Each year the honey bee pollinates crops worth at least $1.4 billion in Canada and $9.3 billion in the United States

Figure 11. Licking and antennating behavior are both part of the retinue response; thus actions induced by the pheromone are also responsible for its dissemination through the colony. In the upper photograph at least one worker bee can be seen licking the queen's abdomen; in the lower photograph several workers make antennal contact. (Photographs courtesy of Kenneth Lorenzen.)

(Robinson, Nowogrodzki and Morse 1989; Scott and Winston 1984). Although hundreds of thousands of bee colonies are moved to crops each year, many crops are still inadequately pollinated. Reduced yields, occasional crop failures and lowered crop quality can all be attributed to this problem (Free 1970; Jay 1986; McGregor 1976; Robinson, Nowogrodzki and Morse 1989).

Since queen pheromone is highly attractive to flying workers, we thought it possible that pollination could be improved by spraying crops with a dilute blend of pheromone. To test this hypothesis, we sprayed blocks of trees or berries with various concentrations of pheromone and monitored the number of workers attracted to the blocks and the crop yields. In almost all of the experiments, up to twice the number of bees visited treated blocks compared to untreated ones.

The effect of pheromone spraying on yield was more variable. In preliminary trials with apple trees we found no increases in yield or improvements in fruit quality. However, pear trees sprayed with pheromone produced fruit that was heavier by an average of 6 percent, which translated to an increase of 30 percent in profits, or about $1,055 per hectare once spraying costs were deducted. Cranberries treated with pheromone yielded about 15 percent more berries by weight, which increased net returns by $4,465 per hectare averaged over two years (Currie, Winston and Slessor 1992; Currie, Winston, Slessor and Mayer 1992). In some cases there was no improvement, whereas in others the yield increase was much greater than these average values. Clearly, if we are able to exercise in the farmer's field the kind of sociochemical control the queen exercises in the nest, we can dramatically expand the economic value of this already beneficial social insect.

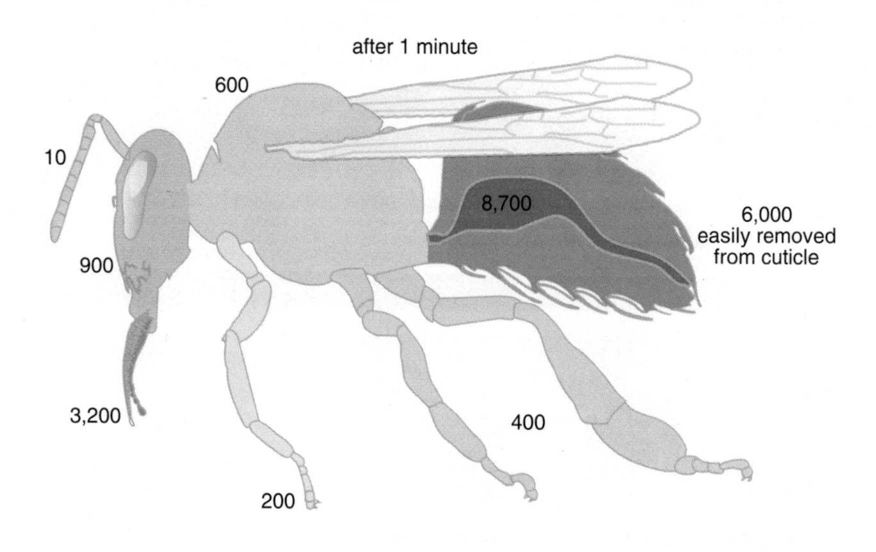

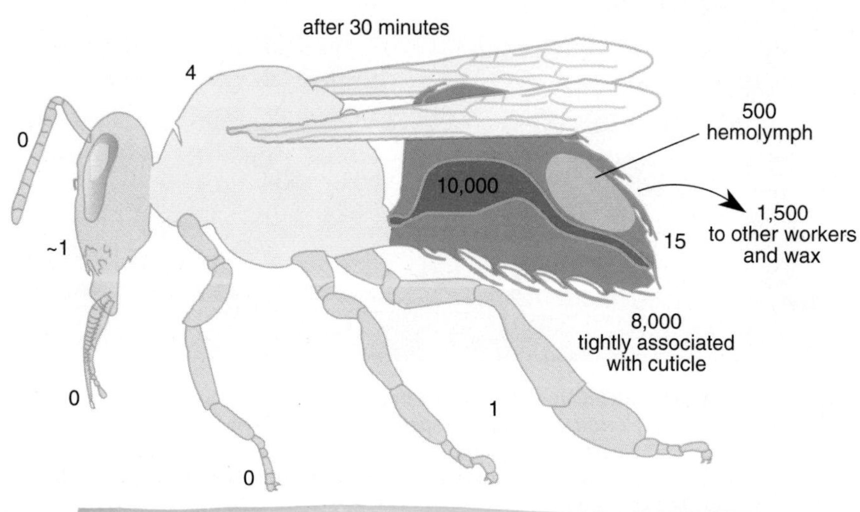

Figure 12. Maps of pheromone on a licking messenger bee illustrate one of the mysteries of pheromone transmission: Most of the pheromone is quickly internalized. The distribution of the pheromone is shown one minute (*upper diagram*) and 30 minutes (*lower diagram*) after the bee contacted the queen. To quantify the internalization of the pheromone by worker bees, 20,000 picograms of radiolabelled 9ODA (the amount a worker typically picks up in her contact with the queen) was applied topically to individual bees. Each bee was then isolated, so that the pheromone could not be transferred to other workers. After a delay, the bee was washed with methanol to dissolve any pheromone on the body surface. The crop and gut were then excised. The amounts of 9ODA in the washes, excised organs and the corpse were determined by liquid scintillation counting. The corpse scintillation count indicated how much pheromone had passed into the body or was so closely associated with the cuticle that it could not be removed by methanol. The pheromone depicted as being external remains available to the colony; it can be picked up by other workers or deposited on the wax comb. The internalized pheromone is effectively removed from circulation.

Acknowledgments

We are grateful to all of our collaborators in the research this article describes, including: J. H. Borden, S. J. Colley, R. W. Currie, H. A. Higo, L.-A. Kaminski, G. G. S. King, K. Naumann, T. Pankiw, E. Plettner, G. D. Prestwich, F. X. Webster, L. G. Willis and M. H. Wyborn. We would also like to acknowledge the technical assistance of P. LaFlamme. Funding for the research came from the Natural Sciences and Engineering Research Council of Canada, the Science Council of British Columbia, the Wright Institute and Simon Fraser University.

Bibliography

Allen, M. D. 1960. The honeybee queen and her attendants. *Animal Behavior* 8:201–08.

Barbier, J., and E. Lederer. 1960. Structure chimique de la "substance royale" de la reine d'abeille (*Apis mellifica*). *Comptes Rendus des Séances de L'Académie de Science* (Paris) 251:1131–35.

Bradshaw, J. W. S., and P. E. Howse. 1984. Sociochemicals of ants. In: *Chemical Ecology of Insects*, W. J. Bell and R. T. Carde (eds). Sunderland, Mass.: Sinauer.

Butler, C. G. 1959. Queen substance. *Bee World* 40:269–75.

Butler, C. G. 1960. The significance of queen substance in swarming and supersedure in honey-bee (*Apis mellifera*) colonies. *Proceedings of the Royal Entomological Society* (London) (A) 35:129–32.

Butler, C. G. 1961. The scent of queen honey bees (*Apis mellifera*) that causes partial inhibition of queen rearing. *Journal of Insect Physiology* 7:258–64.

Butler, C. G., and E. M. Fairey. 1964. Pheromones of the honey bee: biological studies of the mandibular gland secretion of the queen. *Journal of Apicultural Research* 3:65–76.

Callow, R. K., J. R. Chapman and P. N. Paton. 1964. Pheromones of the honey bee: Chemical studies of the mandibular gland secretion of the queen. *Journal of Apicultural Research* 3:77–89.

Callow, R. K., and N. C. Johnston. 1960. The chemical constitution and synthesis of queen substances of honey bees (*Apis mellifera* L.). *Bee World* 41:152–3.

Currie, R. W., M. L. Winston and K. N. Slessor. 1992. Impact of synthetic queen mandibular pheromone sprays on honey bee pollination of berry crops. *Journal of Economic Entomology* (in press).

Currie, R. W., M. L. Winston, K. N. Slessor and D. F. Mayer. 1992. Effect of synthetic queen mandibular pheromone sprays on pollination of fruit crops by honey bees. *Journal of Economic Entomology* (in press).

Duffield, R. M., J. W. Wheeler and G. C. Eickwort. 1984. Sociochemicals of bees. In: *Chemical Ecology of Insects*, W. J. Bell and R. T. Carde (eds). Sunderland, Mass: Sinauer.

Free, J. B. 1970. *Insect Pollination of Crops*. London: Academic Press.

Free, J. B. 1987. *Pheromones of Social Bees*. Ithaca: Cornell University Press.

Higo, H. A., S. J. Colley, M. L. Winston and K. N. Slessor. 1992. Effects of honey bee queen mandibular gland pheromone on foraging and brood rearing. *Canadian Entomologist* 124:409–418.

Hölldobler, B., and E. O. Wilson. 1990. *The Ants*. Cambridge, Mass.: Harvard University Press.

Howse, P. E. 1984. Sociochemicals of termites. In: *Chemical Ecology of Insects*, W. J. Bell and R. T. Carde (eds). Sunderland, Mass: Sinauer.

Jay, S. C. 1970. The effect of various combinations of immature queen and worker bees on the ovary development of worker honey bees in colonies with and without queens. *Canadian Journal of Zoology* 48:168–73.

Jay, S. C. 1972. Ovary development of worker honeybees when separated from worker brood by various methods. *Canadian Journal of Zoology* 50:661–4.

Jay, S. C. 1986. Spatial management of honey bees on crops. *Annual Review of Entomology* 31:49–66.

Juška, A., T. D. Seeley and H. H. W. Velthuis. 1981. How honeybee queen attendants become ordinary workers. *Journal of Insect Physiology* 27:515–19.

Kaminski, L.-A., K. N. Slessor, M. L. Winston, N. W. Hay and J. H. Borden. 1990. Honey bee response to queen mandibular pheromone in a laboratory bioassay. *Journal of Chemical Ecology* 16:841–49.

McGregor, S. E. 1976. *Insect Pollination of Cultivated Crop Plants*. Agricultural Handbook No. 496. Washington, D.C.: U.S. Department of Agriculture.

Naumann, K. 1991. Grooming behaviors and the translocation of queen mandibular gland pheromone on worker honey bees. *Apidologie* 22:523–531.

Naumann, K., M. L. Winston, M. H. Wyborn and K. N. Slessor. 1990. Effects of synthetic honey bee (Hymenoptera: Apidae) queen mandibular gland pheromone on workers in packages. *Journal of Economic Entomology* 83:1271–75.

Naumann, K., M. L. Winston, K. N. Slessor, G. D. Prestwich and F. X. Webster. 1991. The production and transmission of honey bee queen (*Apis mellifera* L.) mandibular gland pheromone. *Behavioral Ecology and Sociobiology* 29:321–32.

Pain, J., M.-F. Hugel and M. C. Barbier. 1960. *Comptes Rendus des Séances de L'Académie de Science* (Paris) 251:1046–8.

Ratnieks, F. L. W., and P. K. Visscher. 1989. Worker policing in the honey bee. *Nature* 342:796–97.

Renner, M., and M. Baumann. 1964. Uber komplexe von subepidermalen drusenzallen (Duftdrusen?) der bienenkonigin. *Naturwissenschaften* 51:68–9.

Robinson, W. S., R. Nowogrodzki and R. A. Morse. 1989. The value of honey bees as pollinators of U.S. crops. *American Bee Journal* 129:411–23, 477–87.

Scott, C. D., and M. L. Winston. 1984. The value of bee pollination to Canadian apiculture. *Canadian Beekeeping* 11:134.

Seeley, T. D. 1979. Queen substance dispersal by messenger workers in honey bee colonies. *Behavioral Ecology and Sociobiology* 5:391–415.

Seeley, T. D., and R. D. Fell. 1981. Queen substance production in honey bee (*Apis mellifera*) colonies preparing to swarm. *Journal of the Kansas Entomological Society* 54:192–96.

Simpson, J. 1958. The factors which cause colonies of *Apis mellifera* to swarm. *Insectes Sociaux* 5:77–95.

Slessor, K. N., G. G. S. King, D. R. Miller, M. L. Winston and T. L. Cutforth. 1985. Determination of chirality of alcohol or latent alcohol semiochemicals in individual insects. *Journal of Chemical Ecology* 11:1659–67.

Slessor, K. N., L.-A. Kaminski, G. G. S. King, J. H. Borden and M. L. Winston. 1988. Semiochemical basis of the retinue response to queen honey bees. *Nature* 332:354–56.

Slessor, K. N., L.-A. Kaminski, G. G. S. King and M. L. Winston. 1990. Semiochemicals of the honey bee queen mandibular glands. *Journal of Chemical Ecology* 16:851–60.

Velthuis, H. H. W. 1970. Queen substances from the abdomen of the honey bee queen. *Zeitschrift fuer vergleichende Physiologie* 70:210–22.

Visscher, P. K. 1989. A quantitative study of worker reproduction in honey bee colonies. *Behavioral Ecology and Sociobiology* 25:247–54.

Webster, F. X., and G. D. Prestwich. 1988. Synthesis of carrier-free tritium-labelled queen bee pheromone. *Journal of Chemical Ecology* 14:957–62.

Willis, L. G., M. L. Winston and K. N. Slessor. 1990. Queen honey bee mandibular pheromone does not affect worker ovary development. *Canadian Entomologist* 122:1093–99.

Wilson, E. O. 1971. *The Insect Societies*. Cambridge, Mass.: Harvard University Press.

Winston, M. L., K. N. Slessor, M. J. Smirle and A. A. Kandil. 1982. The influence of a queen-produced substance, 9HDA, on swarm clustering behavior in the honey bee *Apis mellifera* L. *Journal of Chemical Ecology* 8:1283–88.

Winston, M. L. 1987. *The Biology of the Honey Bee*. Cambridge, Mass.: Harvard University Press.

Winston, M. L., K. N. Slessor, L. G. Willis, K. Naumann, H. A. Higo, M. H. Wyborn and L.-A. Kaminski. 1989. The influence of queen mandibular pheromones on worker attraction to swarm clusters and inhibition of queen rearing in the honey bee (*Apis mellifera* L.). *Insectes Sociaux* 36:15–27.

Winston, M. L., H. A. Higo and K. N. Slessor. 1990. Effect of various dosages of queen mandibular gland pheromone on the inhibition of queen rearing in the honey bee (Hymenoptera: Apidae). *Annals of the Entomological Society of America* 83:234–38.

Winston, M. L., H. A. Higo, S. J. Colley, T. Pankiw and K. N. Slessor. 1991. The role of queen mandibular pheromone and colony congestion in honey bee (*Apis mellifera* L.) reproductive swarming. *Journal of Insect Behavior* 4:649–659.

How Do Planktonic Larvae Know Where to Settle?

In some species the key is a chemical cue, which induces settling through biochemical pathways similar to those operating in the human nervous system

Aileen N. C. Morse

A larva of the red abalone, a large marine snail that clings to rocks in shallow waters off California, hardly seems in control of its fate. Born in a swarm of thousands when a cloud of abalone sperm meets a cloud of abalone eggs, it begins its life adrift. For at least a week and often longer it is carried by currents and tides. Finally, far from its parents and most of its siblings, it settles to the bottom and begins to metamorphose into a juvenile. Its final landing place is where it will live for the first few months of its life.

The process looks random, like the scattering of seeds on the wind. It is not. The abalone larva, though no more than a hundredth of an inch across, is not a passive drifter, and does not necessarily stop at the first place it lands. Its head is surrounded by a fringe of cilia, like the head of a dust mop. By beating these cilia the larva propels itself on a bouncing trajectory through the water column, touching down on the bottom for a few minutes, then launching itself back on the current—again and again, until it happens upon an environment that meets its needs. The crucial factor that makes an environment suitable for the abalone is the presence of coralline red algae, which are plants that grow like a crust on rocks. When the larva touches down on a patch of such algae, it settles for good. Only then does it start to metamorphose, feeding on the algae as it grows.

My colleagues and I at the University of California at Santa Barbara have been studying this remarkable process, and others like it, for 14 years. We are beginning to understand it in some detail at both the cellular and the molecular levels. The abalone larva, we have found, recognizes a unique chemical cue on the surface of the red algae—a small peptide molecule that acts much like the most common neurotransmitter in the human central nervous system. This algal peptide triggers a cascade of chemical reactions in the target sensory cells of the larva; the reactions are similar to those in a human neuron, and they culminate in the electrical "firing" of the sensory cell. In this way, the chemical signal perceived by the larva in its environment is translated into an electrochemical signal recognized by the larval nervous system. Activation of the nervous system turns on the behavioral and cellular processes required for the transition from the arrested larval stage to that of the developing juvenile. Far from being random, the process by which the larvae of the red abalone come to settle on crustose forms of coralline red algae, and only on those algae, is tightly controlled.

Nor is this kind of control unique to the red abalone. We and our colleagues working in several other laboratories have observed similar processes in a number of shallow-water marine invertebrates that attach themselves either permanently or temporarily to a surface close to the ocean floor. Studies such as these are bridging the gulf between two disciplines that deal in vastly different scales: ecology and molecular biology. One of the important questions in ecology is how animals and plants find an environment in which they can successfully survive and reproduce, and how they come to be distributed in the particular way in which they are found in their natural habitats. In the case of some molluscs, in marine worms and in corals, laboratory examination of these complicated processes has begun to answer this question at the molecular level, by first looking to the environment for clues.

Aileen N. C. Morse is Assistant Research Biologist in molecular marine biology at the Marine Science Institute of the University of California at Santa Barbara. Her undergraduate education was obtained at the University of Edinburgh; her Ph. D. was awarded in 1985 by the University of California at Santa Barbara. In addition to her studies of molecular marine ecology in California, she does field work in (and under) the waters of the Caribbean and other tropical seas, and she is an invited participant in the U.S.-India Binational Program on Marine Biological Research. Address: Marine Science Institute, University of California at Santa Barbara, Santa Barbara CA 93106.

Six Million Abalones

Monday is a special day in our laboratory. Each Monday morning Neal Hooker selects two male and two female red abalones from an outside holding tank. The tank is stocked by divers who regularly harvest adult abalones from rocks in the Santa Barbara Channel. For several weeks the abalones have been fattened on kelp, their favorite food, to activate the synchronized development of eggs and sperm in the gonads. The selected abalones have gonads that are ripe and full—green with eggs or white with sperm.

They are brought indoors to a thermostatically controlled larval rearing room. As is true for reproductive processes in many species, maximum efficiency is attained within a very narrow temperature range. We determined experimentally that 15 degrees Celsius is the optimum temperature for development of eggs and sperm, for spawning and fertilization, and for hatching, larval development and subsequent rearing of the red abalone. The males and females are placed in separate dish pans containing seawater, to which we have added a trace of hydrogen peroxide.

Figure 1. Life cycle of the red abalone, a large snail that lives in shallow waters along the California coast, includes a brief interval as a free-swimming larva and a much longer period as a bottom-dwelling juvenile and adult. The abalones breed by releasing clouds of sperm (*white*) and eggs (*green*) directly into the seawater *(1)*. The fertilized eggs grow into larvae that drift and swim in the plankton *(2)*, dispersing from their place of origin and living all the while on a food source derived from the yolk sac. After about seven days the larvae are competent to settle, but they do not actually take up residence on the bottom until they happen upon a suitable environment. In this phase of their life their characteristic activity is "bottom-hopping" *(3)*: they sink to the bottom, linger there briefly, and then swim back to the surface, again and again—until they chance to land on a patch of coralline red algae, a marine plant that encrusts rocks. At this point the abalones settle permanently and begin the irreversible process of metamorphosis. As juveniles *(4)* they graze on the coralline algae; later, as adults *(5)*, they forage on larger plants such as brown kelps.

Within two hours the animals are ready to release their gametes. To protect the highly sensitive sperm and eggs from the possible deleterious effects of oxidation by hydrogen peroxide, the breeding adults are transferred to fresh seawater. Each female releases a cloud of millions of eggs through the respiratory holes that run the length of the shell. Similarly, each male releases hundreds of millions of sperm into the water. Hooker decants the eggs and a small portion of the sperm and mixes them together. By the end of the morning we have six million fertilized abalone eggs.

The hydrogen peroxide trick is one Daniel Morse discovered 14 years ago Divers had reported that in the wild, abalones often spawn in unison. They tend to spend the day in piles of 10 to 50 mature animals. When one abalone spawns, some of its neighbors in the pile have been seen to follow suit. This suggested that animals in the vicinity of a newly released cloud of sperm or eggs may be detecting a chemical "release cue," given off in the cloud. Examination of clouds of sperm and eggs in the laboratory revealed the presence of a particular class of hormones known as prostaglandins. Prostaglandins are "local" hormones; they are short-lived and alter the activities of the cells in which they are synthesized, and in adjoining cells, in a site-specific manner. One of their known functions in higher animals is to mediate physiological reactions that in turn regulate a number of reproductive functions.

It was found that abalones could be induced to spawn in the laboratory by adding prostaglandin to seawater. The prostaglandin is actively taken up by the gonadal tissues; when a particular concentration is reached, the animals spawn. Although this method of stimulating spawning proved highly effective, it was also prohibitively expensive, particularly for the hatchery industries. Hence Morse sought an alternative technique.

In nature, the synthesis of prostaglandins is an endogenous enzymatic process. In the final synthetic step, arachidonic acid is transformed into prostaglandin through the action of an enzyme, in a reaction that also requires the presence of an "active" form of oxygen. It is this process that we exploit in our Monday-morning breeding procedure. Arachidonic acid is always present in gonadal tissues, and it turns out that the active form of oxygen supplied by hydrogen peroxide will initiate its conversion into prostaglandin. In this scheme the abalones produce the spawning trigger themselves.

If the eggs or sperm are not fully developed within the gonad, neither prostaglandin nor hydrogen peroxide are effective inducers of spawning. Our method thus ensures that the eggs and the sperm obtained are fully viable for fertilization and subsequent development.

After the abalones spawn, they cannot spawn again for several months. They are returned to the ocean, to the same spot where they were picked up. Some red abalones in the Santa Barbara Channel have been through our laboratory more than once. They do not suffer from the experience.

The fertilized eggs take about 14 hours to hatch. The free-swimming larvae that emerge take some of the yolk with them and feed on it as they grow. This means they do not have to be supplied with a food source during larval development. Indeed, they require little attention at all—just gently flowing seawater held at 15 degrees Celsius. An antibiotic bath given just after hatching

Figure 2. Male red abalone, stimulated by a small amount of hydrogen peroxide added to the water in a laboratory tank, broadcasts a cloud of sperm. In a 30-minute spawning period it releases hundreds of millions of sperm cells. In both males and females hydrogen peroxide stimulates the production of prostaglandin, a reproductive hormone, which in turn triggers the release of sperm or eggs. This technique for artificially inducing gamete release has made it practical to breed abalones in the laboratory, as well as in commercial hatcheries. (Photograph courtesy of Larry Friesen.)

protects the larvae from destruction by bacteria. Both in the laboratory and in the hatchery, bacteria can thrive in the artificially high densities of animals in rearing chambers. This results in abnormal larval development and greatly reduced numbers of survivors.

During the first seven days following fertilization, larvae undergo a programmed series of developmental changes. On day seven, however, all further larval development is arrested. The larvae have developed to the stage of "competence," meaning they are ready to end their planktonic swimming life-style and settle to the bottom; there they will metamorphose into young juveniles—the first step toward adulthood. After seven days they also are ready for research.

Each Monday, then, our research cycle begins anew: We produce a new batch of larvae, and begin experiments with the batch from the previous week. The ease with which we can induce abalones to synchronously spawn under controlled physiological conditions and thus produce and rear large numbers of uniformly developed larvae is one reason so much of our work has been done with this invertebrate species.

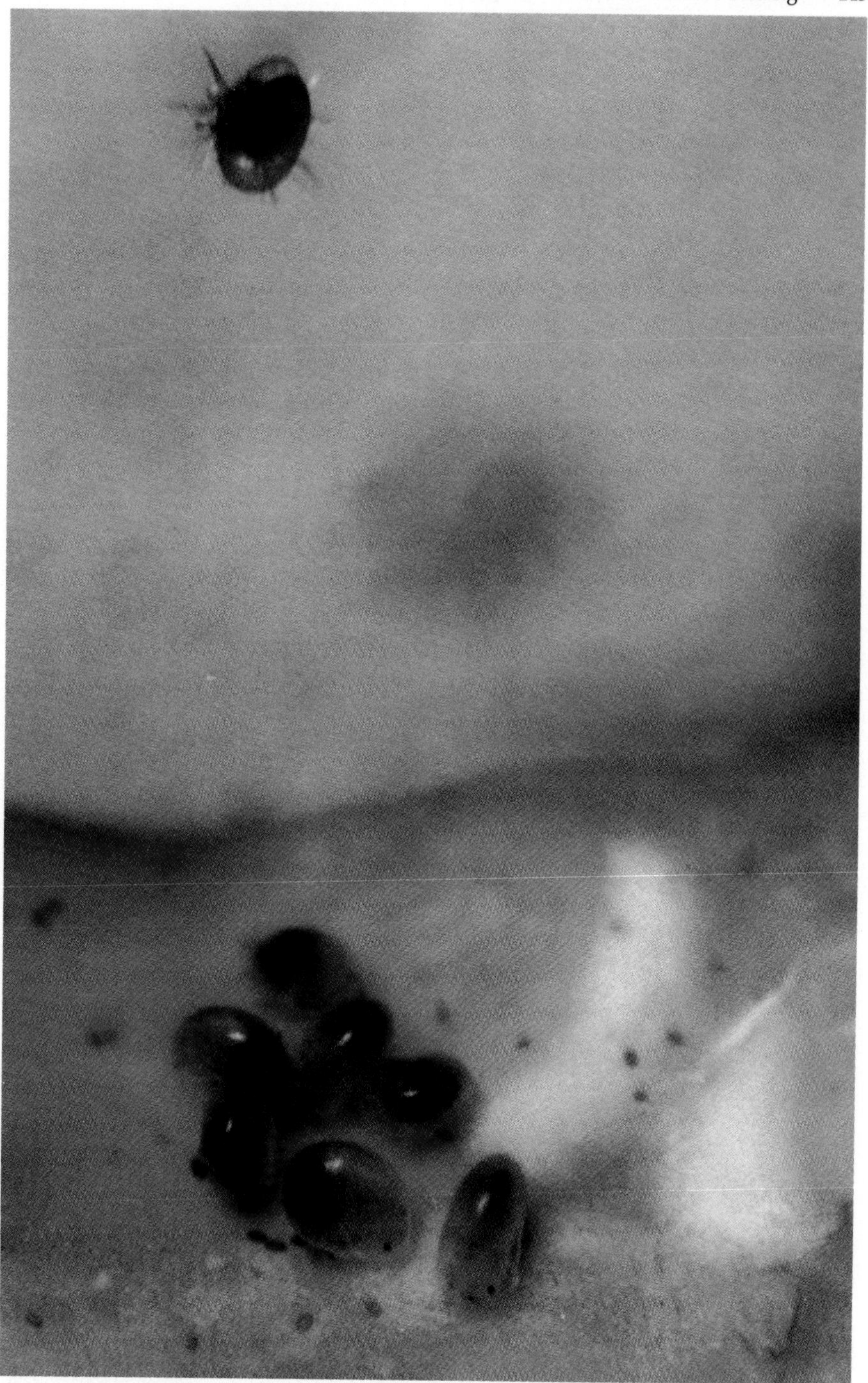

Figure 3. Swimming larva of the red abalone comes in for a landing among some of its siblings on a patch of coralline red algae. Experiments carried out in the author's laboratory have established that the larva recognizes the algae by means of a chemical cue. The molecule that induces permanent settlement is a small peptide, bound to the surface of the algae, related to the human neurotransmitter gamma aminobutyric acid (GABA). The larvae, which are 0.2 millimeter across, settle only in the presence of the natural inducer or various analogues, including GABA itself.

What Are the Larvae Waiting For?

The fact that abalone larvae are developmentally arrested after seven days is another reason for working with this species. Abalones are particularly choosy; their requirements for settling to the bottom and metamorphosing are both stringent and specific. If the larvae are held in clean natural seawater, they continue their normal swimming behavior. They will not settle on to a surface or begin developing into a juvenile abalone. They remain arrested at the stage of competence and can go on for up to a month in this way before they finally use up all their energy source in the yolk sac. This was one of Daniel Morse's earliest observations, and it suggested that abalone larvae have no *endogenous* trigger for the activation of development beyond the larval stage. They obviously have a stringent requirement for an external agent for the induction of settlement and metamorphosis. It was apparent that this agent was not present in the natural seawater in our culture chambers.

We turned to the natural environment for clues to the location and identity of this inducer. We engaged the help and expertise of Mia Tegner of the Scripps Institute of Oceanography. She found that the youngest juvenile abalones she could detect on the ocean floor (they were about two millimeters across) lived almost exclusively on coralline red algae. In the kelp beds along the coast of California, many of the rocks as well as the root-like hold-fasts of the kelp are covered by coralline red algae. The algae form a thin but extensive crust that is permanently cemented to the substrate and is toughened by calcium carbonate deposited between the walls of adjacent cells. Numerous species of algae grow side by side and often overlap to form a patchwork quilt.

In experiments that followed up on Tegner's observation, abalone larvae reared in the laboratory were presented with a variety of coralline red algae. We observed their response in the dissecting microscope. They quickly settled out of the water column and attached themselves to the hard surface of the algae. Remarkably, a single contact event was enough to prevent the resumption of swimming. Metamorphosis began after several hours of contact with the algae. However, only larvae seven days old and older stopped swimming. Younger larvae are not competent to recognize this source of exogenously supplied inducer.

After the larva attaches to the algal surface with its muscular foot, metamorphosis commences. Within 20 hours the larva begins to shed its velum—the fringe of ciliated cells that surround its head and allow it to swim. The result is a dramatic and irreversible commitment to the benthic mode of life: the animal can no longer swim back up into the plankton. The shed cilia are still actively beating; as they "swim" away, many are captured and eaten by the metamorphosing larva.

From now on, the young juvenile lives and moves as a typical snail, creeping along the bottom by stretching and contracting first one lobe and then the other lobe of its oversize foot. The developmental commitment to metamorphosis is irreversible, and proceeds along a genetically preprogrammed path. Several of the genes that are activated at this time, and their corresponding proteins, have been identified and characterized by Marios Cariolou, David Spaulding and Jay Groppe.

After 40 hours, the translucent larval shell has been extended by clearly visible new growth of an opaque, ribbed juvenile shell. Specialized epithelial cells of the mantle—the thin, dark "fringe" lining the shell—deposit the crystals and secrete the organic matrix of the shell. The synthesis of new shell protein can be detected at this time. Internal organogenesis and remodeling have commenced. Rhythmic beating of the fully developed heart is seen through the larval shell at 60 hours. Digestion of food from the algal surface replaces yolk metabolism. This change is accompanied by the synthesis of new digestive enzymes.

For the next few months, as the abalone grows it grazes on the red algae. The red pigment of the algae becomes incorporated into the growing shell. In this way the young animal becomes exquisitely camouflaged from predators, a particular threat at this time in its life. To continue growing, however, it must ultimately seek out richer

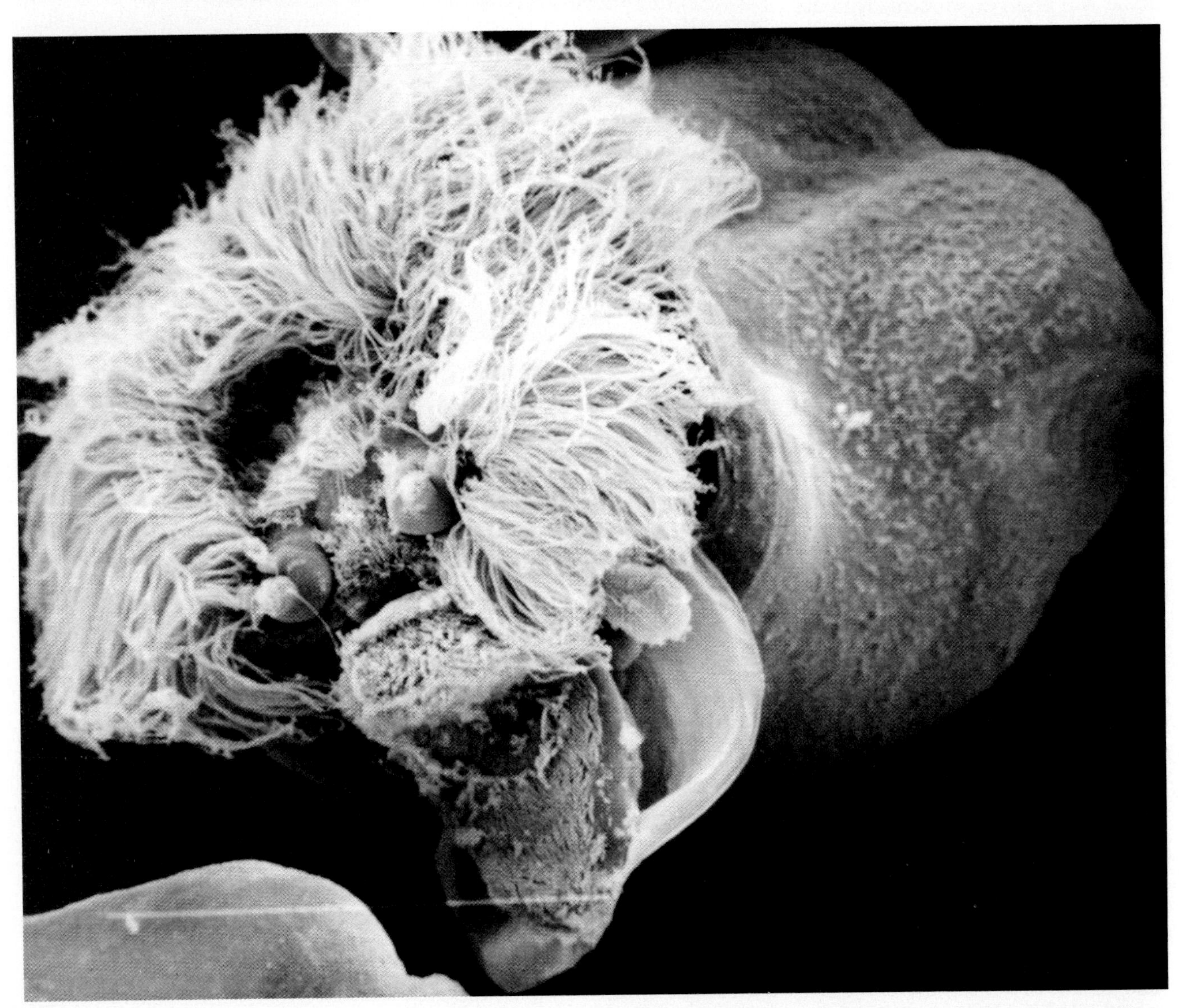

Figure 4. Scanning electron micrograph of an abalone larva shows a thick fringe of long cilia, with which the animal swims. Shorter cilia projecting from the face are the sensory organelles with which the larva detects the peptide that induces metamorphosis. The cilia are shed soon after the animal settles on a patch of red algae, committing the abalone to life as a bottom-dweller.

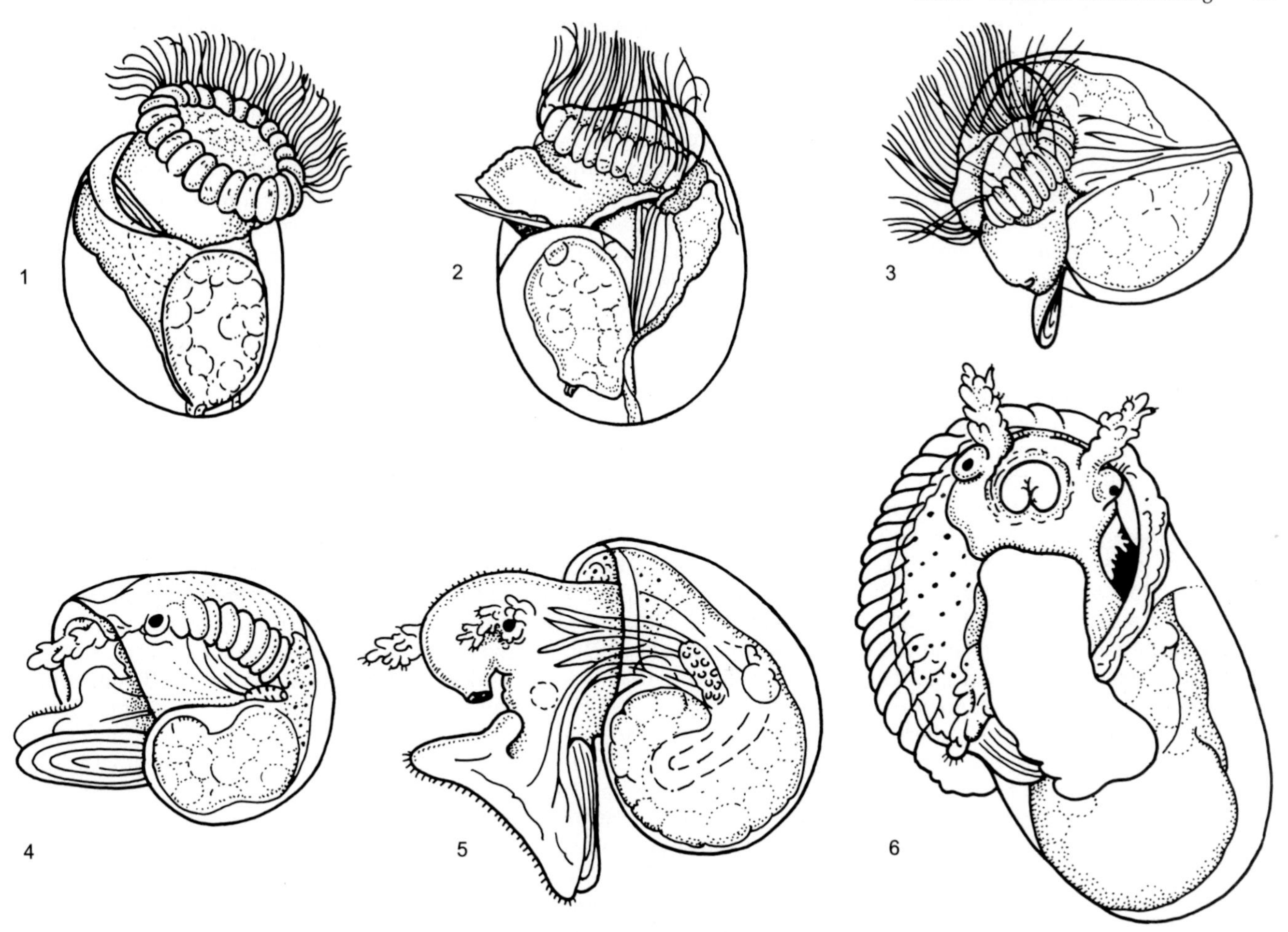

Figure 5. Stages in the development of a young abalone include the crucial transformation from larval to juvenile form. At about 30 hours after fertilization *(1)* and again at 54 hours and 79 hours *(2, 3)*, the abalone larva is still a swimmer, with a prominent mane of cilia. At the time of settlement but before metamorphosis begins *(4)* the abalone's resemblance to other snails is most conspicuous, with a shell that curls up into a symmetrical spiral. At the onset of metamorphosis *(5)*, the animal's foot has enlarged and emerged from the shell, and a creeping mode of locomotion has supplanted swimming. At an age of 12 days *(6)*, the shell has developed an asymmetry, and it will eventually become flattened like one valve of a clamshell. The drawings, based on a series published in 1937 by Doris R. Crofts, show the development of *Haliotis tuberculata*, a species that lives in the English Channel; the timing of development is somewhat different but the stages are similar in *Haliotis rufescens*, the species studied in the author's laboratory.

pastures. Leaving the red algal patch, it creeps off in search of a variety of foliose macroalgae, including the large brown kelps.

We were interested in knowing whether larvae could "recognize" anything other than coralline algae. We presented abalone larvae with other kinds of algae, bacteria and animals commonly found in the natural habitats of juvenile and adult abalone. None of these other organisms induce the larvae to settle and metamorphose; only the coralline red algae have this property. This is what we mean when we say abalone larvae have a *specific* requirement for settlement. As far as we have been able to determine, the natural exogenous cue required for settlement and metamorphosis is associated only with algae of the family Corallinaceae; all members of the family we have tested so far have the agent. The unique availability of this cue on the surfaces of these algae is what determines substrate selection by abalone larvae, and in part is responsible for the observed distribution of young abalone in the ocean.

The fact that abalones are choosy has greatly facilitated our studies of the physiological and molecular processes governing settlement and metamorphosis. Both the stringency of the requirement for an exogenous cue and the recognition of a highly specific cue are necessary to unequivocally interpret assays of larval, cellular and molecular recognition and responsiveness. A number of marine invertebrate larvae have more relaxed and less specific requirements. In these species it has proven difficult to identify the cues involved and to understand the processing of information at the molecular level.

Unmasking the Inducer

Larvae of various species are known to respond to several types of cues, including chemical agents, light and texture. Species whose requirements are both stringent and specific generally recognize the cue by chemosensory means. This has proved to be the case for red abalone larvae. The larvae do not recognize the red algae by sight; they settle equally well in the dark as in the light. Nor do they recognize the algae by their surface texture. What the larvae do recognize is a specific chemical in the algae. This fact was demonstrated in a simple experiment performed more than a decade ago. A sample of the red algae was ground up in seawater to extract the chemical components into solution. Competent larvae responded to this cell-free solution by settling on the bottom and sides of the glass assay vial. Within 24 hours, new growth of the juvenile shell was clearly visible, indicating that the larvae also were metamorphosing. Under these conditions the larvae could not have been responding to the physical surface texture of the algae but had to be detecting an algal chemical cue.

The algae do not release the chemical inducer into the surrounding seawater. Instead, the inducer molecules are tightly complexed to the algal surface. Thus the way an abalone larva finds a home is nothing like the way a male gypsy moth follows a trail of airborne pheromone released by a female; the larva does not swim toward the source of a chemical trail. This process, known as chemotaxis, would not work well over large distances in the turbulent water of the ocean. The chemical trail would soon become too dilute. The abalone larva must make direct contact with the surface of the red algae in order to recognize the chemical cue. It does this by means of the "bottom-hopping" maneuver we described earlier.

But what is the chemical cue? Which of the many molecules in the crude algal extract is it? We separated the individual molecules in the crude extract and used competent larvae to identify the molecule that actively induces them to settle and metamorphose.

Biochemical characterization of the inducer molecule reveals that it is a small, water-soluble peptide—a short chain of amino acids. It cannot be composed of more than 10 amino acids because its apparent molecular weight is only about 1,000, and a typical amino acid has a molecular weight of roughly 100. We are still attempting to ascertain its precise structure. In addition to known amino acids, it contains at least one very unusual amino acid. (This often is the case in small peptides that serve as insect pheromones in land plants.) Based on what we know so far, the molecule appears to represent a new class of compound with, as we have recently discovered, certain exciting properties.

The most intriguing feature of the inducer molecule is its apparent structural and functional resemblance to gamma-aminobutyric acid, or GABA, a neurotransmitter that bridges the gap in nearly half of the known types of chemical synapse in the human brain and spinal cord. Our first clue to this relationship came from the discovery by Daniel Morse that GABA could mimic the effects of the natural algal inducer. Both substances produce the same behavioral and metamorphic response in abalone larvae. This dramatic finding suggested that larvae recognize a chemical environmental cue that is GABA-like. Further evidence came from observing the biological activity of the algal peptide in other animal tissues. The small morphogenetic peptide purified from coralline red algae binds even more tightly than GABA itself to the principal class of GABA receptors purified from mammalian brain cells.

Receptors and Transducers

How do the abalone larvae recognize the GABA-like peptide on the surface of the algae? And how is that chemical signal transduced into an internal signal that prompts the larvae to remain on the algae and metamorphose? Thanks in part to the fact that we have millions of competent larvae to work with each week, we have been able to identify, partially purify and characterize some of the receptors and transducers involved in this process at the molecular level.

The GABA receptors in the human nervous system are cell-surface molecules—large proteins embedded in the plasma membrane. Since the abalone settlement inducer is GABA-like, it was reasonable to suppose that the larval sensory mechanism that specifically detects the environmental cue would also involve cell-surface receptors. Using a radioactively labeled GABA analogue called baclofen, my colleague Henry Trapido-Rosenthal detected receptors that specifically bind this GABA analogue. Because baclofen cannot enter the cells, these labeled receptors must be located on the cell surface.

A number of independent criteria have established that the receptors labeled by baclofen are indeed the ones that recognize the algal inducer. Abalone larvae are induced to settle and metamorphose not only in response to the natural peptide, to GABA and to baclofen, but also in response to a number

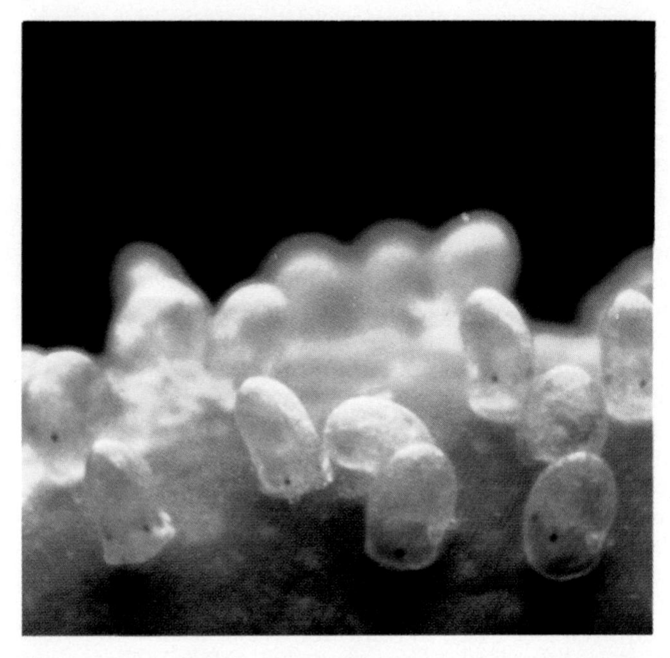

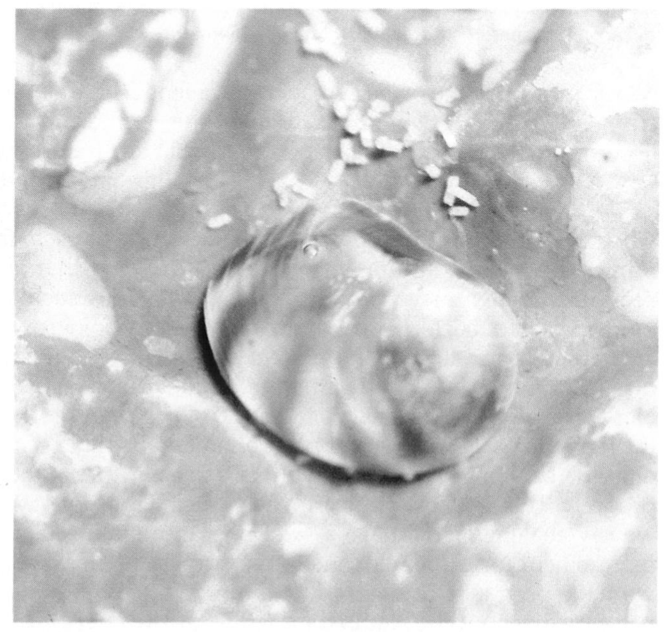

Figure 6. Newly settled juveniles *(left)* feed on diatoms and other microalgae living on the surface of coralline red algae. By "cleaning" the surface in this way, the abalones may be beneficial to the coralline algae. A month after settlement and metamorphosis *(right)*, an abalone has absorbed pigment from the red algae and thereby taken on protective coloration. The white objects above the juvenile abalone are fecal pellets. (Photograph at left courtesy of Robert Sisson.)

Figure 7. Adult red abalones often congregate in heaps of 10 to 50 animals during the day *(photograph above)*, then forage for kelp or other large algae at night. They recognize the kelp with sensory tentacles that protrude from under the shell and through respiratory pores; the pores also allow water to circulate through the animal's gills and are used for the discharge of urine, feces and gametes. Seen from the underside *(photograph below)* an adult abalone consists mainly of a large, muscular foot, with which it creeps along the bottom. The abalone is shown being measured to see if it can legally be harvested. (Photographs courtesy of Bob Evans, Bob Evans Design.)

of known homologous GABA analogues. The relative efficiency of these substances as inducers is correlated with their stereochemical configuration. These analogues were tested for their ability to compete with radioactive baclofen in binding to the identified larval receptors. In all cases, binding of baclofen was reduced, indicating that each of the analogues competes with baclofen for binding to the same receptors. Moreover, their efficiency of competition was found to be directly correlated with their efficiency as inducers in the larval bioassay. GABA and the algal peptide also compete with baclofen for binding to the receptors. The algal peptide has the highest efficiency as both inducer and competitor.

Further proof that Trapido-Rosenthal had identified the right receptors came from his finding that the receptors begin to appear on the abalone larvae before the larvae become competent, and that larvae that do not yet have the receptors are unable to recognize GABA or the algal inducer. In fact, premature exposure

Figure 8. Laboratory assay has been used to identify and isolate agents that induce larval settlement. Each glass vial contains about 200 sibling larvae, all at the stage of competence, so that they are capable of settling and metamorphosing in the presence of suitable chemical signals. Because the larvae in the vial at left are still swimming, one can infer that they have not detected a morphogen; such an inducer is present in the vial at right, where the larvae have settled onto the glass.

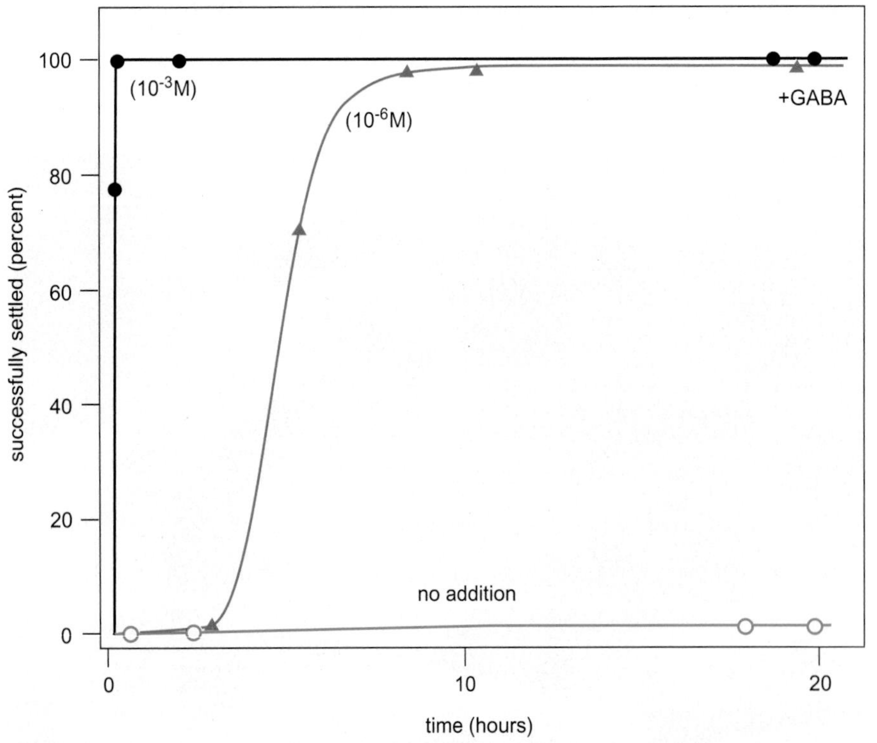

Figure 9. Neurotransmitter GABA is one effective inducer of larval settlement. In the absence of an inducer, almost none of the larvae settle, even after some 20 hours of observation. When GABA is added at a concentration of 10^{-3} mole per liter, all the larvae settle almost immediately—within about seven minutes. At a more realistic concentration of 10^{-6} mole per liter, the response is slower, and so it is several hours before most of the larvae have settled. This lower concentration, however, has proved optimal for the induction of metamorphosis.

to GABA, prior to the development of competence, actually delays the development of the receptors and the ability of larvae to settle and metamorphose in response to GABA or the algal peptide. Eventually, though, the larvae do develop receptors and become responsive; they are only temporarily blinded—or habituated—to GABA. This temporary insensitivity may serve an important function: It may keep the larvae from settling too near their parents and competing with them. Significantly, this process of habituation appears to be confined to the external receptors. The pathway that transduces the external signal within the cell remains fully functional. Habituated larvae can be induced by an internal stimulant of this pathway, which bypasses the normal external receptor.

The function of the internal post-receptor pathway is to convert the chemical signal from the environment into an electrical signal that can be understood and processed by the larval nervous system to activate further development. As we shall see, this transduction involves many of the same basic principles and molecules that are involved in synaptic activation by neurotransmitters in the mammalian central nervous system.

Andrea Baloun Yool and Gregory Baxter demonstrated that the larval receptors, once activated by binding the exogenous inducer, internally activate the enzyme adenyl cyclase. The function of this enzyme is to catalyze the production, from adenosine triphosphate (ATP), of cyclic AMP—a ubiquitous "second messenger," or intracellular signal molecule. When larvae are presented with a molecule that activates adenyl cyclase directly, or one that inhibits the breakdown of cyclic AMP, they settle and metamorphose even in the absence of the natural inducer or GABA. These two results provide substantial evidence, if not solid proof, that cyclic AMP is involved in transducing the external signal to settle. This signal transduction cascade is similar to those found in other neuronal systems.

Baloun Yool and Baxter found that cyclic AMP functions in the target sensory cells of abalone larvae as it does in other cells: acting in concert with calcium ions, it activates the cytoplasmic enzyme protein kinase *A*. This enzyme, in turn, adds a phosphate group to a specific target protein in the cytoplasm—a protein whose identity has yet to be determined. This phosphorylated protein then in some way mediates the opening of channels in the plasma membrane for chloride ions or other anions. The resulting efflux of negatively charged chloride ions or other anions depolarizes the chemosensory cell membrane. The electrochemical signal thus generated can then be propagated by the larval nervous system to activate molecular and cellular development in the larva.

Transduction of the electrical signal is essential, or obligatory, for the larval settlement and metamorphic response. This conclusion is supported by a variety of experiments conducted by Baloun Yool. When larvae are presented with GABA but the outflow of chloride ions is prevented—either by membrane-impermeant molecules that directly block chloride channels or by a high concentration of chloride ions outside the sensory cells—the larvae do not settle. Blockers of sodium and potassium channels are ineffective, demonstrating the specific involvement of chloride channels in the transduction of the metamorphic cue. Larvae do settle, even without GABA, when they are exposed to substances that open chloride channels or to an abnormally low external concentration of chloride ions. Thus the need for the metamorphic environmental cue apparently can be bypassed by directly generating the required electrochemical signal.

The larvae also settle and metamorphose when they are exposed to high external concentrations of positively charged potassium ions in the absence of GABA. In this case, the chemical inducer, its receptor and the associated chloride channel are not involved. The potassium ions effect a depolarization of the membrane by directly causing an opening of potassium channels. Induction by potassium is prevented by membrane-impermeant potassium-channel blockers, but not by the chloride-channel blockers that inhibit the GABA-induced response. The ability of externally provided potassium to circumvent the normal pathway of triggering reactions demonstrates the essential role of depolarization of the chemosensory membrane and the generation of an electrical signal for induction of settlement and metamorphosis. Indeed, after Baloun Yool first developed the potassium method, she and other investigators found that it could be used to trigger settlement in the larvae of other species—molluscs, annelid worms, urchins and bryozoans—whose chemical inducers and receptors they had not yet identified. The finding that this essential depolarization is sensitive to manipulation of the external environment, and to membrane-impermeant blockers, is ample proof that the sensory cells involved are indeed externally accessible.

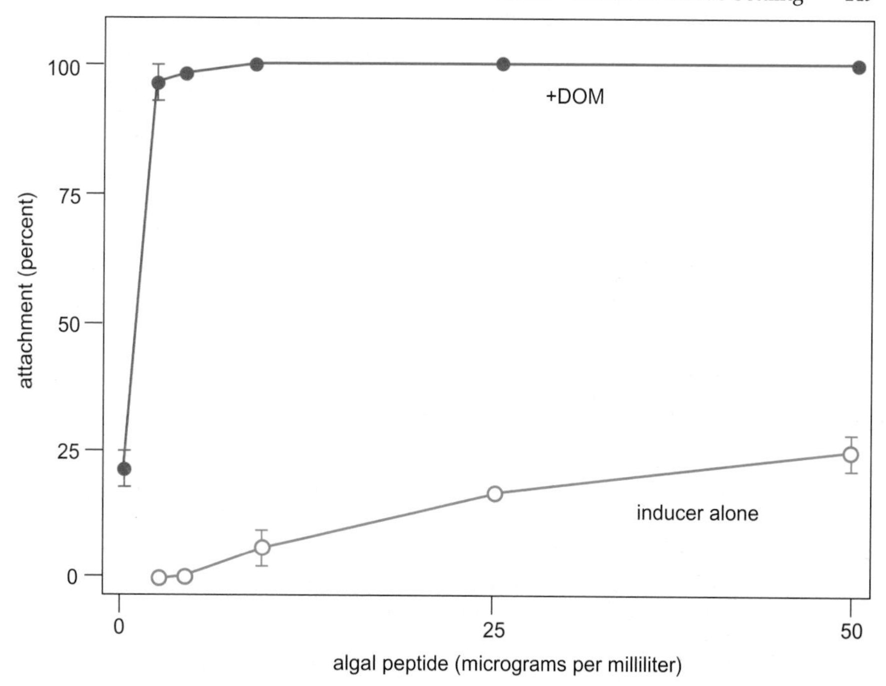

Figure 10. Settlement response is amplified by the amino acid lysine. When the natural inducer peptide, extracted from coralline red algae, is presented in isolation and at low concentration, only a minority of the larvae successfully settle and attach in a glass vial. Adding dissolved organic matter from seawater dramatically increases the rate of settlement. It turns out that lysine is the crucial component of the dissolved organic matter; certain other amino acids that, like lysine, have two amino groups are also effective.

A detailed understanding of the functional relationships between the molecules involved in this cascade will emerge from further molecular dissection and reconstitution of the pathway *in vitro*. We have recently begun to do this. The involvement of cyclic AMP has been confirmed *in vitro*. GABA analogues activate a cyclic-AMP–dependent protein phosphorylation in chemosensory cilia isolated from abalone larvae. Efforts to clone the genes coding for the proteins we have identified *in vivo* are under way. These genetic and molecular analyses will provide information on the structure and function of the proteins involved in the transduction pathways, as well as a basis for comparing the abalone system with its analogues in mammalian and other nervous tissues.

A Regulatory Pathway

While work on the GABA-sensitive receptors was under way, Trapido-Rosenthal discovered that the abalone larva has a second chemosensory pathway involved in settlement and metamorphosis. Competent larvae respond to the amino acid lysine dissolved in seawater, and to other amino acids that, like lysine, have two amino groups. The response to these diamino acids, however, is quite different from the response to the natural inducer or GABA. The larvae do not settle and metamorphose in the presence of lysine alone, but detection of lysine increases the larva's sensitivity to a morphogenetic inducer 100-fold. That is, in the presence of lysine, a much smaller amount of GABA than is normally required is sufficient to induce larvae to settle. This discovery was not entirely surprising: lysine and a number of other amino acids are common modulators of chemosensory systems.

This effect of lysine was first demonstrated *in vivo*, using the larval bioassay. More recently, Baxter confirmed these findings *in vitro*. He identified receptors that specifically bind radioactively labeled lysine. The lysine receptors appear to be on epithelial cilia isolated from the chemosensory cells of the larvae. Circumstantial evidence suggests that these cells may be associated with specialized sensory processes found in the head region of the larva.

Once activated by lysine, the regulatory receptor initiates a cascade of reactions inside the chemosensory cell. We have analyzed the events at the molecular level. This pathway, which we call the regulatory pathway, is entirely independent of the morphogenetic transduction pathway. The larval receptor activated by lysine in turn activates a signal-transducing protein called a *G* protein, which is also a familiar character in many other sensory systems. The *G* protein most probably is situated very close to the receptor in the cell membrane. It activates a membrane-associated enzyme called a phospholipase. The phospholipase cleaves a molecule on the inner surface of the cell membrane, releasing diacylglycerol, a molecule that, like cyclic AMP, is a ubiquitous second messenger. Diacylgycerol activates another membrane-associated enzyme, protein kinase *C*. In turn, this enzyme phosphorylates another protein; we now are now attempting to determine its identity and function.

Although some parts of the regulatory pathway remain mysterious, we do know a great deal more about the *G* protein. Both *in vivo* and *in vitro* studies demonstrate obligatory dependence on the action of this signal-transducing molecule. In fact, the need for lysine can be circumvented by directly stimulating the *G* protein with either cholera toxin or analogues of guanosine triphosphate (GTP). Conversely, *G*-protein inhibitors block the amplification by lysine. These same inhibitors have no effect on metamorphic induction by GABA, thus confirming the independence of the two pathways.

Recently, Lisa Wodicka has obtained a partial nucleotide sequence for the *G* protein. From this sequence she has determined that the larval *G* protein is homologous to G_Q, a new class of *G* protein discovered by Mel Simon's group at Caltech. At present our regulatory pathway ends in mid-air: We cannot yet say how it regulates the morphogenetic pathway, nor whether the two receptors and their independent transduction pathways are located in the same chemosensory cell, or in neighboring cells.

What is the purpose of this facilitatory receptor and its transduction pathway? In what way do they provide "assistance" to larvae in the plankton? At this point we can only speculate. It is possible that amplification of the inducer signal by specific amino acids primes the larvae to settle in potentially favorable conditions. An elevated concentration of lysine and other amino acids in seawater often indicates high levels of nutrient availability, and thus may serve as an indicator of an area in the patchy marine environment where algal food is likely to increase.

Regardless of the adaptive significance of the regulatory mechanism, the observed interaction of the morphogenetic and the regulatory pathways indicates that the abalone larvae may have some ability to fine-tune their choice of a time and place to settle and metamorphose.

Other Animals

Lessons learned from the red abalone also apply to other marine invertebrates. In particular, the same specificity for coralline red algae, and the same dependence on a GABA-like inducer, has

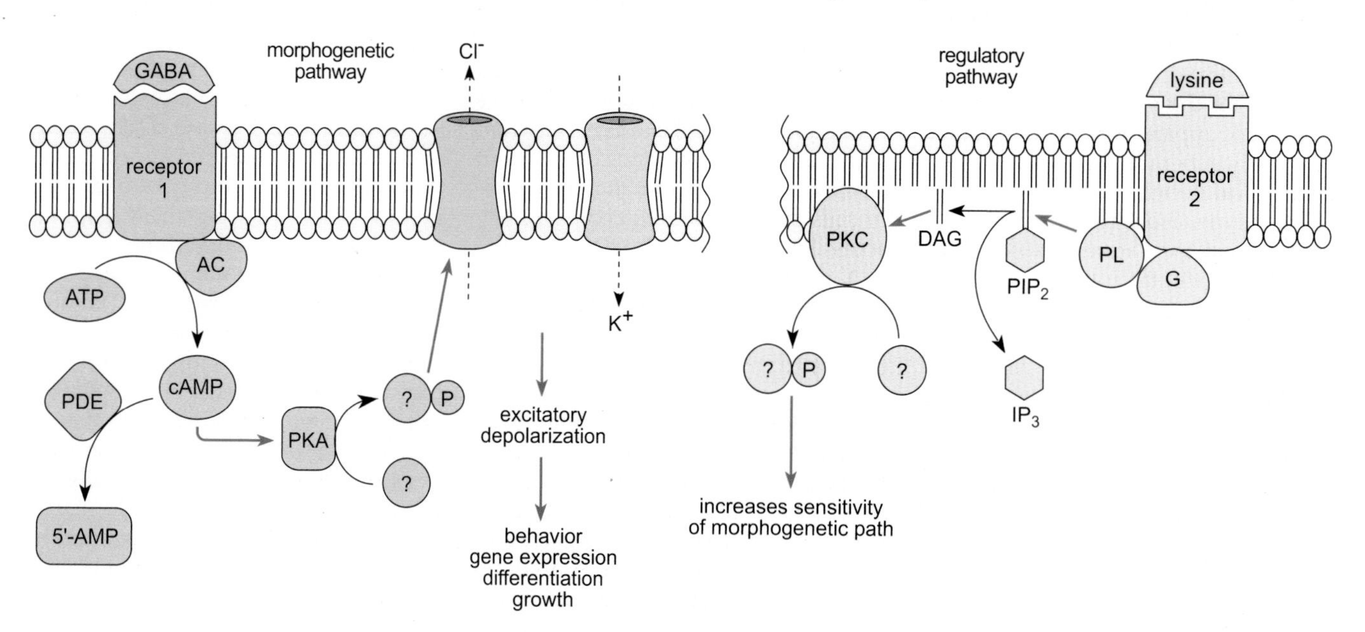

Figure 11. Chemosensory pathways that govern the abalone's settlement response have been traced in molecular detail. The morphogenetic pathway (*blue*) are triggered by the natural inducer peptide or one of its analogues, such as GABA, which binds to a receptor protein (receptor 1) that spans the membrane of a larval sensory cell. Receptor 1 in turn activates the enzyme adenylate cyclase (AC), which catalyzes the synthesis of cyclic adenosine monophosphate (cAMP), a ubiquitous "second messenger" in biochemical systems. The cAMP activates protein kinase *A* (PKA), which then phosphorylates (P) another protein that has not yet been identified. The unknown phosphorylated protein in some way mediates the opening of chloride channels in the cell membrane. Negatively charged chloride ions (Cl^-) flow out of the cell, depolarizing the membrane. Thus the original chemical signal is transformed into an electrical signal, which can be communicated to other cells. In laboratory experiments the need for the chemical morphogen can be bypassed by an excess of potassium ions (K^+) in the seawater. The regulatory pathway (*yellow*), whose action is initiated by lysine or other diamino acids, works in a similar way. Lysine binds to receptor 2, which acts through the *G* protein (G) and phospholipase (PL) to produce the second messenger diacylglycerol (DAG) from phosphatidylinositol bisphosphate (PIP_2). DAG then activates protein kinase *C* (PKC), which phosphorylates another unknown protein. This second phosphorylated protein in some way increases the sensitivity to the morphogenetic inducer about 100-fold. Other components of these reaction pathways include adenosine triphosphate (ATP), phosphodiesterase (PDE), 5′-adenosine monophosphate (5′-AMP), and inositol triphosphate (IP_3). Black arrows represent transformations of one chemical to another through an enzymatic reaction; red arrows indicate activations; dashed arrows show ionic flows across the membrane.

been observed in 12 other species of abalone. Significantly, controlled field studies all have concluded that the youngest juvenile abalones are found exclusively on coralline algae. A number of unrelated invertebrate species, commonly found in the field on coralline algae, may be recognizing the same class of chemical cue. Larvae of these species are induced by coralline algae, by algal extracts and by GABA. More generally, it has been recognized for some time now that larvae in a number of invertebrate phyla recognize particular substrates. Moreover, chemosensory recognition of an inducer appears to be directly correlated with stringency and specificity of larval requirements for the induction of metamorphosis.

In our laboratory Rebecca Jensen has been studying larvae of the honeycomb worm, *Phragmatopoma californica*. This polychaete worm builds itself a protective tube by cementing together sand grains and debris. It lives in aggregations, and it reproduces like the abalone, by broadcast spawning. The larvae drift for an estimated five to eight weeks before settling and metamorphosing. Following work done by Douglas P. Wilson of the Plymouth Laboratory, who studied a related species in Britain, Jensen has shown that honeycomb-worm larvae do not permanently attach and metamorphose until they come into contact with tubes made by adults of their species. In this way the worms build extensive reefs.

To find out what component of conspecific tubes induces the larvae of the honeycomb worm to settle, Jensen first gave clean glass spheres to adult worms, which they incorporated into their tubes. When Jensen removed the spheres, she found they were covered with round plaques of a light brown cement. She also found that the cement-covered glass spheres were enough to induce worm larvae to settle.

Figure 12. Honeycomb worm, *Phragmatopoma californica*, is another organism whose drifting larvae settle in response to a chemical signal. The worm builds itself a protective tube by cementing together sand grains or other debris; its feeding tentacles protrude from the tube (*upper photograph*). Colonies of worms form massive concretions (*lower photograph*), which grow by recruiting new larvae. The signal that induces the larvae to settle is a protein in the cement secreted by the adult worms. (Photographs courtesy of Robert Sisson.)

The cement proved to be a protein stabilized by tough quinone cross-links, similar in structure to the protein of silk. Working with Herbert Waite of the University of Delaware, Jensen has begun to figure out what it is that worm larvae detect in the tubes. The protein has many repeats of a peptide sequence that is rich in dihydroxyphenylalanine, or DOPA, a well-known neurotransmitter. DOPA alone does not seem to be the settlement inducer; a larger portion of the adhesive protein apparently is involved.

Recently, Jensen tested the laboratory-purified inducer for its effectiveness as an inducer in the ocean. She coated plastic surfaces with inducer and immersed them in the ocean, chemical-side down. Downward placement was used to verify that the putative inducer did in fact promote the same behavioral response observed in the laboratory. In response to the inducer-coated panels, the worm larvae had to actively "settle" upward, not simply be paralyzed and fall out of the water. They did attach themselves to downward-facing surfaces and proceeded to metamorphose. The inducer, apparently, is like a powerful fly-paper. This is the first and only time that an inducer of larval metamorphosis identified in the laboratory has been shown to be an effective inducer in the complex environment of the ocean.

The events that take place after the inducer binds to receptors on the sensory cells of *Phragmatopoma* larvae appear to be very similar to those operating in red abalone. In the absence of the natural inducer, compounds that are known to elevate the level of cyclic

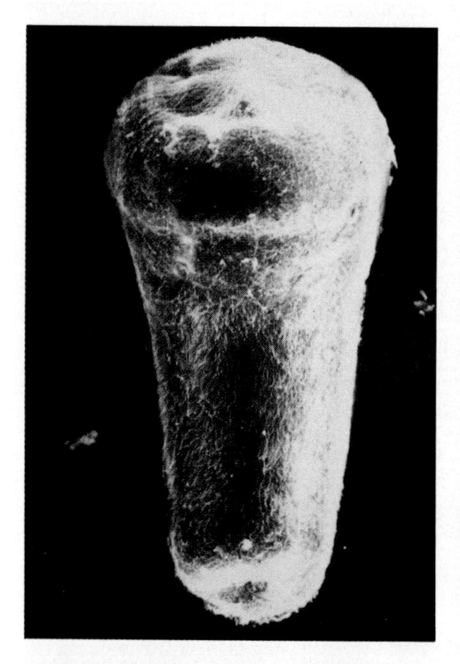

Figure 13. Caribbean lettuce corals of the genus *Agaricia* produce larvae that settle only on a few particular species of coralline red algae, mainly species that tend to grow on the underside of dead corals. The settlement inducer in this case is a polysaccharide in the algal cell wall. Here scanning electron micrographs show a larva in the swimming stage *(left)*; a few hours after exposure to the polysaccharide purified from coralline algae, the larva has cemented itself to a plastic surface and secreted the skeleton of the first polyp stage *(right)*.

AMP in other cells cause the larvae to metamorphose, indicating that cyclic AMP is involved. Artificially increasing the concentration of potassium ions in seawater causes the larvae to metamorphose, even in the absence of the natural inducer. This circumstantial evidence suggests that the natural chemical signal is transduced into an electrical depolarization of the sensory-cell membrane. The components of this transduction pathway must still be pinned down with certainty, but the overall picture appears to be the same.

Daniel Morse, Neal Hooker and I have also been studying the larvae of three species of shallow-water Caribbean "lettuce" corals, the agariciids. Individual coral colonies grow by cloning, but corals also reproduce sexually, giving rise to larvae that disperse in the plankton and colonize new sites on the reef, distant from their parents. A number of field studies have suggested that in the crowded environment of a tropical reef, space is at a premium. Any patch of open space that becomes available is quickly colonized in what has often been thought to be an essentially random manner. We have observed something different, however, about the process of colonization by larvae of lettuce corals.

Agaricia larvae are distinctly choosy about the space in which they settle. They have both a stringent requirement for induction of settlement from the plankton by an exogenous environmental cue, and a specific requirement for a particular cue. These requirements are satisfied by tropical species of coralline red algae. Robert Steneck of the University of Maine has been assisting with the identification of the numerous tropical species tested. Unlike abalone larvae, which will settle on any and all coralline red algae, the coral larvae recognize just a few species. Indeed, the three species of agariciids show different preferences for particular species of coralline algae. These preferences for different substrates may serve to maintain niche diversification of these sympatric species on the reef.

In general, the agariciids tend to prefer algal species that grow on the cryptic underside of piles of dead coral rubble; this may afford them protection from the throng of algal grazers, such as the voracious parrotfish, on the reef. Also unlike abalone larvae, these particular coral species do not have to spend a week in the open ocean as they develop to competence. Instead the lettuce coral larvae are brooded within their parents; they are competent to settle as soon as they are released. Indications are that many of those that settle do so close to their parents and siblings. Genetic analysis of populations in various locations around the same island should answer this definitively.

Like abalone larvae, *Agaricia* larvae are induced to settle by contact with a particular algal chemical complexed to the surface of the algae. Lettuce coral larvae, however, are not recognizing the GABA-like peptide. Instead they are induced to settle and metamorphose by a unique sulfated polysaccharide, a sugar polymer that apparently is present in the cell wall of the algae. We are unsure as to how the coral larva processes this information. We were unable to detect the same kind of transduction pathway as was found in abalone and the honeycomb worm. We suspect that in this more-primitive creature the process of recognition may be mediated by the class of receptors known as lectins. Lectins and the variety of sulfated polysaccharides that they recognize have been widely implicated in highly specific recognition that control differentiation.

Evidence that larvae of bottom-dwelling marine invertebrates have strict requirements for recognition of their settlement substrate is by no means limited to the species studied in our laboratory. Substrate-specificity also has been demonstrated in many other species, although they are not all as selective, nor are their requirements for a chemical inducer as stringent and specific as those of the red abalone, the honeycomb worm or the lettuce coral.

Abalone larvae have provided the most complete model of the larval processes that control the initial events of settlement and metamorphosis in the ocean. It is clear that many sessile invertebrate species that spend their larval phase in the plankton utilize similar cellular and molecular processes. Settlement from the plankton is a *dynamic* process, operating at a molecular level. Coupled to this is transport and delivery of larvae by physical forces such as currents and turbulence. The key factor in determining settlement, however, in any of the many species dependent on an environmental cue is actual detection and subsequent molecular processing of a specific, required cue. The chance of being delivered by physical forces to a recognizable patch on the ocean floor is obviously very low. Estimates place it at far less than 1 in 1,000. Such estimates are based on field studies of recruitment events taking place at some time after the initial settlement event. It so far has proved impossible to monitor initial settlement events in the ocean. Initial settlement events are important factors in determining community structure. Although laboratory studies cannot hope to simulate the complicated interactive effects operating in the ocean, they do show us the potential, the limitations and the outcomes of the responses of larvae to different plants, animals and surfaces likely to be encountered in the ocean.

What these and related studies have shown is that in some instances (and probably many more than we know about), the settlement of larvae is not random. Instead of simply accepting the environment they are dealt by chance, the larvae have evolved complex chemosensory systems for recognizing and selecting an environment that suits them. These systems of receptors and signal transducers may in fact be quite old. They are strikingly similar to ones used by the cells in our own bodies for communication with one another.

My colleagues and I are now trying to test the hypothesis that the chemosensory systems of abalone larvae and the neural transmission systems of the mammalian brain have a common evolutionary origin. By cloning the genes for the various receptors and proteins involved in the signal-transduction pathways and sequencing the DNA, we will be able to see just how homologous they are with their counterparts identified in mammalian brains. If the systems really are as similar as we think they are, we may very well learn lessons of biomedical value.

This research has already had some practical value. Abalone is considered a great culinary delicacy—so much so that the abalone beds off California have been nearly depleted. In the past decade, however, three California companies have been engaged in abalone aquaculture. They use our hydrogen peroxide method to induce spawning of adults, and GABA to induce the larvae to settle. Similar programs are under way in other countries as well. Perhaps the price of abalone may one day return to within the range of the average consumer.

The ocean is a large and complicated place. We have been able to understand one process in one small corner of it. Although this work has focused on the molecular aspects of larval recruitment from the plankton, it is expected that similar analyses will provide insight into other ecologically important biological processes in the ocean.

Acknowledgment

The work described here was supported in part by the National Science Foundation (grant DC 387-18224) and the National Institutes of Health (grant 1-R01 RR06640).

Bibliography

Baloun, A. J., and D. E. Morse. 1984. Ionic control of settlement and metamorphosis in larval *Haliotis rufescens* (gastropoda). *The Biological Bulletin* 167:124–138.

Baxter, G., and D. E. Morse. 1987. G protein and diacylglycerol regulate metamorphosis of planktonic molluscan larvae. *Proceedings of the National Academy of Sciences of the U.S.A.* 84:1867–1870.

Chia, F.-S., and M. E. Rice (eds). 1978. *Settlement and Metamorphosis of Marine Invertebrate Larvae.* New York: Elsevier.

Crofts, D. R. 1937. The development of *Haliotis tuberculata*, with special reference to organogenesis during torsion. *Philosophical Transactions of the Royal Society of London B* 228B:219–268.

Crisp, D. I. 1974. Factors influencing the settlement of marine invertebrate larvae. In *Chemoreception in Marine Organisms* (P. T. Grant and A. M. Mackie, eds.), pp. 177–265. New York: Academic Press.

Jensen, R. A., and D. E. Morse. 1984. Intraspecific facilitation of larval recruitment: gregarious settlement of the polychaete *Phragmatopoma californica* (Fewkes). *Journal of Experimental Marine Biology and Ecology* 83:107–126.

Jensen, R. A., and D. E. Morse. 1988. The bioadhesive of *Phragmatopoma californica* tubes: a silk-like cement containing L-DOPA. *Journal of Comparative Physiology B* 158:317–324.

Jensen, R. A., and D. E. Morse. 1990. Chemically induced metamorphosis of polychaete larvae in both the laboratory and the ocean environment. *Journal of Chemical Ecology* 16:911–930.

Morse, A. N. C. 1988. The role of algal metabolites in the recruitment process. In M.-F. Thompson, R. Sarojini and R. Nagabhushanam, eds., *Marine Biodeterioration*, Oxford & IBH Pub. Co. PVT. Ltd., New Delhi, pp. 463–473.

Morse, A. N. C. 1991. The role of algae in the recruitment of marine invertebrates. In Systematics Association Special Volume Series. Oxford University Press. In press.

Morse, A. N. C., C. A. Froyd and D. E. Morse. 1984. Molecules from cyanobacteria and red algae that induce larval settlement and metamorphosis in the mollusc *Haliotis rufescens*. *Marine Biology* 81:293–298.

Morse, A. N. C., and D. E. Morse. 1984. Recruitment and metamorphosis of *Haliotis* larvae induced by molecules uniquely available at the surfaces of crustose red algae. *Journal of Experimental Marine Biology and Ecology* 75:191–215.

Morse, D. E., and A. N. C. Morse. 1991. Enzymatic characterization of the morphogen recognized by *Agaricia humilis* (scleractinian coral) larvae. *The Biological Bulletin*. In press.

Morse, D. E. 1990. Recent progress in larval settlement and metamorphosis: closing the gaps between molecular biology and ecology. *Bulletin of Marine Science* 46:465–483.

Morse, D. E., H. Duncan, N. Hooker and A. N. C. Morse. 1977. Hydrogen peroxide induces spawning in molluscs, with activation of prostaglandin endo-peroxide synthetase. *Science* 196:298–300.

Morse, D. E., N. Hooker, H. Duncan and L. Jensen. 1979. γ-aminobutyric acid, a neurotransmitter, induces planktonic abalone larvae to settle and begin metamorphosis. *Science* 204:407–410.

Morse, D. E., N. Hooker, A. N. C. Morse and R. A. Jensen. 1988. Control of larval metamorphosis and recruitment in sympatric agariciid corals. *Journal of Experimental Marine Biology and Ecology* 116:193–217.

Strathman, M., and M. I. Simon. 1990. G protein diversity: A distinct class of α subunits is present in vertebrates and invertebrates. *Proceedings of the National Academy of Sciences of the U.S.A.* 87:9113–9117.

Trapido-Rosenthal, H. G., and D. E. Morse. 1985. *L*-α, ω-diamino acids facilitate GABA induction of larval metamorphosis in a gastropod mollusc (*Haliotis rufescens*). *Journal of Comparative Physiology B* 155:403–414.

Trapido-Rosenthal, H. G., and D. E. Morse. 1986a. Availability of chemosensory receptors is down-regulated by habituation of larvae to a morphogenetic signal. *Proceedings of the National Academy of Sciences of the U.S.A.* 83:7658–7662.

Trapido-Rosenthal, H. G., and D. E. Morse. 1986b. Regulation of receptor-mediated settlement and metamorphosis in larvae of a gastropod mollusc (*Haliotis rufescens*). *Bulletin of Marine Science* 39:383–392.

Wodicka, L. M., and D. E. Morse. 1991. cDNA sequences reveal mRNAs for two Gα signal transducing proteins from larval cells. *The Biological Bulletin*. In press.

Yool, A. J., S. M. Grau, M. G. Hadfield, R. A. Jensen, D. A. Markell and D. E. Morse. 1986. Excess potassium induces larval metamorphosis in four marine invertebrate species. *The Biological Bulletin* 170:255–266.

Hunter-Gatherers of the New World

Kim Hill
A. Magdalena Hurtado

Foraging peoples live by hunting, fishing, and collecting wild plants and insects. In the last 20 years anthropological research among foragers, or hunter-gatherers, as they are often called, has become increasingly important, for three reasons. First, the opportunity to study humans as foragers has been disappearing at an extremely rapid rate in the last 100 years. Because of worldwide economic trends, the transition from foraging to other forms of subsistence may be irreversible. Thus, we are likely to be the last generation to witness our fellow humans living in a way that was typical of most of human history.

Second, modern foragers live in relatively small groups (usually 15 to 100 individuals) in which subsistence activities produce immediate results, and a limited number of behavioral options are open to band members. This means that the ability to study direct links between ecological or social variables and behavioral patterns is generally greater in foraging societies than in more complex human settings.

Third, because our hominid ancestors spent all but the last 10,000 years (less than one percent of the time span of hominid history) living in small groups that subsisted on wild resources, we should be able to learn a great deal about human history and the evolution of human traits by studying modern foragers. Virtually all modern human anatomical traits evolved when foraging was universal, and the human nervous system and physiological mechanisms that generate behavior also evolved while humans lived by hunting and gathering.

Observations of the Ache, a foraging people in Paraguay, indicate that no single pattern of behavior is typical of the hunter-gatherer way of life

Stimulated by these factors, and strongly influenced by the seminal work of Richard Lee on the !Kung San as well as a provocative conference on "Man the Hunter" (Lee and DeVore 1968), a number of ecologically oriented research projects on modern foragers have been carried out in the past two decades: the Harvard Kalahari project and other !Kung San studies (for example, Lee and DeVore 1976; Tanaka 1980), research in Australia and Indonesia (Meehan 1977; Jones 1980; Griffin and Estokio-Griffin 1985), research in the Arctic and sub-Arctic (Binford 1978; Smith 1985), the Harvard Pygmy Project and other pygmy studies (Hart 1978; Harako 1981; Bailey and Peacock 1988), and the Utah hunter-gatherer project on Ache and Hiwi foragers, as well as the Utah-UCLA Hadza project (O'Connell et al. 1988; Hawkes et al. 1989).

The most important lesson that can be derived from these and numerous other forager studies is that very few "typical" hunter-gatherer patterns emerge. Instead, groups vary in almost every parameter that has been measured: composition of diet, food-sharing, men's and women's work patterns, subsistence strategies, childcare, settlement patterns, marriage systems, and fertility and mortality. The nonspecialist anthropological audience has been slow to appreciate the importance of this variability. Instead, the !Kung San studies of Lee are often cited as the typical hunter-gatherer pattern. This is almost certainly due to the outstanding quality of Lee's work and to the supposed appropriateness of the African savanna as an ecological context for understanding earlier hominids. Indeed, the !Kung study remains a cornerstone of modern forager research, but we must learn to build upon it if we are to develop our understanding of human evolutionary history.

One of the most useful approaches currently employed in forager studies is that of behavioral ecology, or the study of behavior from an evolutionary perspective. This approach assumes that behavioral patterns are generally adaptive and that variations are due to differences in the costs and benefits to fitness in each environmental and social context.

Much of the work presented in this paper was stimulated by four main issues in hunter-gatherer studies. First, what are the likely causes and consequences of the major dietary and technological shifts observed in the archaeological record, and what can these tell us about the diet of our hominid ancestors? Next, how and why is food shared among current foragers, and what are the implications of these patterns for the evolution of group living, settlement patterns, and the sexual division of labor in hominids? Third, what can activity profiles of modern foragers tell us about our past? Is the foraging way of life one of ease and leisure or a difficult one requiring hard work to survive? Fourth, in the area of demographics, what are the basic trends in fertility and mortality that characterize modern foragers?

In addition to these issues, most researchers are now concerned with the relationship between ecological constraints and different behavioral patterns. An understanding of this relationship should make it possible to build models that specify how independent variables will affect the behavior observed in any human community. Only through this uniformitarian approach can we hope to know what our ancestors did in the unobservable past.

This paper reports some of the results obtained in eight years of work with Ache foragers of Paraguay and describes preliminary findings from an ongoing project with the Hiwi foragers of the Venezuelan savanna. In all these studies we emphasize the variation through time and across groups of individuals, in order to derive likely explanations for differences observed between foraging groups.

The Ache of Paraguay

The Ache are a native population of Paraguay who until recently were full-time nomadic foragers (Clastres 1972; Melia et al. 1973). They consist of four independent groups; we have studied primarily the northernmost group, the last to make permanent peaceful contact (in the 1970s). This group was made up of 10 to 15 small bands that had no specific territories but roamed over an area of about 18,500 km^2 in eastern Paraguay. Each band had a smaller home range, but adults generally knew the entire area covered by the group. Recall by informants suggests the median size of a band was 48 people; the range on a given day was from 3 to 160.

The region is mainly neotropical, semideciduous evergreen forest, with a tree canopy about 20 m high and undergrowth more dense than that observed in many other primary tropical forests of South America. Since 1975, much of the area has been cut for agriculture and cattle pasture, but sizable pieces of primary forest (about 2,400 km^2) still exist near the Ache settlements. Transects measuring mammalian densities suggest a crude biomass of only about 400 kg/km^2 for the most commonly encountered mammals. This is about half the crude biomass measured for the same species in Barro Colorado, Panama (Eisenberg and Thorington 1973), and most forests of South America that have been studied show considerably higher species diversity.

Before contact, Ache bands foraged in this area and moved campsite frequently while hunting and gathering. Now they live primarily at agricultural mission stations but still spend about 25% to 35% of their time on overnight trips back in the forest, foraging for subsistence. We have monitored the Ache diet almost from the point of first contact to the present and are able to reconstruct the traditional diet to some extent by means of observations on forest trips (Hill 1983).

The data from short-term trips (with a range of 4 to 15 days) suggest that as foragers the Ache eat an astounding 3,700 calories per person per day (Hill et al. 1984). (By contrast, active adult Americans consume about 2,700 calories per day.) When the Ache are living in the forest, an average of 56% of their calories come from mammalian meat (ranging from 46% to 66%, depending on the season), with honey making up 18% (range 6–30%) and plants and insects providing an average of 26% (range 15–49%).

On a foraging trip, camp members rise early, eat whatever is left over from the previous day, and set out in search of food. Men lead the way, carrying only bows and arrows, and women and children follow, the women carrying young children and the family's possessions in a woven basket. Some men walk with their wives and carry children on their shoulders. Ache foragers do not walk on trails but break a new path through the forest each day. Usually the leaders set out in the direction of an area known or thought to contain important food resources.

After walking together for about an hour, the two sexes separate, with men walking further and more rapidly in search of game, and women and children slowly progressing in the general direction the men have set out. Men generally eat very little during the day, but women and children sometimes collect and eat fruits and insects while men hunt, and women often process palm trunks for their starchy fiber near the end of the day. This snacking usually accounts for less than 5% of all food consumed (Hill et al. 1984).

All camp members come together again at the end of the day, when they clear a small camp in the underbrush, build fires, and prepare and share food extensively. Evening is considered the most pleasant time, with band members enjoying their only large meal of the day, and joking and singing in the night. While in the forest the Ache sleep on the ground or on palm-leaf mats in a small circle. They build palm-leaf huts to sleep in only if it begins to rain. The next morning the band moves on again in search of food unless there is heavy rainfall throughout the day.

A model for food choice

The tropical forests of Paraguay are believed to contain several hundred species of edible mammals, birds, reptiles, amphibians, and fish, but the Ache have been observed to exploit only about 50 of them. Similarly, the forest holds hundreds of edible fruits and insects, yet the Ache exploit only about 40 of these. Over 98% of the total calories in the diet we observed between 1980 and 1983 were supplied by only 17 different resources. What can account for the fact that the Ache seem to ignore many edible resources in their environment?

The optimal diet model (MacArthur and Pianka 1966; Emlen 1966) was developed to predict which of an array of resources will be exploited if organisms attempt to maximize rates of food acquisition. In order to maximize the overall rate of return, foragers should attempt to obtain a resource only when the expected return rate is higher than what they can obtain on average if they

Kim Hill is an assistant professor in the Department of Anthropology and a member of the Evolution and Human Behavior program at the University of Michigan. He received his B.A. in biology and his Ph.D. in anthropology from the University of Utah. A. Magdalena Hurtado is a research scientist at the Instituto Venezolano de Investigaciones Científicas and in the Department of Anthropology at the University of Michigan. She received her B.A. in anthropology from SUNY-Purchase and her Ph.D. in anthropology from the University of Utah. Address: Department of Anthropology, 1054 L.S.A., University of Michigan, Ann Arbor, MI 48109–1382.

Figure 1. When in the forest the Ache eat well, taking in as much as 3,700 calories per day. The fruit *Rheedia brazilense (above),* which ripens in January, is a favorite food, as is *Cebus apella,* the capuchin monkey *(right).* (Photo above by K. Hill; photo at right by K. Hawkes.)

ignore it and continue to search for other resources.

Between 1980 and 1986, using Ache data, we were able to test the prediction that each resource exploited should be characterized by a higher caloric value than that observed for foraging overall. Our data show that all 16 items observed to be exploited in 1980, and 25 of 26 items exploited in 1981–83, are characterized by higher return rates than a forager could expect to obtain if he ignored the item and continued foraging (Hawkes et al. 1982; Hill et al. 1987). Whereas return rates for a whole day of foraging worked out to about 1,250 cal/h for men and about 1,090 cal/h for women, the average rate for any particular food obtained was, for men, about 3,500 cal per hour of foraging, and for women about 2,800 cal per hour of foraging. Although we were unable to measure resources *not* taken by the Ache, experience suggests that many would indeed be characterized by low returns (small fruits, birds, insects, reptiles). Thus, Ache foragers apparently do behave as if they chose to exploit only those resources that would increase their overall rate of food acquisition.

The model may be particularly useful for understanding subsistence changes that occur through time, or as a result of changes in technology. For example, in 1980 a few Ache hunters acquired shotguns, which raised their overall return rate from 910 cal/h (with bow and arrow) to 2,360 cal/h. Because some of the game taken by Ache men is characterized by return rates below the new 2,360 cal/h but above the rate of bow-and-arrow hunting, shotgun hunters should ignore some low-return animals that are taken by bow hunters. This prediction was generally met by observations in the field (Hill and Hawkes 1983). Most notably, shotgun hunters spent less than 2% of their time pursuing capuchin monkeys (with a return rate of 1,215 cal/h), whereas bow hunters spent over 13% of their time chasing capuchin monkeys *on the same foraging trips.* Several times shotgun hunters were observed to leave monkey hunts and continue searching for other, more profitable game.

Continued work with the Ache has pointed out both the utility and some shortcomings of models derived from optimal foraging theory (Hill et al. 1987; Hill 1988). Limitations that we have noted are, first, that some items that must be processed extensively may be exploited even when the resulting return rates are low, if processing normally takes place during times that foraging is not possible. Second, a short-term risk of imminent starvation or a reduction in the variance of daily food intake may lead to foraging behavior not predicted by simple models. Third, the biological value of foods is probably not reducible to calories when food types differ greatly.

A balance of nutrients is likely to be especially important in decisions in which the forager faces a choice between foods high in carbohydrates (such as plants) and those high in protein and lipids (animals and insects). From observations on the Ache, it appears that the sexual division of labor and the foraging strategy of males in general can be predicted by optimal foraging models only if the higher value of foods rich in proteins and lipids over carbohydrates is taken into account (Hill et al. 1987). The most important lesson from the Ache studies, however, is that simple models based on the assumption that individuals will attempt to maximize their rate of food acquisition while foraging are indeed useful for predicting subsistence patterns.

Food-sharing

The extensive sharing of food has been reported for many foraging peoples, but until recently there have been no quantitative studies that allow us to determine exactly what the sharing pattern looks like and how it varies from one foraging group to another. Understanding this variation is crucial, because food-sharing has been postulated to be critical in shaping the unique character of human sociality (e.g., Washburn and Lancaster 1968; Isaac 1978; Lancaster 1978; Kithara-Frisch 1982; Zihlman 1983).

The Ache share food throughout the band. Women who share vegetable items are usually praised, and young children are taught that stinginess is the worst trait a person can have. All hunters, regardless of status or hunting success, give up their kills to be distributed by others, and they almost never eat from their own kill. Nevertheless, there is some interesting variation in the way different resources are shared.

Although about 75% of all food consumed in an Ache band is acquired by a person outside the consumer's nuclear family, different resources are not all shared to the same extent (Kaplan et al. 1984). Game items are shared most, with more than 90% of the meat a hunter acquires being consumed by individuals not in his immediate family. Honey is shared somewhat less, and plant and insect foods least (Kaplan and Hill 1985).

Statistical analyses show that wives, children, and siblings receive no more of the meat or honey acquired by a man (their husband, father, or brother) than would be predicted by random chance if all the food were simply divided up among band members. This clearly contradicts a common assumption that food is always shared preferentially with close kin. However, the husband, children, and siblings of a woman were found to consume more of the food she collected than would be predicted by chance (Kaplan et al. 1984).

Further analyses suggest food-sharing among the Ache may serve to reduce daily variance in consumption. Food-sharing should be most beneficial if, by their own foraging, individuals acquire more than they can eat on some days and nothing on other days—a pattern that, indeed, characterizes Ache men's hunting returns (Hill and Hawkes 1983; Kaplan et al. 1989). If sharing is a strategy to reduce daily variance in food intake, and if different types of food (e.g., vegetables and meat) are not interchangeable, we should find a correlation between the daily variance in acquisition of a type of food and the extent to which it is shared. Our data from the Ache confirm this relationship. Moreover, the absolute reduction in variance of daily food intake is high enough to be biologically significant. During the time of our study, the average nuclear family reduced its daily variance from 13,243 cal to 4,863 cal by sharing food. A simple model suggests that food-sharing may lead to an 80% increase in nutritional status (Kaplan and Hill 1985).

Differences in observed patterns of food-sharing across modern foragers may therefore be partially due to differences in the daily variance of major food types acquired. This hypothesis leads to useful predictions: for example, among foragers that are able to store food, we might expect very little sharing, because storage reduces variance in the availability of food. Additionally, the data suggest that food-sharing and associated social patterns may not have arisen in hominid history until our ancestors began to use subsistence strategies that produced a high daily variance in food acquisition.

The division of labor

One of the major issues in the study of modern foragers has been just what determines how much time they spend in different activities. Some researchers see foragers as members of an original "affluent society" (e.g., Sahlins 1972), in which work effort is low because "needs" are few and easily met. Others (for example, Hawkes et al. 1985) have questioned this generalization. Early quantitative work with the !Kung San of Africa (Lee 1968) tended to support the low-effort model (which was partially derived from !Kung data) and has led more recent workers to monitor carefully how much time is spent in subsistence work and what other activities are important throughout the day. In addition, because a marked division of labor along sexual lines is an important characteristic distinguishing humans from other primates, it is of interest to describe the range of activities specifically for men and for women.

Our collected data—some 63 days' worth of focal

Figure 2. The division of labor along sexual lines is clearly marked among the Ache, with men spending almost 7 hours per day in hunting and in processing both meat and other foods. Resting, socializing, and light work together account for about another 4.5 hours per day. (Photo by K. Hill.)

studies each on men and women and 1,055 person-days of subsistence studies—show that in the forest Ache men spend about 6.7 h/day in subsistence activities (searching, acquiring resources, and processing food) and another 0.6 h/day working on the tools used in subsistence activities. Men also spend about 4.5 h resting, socializing, or in light activities each day (Hill et al. 1985). Women spend about 1.9 h in subsistence activities, 1.9 h moving camp, and about 8 h in light work or childcare (Hurtado et al. 1985). The contrast between the genders may not be surprising in light of the finding that men provide 87% of the energy supplied in the Ache diet and close to 100% of the protein and lipid consumed.

These data contradict the simple generalization that foragers spend little time in subsistence work. The Ache spend more than twice as much time in procuring, processing, and transporting food as !Kung men and women, who take 3.1 and 1.8 h/day, respectively, for such activities (Lee 1979). We developed a model which assumes that foragers will spend time in those activities which lower the mortality rates of their children and increase their own reproductive rates (Hawkes et al. 1985; Hill 1983; Hurtado 1985). This model explicitly rejects the notion, based on a concept of "limited needs" (Hawkes et al. 1985), that foragers work few hours per day because they do not need or want any more food.

What can account for the fact that the Ache seem to ignore many edible resources in their environment?

Among the Ache, who eat much better than most foragers who have been studied, there is still evidence that the acquisition of more food than usual results in higher rates of successful reproduction (Hill and Kaplan 1988). In virtually all primitive, peasant, and economically underdeveloped populations studied, it has been shown that a greater intake of food than usual leads to lower rates of child mortality (Behm 1983; Chen 1983). Many of these populations spend little time each day in work, relative to well-nourished modern Americans (e.g., Minge-Klevana 1980). Given the severe effects of dietary stress, however, it is unlikely that these populations simply choose to spend little time in subsistence work. More plausibly, they limit the hours they spend in food acquisition because opportunities are poor or entail at times an increased risk of injury or death, or because time spent in alternate activities has a greater effect on reproduction and children's survival.

The Ache patterns of time allocation provide some support for these generalizations. Both sexes work less on days when opportunities for food acquisition are poor. Men hunt fewest hours on days of poor weather (Hill et al. 1985), and women forage for fewer hours when no high-quality fruits are in season (Hurtado et al. 1985). Both sexes also reduce the time spent in subsistence activities when other important needs arise. Men spend fewer hours hunting when they have children to take care of (Hill et al. 1985), and women spend less time in foraging when they are nursing and caring for an infant than do women without an infant (Hurtado 1985; Hurtado et al. 1985). These data suggest that the sexual division of labor in hominids may have arisen when opportunities for obtaining new foods were ignored by females because they resulted in higher rates of child mortality.

Figure 3. Women spend about 8 hours per day in light work, childcare, or a combination of the two, as with the woman here shown weaving a basket while her son looks on. Women also spend almost 2 hours per day moving camp, transporting their family's possessions in a basket of this kind. (Photo by K. Hawkes.)

Mortality and fertility

Until Howell's study of !Kung demography, very little was known about mortality and fertility in foraging societies. Howell (1979) suggests that mortality rates among the !Kung are not significantly different from those observed currently in underdeveloped countries: high in infancy and early childhod and low in middle age, with about a third of the population surviving to old age. However, as compared with other populations that do not practice contraception, fertility rates are low among the !Kung, with a mean of 4.7 live births to each woman, owing primarily to an interval between live births that is close to four years. Our major goal in studying the demography of the Ache, and later the Hiwi of Venezuela, was to determine how similar their patterns of mortality and fertility were to those reported for the !Kung and to examine some of the possible determinants and implications of each. Comparison of the Ache with the !Kung is particularly interesting, because the !Kung are foragers in a dry desert, whereas the Ache live in the tropical forest, under very different ecological constraints.

Our data on mortality and reproduction come from interviews with all currently living Ache men and women. This gave us reproductive histories of 166 women, 65% of which were cross-checked at least once. Preliminary demographic data on the Hiwi are based on reproductive interviews with 35 men and women.

Mortality rates for the Ache look surprisingly similar to those reported for the !Kung, except in the highest age classes (where the !Kung sample size is very small). For both populations, about 20% of all children die before reaching one year of age, and about 60% survive to the age of 15. By contrast, only 48% of Hiwi children survive to age 15. The major causes of death differ for children in the three populations. Whereas 90% of !Kung children's deaths were reported as due to illness, 62% of the Hiwi and only 32% of the Ache children die from illness. The single greatest cause of death among Ache children is homicide, which accounts for 31% of the mortality in the Ache and 14% in the Hiwi, but only 4% in the !Kung.

Differences in causes of adult mortality among Ache, Hiwi, and !Kung foragers show similar trends.

The opportunity to study humans as foragers has been disappearing extremely rapidly in the last 100 years

Among Ache adults, warfare and accidents (snakebite, fall out of tree, attack by jaguar, etc.) account for 73% of all adult deaths. Among the Hiwi, warfare and accidents account for 39%, and among the !Kung only 11%. For the !Kung, illness is the cause of 88% of adult mortality, whereas for the Hiwi the proportion is 56%, and for the Ache it is only 17%. The data show that although overall mortality rates are similar, the Ache and Hiwi are more prone to accidental and violent death, whereas !Kung foragers of all ages are most likely to die from illness.

Fertility rates also differ significantly. The mean total lifetime fertility (number of live births) for women who have completed their childbearing years is 4.7 among the !Kung, 5.1 among the Hiwi, and 7.2 among the Ache. The average interval between births is 48 months for the !Kung and 38 months for the Ache.

Three factors are important in explaining the variation in birth intervals among populations: first, nursing patterns may differ in ways that result in a longer period of nonreproductive cycling among some populations. Second, greater levels of activity may lower the fertility of some populations of women who work harder. Third, nutritional intake may vary, with some populations who eat better more likely to become pregnant even when nursing and activity patterns are the same. In particular, nutritional differences between the !Kung and the Ache may account for more of the observed difference in birth intervals than do nursing patterns. As for Hiwi women, preliminary data show them to be intermediate between the !Kung and Ache in their interval between births, as in their caloric consumption and body weight.

A top priority for research

Comparison of the Ache to other foragers reveals considerable variation between groups for most parameters. No single group of hunter-gatherers can be considered as typical, and no group can legitimately be used as an analogue for understanding our ancestors. Instead, careful testing of hypotheses should make it possible to explain the observed variation and eventually to reconstruct hominid behavioral patterns under a variety of conditions.

The amount of attention directed to hunter-gatherers around the world has increased dramatically in the past 20 years, but unfortunately the opportunities for

Figure 4. Like other foraging peoples throughout the world, the Ache face losing their way of life as their territory is threatened by the development of the modern state within which they reside. Opportunities to study the rich variety of behavior associated with hunting and gathering may soon become scarce. (Photo by K. Hill.)

study are decreasing even more rapidly, despite new reports of full-time foragers in South America, Madagascar, New Guinea, and Indonesia. Studies on these groups and the remaining foragers in South America, Africa, Australia, and Asia should be given top priority if we hope to learn more before this opportunity disappears forever. Similarly, studies of groups that spend some time trekking nomadically, living mainly off wild resources, can give us important insights.

The increasing trend toward quantitative studies that pay careful attention to ecological constraints on the lives of foragers is not without its strong critics—those who believe that responsible anthropologists should focus entirely on issues of political rather than scientific content. It is true that foraging peoples have lost their traditional territories at an alarming rate. They are almost universally incorporated into the lowest economic strata of the modern states in which which they reside, and often their health and well-being deteriorate rapidly on contact with modern society. We have a responsibility to help and protect these people, who have given us the opportunity to understand our origins better. However, we must not forget that we also owe it to our children, and to the children of the foragers who remain, to learn what we can about this once-universal human way of life in the short period left to us. Scientific and humanistic goals need not be mutually exclusive. Time spent residing with present-day foragers can be used productively both to help and to learn about the people who are so kind as to allow our presence as their guests. Let us hope that they remember us as having done all we could to improve their situation and preserve the knowledge of their rapidly disappearing way of life.

References

Bailey, R., and N. Peacock. 1988. Efe pygmies of northeast Zaire: Subsistence strategies in the Ituri forest. In *Uncertainty in the Food Supply*, ed. D. du Garinne and G. A. Harnson. Cambridge Univ. Press.

Behm, H. 1983. Final report on the research project on infant and child mortality in the Third World. In *Infant and Child Mortality in the Third World.* Paris: CICRED.

Binford, L. R. 1978. *Nunamuit Ethnoarchaeology.* Academic Press.

Chen, L. 1983. Child survival: Levels, trends, and determinants. In *Determinants of Fertility in Developing Countries: A Summary of Knowledge.* Nat. Acad. Sci.

Clastres, P. 1972. The Guayaki. In *Hunters and Gatherers Today*, ed. M. Bicchieri. Holt, Rinehart & Winston.

Eisenberg, J., and R. Thorington, Jr. 1973. A preliminary analysis of a neotropical mammal fauna. *Biotropica* 5:150–61.

Emlen, J. M. 1966. The role of time and energy in food preference. *Am. Naturalist* 100:611–17.

Griffin, P., and A. Estokio-Griffin. 1985. *The Agta of Northeastern Luzon: Recent Studies.* Manila, Philippines: San Carlos Publ.

Harako, R. 1981. The cultural ecology of hunting behavior among Mbuti pygmies in the Ituri forest, Zaire. In *Omnivorous Primates*, ed. R. S. O. Harding and G. Teleki. Columbia Univ. Press.

Hart, J. 1978. From subsistence to market: A case study in Mbuti net hunters. *Human Ecol.* 6:325–53.

Hawkes, K., K. Hill, and J. O'Connell. 1982. Why hunters gather: Optimal foraging and the Ache of eastern Paraguay. *Am. Ethnol.* 9: 379–98.

Hawkes, K., J. O'Connell, K. Hill, and E. Charnov. 1985. How much is enough? Hunters and limited needs. *Ethol. Sociobiol.* 6:3–15.

Hawkes, K., J. O'Connell, and N. Blurton-Jones. 1989. Hardworking Hadza grandmothers. In *Comparative Socioecology of Mammals and Man*, ed. R. Foley and V. Standen. Basil Blackwell.

Hill, K. 1983. Adult male subsistence strategies among Ache hunter-gatherers of eastern Paraguay. Ph.D. diss., Univ. of Utah.

———. 1988. Macronutrient modifications of optimal foraging theory: An approach using indifference curves applied to some modern foragers. *Human Ecol.* 16:157–97.

Hill, K., and K. Hawkes. 1983. Neotropical hunting among the Ache of eastern Paraguay. In *Adaptive Responses of Native Amazonians*, ed. R. Hames and W. Vickers, pp. 139–88. Academic Press.

Hill, K., and H. Kaplan. 1988. Tradeoffs in male and female reproductive strategies among the Ache: Parts 1 and 2. In *Human Reproductive Behavior*, ed. L. Betzig, P. Turke, and M. Borgerhoff-Mulder. Cambridge Univ. Press.

Hill, K., K. Hawkes, A. Hurtado, and H. Kaplan. 1984. Seasonal variance in the diet of Ache hunter-gatherers in eastern Paraguay. *Human Ecol.* 12:145–80.

Hill, K., H. Kaplan, K. Hawkes, and A. Hurtado. 1985. Men's time allocation to subsistence work among the Ache of eastern Paraguay. *Human Ecol.* 13:29–47.

———. 1987. Foraging decisions among Ache hunter-gatherers: New data and implications for optimal foraging models. *Ethol. Sociobiol.* 8: 1–36.

Howell, N. 1979. *Demography of the Dobe !Kung.* Academic Press.

Hurtado, A. 1985. Women's subsistence strategies among Ache hunter-gatherers of eastern Paraguay. Ph.D. diss., Univ. of Utah.

Hurtado, A., K. Hawkes, K. Hill, and H. Kaplan. 1985. Female subsistence strategies among Ache hunter-gatherers of eastern Paraguay. *Human Ecol.* 13:1–28.

Isaac, G. 1978. The food-sharing behavior of proto-human hominids. *Sci. Am.* 238:90–108.

Jones, R. 1980. Hunters in the Australian coastal savanna. In *Human Ecology in Savanna Environments*, ed. D. Harris. Academic Press.

Kaplan, H., K. Hill, K. Hawkes, and A. Hurtado. 1984. Food-sharing among the Ache hunter-gatherers of eastern Paraguay. *Curr. Anthropol.* 25:113–15.

Kaplan, H., and K. Hill. 1985. Food-sharing among Ache foragers: Tests of explanatory hypotheses. *Curr. Anthropol.* 26:223–45.

Kaplan, H., K. Hill, and A. Hurtado. 1989. Risk, foraging and food-sharing among the Ache. In *Risk and Uncertainty in the Food Supply*, ed. E. Cashdan. Westview Press.

Kitahara-Frisch, J. 1982. Nature of the basic early hominid dietary adaptation. *Life Sci. Inst. Ann. Rev.*, pp. 165–83.

Lancaster, J. B. 1978. Carrying and sharing in human evolution. *Human Nat.* 1:83–89.

Lee, R. B. 1968. What hunters do for a living, or how to make out on scarce resources. In *Man the Hunter*, ed. R. B. Lee and I. DeVore. Aldine.

———. 1979. *The !Kung San: Men, Women and Work in a Foraging Society.* Cambridge Univ. Press.

Lee, R. B., and I. DeVore. 1968. *Man the Hunter.* Aldine.

———. 1976. *Kalahari Hunter-Gatherers.* Harvard Univ. Press.

MacArthur, R. H., and E. R. Pianka. 1966. On the optimal use of a patchy environment. *Am. Naturalist* 100:603–09.

Meehan, B. 1977. Hunters by the seashore. *J. Human Evol.* 6:363–70.

Melia, B., L. Miraglia, M. Munzel, and C. Munzel. 1973. *La Agonia de los Ache-Guayaki.* Asunción: Univ. Católica.

Minge-Klevana, W. 1980. Does labor time increase with industrialization? A survey of time allocation studies. *Curr. Anthropol.* 21:279–98.

O'Connell, J., K. Hawkes, and N. Blurton-Jones. 1988. Hadza scavenging: Implications for Plio-Pleistocene hominid subsistence. *Curr. Anthropol.* 29:356–63.

Sahlins, M. 1972. *Stone Age Economics.* Aldine.

Smith, E. A. 1985. Inuit foraging groups: Some simple models incorporating conflicts of interest, relatedness, and central place sharing. *Ethol. Sociobiol.* 6:37–57.

Tanaka, J. 1980. *The San Hunter-Gatherers of the Kalahari.* Univ. of Tokyo Press.

Washburn, S., and C. S. Lancaster. 1968. The evolution of hunting. In *Man the Hunter*, ed. R. B. Lee and I. DeVore, pp. 293–303. Aldine.

Zihlman, A. L. 1983. A behavioral reconstruction of *Australopithecus*. In *Hominid Origins*, ed. K. J. Reichs, pp. 207–38. Washington, DC: Univ. Press of America.

Thomas D. Seeley

The Ecology of Temperate and Tropical Honeybee Societies

Ecological studies complement physiological ones, offering a new perspective on patterns of honeybee adaptation

In his masterful synthesis of insect sociology, Wilson (1971) identified three overlapping stages in the history of research on social insects: a natural history phase, a physiology phase, and a population-biology or ecology phase. Until the mid-1970s, most students of the social insects focused on the first two stages, either describing the diverse societies of termites, ants, wasps, and bees or experimentally analyzing the physiological bases of their social systems. Quite recently, ecological studies of the social insects have also flourished. As a result, we are beginning to understand the properties of social insect colonies as adaptations fixed by natural selection.

Throughout the history of insect sociology the premier object of study has been a single species of social bee: *Apis mellifera*, one of four living honeybee species. Its preeminence stems from several factors. First, honeybee societies rank among the most complex of all insect societies, with such advanced features as strong dimorphism of queen and worker, elaborate division of labor by age, precise control of nest temperature, and a remarkable system of communication based on a dance language (Michener 1974). Ease of study also favors research on honeybees. Not only are their colonies easily maintained in man-made hives of the sort used by beekeepers, but they will even live in glass-walled observation hives, enabling humans to peer into the heart of their society.

Still another factor is surely man's ancient fascination with the honeybee. The great insect sociologist William Morton Wheeler (1923) expressed this accurately when he wrote:

> Its sustained flight, its powerful sting, its intimacy with flowers and avoidance of all unwholesome things, the attachment of the workers to the queen—regarded throughout antiquity as a king—its singular swarming habits and its astonishing industry in collecting and storing honey and skill in making wax, two unique substances of great value to man, but of mysterious origin, made it a divine being, a prime favorite of the gods, that had somehow survived from the golden age or had voluntarily escaped from the garden of Eden with poor fallen man for the purpose of sweetening his bitter lot. [p. 91]

When the human fascination with the honeybee turned scientific, man began to describe and experimentally analyze the interwoven phenomena of colony life cycle, caste structure, and communication codes that make up the social organization of bees. This approach gained strong impetus from the highly crafted studies of the Nobel Laureate Karl von Frisch (1967, 1971) and continues apace today. We now understand in fair detail such topics as honeybee caste determination, sensory physiology, nest micrometeorology, and communication among colony members. In stark contrast, the ecology of honeybee societies remains a largely uncharted area of study. In short, we know a great deal about how honeybee societies work, but remarkably little about the pressures of natural selection that have shaped them.

The honeybee's dance language provides a clear example of this imbalance between mechanistic and functional knowledge. Over the past 60 years, aided by more than 40 graduate students, von Frisch (1967) has assembled a detailed picture of the physiological mechanisms of the behaviors that unfold when a scout bee discovers a rich patch of flowers, flies back to her nest, and recruits nestmates to gather the food by using the dance language. However, the precise ecological significance of this finely tuned system of communication has until recently remained a matter of speculation. Only in the past three years have researchers begun to analyze in a systematic way how the dance language helps colonies living in nature collect their food (Visscher and Seeley 1982; Seeley, in press).

In this essay I will describe some recent developments in the ecological study of honeybee social behavior, drawing in particular on a general program of behavioral-ecological research I have conducted over the past eight years. Most of my research deals with colonies of *A. mellifera* living in the northeast region of the United States. *A. mellifera* occurs as a native in Europe, western Asia, and Africa, but has been introduced throughout the world by man. A smaller portion of the research program considers the ecologies of the three other honeybee species: *A. florea*, *A. cerana*, and *A. dorsata*. Except for *A. cerana*, whose range extends north into China and Japan in eastern Asia, these three species are found only in southern Asia.

Colonies living in nature

Wild colonies of honeybees have been hunted by the peoples of Asia, Africa, and Europe for hundreds if not thousands of years (Crane 1975).

Thomas D. Seeley is Assistant Professor of Biology at Yale University. He is a graduate of Dartmouth College and Harvard University. His scientific interests center on the social behavior of insects, especially the honeybee. The area of research described in this article will be reviewed more fully in his book Honeybee Ecology, *to be published by Princeton University Press. Address: Department of Biology, Yale University, New Haven, CT 06511.*

Thus man has long had contact with honeybee colonies in nature, but with the intention of robbing them of their beeswax and honey, not studying them. When, in the 1500s, Europeans turned from exclusively plundering bees to also examining the fundamental facts of honeybee life, they observed bees living in man-made hives. To the present day, virtually all scientific research on honeybee biology has been conducted with colonies occupying beehives placed in locations that are convenient for scientists. For example, much of von Frisch's pioneering research on honeybee communication was carried out in the courtyard of the Munich Zoological Institute, a converted monastery in the heart of a city.

An important first step in studying honeybee ecology, therefore, was to describe the nests and life history of wild colonies of *A. mellifera*. This required turning away from colonies living in man-made hives in ecologically disturbed habitats and instead studying colonies inhabiting hollow trees in forests—in this case colonies found in the countryside near Ithaca, New York (Fig. 1). We located wild colonies either through information from local residents or by beelining, the old bee hunter's technique of inducing bees to forage from a comb filled with sugar water and then tracing the bees back along their flight lines until the bee tree is reached (Edgell 1949). Some of the bee trees were felled to collect the colonies and dissect their nests. Others were left standing for long-term observations of colony mortality. To determine the distinctive reproductive patterns of wild colonies, we simultaneously studied reproduction in colonies living in man-made hives the same size as the tree cavities occupied by wild colonies.

When allowed to remain in hollow trees in the forest, honeybee colonies live quite differently than they do when they inhabit a beekeeper's or bee researcher's standard beehives. Whereas a beekeeper desires a large, nonreproducing colony capable of stockpiling a vast quantity of honey (much more honey, in fact, than the colony would ever need), wild colonies grow to only one-third to one-half the population size, sequester only as much honey as they need, and devote their remaining

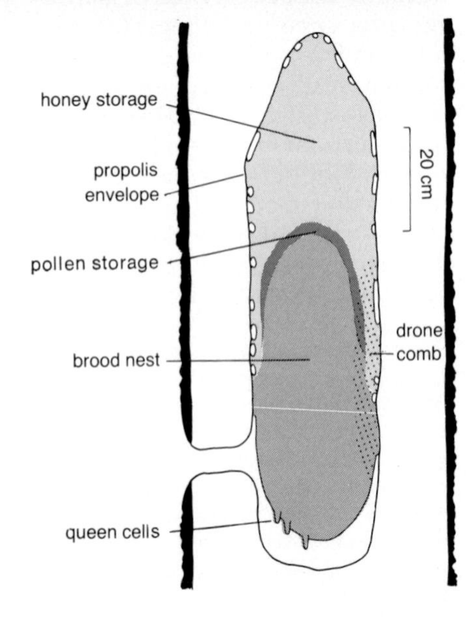

Figure 1. A knothole that serves as the entrance to an *Apis mellifera* nest is visible high up in the left fork of this intact bee tree in central New York State (*above*). A cross-section through a typical nest (*right*) shows what lies beyond the entrance. Layers of vertical comb nearly fill the tree hollow. Honey is stored in the upper region, pollen is packed in a narrow band directly below, and brood is reared in the lowermost portion of each comb. Queen cells house the new queens before swarming; drones occupy special cells at the edge of the nest. A polished layer of propolis, or tree resins, seals the nest cavity. Before a swarm occupies such a cavity, scouts carefully check its suitability by measuring the volume of the cavity and noting the size of the entrance opening, its compass orientation, and its proximity to the floor of the cavity. (All photographs by the author; sketch after Seeley and Morse 1976.)

energies to colony reproduction, swarming nearly every year (Seeley and Morse 1976; Seeley 1978; Winston 1980). These differences between wild colonies and beekeeper's colonies do not reflect genetic differences between the two colony types. Rather, they reflect the beekeeper's practice of providing abundant hive space (usually more than 160 liters), so that colonies rarely become overcrowded and thus are rarely stimulated to swarm. Wild colonies, in contrast, live in cavities that are usually smaller than 80 liters in volume; they soon outgrow this space and so swarm annually (Seeley 1977).

The two types of colonies also differ strikingly in their patterns of survival. Because beekeepers provide their colonies with snug hives, preconstructured combs, and supplementary food in lean times, only about 10% of domestic colonies perish annually. Wild colonies face a much harsher existence. Each colony must search the forest for a tree cavity, construct energetically expensive beeswax combs, and laboriously collect from millions of flowers the tens of kilograms of pollen and honey consumed by a colony yearly (Ribbands 1953). If in a given summer the plants yield abundant food, wild colonies thrive and multiply; if not, many starve. In their first year fully 75% of wild colonies starve, but if a colony survives this critical period, mortality drops to about 20% per year (Seeley 1978). Evidently selection is most intense, and hence evolution potentially fastest, at the stage of colony founding. It is therefore perhaps not surprising that honeybee colonies possess the complex adaptations for colony founding discussed below.

The challenge of winter

The finding of greatest ecological meaning from our initial descriptive studies is that there is strong seasonality in colony mortality. The vast majority of wild colonies perish during the winter; 77% and 90% of the deaths of first-year and established colonies, respectively, occur during the winter (Seeley 1978). Death is almost always due to starvation. It appears that the paramount ecological challenge faced by honeybee colonies in cold temperate regions of the world is winter survival. Much of the honeybee's biology can be understood as adaptations to this particular problem.

Unlike other social insect species in cold climates, honeybee colonies do not overwinter either as dormant, solitary queens or as inactive colonies in refuge deep underground, but as full, active colonies in self-heated nests. To achieve this, each colony contracts in autumn into a tight cluster of bees whose surface temperature is maintained above about 10°C. Heat is generated within the cluster by microvibrations produced by the bees' massive flight muscles. The colony's stored honey provides the fuel (Johansson and Johansson 1979; Seeley and Heinrich 1981).

The reason that honeybees practice such an exotic technique of overwintering is largely historical. Honeybees apparently arose in the Old World tropics and only later penetrated temperate regions (Wilson 1971; Michener 1974). All honeybee species living in the tropics today show a basic ability to control the central, nursery region of the nest at 30 to 36°C year round (Morse and Laigo 1969; Darchen 1973; Akratanakul 1977 diss.). When *A. mellifera* expanded out of the tropics into regions with harsher climates, its colonies adapted by simply refining their preexisting methods of controlling nest microclimate, rather than evolving wholly new physiological techniques for surviving periods of low body temperature.

Success in maintaining a warm microenvironment inside the nest and thus in overwintering requires that a colony occupy a well-sheltered tree cavity which tightly encloses the bees and their honey-filled combs, and that it possess the 20 or more kilograms of honey needed for winter heating fuel. Bees fulfill the first requirement through their elegant process of nest site selection (Lindauer 1955, 1961; Seeley et al. 1980; Seeley 1982). After leaving the parent nest and settling nearby, a honeybee swarm sends out a few hundred scout bees who search the surrounding forest for potential nest sites; from among the twenty or so sites found they select the single best site for their colony's future home.

A number of investigators (Seeley and Morse 1978a,b; Jaycox and Parise 1980; Gould 1982; Rinderer et al. 1982) have analyzed the preferences of house-hunting bees by setting out series of paired nest-boxes differing in a single variable such as entrance height or cavity volume and noting the patterns of occupation by wild swarms. Scouts reject cavities that have a volume of less than about 15 liters, undoubtedly because such small cavities cannot enclose enough honey-filled combs to fuel a colony through a winter. They measure the size of the cavity by walking about inside it and integrating the sensations thus experienced into a perception of volume (Seeley 1977).

Scouts also reject cavities whose entrance holes are larger than about 70 cm^2, probably because such sites are excessively drafty. Small openings in the walls of the cavity do not disqualify a site, because bees can plug these with tree resins. The bees' preference for an entrance opening that lies at the floor of the cavity and that faces south is probably also an adaptation for overwintering (Avitabile et al. 1978). An entrance located at the bottom of the cavity may help minimize convectional heat loss, and a southern exposure ensures a sun-warmed platform from which bees can make the critical cleansing flights to eliminate accumulated body wastes on occasional mild days in midwinter.

Fulfilling the second requirement for overwintering, sufficient honey stores, is fostered by several unique elements of honeybee social organization. One is the dance language, a system of communication whereby a scout that has discovered a rich food source recruits nestmates to help gather her find by performing inside the nest a miniature reenactment of a flight out to the flowers (von Frisch 1967; Michener 1974; Gould 1976). To understand exactly how this ability to communicate helps colonies collect the many kilograms of honey needed for winter, Visscher and I (1982) recently monitored the foraging behavior of a honeybee colony living in a deciduous forest in central New York State (Fig. 2). We envisaged the colony as a gigantic amoeba fixed on its nest site but able to send pseudopods—i.e., groups of foragers—out across the forest to patches of flowers rich in nectar and pollen. Of course, it is utterly impossible to follow a colony's entire force of 10,000 or more foragers by observing them directly; they are spread across some 25 km^2 of forest. However, we suspected that one could infer where a colony is foraging by tapping into the bees' own communication system—that is, by reading the dances by which foragers were recruited.

The heart of our experimental procedure, therefore, was to establish a normal-sized colony in a glass-walled observation hive and to sample, record, and translate the colony's dances. In each dance a bee walks briskly and repeatedly across one of the nest's vertical combs, shaking her abdomen back and forth to attract the attention of the surrounding bees. The duration of these so-called "waggle runs" is proportional to the distance to the patch of flowers being advertised. Nearby sites are indicated by waggle runs lasting less than one second, distant sites by runs of up to several seconds. The direction of the patch of flowers is encoded in the angle of the dance. Waggle runs in which the bee walks straight up the vertical comb indicate flower patches directly in line with the sun, whereas waggle runs performed 90° to the right of this indicate patches 90° to the right of the sun, and so forth. Whether the patch being advertised yields pollen or nectar is indicated by the presence or absence of loads of pollen on the dancer's hind legs.

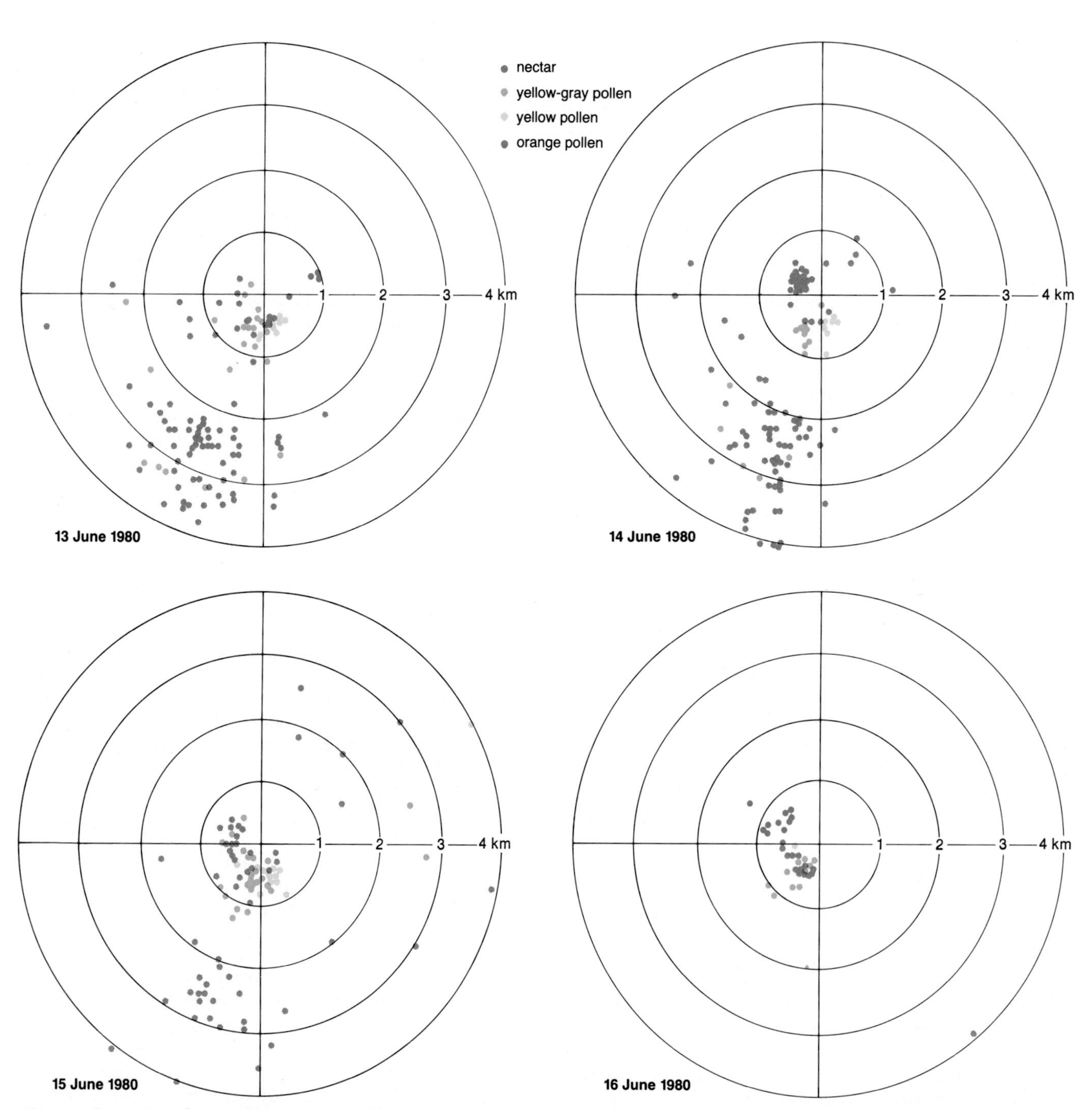

Figure 2. By reading the recruitment dances of foragers, it was possible to map the location of food sources on consecutive days and thus to study the overall dynamics of a colony's foraging activity. The nature of the food source was determined by observing whether returning foragers unloaded nectar (*gray dots*) or pollen (*colored dots*); recording the color of the pollen brought back in each case gave a further key to the distribution of the foragers. The pattern that emerges is one of a shifting mosaic of food sources, as the honeybee colony uses information gained from the searches of thousands of foragers to help direct its forces to the most rewarding flower patches. (After Visscher and Seeley 1982.)

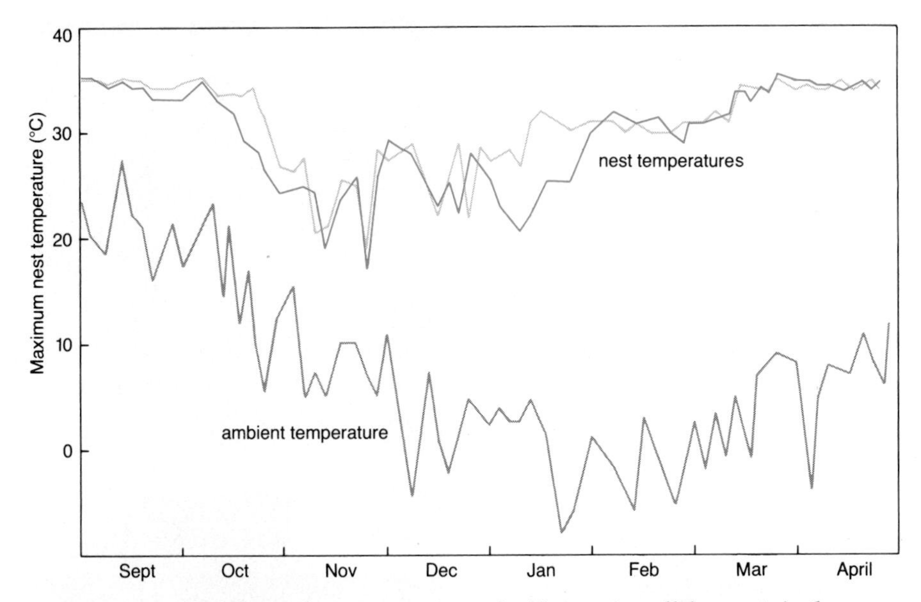

Figure 3. Records of the maximum temperatures inside two *A. mellifera* nests in the course of one winter show a remarkable ability to control the climate within the nest. When colonies cease brood rearing in October, they lower the nest temperature from about 34°C to around 20°C to conserve fuel. The return to temperatures of about 30°C in December and January marks the midwinter resumption of brood rearing.

By carefully timing the duration and measuring the angles of over 100 dances each day, we were able to plot the colony's forage sites as shown in Figure 2. Comparison of the maps for consecutive days revealed the dynamics of the colony's foraging. We discovered that the colony routinely foraged several kilometers from its nest; 95% of the foraging took place within 6 km, and the median distance traveled was 1.7 km. The colony frequently—at least once daily—adjusted the distribution of its foragers among the patches, and it worked relatively few patches each day. On average, 9.7 patches accounted for 90% of the colony's foraging in a given day. In short, the overall pattern was one of exploitation of a vast, rapidly changing mosaic of flower patches. Evidently the dance language is one component of a strategy that enables the colony to forage efficiently on patches of flowers which are widely dispersed and which vary greatly in profitability. By pooling information gained from the reconaissance of thousands of foragers, the colony is able to monitor the flower patches around the nest continuously. By also somehow identifying which patches are best (Boch 1956) and using the dance language to direct foragers to these patches, colonies appear to be able to keep their forces consistently focused on a few top-quality patches of flowers. The result is highly efficient food collection based on what might be called an "information center" strategy of foraging.

To stockpile sufficient winter stores, colonies need not only efficient foraging techniques but also adequate time to collect food. Time is especially valuable for newly founded colonies. Within the short summer season of temperate regions, each first-year, or "founder," colony must find a proper homesite, build from scratch a nest of costly beeswax combs, rear young that can survive the winter, and gather provisions for winter. All that a swarm takes from the parent colony to help establish the daughter colony, besides the workers themselves, is one stomachful of honey per worker (Combs 1972).

Apparently colonies have been selected to swarm as early as possible in late spring or early summer to allow maximum time for a founder colony to become established. Around Ithaca, swarming starts by mid-May, peaks in early June, and is largely completed by early July (Fell et al. 1977). A similar peak period for swarming in late spring and early summer has been observed in other north temperate locations (Simpson 1959; Martin 1963). The importance of the swarming date was demonstrated by experiments conducted in Ithaca over three years in which standard-sized swarms were placed in empty hives on May 20 and June 30 (20 days before and after the median date for swarming, respectively) and patterns of survival through the following winter were noted. Although, as expected, the probability of surviving to the next spring was low for both groups, it was significantly higher for colonies that swarmed early than for those that swarmed late: 0.55 and 0.33, respectively (Seeley and Visscher, unpubl.).

Colonies cannot divide by swarming until they have grown large enough to ensure that both parent and daughter colonies will have a sufficiently large worker force. Thus selection favors rapid growth of the colony in the spring. This may explain why colonies, having ceased rearing brood in October, resume this activity in mid-December, shortly after the winter solstice (Jeffree 1956; Avitabile 1978). What enables colonies to perform the remarkable feat of raising young in midwinter is again their outstanding capacity to regulate nest temperature. As shown in Figure 3, a colony can maintain the core of its winter cluster at a cozy 32 to 34°C, the brood-nest temperature, even when the temperature outside the nest cavity is 0°C or colder (Simpson 1961; Seeley and Heinrich 1981).

The theme that emerges from these ecological studies of *A. mellifera* in North America is one of adaptation to a harsh physical environment, especially the long, freezing winters of north temperate regions. Adaptations include the choice of nest sites providing adequate shelter, the ability to regulate nest temperature well above the ambient temperature for months on end, storage of tens of kilograms of honey for heating fuel, and even the timing of the basic annual cycle of rearing brood and swarming, which helps founder colonies become established during the short summers of the temperate zone.

Predation in the tropics

Paradoxically, although the studies described so far reveal a pattern of adaptation to a challenging physical environment, honeybees are essentially tropical insects, most of whose evolution occurred in mild climates. Thus most of the features of honeybee biology discussed above (the foraging strategy being an exception)

Figure 4. Workers of the three honeybee species found in tropical Asia show substantial differences in size. The smallest worker, that of *A. florea* (*top*), differs by approximately 30% in head width from the *A. cerana* worker (*center*), which in turn differs 40% from the largest worker, *A. dorsata* (*bottom*). An important related difference is stinger size; larger stingers such as those possessed by *A. dorsata* workers produce several times the pain of smaller stingers.

are secondary adaptations evolved by *A. mellifera* in expanding out of the tropics. To understand the primary evolution of the honeybee societies, therefore, one must study these bees in the tropics.

I chose to study the tropical honeybees living in Southeast Asia (Seeley et al. 1982). Here one finds three sympatric honeybee species—*A. florea*, *A. cerana*, and *A. dorsata*—a situation which provides a wonderful opportunity to analyze the ecological forces shaping insect societies through a comparison of species. The logic of this approach is that different species have evolved in relation to different ecological pressures, and that comparisons among closely related species can reveal how differences in their traits reflect differences in their ecologies (Lack 1968; Clutton-Brock and Harvey 1979; Krebs and Davies 1981). Because the species being compared are closely related phylogenetically, one can assume that their differences reflect recent adaptive divergence within a monophyletic group rather than long-held historical differences among phylogenetically distant species. One special value of comparative studies is that they can reveal how entire constellations of traits in a given species—morphological, behavioral, and physiological—are interrelated and are influenced by the same ecological pressures. In other words, the comparative approach can reveal the overall adaptive strategy of a species.

The three Asian honeybee species are ideally suited to this type of study. Their close phylogenetic relationship is indicated by the numerous anatomical and behavioral traits (such as the dance language) by which insect systematists have grouped them in the genus *Apis* (Michener 1974; Winston and Michener 1977). Equally important, however, is the fact that despite this close relationship the three species exhibit a curious array of differences in behavior and ecology and so provide rich material for comparison.

Although little was known about these bees, a few papers (Tokuda 1924; Roepke 1930; Lindauer 1956; Morse and Laigo 1969) reported enough to allow a preliminary comparison of the three species before fieldwork was begun (Table 1). The available information suggested that each species possesses a different strategy of colony defense. Specifically, it appeared that *A. florea* colonies defend themselves against visual predators by building low, widely dispersed, hidden nests, and that they use sticky barriers for protection against ants; that *A. cerana* colonies rely on a strategy of nesting inside cavities for protection against

Figure 5. Nest sites of the three Asian honeybee species differ widely. This typical *A. florea* nest (*top*) is built low on a slender branch of a small tree; a protective blanket of bees covers the nest comb. About 20 cm wide, the nest was originally hidden by vegetation. *A. cerana* colonies characteristically nest in cavities with small entrance openings (*center*). Here the nest was constructed in the cavity of a tree, which has been cut open to expose the interior. The bees have been killed, and most have dropped off the combs. The two taller combs at the right each measure about 30 cm. Colonies of *A. dorsata* typically nest in tall trees (*bottom*). The two nests shown here, each about 100 cm wide, are located about 20 m high in a dipterocarp tree. As in *A. florea* nests, bees cover the nest comb.

Figure 6. An *A. florea* colony salvages wax from a nest it was forced to vacate by a predator's attack. The loads of wax visible on each bee's hindlegs will be used to build comb at a new site. All three species of Asian honeybees suffer from intense predation that causes frequent shifts in nest sites.

large predators, and on direct attacks by their relatively large workers to repel smaller enemies such as ants; and that *A. dorsata* colonies counter predators by nesting in lofty aggregations in the treetops and by launching powerful stinging attacks against intruders capable of penetrating to such heights. It seemed likely that two factors set the stage for this radiation in defense strategies: differences in the size of the workers (Fig. 4), which could strongly influence the effectiveness of defense by direct fighting, and differences in the type of nest site (tree cavity or open air), which might profoundly influence a colony's basic defense situation. Presumably these differences could be traced to interspecific competition for food and nest sites.

To test this hypothesis, Robin Hadlock Seeley and I assembled basic information on the natural history of the three species by observing them living under natural conditions. Our study was conducted in the mountains of northeast Thailand, in and around the Khao Yai National Park. This park comprises 2,000 km^2 of tropical semievergreen rain forest—apparently an ideal habitat for honeybees, since it contains abundant colonies of all three species (Fig. 5).

We found that the bees in this forest suffer severe predation, probably because their brood- and honey-filled nests are bonanzas for numerous species. To quantify the intensity of this predation, we monitored groups of colonies for several months, inspecting each colony twice monthly. We found that approximately 25% of all *A. florea* and 10% of all *A. cerana* colonies are forced each month to abandon their nests because of a predator's attack (Fig. 6). These figures imply that on average each colony's nest is destroyed every four months in the case of *A. florea*, and every ten months for *A. cerana*. Comparable data could not be collected for *A. dorsata* colonies because they migrate between the highlands and the lowlands, following the rains and the flowers (Koeniger and Koeniger 1980), and so were absent from the area during much of our seven-month stay. However, it is clear from attacks we witnessed on *A. dorsata* colonies and destroyed nests we encountered that colonies of this species are also constantly threatened by predators.

What predators cause these high rates of nest destruction? A number of vertebrates—tree shrews (*Tupaia glis*), rhesus monkeys (*Macaca mulatta*), Eurasian honey buzzards (*Pernis apivorus*), and Malayan honey bears (*Helarctos malyanus*)—together with giant social wasps (*Vespa* spp.) are the principal predators capable of mounting massive, overwhelming attacks which can force colonies to abandon their nests. Other predators, such as agamid lizards and weaver ants (*Oecophylla smaragdina*), impose a chronic stress on colonies by extracting a few workers daily but rarely drive colonies from their nests.

As we came to know the three Asian honeybee species during seven months of fieldwork, we found that our initial hypothesis neatly meshed with our observations. Most of the differences in colony design among the three species do seem to reflect three different strategies of colony defense.

A. florea colonies build their defense around avoiding detection by predators. To this end, the bees build small, widely dispersed nests usually set deep in dense, shrubby vegetation along forest margins. One dramatic demonstration of this emphasis on nest concealment came in the dry season, when most *A. florea* colonies lost their cover as plants shed their leaves. Once exposed, colonies waited about two weeks,

Table 1. Preliminary comparison of the three honeybee species in tropical Asia

Property	*A. florea*	*A. cerana*	*A. dorsata*
Worker size (length)	small	medium	large
Nest site	open	cavity	open
Nest height	low (<5 m)	?[a]	high (>10 m)
Nest visibility	hidden?	?	conspicuous?
Nest dispersion	solitary	solitary	aggregated
Sticky barrier	present	absent	absent
Colony population	small (<5,000)	?	large (>20,000)
Colony aggressiveness	timid?	?	fierce

SOURCES: Tokuda 1924; Roepke 1930; Lindauer 1956; Morse and Laigo 1969
[a] Question marks indicate that the characteristic is uncertain or unknown.

until their brood had matured, and then moved a hundred meters or so to a new, well-hidden site. To help minimize the cost of switching nest sites, each colony transported all its stored honey to the new nest and even salvaged much of the old nest's wax for use in building the new comb.

We experimentally tested whether these shifts in nest site served to conceal nests from predators or simply to keep them out of the burning tropical sun by clipping away the vegetation surrounding several nests but leaving vegetation overhead to provide shade. Other nests that were similarly disturbed but left surrounded by vegetation served as controls. All colonies whose nests were exposed abandoned their nests within two weeks, whereas after eight weeks only one control colony had shifted nest sites. Evidently loss of concealment does trigger a shift in nest site by *A. florea* colonies. Reliance on nest concealment makes good sense for these bees. Being built in the open, their nests are basically vulnerable, and because their workers are small and so deliver only relatively painless stings, colonies usually cannot repel large invaders. Their best option seems to be to prevent discovery of the nest in the first place.

Figure 7. *A. florea* workers protect their colony from the weaver ant (*Oecophylla smaragdina*), an important predator, by plastering a sticky band of plant resin around the branch leading to their nest. Here four trapped weaver ants – one partially hidden by the branch – show the effectiveness of this strategy. *A. florea* evades larger predators by concealing its nest.

Nest concealment is not the sole line of defense mounted by *A. florea* colonies. In fact, one of their most important predators, the weaver ant, whose colonies contain hundreds of thousands of workers (Hölldobler 1970), easily locates *A. florea* nests. Because of their diminutive size, *A. florea* workers are no match for this large ant in direct fighting, and colonies rely instead on sticky bands of plant resin plastered around the branch supporting the nest (Fig. 7). The vulnerability of colonies without such sticky bands was easily demonstrated by bending a leaf bearing a weaver ant onto the nest. The bees would immediately retreat, but the ant always managed to yank a bee off the nest and kill it.

A. cerana colonies have evolved a different strategy of colony defense, one that centers on their habit of nesting inside cavities such as caves and hollow trees. These colonies are quite conspicuous (at least to humans, but probably also to most other vertebrates), because their medium-sized workers stream in and out of low, clearly visible entrance holes. Their defense thus relies little upon nest concealment. Instead, *A. cerana* colonies make it difficult for large predators to gain access to their nests by selecting cavities whose largest entrance opening is usually a narrow crack or hole no bigger than 40 cm^2.

When we simulated attacks on these colonies by scraping at the entrances of nests, the guard bees simply retreated to safety inside their nest cavities. In our long-term observations of these colonies we frequently encountered evidence of attacks by predators in the form of signs of digging and scratch marks around the nest entrance, but apparently the predators could not reach the walled-in colonies and the bees were unharmed. Adaptations related to this defense include multiple combs (rather than a single comb, as in the nests of *A. florea* and *A. dorsata*), to fit within the limited space of the cavity, small and nonaggressive colonies (since massive counterattacks are not needed), and probably special techniques of ventilation to prevent asphyxiation within the enclosed nest cavity (Seeley 1974). Small predators such as ants can enter these nests easily, but *A. cerana* workers are large enough to fend off such enemies by direct fighting.

The defense strategy of *A. dorsata* colonies follows a still different design: they nest high—usually above 15 m—in the tallest trees in the forest, thereby evading all predators not skilled in climbing or flying, and launch massive stinging attacks with their sizable colonies of relatively large, ferocious workers against whatever predators can reach their nests. No attempt is made to conceal their huge nests, which can measure a meter or more across and which appear as conspicuous dark objects hanging in the treetops. When we simulated an attack by the Eurasian honey buzzard, a specialist feeder on brood of social wasps and bees, by swinging a black binocular case near an *A. dorsata* colony without striking it, over 100 guards immediately chased the case in a comet's-tail formation while it "flew" back and forth within 5 m of the nest. When we retrieved the case after three minutes of attack we counted 72 stings deeply embedded in its leather cover.

One striking feature of *A. dorsata* is their habit of nesting in groups: a single tree can contain 10 or more *A. dorsata* colonies. At first we thought that this was also an adaptation to colony defense, since grouped colonies could pool their forces. However, it seems probable that this phenomenon is merely a by-product of the fact that colonies converge on a few especially tall trees in choosing nest sites. When I climbed up into an aggregation of nests and attacked one, the grouped colonies did not unleash a concerted counterattack. Only the colony attacked fought back with hundreds of stinging bees.

In general, our observations in the forests of Thailand support the hypothesis that the honeybees of tropical Asia have strongly diverged in their strategies of colony defense, and that this divergence underlies the initially puzzling array of differences among colonies of the three species. In each species the colony's system of defense involves numerous interwoven lines of adaptation, including choice of nest site, nest architecture, population size, and worker morphology, physiology, and behavior. It seems clear that predation has been a pervasive and powerful force in the evolution of these bee societies.

In this discussion I have tried to show how an ecological approach to the honeybee societies reveals the adaptive theme of each species. I have also tried to demonstrate that a synthesis of physiological and ecological approaches provides a broad framework for understanding the social behavior of honeybees, and that, when combined, the two approaches are mutually inspiring. A clear example of this synergism comes from the study of the foraging strategy of honeybee colonies. Here we rely on being able to interpret the bees' dance language, but at the same time find that understanding how this communication functions in nature raises new physiological questions, such as how colonies identify the most profitable flower patches. In essence, physiological studies have paved the way for precise ecological investigations, and in turn, the fledgling ecological approach illuminates pathways into new areas of physiological investigation.

References

Akratanakul, P. 1977. The natural history of the dwarf honey bee, *Apis florea* F., in Thailand. Ph.D. diss., Cornell Univ.

Avitabile, A. 1978. Brood rearing in honeybee colonies from late autumn to early spring. *J. Apic. Res.* 17:69–73.

Avitabile, A., D. P. Stafstrom, and K. J. Donovan. 1978. Natural nest sites of honeybee colonies in trees in Connecticut, USA. *J. Apic. Res.* 17:222–26.

Boch, R. 1956. Die Tänze der Bienen bei nahen und fernen Trachtquellen. *Z. vergl. Physiol.* 38:136–67.

Clutton-Brock, T. H., and P. H. Harvey. 1979. Comparison and adaptation. *Proc. Roy. Soc. London B* 205:547–65.

Combs, G. F. 1972. The engorgement of swarming worker honeybees. *J. Apic. Res.* 11:121–28.

Crane, E. 1975. *Honey: A Comprehensive Survey.* Heinemann.

Darchen, R. 1973. La thermorégulation et l'ecologie de quelques espèces d'abeilles sociales d'Afrique (Apidae, Trigonini et *Apis mellifica* var. *adansonii*). *Apidologie* 4:341–70.

Edgell, G. H. 1949. *The Bee Hunter.* Harvard Univ. Press.

Fell, R. D., J. T. Ambrose, D. M. Burgett, D. DeJong, R. A. Morse, and T. D. Seeley. 1977. The seasonal cycle of swarming in honeybees. *J. Apic. Res.* 16:170–73.

Gould, J. L. 1976. The dance-language controversy. *Quart. Rev. Biol.* 51:211–44.

______. 1982. Why do honey bees have dialects? *Beh. Ecol. Sociobiol.* 10:53–56.

Hölldobler, B. 1979. Territories in the weaver ant (*Oecophylla longinoda* [Latreille]): A field study. *Z. Tierpsychol.* 51:201–13.

Jaycox, E. R., and S. G. Parise. 1980. Homesite selection by Italian honeybee swarms, *Apis mellifera ligustica* (Hymenoptera: Apidae). *J. Kansas Ent. Soc.* 53:171–78.

Jeffree, E. P. 1956. Winter brood and pollen in honeybee colonies. *Insectes Soc.* 3:417–22.

Johansson, T. S. K., and M. P. Johansson. 1979. The honey bee colony in winter. *Bee World* 60:155–70.

Koeniger, N., and G. Koeniger. 1980. Observation and experiments on migration and dance communication of *Apis dorsata* in Sri Lanka. *J. Apic Res.* 19:21–34.

Krebs, J. R., and N. B. Davies. 1981. *An Introduction to Behavioral Ecology.* Sinauer.

Lack, D. M. 1968. *Adaptations for Breeding in Birds.* Methuen.

Lindauer, M. 1955. Schwarmbienen auf Wohnungssuche. *Z. vergl. Physiol.* 37:263–324.

______. 1956. Über die Verständigung bei indischen Bienen. *Z. vergl. Physiol.* 38:521–57.

______. 1961. *Communication among Social Bees.* Harvard Univ. Press.

Martin, P. 1963. Die Steuerung der Volksteilung beim Schwärmen der Bienen. Zugleich ein Beitrag zum Problem der Wanderschwärme. *Insectes Soc.* 10:13–42.

Michener, C. D. 1974. *The Social Behavior of the Bees: A Comparative Study.* Harvard Univ. Press.

Morse, R. A., and F. M. Laigo. 1969. *Apis dorsata* in the Philippines. *Mono. Philipp. Assoc. Entomol.* 1:1–96.

Ribbands, C. R. 1953. *The Behaviour and Social Life of Honeybees.* London: Bee Research Association.

Rinderer, T. E., K. W. Tucker, and A. M. Collins. 1982. Nest cavity selection by swarms of European and Africanized honeybees. *J. Apic. Res.* 21:98–103.

Roepke, W. 1930. Beobachtungen an indischen Honigbienen, insbesondere an *Apis dorsata* F. *Meded. Landbouw. Wageningen* 34:1–28.

Seeley, T. D. 1974. Atmospheric carbon dioxide regulation in honey-bee (*Apis mellifera*) colonies. *J. Insect Physiol.* 20:2301–05.

______. 1977. Measurement of nest cavity volume by the honeybee (*Apis mellifera*). *Beh. Ecol. Sociobiol.* 2:201–27.

______. 1978. Life history strategy of the honey bee, *Apis mellifera*. *Oecologia* 32:109–18.

______. 1982. How honeybees find a home. *Sci. Am.* 247:158–68.

______. In press. Division of labor between scouts and recruits in honeybee foraging. *Beh. Ecol. Sociobiol.*

Seeley, T. D., and B. Heinrich. 1981. Regulation of temperature in the nests of social insects. In *Insect Thermoregulation*, ed. B. Heinrich, pp. 159–234. Wiley.

Seeley, T. D., and R. A. Morse. 1976. The nest of the honey bee (*Apis mellifera* L.). *Insectes Soc.* 23:495–512.

______. 1978a. Nest site selection by the honey bee, *Apis mellifera*. *Insectes Soc.* 25:323–37.

______. 1978b. Dispersal behavior of honeybee swarms. *Psyche* 84:199–209.

Seeley, T. D., R. A. Morse, and P. K. Visscher. 1980. The natural history of the flight of honeybee swarms. *Psyche* 86:103–13.

Seeley, T. D., R. H. Seeley, and P. Akratanakul. 1982. Colony defense strategies of the honeybees in Thailand. *Ecol. Mono.* 52:43–63.

Seeley, T. D., and P. K. Visscher. Unpubl. Survival of honeybee colonies in cold temperate climates: The critical timing of brood rearing and colony reproduction.

Simpson, J. 1959. Variation in the incidence of swarming among colonies of *Apis mellifera* throughout the summer. *Insectes Soc.* 6: 85–99.

______. 1961. Nest climate regulation in honeybee colonies. *Science* 133:1327–33.

Tokuda, Y. 1924. Studies on the honeybee, with special reference to the Japanese honeybee. *Trans. Sapporo Nat. Hist. Soc.* 9:1–27.

Visscher, P. K., and T. D. Seeley. 1982. Foraging strategy of honeybee colonies in a temperate deciduous forest. *Ecology* 63:1790–1801.

von Frisch, K. 1967. *The Dance Language and Orientation of Bees.* Harvard Univ. Press.

______. 1971. *Bees: Their Vision, Chemical Senses, and Language.* Cornell Univ. Press.

Wheeler, W. M. 1923. *Social Life among the Insects.* Harcourt, Brace and Co.

Wilson, E. O. 1971. *The Insect Societies.* Harvard Univ. Press.

Winston, M. L. 1980. Swarming, afterswarming, and reproductive rate of unmanaged honeybee colonies (*Apis mellifera*). *Insectes Soc.* 27:391–98.

Winston, M. L., and C. D. Michener. 1977. Dual origin of highly social behavior among bees. *PNAS* 74:1135–37.

Part IV
The Evolutionary Basis of Reproductive Behavior

This fourth and final section contains seven articles that illustrate how behavioral ecologists analyze questions about reproductive behavior. All the authors employ an adaptationist approach, with its central tenet that among the major forces of evolution—mutation, migration, drift, and natural selection—only selection leads to adaptation. This being so, we expect individuals to behave in ways that enhance the transmission of their genes, an unconscious goal that is generally achieved by maximizing lifetime reproduction. Tests of this prediction involve experimental manipulations, long-term observations of individual animals in their natural habitats, and comparisons among related species that differ in particular ways and among unrelated species that are behaviorally similar.

Randy Thornhill and Darryl Gwynne employ all these approaches in attempting to understand why males and females nearly always differ dramatically in the ways they go about reproducing. Generally males are extremely eager to court and mate, but expend relatively little time and energy on rearing young, whereas females devote much more effort to choosing their mate and rearing their offspring. These "traditional" sex roles are far from universal, however, as is seen in numerous insects. For example, female Mormon crickets often compete vigorously for access to mates, while the males are sexually coy, sometimes rejecting willing partners. Thornhill and Gwynne believe that, in general, whichever sex invests the least in offspring will exhibit "male-like" behavior. Thus males behave in "traditional" fashion when they invest less than females, but when the tables are turned and males invest more, they behave like "traditional" females.

Try to develop ontogenetic or functional hypotheses to account for sex role differences that are alternatives to those proposed by Thornhill and Gwynne. For example, is the idea that females learn how to behave by watching their mother sufficient to explain sex roles in crickets and other species? And, might it be the relative *rate* at which males and females can produce offspring, rather than relative investment, that determines sex roles? Can you uncover the logic behind this alternative, and the relationship between the parental-investment and reproductive-rate hypotheses?

Thornhill and Gwynne's article reveals that sexual reproduction is not always a cooperative affair. When male and female reproductive interests differ, an evolutionary "battle of the sexes" ensues. One outcome, according to William Eberhard, is the varied and bizarre morphology of animal genitalia. Eberhard proposes that males use "copulatory courtship behavior" to encourage choosy females to accept their sperm and to use the sperm eventually to fertilize their eggs. According to this hypothesis, through evolutionary time female choosiness has resulted in the elaboration of male intromittent organs and behaviors enabling males to deliver effective sensory messages to females.

Eberhard's article presents numerous adaptationist hypotheses to explain the shapes and sizes of intromittent organs, their spines, whorls, and horns, and the speed and frequency of copulations. A useful exercise would be to list these alternative hypotheses and then to develop key predictions that follow from each, predictions that would enable you to discriminate among the competing ideas.

But let us imagine that Eberhard is correct about the role of sexual selection in molding penises and other devices associated with copulatory courtship. Many questions arise from this approach. For example, under what circumstances would *females* be expected to possess penis-like structures? Why do males in many species of rodents and ungulates have bones in their penises, whereas most primates lack penis bones? Why do male birds generally lack penises altogether, whereas most male mammals have large and elaborate organs? And why are there exceptions among the birds, with intromittent organs being found in ostriches, rheas, tinamous, ducks, geese, and swans?

The different kinds of mating decisions made by males and females generate a diversity of mating systems. In fishes, this diversity reaches a level of complexity unknown among birds and mammals. As Robert Warner shows, in reef fishes, "maleness" and "femaleness" are often transitory. Individuals in many species start out their reproductive lives as females but later switch to being brilliantly-colored males. In other species, individuals can change from being male to female, while in a few deep-water fishes individuals produce both sperm and eggs throughout life. Warner's article develops a general theory to explain how sexual selection and mating system diversity underlie the evolution of different kinds of hermaphroditism among fishes.

Warner's article offers an opportunity to identify and contrast hypotheses for sex reversal at multiple levels of analysis (see the article in Part I by Holekamp and Sherman). Given the fitness benefits that Warner believes derive from the ability to switch sex under appropriate circumstances, why do you suppose sex reversal is a relatively rare phenomenon, even in fishes, and nonexistent in birds, mammals, reptiles, and insects? Can you develop a non-adaptationist explanation for the irregular distribution of sex reversal? What are the reproductive costs of hermaphroditism and sex reversal?

The product of reproductive behavior is progeny, and in most species parents are solicitous toward their young. But in a surprising number of predatory birds, parents tolerate the elimination of one chick by another. Indeed, the parents rarely interfere in siblicidal aggression, even when one chick is near death. Adaptationists are justifiably puzzled by siblicide because it seems impossible to explain via natural selection. In fact, some people have argued that the apparently "stupid" behavior of the parent birds occurs because of a physical constraint on brain size. Since birds must economize on weight if they are to get off the ground, they can afford to carry only a limited amount of brain tissue—even if this means that they are cognitively handicapped.

The paper by Douglas Mock, Hugh Drummond, and Christopher Stinson will enable you to contrast the "bird-brained" argument with an adaptationist alternative. Mock and his colleagues adopt what is, at first sight, a curious hypothesis: that there are important fitness *benefits* for siblicidal individuals and their parents. These benefits may arise when food shortages make it impossible for parents to provision a full clutch all the way to fledging. The authors test their idea using a clever combination of comparative and experimental approaches. Consider how examples of convergent and divergent evolution were used in this article to generate comparative tests. It might also be worthwhile to think about the use and interpretation of cross-fostering experiments in studying behavior. Finally, it is interesting to consider why, if food is the important limiting factor, siblicidal chicks never eat their departed siblings.

The fierce selfishness of nestling egrets, boobies, and black eagles represents an extreme form of uncooperative behavior. At the other end of the spectrum, students of animal behavior can point to species in which some individuals give up opportunities to reproduce in order to assist others in doing so. Consider a jackal pup born on the Ndutu Plains of central Kenya. If it survives the many hardships of puppyhood it may remain on its natal territory into adulthood and, rather than loafing about, help feed and protect its parents' new litter.

Delaying personal reproduction to aid siblings is as much a puzzle for adaptationists as is the destruction of siblings. How can behavior that reduces reproductive opportunities enhance fitness? Patricia Moehlman provides an evolutionary explanation by noting that if a helpful yearling jackal can *increase* the number of its siblings that survive to reproduce, it is actually promoting the propagation of its own genes. Since full siblings have the same parents, they share a high proportion of genes that are identical by descent. For an individual, "creating" extra siblings (i.e., by helping them to survive) has fundamentally the same genetic consequences as creating and nurturing one's own offspring.

This argument may also help explain even more extreme cases of reproductive self-sacrifice. In eusocial species some individuals typically spend their whole lives as nonbreeding helpers, unlike jackals, which eventually disperse to start their own families. Most eusocial creatures are insects: the ants, the termites, many bee and wasp species, and a few beetles, aphids, and thrips. In these creatures, workers, which are sometimes physiologically sterile, spend their lives helping their mother produce additional siblings, some of which will reproduce and pass on the genes they share in common with the workers that raised them.

Prior to 1981 no eusocial vertebrate was known. But then the naked mole-rat, a buck-toothed, burrowing, East African rodent burst on the scene. As Rodney Honeycutt's article illustrates, naked mole-rats are wonderfully odd animals that live in huge colonies within which only one female and one to three males reproduce. The nonbreeders work to forage and to excavate, maintain, and defend the colony's vast subterranean burrow system. Like ants and termites, as well as other less altruistic species such as jackals, naked mole-rat workers appear to direct their assistance toward individuals that share a relatively high proportion of their genetic material through common descent.

But the discovery that helpers, whether permanently sterile or not, direct their assistance to close relatives does not provide a complete explanation for the evolution of alloparental behavior. After all, full siblings are on average as closely related as parents are to offspring in *all* diploid species—yet the taxonomic distribution of helping is spotty. Consider how both Moehlman and Honeycutt test the hypothesis that special ecological and demographic factors are required for the evolution of reproductive self-sacrifice. What are the similarities and differences in the ecological pressures these two authors believe are responsible for the evolution of natal philopatry and helping in silver-backed jackals and naked mole-rats? The evolution of cooperative breeding in vertebrates and eusociality in insects are usually discussed as separate topics. Should this continue? Might some of the same ecological pressures that favored helping in jackals and mole-rats also apply to the evolution of eusociality in honey bees (see the articles by Winston and Slessor and by Seeley)?

The final article, by David Buss, also deals with a highly social species that is capable of great cooperation. Buss raises the intriguing issue of how human beings choose their spouses. He mailed questionnaires to hundreds of American men and women, and discovered that the sexes value similar attributes in potential partners and that individuals tend to marry someone who shares many of their attitudes and characteristics. The surveys also revealed an interesting difference between the sexes: men are more concerned with the physical attractiveness of women, while women attach greater importance to the earning capacity of men.

How might an evolutionary approach might be applied to Buss' findings? First, try to develop several hypotheses about how positive assortative mating (like marries like) might enhance individual fitness. Next propose some alternative *non*-adaptationist explanations for why individuals tend to marry those who share their views and characteristics. Finally, can you account for the sexual differences in the importance men and women attach to physical appearance and earning capacity in adaptationist and non-adaptationist terms?

Adaptationist approaches to human behavior are controversial because many people feel that our behavior is "culturally determined" and therefore independent of biology. According to this school of thought, the great diversity of the world's cultures implies that humans can adopt any behavior they choose. What sorts of criticisms would these individuals level against Buss's study and interpretations (and against the study of Hill and Hurtado in Part III)? In order to test the "arbitrary culture" hypothesis, what predictions can be made about the characteristics men and women prefer in mates among different cultures? How might we discriminate this hypothesis from the one Buss champions?

Taken together, the seven articles in this section illustrate how behavioral biologists have attempted to make evolutionary sense of some important puzzles about reproductive behavior. Why do the sexes differ in eagerness to mate and criteria for choosing partners? Why do some species exhibit reversals of sex roles and even of individuals' sex? Why do parents tolerate, and sometimes even encourage, the murder of certain offspring? Why do some individuals forego reproduction in favor of assisting others to reproduce? The successes behavioral ecologists have had in answering these questions by studying Mormon crickets, masked boobies, naked mole-rats, and even humans testify to the range and power of an evolutionary approach to animal behavior.

Randy Thornhill
Darryl T. Gwynne

The Evolution of Sexual Differences in Insects

The ultimate cause of sexual differences in behavior may be the relative contribution of the sexes to offspring

Evolutionary biologists strive to understand the diversity of life by the study of the evolutionary processes that produced it. Among the more fundamental of these processes is sexual selection, which has probably been a major form of selection in the evolutionary background of all organisms with two sexes. An important question for evolutionary biologists, as one theorist (Williams 1975) has put it, is "Why are males masculine and females feminine and, occasionally, vice versa?" This question focuses on evolutionary causation by natural or sexual selection rather than on the proximate causes of sexual differences, such as genetic influences, hormones, or development.

Sexual selection is distinguished from natural selection in terms of how the differential in the reproduction of individuals is brought about (Darwin 1874). Sexual selection is differential reproduction of individuals in the context of competition for mates. Natural selection is differential reproduction of individuals due to differences in survival. Since reproduction is necessary for selection to act, natural selection also includes differential reproduction of individuals in the context of reproductive acts, such as obtaining a mate of the right species, proper fertilization, and so on. Although both forms of selection involve competition between individuals for genetic representation, competition for mates is a key factor for distinguishing sexual from natural selection. The competition among members of one sex (usually males) for the opposite sex may take the form of attempting to coax choosy individuals to mate, leading to intersexual selection, or may involve striving to obtain access to already receptive individuals who are willing to mate with any individual of the opposite sex, leading to intrasexual selection.

In his treatise on sexual selection, Darwin (1874) compiled an encyclopedic volume of comparative support for the crucial role of the process in the evolution of morphological and behavioral traits important in sexual competition. Current studies of sexual selection involve several approaches. Some researchers seek to describe the consequences and nature of sexual selection by the study of behavioral and morphological traits important in sexual competition as well as by observing the types of mating associations (monogamy and polygyny, for example) in animals and plants (e.g., Bradbury and Vehrencamp 1977; Emlen and Oring 1977; Thornhill and Alcock 1983). Another approach is to measure the intensity of sexual selection, focusing on variation in the reproductive success of individuals in nature (e.g., Payne 1979; Wade and Arnold 1980; Thornhill 1981, 1986; Gwynne 1984a). Other studies are attempting to elucidate how sexual selection works, and there are several competing hypotheses (reviewed in Thornhill and Alcock 1983). Subtle forms of female mate choice and male-male competition for females have been discovered and are under investigation (Parker 1970; Thornhill 1983; Smith 1984). The area of study we will address in this paper concerns factors that control the operation of sexual selection—factors that govern the extent to which one sex competes for the other.

The evidence of sexual selection in nature raises a number of questions about these factors. When reproducing, why are males usually more competitive and less discriminating of mates than females and thus subject to greater sexual selection? And why in a few exceptional species is the intensity of sexual selection on the sexes apparently reversed, with females competing for males and males discriminating among mates? Moreover, why does the extent of sexual selection vary in the same species?

We will address these questions using evidence from insects, which, as one of the most diverse groups of animals, exhibit a variety of different reproductive biologies and thus provide a wealth of comparative information with which to examine the theory of sexual selection. We will first discuss theory concerning the control of sexual selection and then examine the theory in light of what is known about insects.

Control of sexual selection

Bateman (1948) argued that males typically are more sexually competitive than females primarily because of the sexual asymmetry in gamete size. He noted that female fertility is limited by the production of large,

Randy Thornhill is Professor of Biology at the University of New Mexico. He received his B.S. and M.S. from Auburn University and his Ph.D. (1974) from the University of Michigan. His research interests include sexual selection, the evolution of sexual differences, insect mating systems, and human social behavior. Darryl Gwynne attended the University of Toronto and Colorado State University (Ph.D. 1979). He was a postdoctoral fellow at the University of New Mexico and is currently a Research Fellow at the University of Western Australia. His interests include sexual selection and communication in insects. *The research reported here was supported by grants from the NSF, a Queen Elizabeth II Fellowship (Australia), and the Australian Research Grants Scheme.* *Address for Dr. Thornhill: Department of Biology, University of New Mexico, Albuquerque, NM 87131.*

Figure 1. The female in this mating pair of hangingflies (*Hylobittacus apicalis*) is feeding on a blow fly captured and provided to her by the male. This nuptial feeding by the male enhances the female's fecundity, reduces the risks that must be taken by the female to obtain food, and therefore promotes the male's reproductive success. (From Thornhill 1980.)

costly gametes; for the same amount of reproductive effort, a male produces vastly more gametes. Therefore, the reproductive success of males is limited by their success at inseminating females and not, as in females, by their ability to produce gametes. Bateman's work with fruit flies, *Drosophila melanogaster,* also demonstrated empirically that the intensity of sexual selection is greater on males than on females, and the difference is due to greater variation in mating success among males.

Williams (1966) and Trivers (1972) built a more comprehensive theory than Bateman in noting that what controls the intensity of sexual selection and explains the evolution of sexual differences in reproductive strategy is not just prezygotic investment by the sexes in gametes but all goods and services that contribute to the next generation—that determine the number and survival of offspring. The large disparity in gamete size itself predicts neither the reduced sexual differences seen in monogamous species nor the competitive females and choosy males seen in species with reversals in sex roles.

Material contributions by each sex to the next generation determine the reproductive rate of the population; therefore, sexual competition is ultimately for these contributions because the greater the amount obtained, the higher the reproductive success of a competitor (Thornhill 1986). Simply put, when one sex (usually the females) contributes more to the production and survival of offspring, sexually active members of this sex are in short supply and thus become a limiting resource for the reproduction of the opposite sex; and the extent of sexual competition in the sex contributing least should correspond with the degree to which the opposite sex exceeds it in this contribution. Furthermore, the sex with the greater contribution is subject to a greater loss of fitness if it makes an improper mate choice, because its contribution represents a large fraction of its total reproductive contribution. This asymmetry, coupled with the availability of the sex investing less, is expected to favor mate choice by the sex contributing more.

An important point to note is that not all forms of effort expended by the sexes in reproduction are expected to control the extent of sexual selection. Energetic expenditures by males that are used in obtaining fertilizations but that do not allow greater reproduction by females or do not promote the fitness of offspring are excluded (Trivers 1972; Thornhill and Alcock 1983; Gwynne 1984b).

Male efforts that are expected to control the extent of sexual selection encompass an array of resources that are of material benefit to females, offspring, or both. Such resources include nutrition, protection, and, under certain conditions, genes in spermatozoa. Bateman argued persuasively that female reproductive success is rarely limited by sperm per se. However, a high variance in the genetic quality of males available as mates may result in female competition for the best mates if genetic variation among males affects offspring survival and if males of high genetic quality limit female reproductive success. Although there is little information with which to examine the influence of high-quality sires on the operation of sexual selection, there are a number of studies, which we will now discuss, that examine sexual selection in species in which males provide immediate, material services.

Contributions to offspring

Both males and females can provide for offspring in a variety of ways. Females invest directly in their offspring both through the material investment in eggs and zygotes as well as through maternal care. The better-known examples of male investment concern direct paternal care of offspring; this has been observed in a large number of species in a variety of animal groups (Ridley 1978). However, males can also contribute indirectly to offspring by providing benefits to their mates, both before and after mating. Examples include the nutritional benefits of "courtship feeding," which is observed in certain birds (Nisbet 1973) and insects, such as the hangingfly *Hylobittacus apicalis* shown in Figure 1 (Thornhill 1976). Protection of the mate is another example of such benefits (Gwynne 1984b; Thornhill 1984). The nature of selection leading to the evolution of benefit-providing males is still poorly understood (Alexander and Borgia 1979; Knowlton 1982). But regardless of the evolved function of the phenomenon, with the evolution of benefit-providing males there may be a change in the action of sexual selection on the sexes.

Contributions by one sex that affect the number and survival of offspring potentially limit the reproduction of the opposite sex. Thus the relative investment of the sexes in these sorts of reproductive efforts should determine the extent to which each sex competes for the opposite sex. This hypothesis can be tested by comparing species or populations with differing investment patterns or by directly manipulating resources that limit the reproduction of the population. Examples we will review use these methods.

There have been few attempts to estimate the extent of sexual selection on the sexes in nature. However, variance in the reproductive success of the sexes has been estimated for *Drosophila melanogaster* (Bateman 1948), a damselfly (Finke 1982), red-winged blackbirds (Payne 1979), and red deer (Clutton-Brock et al. 1982). In these species, the parental contribution by the male is smaller than that of the female; as predicted, all species show greater sexual selection on the males.

Observed sexual differences, the consequences of sexual selection, typically serve as evidence for the relative intensity of sexual selection on the sexes in the evolutionary past. In most species, females provide a large amount of parental contribution and males little, and it is primarily the males that show secondary sexual traits of morphology and behavior which function in competition for mates. It is also well known that sexual differences are greatly reduced under monogamy. This is as expected, because both sexes of monogamous species engage in similar levels of parental care. However, for this comparative test to succeed, a sex reversal in the courtship and competitive roles should be observed in species in which males provide a greater portion of the total contribution affecting offspring number and survival.

Parental care provided only by the male is found throughout the vertebrates, particularly in frogs and toads, fishes, and birds, and is likely to represent a limiting resource for female reproduction. In certain seahorses and pipefishes (Syngnathidae), males care for eggs in a specialized brood pouch (Breder and Rosen 1966). In these fishes, male parental care appears to limit female reproductive success, and females are larger and more brightly colored than males, as well as being more competitive in courtship (Williams 1966, 1975; Ridley 1978). In frogs of the genus *Colostethus,* it is the male in one species and the female in another that provide parental care by carrying tadpoles on their backs. In both species, as predicted by theory, it is the sex emancipated from parental duties that defends long-term mating territories and has a higher frequency of competitive encounters (Wells 1980). In species of birds in which males provide most of the parental care, the roles in courtship behavior are reversed; females compete for mates and sometimes are the larger or more brightly colored sex (reviewed by Ridley 1978).

Exclusive paternal care of eggs or larvae is restricted to about 100 species of insects, all of which are within the order Hemiptera, or true bugs (Smith 1980). In the giant water bugs (Hemiptera: Belostomatidae), females adhere eggs to the wing covers of their mates, and the males aerate the eggs near the water surface and protect them from predators. For *Abedus herberti,* Smith (1979) provides evidence that male back space is a limiting resource for females and that male parental care is essential for offspring survival; females actively approach males during courtship, and males reject certain females as mates.

Although direct investment in offspring through parental care is uncommon within the insects, indirect paternal contributions, with males supplying the females with nutrition or other services such as guarding, is widespread in a number of taxa.

The guarding of the female by the male after mating is usually thought of as functioning to prevent other males from inseminating the female (Parker 1970). However, an alternative hypothesis is that guarding evolved in the context of supplying protection for the female and that it thereby enhances male reproductive success (Gwynne 1984b; Thornhill 1984). Male guarding is known to benefit the female in several species: in damselflies (*Calopteryx maculata*), guarding by males after mating allows females to oviposit undisturbed by other males (Waage 1979); in waterstriders (*Gerris remigis*), harassment of guarded females by other males is similarly reduced, allowing females much longer periods during which to forage for food (Wilcox 1984). At present there is no information for these species concerning whether certain males protect females better than others, which would lead to female competition for more protective males. If female competition occurs, selection should favor mate choice by males.

Mate choice by males has been observed in species in which males provide protection or other services to females. In brentid weevils (*Brentus anchorago*) males prefer large females as mates and are known to assist their mates in competition for oviposition sites by driving away nearby ovipositing females (Johnson and Hubbell 1984). Male lovebugs (*Plecia nearctica*), so named for their two- to three-day-long periods of copulation, also prefer to mate with large females (Hieber and Cohen 1983). Lengthy copulation in this species may be beneficial for females in that copulating pairs actually fly faster than unattached lovebugs (Sharp et al. 1974). Similar benefits may be obtained by paired amphipod crustaceans (*Gammarus pulex*); pairs in which males are larger than females have a superior swimming performance that minimizes the risk of being washed downstream (Adams and Greenwood 1983). Perhaps these sorts of services supplied by male crustaceans explain the presence of male choice of mates seen in certain groups (e.g., Schuster 1981).

Males can supply nutrition in several ways. Our research has dealt with courtship feeding, where food items such as prey or nutritious sperm packages (spermatophores) are eaten by females, and we discuss this behavior in detail below for male katydids and scorpionflies. There are also more subtle forms of contribution; in several insect species spermatophores or other ejaculatory nutrients are passed into the female's genital tract at mating (Thornhill 1976; Gwynne 1983; Thornhill and Alcock 1983). A number of researchers have done some interesting work on a similar phenomenon in crustaceans. Electrophoretic studies of proteins in the ovaries and the male accessory gland of a stomatopod shrimp (*Squilla holoschista*) strongly suggest that a specialized protein from the male's accessory glands is transferred with the ejaculate into the female's gonopore and then is translocated intact into the developing eggs; females of this species usually initiate mating and will mate repeatedly (Deecaraman and Subramoniam 1983). In a detailed study of another

Figure 2. A female katydid (*Requena verticalis*) just after mating shows the large spermatophore that has been attached by the male to the base of her ovipositor (*top*). The female grasps the nutritious spermatophylax (*middle*) and eats it (*bottom*), leaving the sperm ampulla portion of the spermatophore in place. After insemination, the ampulla is also eaten. The nutrients in the spermatophore represent a considerable material contribution by the male in the reproductive success of the female. (Photos by Bert Wells.)

Figure 3. While this female Mormon cricket (*Anabrus simplex*) is atop the male, the male apparently weighs his potential mate and will reject her if she is too light. Males select mates among females that compete for access to them, preferring females that are larger and therefore more fecund. This represents a reversal of the sex roles much more commonly found in nature. (Photo by Darryl Gwynne.)

stomatopod, *Pseudosquilla ciliata,* Hatziolos and Caldwell (1983) report a reversal in sex roles, with females courting males that appear reluctant to mate; in the absence of obvious male parental contribution, these researchers cite work with insects in suggesting that male *Pseudosquilla* may provide valuable nutrients in the ejaculate.

Studies with several butterfly species have used radiolabeling to show that male-produced proteins are incorporated into developing eggs as well as into somatic tissues of females (e.g., Boggs and Gilbert 1979). Lepidopteran spermatophores potentially represent a large contribution by the male (up to 10% of body weight), and proteins ingested by females are likely to represent a limiting resource for egg production in these insects that feed on nectar as adults (Rutowski 1982). Preliminary experiments by Rutowski (pers. com.) with alfalfa butterflies (*Colias* spp.) indicate that females receiving larger spermatophores lay more eggs. Consistent with theory, there is evidence of males choosing females and of competition by females for males. For example, in the checkered white butterfly (*Pieris protodice*), males prefer young, large (and thus more fecund) females to older, smaller individuals (Rutowski 1982). And in *Colias* certain females were observed to solicit courtship by pursuing males; these females may have had reduced protein supplies, as they were shown to have small, depleted spermatophores in their genital tracts (Rutowski et al. 1981). Although there is variation between species of butterflies in the size of the male spermatophore, this variation apparently does not result in large differences between species in the male contribution; a review of the reproductive behavior of several butterfly species did not show consistent differences in courtship when species with small spermatophores were compared to those with large spermatophores (Rutowski et al. 1983). However, as shown by Marshall (1982) and confirmed by our studies described below, spermatophore size is not always a useful measure of the importance of the male nutrient contribution.

Reversal of courtship roles in katydids

Katydids (Orthoptera: Tettigoniidae) are similar to butterflies in that males transfer spermatophores to their mates, and, as shown by radiolabeling, spermatophore nutrients are used in egg production (Bowen et al. 1984). In contrast to the mated female butterfly, the katydid female ingests the spermatophore by eating it (Fig. 2). The spermatophore consists of an ampulla which contains the ejaculate and a sperm-free mass termed the spermatophylax (Gwynne 1983). Immediately after mating, the female first eats the spermatophylax; while this is being consumed, insemination takes place, after which the empty sperm ampulla is also eaten (Gwynne et al. 1984). However, the katydid spermatophylax appears not to function as protection of the ejaculate from female feeding. The spermatophylax of the katydid *Requena verticalis* is more than twice the size necessary to allow the transfer both of the spermatozoa and of substances that induce a four-day nonreceptive period in females (Gwynne, unpubl.).

Spermatophore nutrients are important to the reproductive success of the female katydid. Laboratory experiments have shown that as consumption of spermatophylax increases, both the size and the number of eggs that females subsequently lay also increase (Gwynne 1984c). Furthermore, the increase in the size of eggs appears to be determined only by male-provided nutriment; an increase in protein in the general diet increases egg number but does not affect egg size (Gwynne, unpubl.).

The size of the spermatophore produced by male katydids varies from less than 3% of male body weight in some species to 40% in others (Gwynne 1983). Differences in the size of the male contribution conform with the predictions of sex-difference theory: in two species of katydids that make very large investment in each spermatophore (25% or more of male weight) and that have been examined in detail—the Mormon cricket (*Anabrus simplex*) and an undescribed species (*Metaballus* sp.) from Western Australia—there is a complete reversal in sex roles, with females competing aggressively for access to males that produce calling sounds, and males selecting mates, preferring large, fecund females (Gwynne 1981, 1984a, 1985). Figure 3 illustrates this reversal in the Mormon cricket. There is no evidence of such a reversal of courtship roles in species with smaller spermatophores; in these species males compete for mating territories and females select mates (Gwynne 1983).

It is evident, however, that a complete estimate of the contribution to offspring requires more than a simple measure of the relative contribution by the sexes to offspring such as the weight of the spermatophore relative to the weight of a clutch of eggs. Both species of katydids showing a role reversal in courtship behavior also had populations that showed no evidence of the reversed roles. For the Mormon cricket, the simple measure of relative contribution did not show a higher contribution by males at the sites of role reversal (Gwynne 1984a). However, these sites had very high population densities, with individuals of both sexes competing vigorously for food in the form of dead arthropods and certain plants. These observations suggest the hypothesis that the limited food supplies at these sites resulted in few spermatophores being produced and that spermatophore nutrients were thus a limiting resource for female reproduction. Food did not appear to be scarce at sites of low population density where the reversal in courtship roles was not observed. Support for the hypothesis that food is a limiting resource at high-density sites came from dissections of the reproductive accessory glands that produce the spermatophore in a sample of males from each of the sites. Only the few calling males at sites of high density had glands large enough to produce a spermatophore, whereas most males at the low-density site had enlarged glands. This difference between the males at the high- and low-density sites was not a result of a higher number of matings by males at the high-density site.

Differences between individuals from the two sites indicate that sexual selection on females at high-density sites was intense compared to the low-density site. (Sexual selection is measured by variance in mating success; see Wade and Arnold 1980.) Some females were very successful at obtaining spermatophores. These tended to be large females that were preferred by males as mates. The evolutionary consequences of the apparently greater sexual selection on high-density females was not only aggressive female behavior in competition for calling mates but also a larger female body size at this site relative to males. This sexual dimorphism was not seen at sites of low density.

Variation in the expression of sexual differences within the same katydid species suggested that behavior might be flexible; that is, females become competitive and males choosy when they encounter certain environments. This hypothesis was examined using the undescribed *Metaballus* species of katydid from Western Australia, which is similar to the Mormon cricket in that only certain populations show female competition for mates and show males that reject smaller, less fecund females. In this species, discriminating males call females by producing a broken "zipping" song from deep in the vegetation. Sites where courtship roles are reversed consist of mainly the zipping male song, whereas at sites of male competition, males produce continuous songs that appear to be louder. An experiment was conducted which involved shifting a number of males and females from a site where role reversal was not observed to one in which it was noted. The behavior of the males that were moved to the role-reversed site changed to resemble that of the local males: their song changed from a continuous to a zipping song, the duration of courtship increased (possibly to assess the quality of their mates), and they even rejected females as mates. Thus, sexual differences in behavior are plastic; courtship roles of the sexes appear to be dependent on the environment encountered.

For the Mormon cricket, it is likely that the relative contributions of the sexes is the factor controlling sexual selection. A comparison of the weights of spermatophores and egg clutches is undoubtedly a poor estimate of relative contribution by the sexes; spermatophores seem to be important to the reproduction of females at both sites (Gwynne 1984a). However, if food supplies limit spermatophore production at sites where reversals in courtship roles are observed, and if females cannot obtain spermatophore nutrients from other food sources, then spermatophores are likely to have a greater influence on female fecundity and thereby are more valuable to female reproduction at these sites. Thus, the total contribution from the males at these sites is probably larger than that of the females.

Nuptial feeding in scorpionflies

Most of the evidence supporting the hypothesis that the relative contribution of the sexes to offspring is an important factor controlling the extent of sexual selection has been derived from comparisons between or, in the katydid work, within species. In contrast, studies were conducted in which the relative contribution of males was manipulated to determine its effect on the extent of sexual selection (Thornhill 1981, 1986). This research has focused on scorpionflies of the genus *Panorpa* (Panorpidae), in which males use either dead arthropods or nutritious products of salivary glands to feed their mates.

Males must feed on arthropod carrion, for which they compete through aggression, before they can secrete a salivary mass. Males in possession of a nuptial gift release pheromone that attracts conspecific females from some distance. Females can obtain food without male assistance, but doing so is risky because of exposure to predation by web-building spiders. Movement in the habitat required to find dead arthropods exposes females and males to spider predation, and dead arthropods unattended by males are frequently found in active spider webs. The gift-giving behavior of males is an important contribution, because dead arthropods needed by females to produce eggs are limited both in the absolute sense and in terms of the risks in obtaining them.

In a series of experiments, individually marked male and female *Panorpa latipennis* were placed in field enclosures, and variances in mating success of the sexes were determined in order to estimate the relative intensity of sexual selection. Dead crickets taped to vegetation represented the resource that males defended from other males and to which females were attracted. In one experiment, three treatments were established in which equal numbers of males and females were added to each enclosure and the number of dead crickets varied—two, four, or six crickets per enclosure. As predicted, competition among males was greatest in the enclosures with two crickets; the intensity of sexual selection, calculated by variance in male mating success, was greatest in this treatment and was lowest in the treatment with six crickets.

Variance in female mating success was low and was not significant across cricket abundances over the seven days of the experiment. Sexual selection on females probably often arises from female-female competition for the best mates regardless of the number of mates. *Panorpa* females prefer males that provide large, fresh nuptial gifts of dead arthropods over males that provide salivary masses, and males only secrete saliva when they cannot compete successfully for dead arthropods (Thornhill 1981, 1984). This female mating preference is adaptive in that females mating with arthropod-providing males lay more eggs than females mating with saliva-providing males. Thus an accurate measure of sexual selection on female *Panorpa* would include the variation in egg output by females in relation to the resource provided by the mates of females. This information is not available at present.

However, the results on males from this experiment clearly support the hypothesis that sexual selection is determined by the relative contribution of the sexes. As food is a limiting resource for reproduction by female scorpionflies, the total contribution of food by males in enclosures with more crickets was greater than in enclosures with fewer crickets, and the intensity of sexual selection on males declined as males contributed relatively more.

Such studies of the factors controlling the operation of sexual selection are important for two major reasons. The first is simply that sexual selection has been such an important factor in the evolution of life. Sexual selection seems inevitable in species with two sexes because, as Bateman (1948) first pointed out, the relatively few large female gametes will be the object of sexual competition among the males, whose upper limit to reproductive success is set by the number of ova fertilized rather than by production of the relatively small, energetically cheap sperm. The role of sexual selection in the history of life can best be explored when such controlling factors are fully understood. The second reason is related to the first: the difference in the operation of sexual selection on the sexes may ultimately account for all sexual differences. Only sexual selection acts differently on the sexes per se (Trivers 1972). Natural selection may act on and may even magnify sexual differences in behavior and morphology, but probably only after these differences already exist as a result of the disparate action of sexual selection.

The insight of Williams (1966) and Trivers (1972) is that the relative contribution of materials and services by the sexes in providing for the next generation is the most important factor controlling the operation of sexual selection. In insects, contributions supplied by males to their mates include not only the paternal care of young, a well-studied phenomenon in vertebrates, but also other services such as courtship feeding, subtle forms of nutrient transfer via the reproductive tract, and "beneficial" guarding of mates.

References

Adams, J., and P. J. Greenwood. 1983. Why are males bigger than females in precopula pairs of *Gammarus pulex? Behav. Ecol. Sociobiol.* 13:239–41.

Alexander, R. D., and G. Borgia. 1979. On the origin and basis of the male-female phenomenon. In *Sexual Selection and Reproductive Competition in the Insects*, ed. M. S. Blum and N. A. Blum, pp. 417–40. Academic.

Bateman, A. J. 1948. Intrasexual selection in *Drosophila. Heredity* 2:349–68.

Boggs, C. L., and L. E. Gilbert. 1979. Male contribution to egg production in butterflies: Evidence for transfer of nutrients at mating. *Science* 206:83–84.

Bowen, B. J., C. G. Codd, and D. T. Gwynne. 1984. The katydid spermatophore (Orthoptera: Tettigoniidae): Male nutrient investment and its fate in the mated female. *Aust. J. Zool.* 32:23–31.

Bradbury, J. W., and S. L. Vehrencamp. 1977. Social organization and foraging in emballonurid bats. III. Mating systems. *Behav. Ecol. Sociobiol.* 2:1–17.

Breder, C. M., and D. E. Rosen. 1966. *Modes of Reproduction in Fishes*. Nat. Hist. Press.

Clutton-Brock, T. H., F. E. Guinness, and S. D. Albon. 1982. *Red Deer: Behavior and Ecology of Two Sexes*. Univ. of Chicago Press.

Darwin, C. 1874. *The Descent of Man and Selection in Relation to Sex*, 2nd ed. New York: A. L. Burt.

Deecaraman, M., and T. Subramoniam. 1983. Mating and its effect on female reproductive physiology with special reference to the fate of male accessory sex gland secretion in the stomatopod, *Squilla holoschista. Mar. Biol.* 77:161–70.

Emlen, S. T., and L. W. Oring. 1977. Ecology, sexual selection, and the evolution of mating systems. *Science* 197:215–22.

Finke, O. M. 1982. Lifetime mating success in a natural population of the damselfly *Enallagma hageni* (Walsh) (Odonata: Coenagrionidae). *Behav. Ecol. Sociobiol.* 10:293–302.

Gwynne, D. T. 1981. Sexual difference theory: Mormon crickets show role reversal in mate choice. *Science* 213:779–80.

———. 1983. Male nutritional investment and the evolution of sexual differences in the Tettigonidae and other Orthoptera. In *Orthopteran Mating Systems: Sexual Competition in a Diverse Group of Insects*, ed. D. T. Gwynne and G. K. Morris, pp. 337–66. Westview.

———. 1984a. Sexual selection and sexual differences in Mormon crickets (Orthoptera: Tettigoniidae, *Anabrus simplex*). *Evolution* 38:1011–22.

———. 1984b. Male mating effort, confidence of paternity, and insect sperm competition. In Smith 1984, pp. 117–49.

———. 1984c. Courtship feeding increases female reproductive success in bushcrickets. *Nature* 307:361–63.

———. 1985. Role-reversal in katydids: Habitat influences reproductive behavior (Orthoptera: Tettigoniidae, *Metaballus* sp.). *Behav. Ecol. Sociobiol.* 16:355–61.

Gwynne, D. T., B. J. Bowen, and C. G. Codd. 1984. The function of the katydid spermatophore and its role in fecundity and insemination (Orthoptera: Tettigoniidae). *Aust. J. Zool.* 32:15–22.

Hatziolos, M. E., and R. Caldwell. 1983. Role-reversal in the stomatopod *Pseudosquilla ciliata* (Crustacea). *Anim. Behav.* 31:1077–87.

Hieber, C. S., and J. A. Cohen. 1983. Sexual selection in the lovebug, *Plecia nearctica*: The role of male choice. *Evolution* 37:987–92.

Johnson, L. K., and S. P. Hubbell. 1984. Male choice: Experimental demonstration in a brentid weevil. *Behav. Ecol. Sociobiol.* 15:183–88.

Knowlton, N. 1982. Parental care and sex role reversal. In *Current Problems in Sociobiology*, ed. King's College Sociobiology Group, pp. 203–22. Cambridge Univ. Press.

Marshall, L. D. 1982. Male nutrient investment in the Lepidoptera: What nutrients should males invest? *Am. Nat.* 120:273–79.

Nisbet, I. C. T. 1973. Courtship-feeding, egg size, and breeding success in common terns. *Nature* 241:141–42.

Parker, G. A. 1970. Sperm competition and its evolutionary consequences in the insects. *Biol. Rev. Cambridge Philos. Soc.* 45:525–67.

Payne, R. B. 1979. Sexual selection and intersexual differences in variance of breeding success. *Am. Nat.* 114:447–66.

Ridley, M. 1978. Paternal care. *Anim. Behav.* 26:904–32.

Rutowski, R. L. 1982. Mate choice and lepidopteran mating behavior. *Fla. Ent.* 65:72–82.

Rutowski, R. L., C. E. Long, and R. S. Vetter. 1981. Courtship solicitation by *Colias* females. *Am. Midl. Nat.* 105:334–40.

Rutowski, R. L., M. Newton, and J. Schaefer. 1983. Interspecific variation in the size of the nutrient investment made by male butterflies during copulation. *Evolution* 37:708–13.

Schuster, S. M. 1981. Sexual selection in the Socorro Isopod *Thermosphaeroma thermophilum* (Cole) (Crustacea: Peracarida). *Anim. Behav.* 29:698–707.

Sharp, J. L., N. C. Leppala, D. R. Bennett, W. K. Turner, and E. W. Hamilton. 1974. Flight ability of *Plecia nearctica* in the laboratory. *Ann. Ent. Soc. Am.* 67:735–38.

Smith, R. L. 1979. Paternity assurance and altered roles in the mating behaviour of a giant water bug, *Abedus herberti* (Heteroptera: Belostomatidae). *Anim. Behav.* 27:716–25.

———. 1980. Evolution of exclusive postcopulatory paternal care in the insects. *Fla. Ent.* 63:65–78.

———, ed. 1984. *Sperm Competition and the Evolution of Animal Mating Systems*. Academic.

Thornhill, R. 1976. Sexual selection and paternal investment in insects. *Am. Nat.* 110:153–63.

———. 1980. Sexual selection in the black-tipped hangingfly. *Sci. Am.* 242:162–72.

———. 1981. *Panorpa* (Mecoptera: Panorpidae) scorpionflies: Systems for understanding resource-defense polygyny and alternative male reproductive effort. *Ann. Rev. Ecol. Syst.* 12:355–86.

———. 1983. Cryptic female choice in the scorpionfly *Harpobittacus nigriceps* and its implications. *Am. Nat.* 122:765–88.

———. 1984. Alternative hypotheses for traits believed to have evolved in the context of sperm competition. In Smith 1984, pp. 151–78.

———. 1986. Relative parental contribution of the sexes to offspring and the operation of sexual selection. In *The Evolution of Behavior*, ed. M. Nitecki and J. Kitchell, pp. 10–35. Oxford Univ. Press.

Thornhill, R., and J. Alcock. 1983. *The Evolution of Insect Mating Systems*. Harvard Univ. Press.

Trivers, R. L. 1972. Parental investment and sexual selection. In *Sexual Selection and the Descent of Man, 1871–1971*, ed. B. Campbell, pp. 136–79. Aldine.

Waage, J. K. 1979. Adaptive significance of postcopulatory guarding of mates and nonmates by *Calopteryx maculata* (Odonata). *Behav. Ecol. Sociobiol.* 6:147–54.

Wade, M. J., and S. J. Arnold. 1980. The intensity of sexual selection in relation to male sexual behaviour, female choice, and sperm precedence. *Anim. Behav.* 28:446–61.

Wells, K. D. 1980. Social behavior and communication of a dendrobatid frog (*Colostethus trinitatis*). *Herpetologica* 36:189–99.

Wilcox, R. S. 1984. Male copulatory guarding enhances female foraging in a water strider. *Behav. Ecol. Sociobiol.* 15:171–74.

Williams, G. C. 1966. *Adaptation and Natural Selection*. Princeton Univ. Press.

———. 1975. *Sex and Evolution*. Princeton Univ. Press.

Animal Genitalia and Female Choice

William G. Eberhard

When I was a senior in college I took a course in ichthyology and learned to enjoy thumbing through taxonomic drawings, which displayed the fascinating theme-and-variations patterns that are so common in nature. The various species in a genus were basically similar, but each had a set of seemingly senseless and often surprising and aesthetically pleasing differences. Later that year I became interested in spiders, and I can still remember my disappointment upon finding that similar drawings of whole spiders did not accompany papers on spider taxonomy. Instead, illustrations in spider papers were limited to male and female genitalia, which are generally extremely complex structures lacking the elegant sweep of fish profiles. Even closely related species of spiders can usually be distinguished by the genitalia alone.

This was my first encounter with a major pattern in animal evolution: among closely related species that employ internal fertilization, the genitalia—especially male genitalia—often show the clearest and most reliable morphological differences. For some reason, the genitalia of most spiders have evolved rapidly, becoming distinct even in recently diverged lines. In contrast, animals that employ external fertilization, such as most fish, do not have species-specific genital morphology.

These trends are widespread. Groups in which intromittent genitalia (for placing gametes inside the mate) are often useful for distinguishing species include flatworms, nematodes, oligochaete worms, insects, spiders, millipedes, sharks and rays, some lizards, snakes, mites, opilionids, crustaceans, molluscs, and mammals (including rodents, bats, armadillos, and primates). In contrast, groups that employ external fertilization all lack species-specific genitalia; they include echinoderms, most polychaete worms, hemichordates, brachiopods, sipunculid worms, frogs, birds, a few insects, and most fish. In such cases, both males and females have only a simple opening through which gametes are released. Even within groups that have recently switched from external to internal fertilization, for example guppies, whose males use a modified anal fin to introduce sperm into the female, the intromittant organs are often useful for distinguishing species.

Rapid evolutionary divergence of male genitalia may be explained by the ability of females to choose the paternity of their offspring

Rapid and divergent genital evolution also occurs in species in which the male, rather than penetrating the female himself, introduces a spermatophore, or package of his sperm, into the female. In many octopuses, squids, scorpions, some pseudoscorpions, some snails and slugs, some arrow worms, and pogonophoran worms, it is the spermatophore, rather than male genital structures, that is morphologically complex and species-specific.

Why do male mating structures possess such a bewildering diversity of forms? Surely the transfer of a small mass of gametes does not require the elaborate genital structures carried by the males of many groups. Two explanations were proposed some time ago: lock-and-key and pleiotropy. Neither is particularly convincing. According to the lock-and-key hypothesis, females have evolved under selection favoring those individuals that avoided wasting eggs by having them fertilized by sperm of other species. Elaborate, species-specific female genitalia (locks) admit only the genitalia of conspecific males (keys), enabling females to avoid mistakes in fertilization.

Originally proposed nearly 150 years ago for insects (see Nichols 1986), the lock-and-key idea fell into disrepute when it was established that locks are too easily picked. Studies of groups in which females have complex genitalia showed that the female genitalia could not exclude the genitalia of males of closely related species (for a summary of evidence see Shapiro and Porter 1989). The lock-and-key hypothesis is inapplicable in many other groups in which the female genitalia are soft and mechanically incapable of excluding incorrect keys while the male genitalia or spermatophores are nevertheless species-specific in form (flatworms, nematodes, arrow worms, annelid worms, sharks and rays, guppies and their relatives, snakes, lizards, snails and slugs, octopuses and squids, and many insects).

A species-isolation function, whether mechanical or otherwise, is improbable for several reasons. In some

William Eberhard is a member of the staff of the Smithsonian Tropical Research Institute and a professor at the University of Costa Rica. He received both undergraduate and graduate degrees from Harvard University. His research interests include the behavior and ecology of web-spinning spiders, functional morphology of beetle horns and earwig forceps, evolutionary interactions between subcellular organelles and plasmids and the cells that contain them, and the evolution of animal genitalia. Address: Escuela de Biología, Universidad de Costa Rica, Ciudad Universitaria, Costa Rica.

Figure 1. The male of this pair of *Altica* beetles performs complex courtship behavior after copulation has begun. Within the first two minutes of copulation, he inflates inside the female a sac which emerges from the tip of the brown, cylindrical basal portion of his genitalia *(visible in the top photo)*, and passes his sperm into the female. During the rest of the approximately 20-minute copulation, he periodically thrusts forward as far as he can *(middle photo)* in a more or less stereotyped pattern. The thrusts do not move his genitalia deeper into the female's genital tract; rather they stretch the walls of the entire basal portion of the tract as it is displaced forward within her body. Between bouts of thrusting the male often rubs his rear tarsi gently but persistently near the tip of the female's abdomen *(bottom photo)*. After he has withdrawn his genitalia, the male sometimes gives her additional, more vigorous rubs with his hind legs. (Photos by D. Perlman.)

groups with species-specific genitalia, males and females exchange species-specific signals during courtship, and probably seldom if ever reach the point of making genital contact with members of other species. For example, some female moths attract males with species-specific blends of pheromones, and the males, after finding the females, court them with additional species-specific pheromones before beginning to copulate. Yet the male moths have species-specific genitalia (Baker and Cardé 1979). Species-specific genitalia have even evolved in situations in which mistaken, cross-specific matings are essentially impossible, for example in island-dwelling species isolated from all close relatives, or parasitic species that mate on hosts which never harbor more than one species of the parasite. In some of these groups, such as the pinworms of primates, male genitalia provide the best morphological characteristics known to distinguish closely related species (Inglis 1961).

The alternative hypothesis, pleiotropy, is no more satisfying. It holds that genital characteristics are chance effects of genes that code primarily for other characteristics, such as adaptations to the environment. But this idea fails to explain why incidental effects should consistently occur on genitalia and not elsewhere. Nor does it explain why incidental effects fail to occur in species employing external rather than internal fertilization. It also cannot account for the genital morphology of a number of groups, such as spiders and guppies, in which organs (e.g., a pedipalp or anal fin) other than the primary male genitalia acquire the function of introducing sperm into the female; these other organs consistently become subject to the putative "incidental" effects while the primary genitalia do not.

So, until recently, a pattern widespread in animal evolution was left without a plausible explanation. Recently, however, a resurgence of interest in Darwin's ideas on sexual selection and advances in evolutionary theory to encompass male-female conflicts have stimulated new hypotheses.

Two mechanisms proposed by Darwin are potentially involved: male-male competition and female choice. Recent hypotheses are that male genitalia function to remove or otherwise supersede sperm introduced in previous matings of the female (Waage 1979; Smith 1984), and that male genitalia are often used as "internal courtship devices"—inducing the female to use a male's sperm—and thus are under sexual selection by female

Figure 2. The penis of *Notomys mitchelli*, an Australian rodent, is extremely elaborate. The movements of such an organ certainly do not go unnoticed by the female while it is within her. (From Breed 1986; courtesy of W. G. Breed.)

choice (Eberhard 1985). While sperm removal and displacement have been documented in several cases, these are unlikely to be general explanations for the trend of genitalia to evolve and diverge rapidly, because males of many groups with species-specific genitalia do not penetrate deep enough into the female to reach sites where sperm from previous copulations are stored. In the remainder of this article I present some of the evidence supporting the female-choice hypothesis, and show how it calls into question some basic and intuitively "obvious" notions about animal behavior and morphology.

Copulatory courtship and genital stimulation

The obvious function of a male's courtship behavior is to induce the female to mate. Yet males of many species of insects appear to court the female even after they have achieved genital coupling. Their behavior includes typical courtship movements such as waving antennae or colored legs, stroking, tapping, rubbing, or biting the female's body, buzzing the wings in stereotyped patterns, rocking the body back and forth, and singing (Eberhard, unpubl.). A survey of studies of copulation behavior in insects showed that, in just over one-third of 302 species, the male performs behavior apparently designed to stimulate the female (Eberhard, unpubl.). In some species, such as the beetles shown in Figure 1, the male combines movements of body parts such as antennae and hind feet with more or less stereotyped movements of his genitalia. Both genital and non-genital behavior patterns differ in closely related species.

Many male mammals move their genitalia in and out of the female in more or less stereotyped movements prior to insemination (Dewsbury 1972). Some also perform post-ejaculatory intromissions which differ from the earlier ones and which increase the likelihood that the mating will result in the female becoming pregnant (Dewsbury and Sawrey 1984). Male goldeneye ducks perform four different displays after copulation (Dane and van der Kloot 1962).

Other observations also indicate that male genitalia themselves perform copulatory courtship. A study by Lorkovic (1952) showed that the genitalic "claspers" of some male butterflies are rubbed gently back and forth on the sides of the female's abdomen during copulation. Some snails thrust genitalic darts into the female during courtship or copulation (Fretter and Graham 1964), while some moths have elongate, sharp-pointed scales on their penes which are designed to fall off inside the female (Busck 1931), probably delivering stimuli to the female after the male has left. In a variety of groups, ranging from *Drosophila* flies to mice and sheep, copulatory behavior persists even after the male has exhausted his supply of sperm (Dewsbury and Sawrey 1984). In marmosets, male stimulatory effects during copulation have been documented by showing that some female responses disappear when the female's reproductive tract is anesthetized (Dixson 1986). Male cats and some male rodents have backwardly directed spines on their penes which make stimulation of the female inevitable (Fig. 2); ovulation in female cats is known to be induced by mechanical stimulation of the vagina (Greulich 1934; DeWildt et al. 1978). These stimulation devices probably represent mechanical equivalents of the visual displays of erect, brightly colored male penes in some lizards and primates (Bohme 1983; Eckstein and Zuckerman 1956; Hershkovitz 1979).

In some groups, mechanical stimulation and sperm transfer are particularly clear because they are performed separately. Male spiders and millipedes generally transfer sperm to modified structures on their pedipalps or legs, then use these secondary genitalia to introduce sperm into the female. In some species copulation always occurs first with the secondary genitalia empty; the male then withdraws, loads the secondary genitalia with sperm, and copulates again (Austad 1984). A number of species, including beetles, wasps, and rodents, perform a series of preliminary or extra intromissions which apparently do not result in sperm transfer (Cowan 1986; Schincariol and Freitag 1986; Dewsbury, in press). Mallards, one of the few bird groups having intromittent organs, frequently copulate during pair formation, six months before egg-laying, when the male gonads are repressed and sperm are not produced (McKinney et al. 1984).

Copulatory courtship behavior is understandable given the perspective that copulation is only one of a series of events which must occur if a male is to sire offspring. The female must remain still, or at least not actively attempt to terminate copulation prematurely. In rats and fleas, for instance, genitalic stimulation increases a female's tendency to stand still (Rodriguez-Sierra et al. 1975; Humphries 1967). The male's sperm must be transported to the storage site or fertilization site; this process seems to depend to a large extent on female peristalsis or other transport movements rather than on the motility of the sperm (see Overstreet and Katz 1977 on mammals; Davey 1965 on insects). Females of many species have, associated with their sperm-storage organs, glands which must be activated to help keep sperm alive and healthy. In some mammals and arthropods, sperm must be "activated" once inside the female in order to become capable of fertilization (Hamner et al. 1970; Leopold and Degrugillier 1973; Brown 1985). In some species, such as roaches and cats, ovulation and maturation of the eggs are induced by mechanical stimuli associated with copulation (Roth and Stay 1961; Greulich 1934). Brood care in earwigs is thought to be induced by mating (Vancassel cited in Lamb 1976).

Another critical female response often associated with copulation is the lack of further sexual receptivity. Sperm from subsequent matings can offer dangerous competition because it is extremely rare for internal fertilization of eggs to occur immediately following copulation. In some bees and wasps, postcopulatory courtship appears to reduce the frequency of remating by females (van den Assem and Visser 1976; Alcock and Buchmann 1985). Stimuli from both copulation (without spermatophore transfer) and the spermatophore itself reduce further sexual receptivity in the female butterfly *Pieris rapae* (Obara et al. 1975; Sugawara 1979). Finally, in mice, copulation can inhibit transport of sperm from previous matings (Dewsbury 1985). In sum, a male's reproductive success can be greatly affected by his ability to induce females to perform any of several critical post-coupling activities. Copulatory courtship behavior by males probably serves this end.

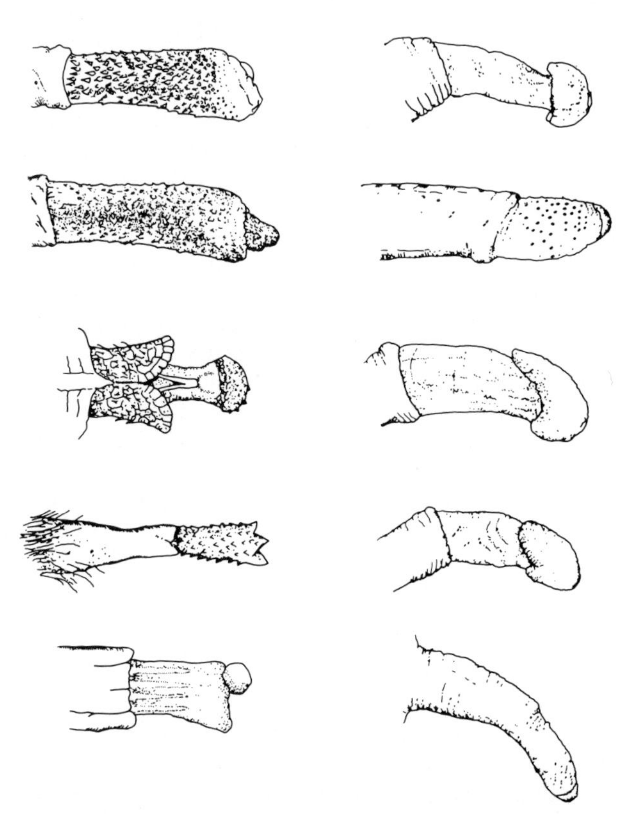

Figure 3. As predicted by the female-choice theory, more elaborate penis morphology occurs in species in which the female is more likely to mate with more than one male, thus being able to choose the father of her offspring. The male genitalia above are from different primate species. Those on the right belong to species in which a single male usually monopolizes the matings of a female during a period of estrous; those on the left belong to species in which receptive females can be mated by more males. (On the left proceeding down: ***Galago crassicaudatus, G. garnettii, Arctocebus calabarensis, Euoticus elegantulus, Nycticebus coucang.*** **On the right proceeding down:** ***Colobus guereza, Callithrix jacchus, Mandrillus sphinx, Erythrocebus patas, Saguinus oedipus.*****)(After Dixson 1987; drawn to different scales.)**

Number of mates and coyness

The female-choice hypothesis predicts that genital morphology of males should be under stronger sexual selection in those species in which females mate with more males. Dixson (1987) recently tested this prediction using a sample of 130 primate species. His data show, even when corrected for possible effects of common ancestors, that males of species in which sexually active females are not monopolized by single males have relatively longer genitalia, more highly developed hard spines on the penis, more complex shapes at the tip of the penis, and a more developed baculum (penis bone)(Fig. 3). In addition, males of species in which receptive females are not monopolized display more elaborate copulatory behavior, with prolonged and multiple intromissions. A similar trend occurs in *Heliconius* butterflies; males of species in which females remate more often tend to have more distinctive genitalia (Eberhard 1985).

If reproductive processes in females are triggered by male stimuli after coupling, then sexual selection theory predicts that it will be advantageous for a female to avoid having each copulation result in the fertilization of all her available eggs. This is because females able to favor males proficient at stimulation will have sons that are superior reproducers because the sons will be, on average, proficient stimulators (Fisher 1958). This type of selection can, in theory, give rise to a runaway process in which males develop increasingly elaborate apparatus and females become increasingly discriminating. Along with the probable advantages to females of controlling the timing of fertilization, female choice may help explain the tortuous and complex morphologies of many female reproductive tracts (Fig. 4, 5) and the rarity of designs in which males simply place their sperm at the site of fertilization. Female genitalia may be designed not only to facilitate fertilization, but also to prevent it under certain circumstances.

Perhaps the most dramatic and well-documented case supporting this idea is that of bedbugs and their allies (Carayon (1966). Some male bedbugs have evolved a hypodermic penis which can be inserted at a variety of sites on the female's body; the sperm are injected into her blood, and they migrate to the ovaries where they accumulate in huge masses even after only a single copulation. In some groups the females have responded by evolving a new genital system, complete with an opening on the top of the abdomen, a storage organ, and ducts to the oviduct. This new system would seem unnecessary for sperm transport, and both its developmental origin and mode of action suggest that it serves instead to selectively prevent fertilization. The cells of the new female system are derived from types used to combat infections, and only a very small proportion of the sperm that enter ever reach the ovaries.

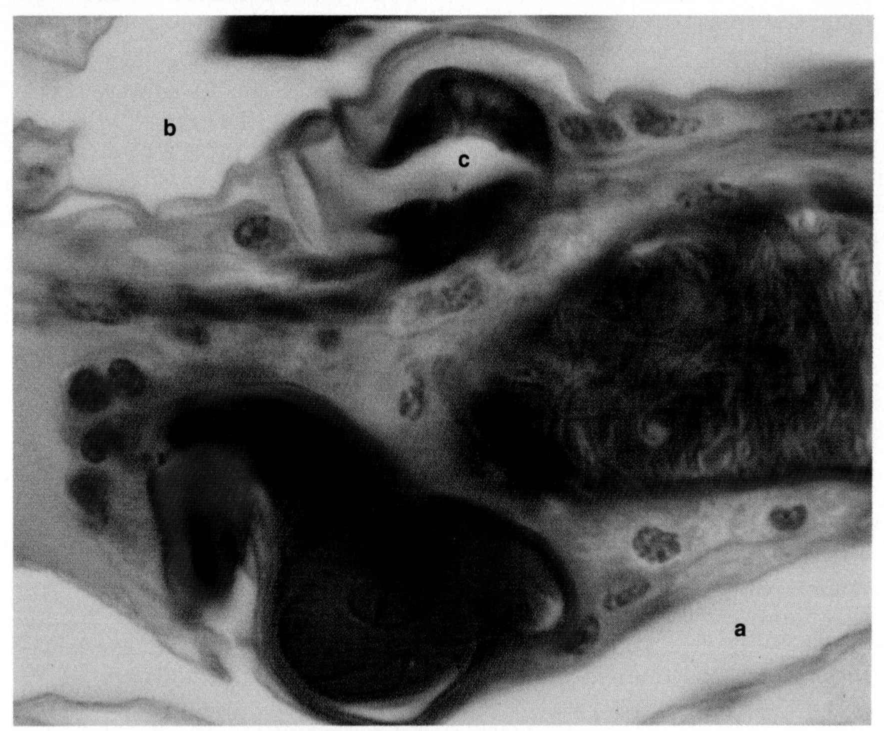

Figure 4. This transverse section of a portion of the female reproductive organs of the velvet water bug, *Hebrus ruficeps*, makes it clear that the female has control over the fate of sperm inside her. The male inserts his phallus along the shaft of her ovipositor into her genital chamber *(a)*, where he then everts a sac at the tip of the phallus into a farther chamber, the gynatrial sac *(b)*. He depends on the female to draw his sperm—each of which is longer than her entire body—out of his genitalia with a special set of dilator muscles. The female must then pump the sperm into the proximal end of her storage organ, the spermatheca, using the spermathecal pump *(c)* and spermathecal muscles. (From Heming-van Battum and Heming 1986; courtesy of B. Heming.)

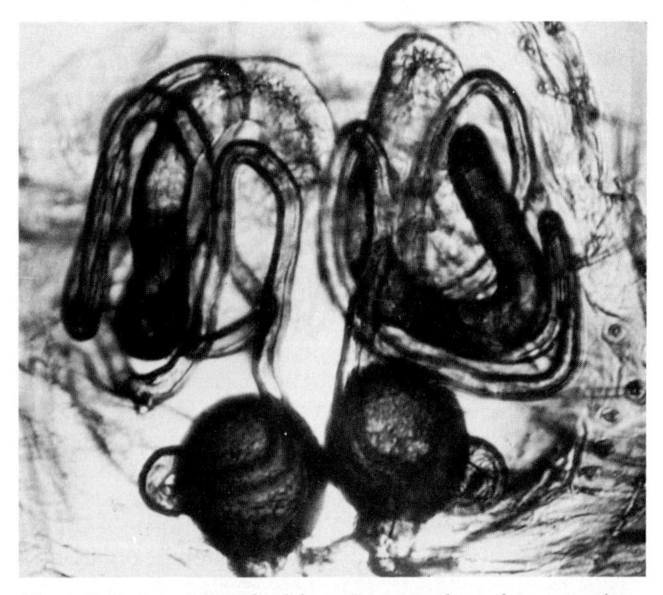

Figure 5. Tortuous ducts lead from the external copulatory opening *(circular forms at bottom)* to the internal sperm storage cavity in a female *Goyenia sylvatica* spider. The male has a pair of long, slender structures which presumably traverse at least part of this maze of ductwork. Females may test the mechanical properties of male genitalia with such hard-to-reach structures. (Photo by R. Foster.)

Such complicated coyness is not limited to insects. In pigs and several rodents, the rapid buildup of sperm in the uterus after mating is followed by the arrival of leucocytes which engulf and digest the sperm (Overstreet and Katz 1977). Other female mammals have barriers to sperm transport: as much as 80% of the sperm in the ejaculate of a rabbit is expelled from the vagina within minutes of coitus; an additional 15% probably does not migrate upward from the vagina; most of the remaining 5% is lost or retained along the way (Overstreet and Katz 1977). The cervix of a female human moves away from rather than toward the vagina during sexual arousal (Masters and Johnson 1966). Among males, some individuals are apparently better than others at overcoming these female barriers (Overstreet and Katz 1977).

Reconsidering old ideas

These observations suggest that the concept of courtship needs to be expanded to include behavior designed to induce the female not only to accept a male's copulation, but also to accept, care for, and use his sperm rather than that of other males. By using the sperm of some males and not others with whom she has copulated, the female can exercise what Thornhill (1983) has termed "cryptic female choice." If this is the case, then evolutionary biologists cannot measure the selective effects of male behaviors and morphologies by simply counting copulations. A male's reproductive success is not necessarily a simple function of the number of females with whom he copulates (Fig. 6).

Recognizing the possible role of sexual selection forces one to rethink the functions of intromission and copulation and the evolutionary origin of external male genitalia. Consider the fact that intromittent organs are more common in males than females. If mating is a cooperative effort by the male and female to fertilize eggs, and if intromittent genitalia have evolved because of the advantage of protecting gametes and zygotes from environmental vicissitudes (both generally accepted notions), then why is it that the male—the sex with relatively small, energetically cheap gametes—nearly always has the intromittent organ even though internal fertilization has evolved many times independently?

Use of a classic comparative technique—looking for an exception to the rule and checking that case for other unusual traits—suggests that sexual selection is involved. A well-documented exception occurs in sea horses and their relatives. The female sea horse has a functional penis, called an ovipositor, which she inserts into the male to transfer her eggs into his pouch, where they are fertilized. Sea horses are unusual in another respect: the males make large investments in the offspring, brooding the eggs and in some cases nourishing the offspring in the pouch before they leave. Females of some species are limited reproductively by a lack of access to males who will brood their eggs (Berglund et al. 1989). The coincidence of a reversal both in sexual morphology and in which of the two sexes represents a "bottleneck" or limiting resource in the reproduction of the other, is expected if sexual selection is involved. In more typical cases, male reproduction is limited by the number of mates, whereas female reproduction is limited not by males but by other factors such as food availability. Sexual selection theory predicts that the sex whose reproduction is more limited by access to the other sex will be selected to distribute its gametes to as many members of the opposite sex as possible. The sea horses thus suggest that intromittent organs in general are competitive devices for placing gametes in favorable sites and inducing even not completely willing partners to accept and use the gametes.

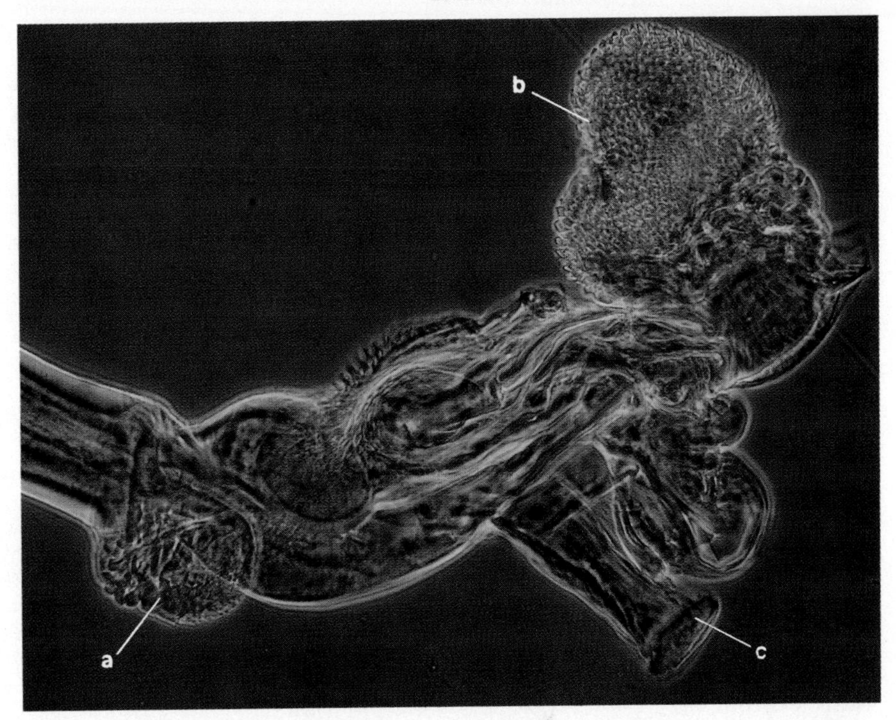

Figure 6. The complex tip of the genitalia of a male medfly is at the end of a long, snakelike structure which is about 40% of the length of the male's body. He inserts his genitalia by folding them and threading them up the shaft of the female's ovipositor. Then he gradually unfolds them inside her. They must go still deeper, however, and the male apparently uses an inching movement of an expandable finger-like structure *(a)* that pulls them farther inward. When the tip finally reaches the upper end of the vagina, it must hit a deeper target; the male uses a second expandable sac *(b)* to drive a tubular structure *(c)* into a cone in the wall of the vagina, and from there into a cylindrical organ attached to the cone's tip. The female has muscles in the ducts leading to the sperm storage sites (spermathecae), and apparently these must contract in order for the sperm to be pumped into the spermathecae. (Photo by M. Vargas and H. Camacho).

This implies that the notion that internal fertilization evolved to protect gametes and zygotes (Hinton 1964) may have to be reevaluated. Protection in some cases may have been an incidental consequence of competitive maneuvers for placing gametes in positions where they were more likely to achieve fertilization (Parker 1970). Sexual selection in this context may have had other important incidental consequences. Perhaps some lineages were able to move from aquatic to terrestrial habitats partly because ancestral males had evolved intromittent organs.

Another classical notion that must be modified is the distinction, made by Darwin himself, between primary sexual characteristics, which are supposed to function strictly for gamete transfer, and secondary sexual characteristics, such as horns or bright colors, which function in the context of sexual selection (as in fights between males over females or the attraction of females). Correcting Charles Darwin on a point of natural selection theory is an event so unusual that it inspires trepidation, even when the point in question is only a footnote in one of his books.

Good genes or propaganda?

Studies of genital function inspired by new evolutionary theory may in turn help advance the theory. Evolutionary theorists are currently divided over whether or not female choice necessarily favors males that are superior in other contexts, as in the ability to resist parasites or capture more prey (Bradbury and Andersson 1987). The "good genes" hypothesis interprets elaborate morphologies such as the tails of peacocks, which are classically associated with sexual selection by female choice, as indicators of overall male quality. Females may judge the health of the male by his ability to produce such extravagant features (Zahavi 1987; Hamilton and Zuk 1982). On the other hand, as Ronald Fisher argued over 50 years ago, male signaling ability per se should be advantageous once females begin to use male signals as criteria for mating, or, as we have seen here, as triggers for other postcoupling processes leading to fertilization. A male with a greater ability to capture the female's attention and induce her to accept and use his sperm for fertilization could be favored whether or not that talent for "advertisement" is correlated with other aspects of male fitness. The ability to produce effective signals would, in and of itself, be a component of the male's fitness. Clarification of the causes of genital evolution may help resolve this debate, because typical genital characteristics probably have little or no relationship with overall male fitness in other contexts. If the female-choice hypothesis for genital evolution is correct, then the choosing of males on the basis of advertisement rather than good genes must be widespread.

The study of genitalia probably represents the last major frontier of an old and very successful branch of biology: functional morphology. It can be difficult to observe genitalia in action, and special techniques such as freezing pairs at different stages of copulation are often needed. However, the payoffs in increased understanding that result when genitalic events are included in studies of the selective consequences of different male and female behaviors are already evident (Simmons 1986; 1987). In addition to cryptic female choice, direct competition between males can occur inside females, as with the removal or repositioning of sperm from previous copulations in odonates (Waage 1986; Siva-Jothy 1988). Other possibilities are chemical trickery, in which sperm from previous males are induced to leave the storage site and become diluted by the sperm of a second male (Gilbert et al. 1984), and shuffling or selective discarding of sperm by the female as a result of genital stimulation (Otronen 1988). Studies of such phenomena, combined with more information on unknown behavior of the sperm inside the female, will probably constitute a major area of interest in evolutionary studies of behavior and morphology. There are many exciting surprises in store.

References

Alcock, J., and S. L. Buchmann. 1985. The significance of post-insemination display by male *Centris pallida* (Hymenoptera: Anthophoridae). *Zeitschrift für Tierpsychologie* 68:231–43.

Assem, J. van den, and J. Visser. 1976. Aspects of sexual receptivity in the female *Nasonia vitripennis* (Hymenoptera: Pteromalidae). *Biol. Behav.* 1:37–56.

Austad, S. N. 1984. Evolution of sperm priority patterns in spiders. In *Sperm Competition and the Evolution of Animal Mating Systems,* ed. R. L. Smith, pp. 223–49. Academic Press.

Baker, T. C., and R. T. Cardé. 1979. Courtship behavior of the oriental fruit moth *(Grapholitha molesta)*: Experimental analysis and consideration of the role of sexual selection in the evolution of courtship pheromones in the Lepidoptera. *Ann. Entom. Soc. Am.* 72:173–88.

Berglund, A., G. Rosenquist, and I. Svensson. 1989. Reproductive success of females limited by males in two pipefish species. *Am. Natur.* 133:506–16.

Bohme, T. 1983. The Tucano Indians of Colombia and the lizard *Plica plica*: Ethnological, herpetological and ethological implications. *Biotropica* 15:148–50.

Bradbury, J., and M. Andersson. 1987. *Sexual Selection: Testing the Alternatives.* Wiley.

Breed, W. G. 1986. Comparative morphology and evolution of the male reproductive tract in the Australian hydromysine rodents (Muridae). *J. Zool.* A209:607–29.

Brown, S. 1985. Mating behavior of the golden orb-weaving spider, *Nephila clavipes*: II. Sperm capacitation, sperm competition, and fecundity. *J. Comp. Psychol.* 99:167–75.

Busck, A. 1931. On the female genitalia of the microlepidoptera and their importance in the classification and determination of these moths. *Bull. Brooklyn Entom. Soc.* 26:119–216.

Carayon, J. 1966. Traumatic insemination and the paragenital system. In *Monograph of Cimicidae,* ed. R. Usinger, pp. 81–166. Entom. Soc. Am.

Cowan, D. P. 1986. Sexual behavior of eumenid wasps (Hymenoptera: Eumenidae). *Proc. Entom. Soc. Wash.* 88:531–41.

Dane, B., and W. van der Kloot. 1962. Analysis of the display of the goldeneye duck *(Bucephala clangula* L.) *Behavior* 22:282–328.

Davey, K. G. 1965. *Reproduction in Insects.* Oliver and Boyd.

DeWildt, D. E., S. C. Guthrie, and S. W. J. Seager. 1978. Ovarian and behavioral cyclicity of the laboratory-maintained cat. *Hormones and Behav.* 10:251–57.

Dewsbury, D. 1972. Patterns of copulation behavior in mammals. *Quart. Rev. Biol.* 47:1–33.

———. 1985. Interactions between males and their sperm during multi-male copulatory episodes of deer mice *(Peromyscus maniculatus). Animal Behav.* 33:1266–74.

———. In press. Copulatory behavior as courtship communication. *Ethology.*

Dewsbury, D., and D. K. Sawrey. 1984. Male capacity as related to sperm production, pregnancy initiation, and sperm competition in deer mice *(Peromyscus maniculatus). Behav. Ecol. Sociobiol.* 16:37–47.

Dixson, A. F. 1986. Genital sensory feedback and sexual behaviour in male and female marmosets *(Callithrix jacchus). Physiol. Behav.* 37:447–50.

———. 1987. Observations on the evolution of genitalia and copulatory behavior in primates. *J. Zool.* 213:423–43.

Eberhard, W. G. 1985. *Sexual Selection and Animal Genitalia.* Harvard Univ. Press.

Eckstein, P., and S. Zuckerman. 1956. Morphology of the reproductive tract. In *Marshall's Physiology of Reproduction,* ed. A. S. Parkes, vol. 1, pp. 43–155. Longmans, Green and Co.

Fisher, R. 1958. *The Genetical Theory of Natural Selection.* Dover.

Fretter, V., and A. Graham. 1964. Reproduction. In *Physiology of Mollusca,* ed. K. M. Wilbur and C. M. Yonge, pp. 127–64. Academic Press.

Gilbert, D. G., R. C. Richmond, and K. B. Sheehan. 1984. Studies of esterase 6 in *Drosophila melanogaster.* V. Progeny production and sperm use in females inseminated by males having active or null alleles. *Evolution* 38:24–37.

Greulich, W. W. 1934. Artificially induced ovulation in the cat *(Felis domestica). Anat. Rec.* 58:217–23.

Hamilton, W. D., and M. Zuk. 1982. Heritable true fitness and bright birds: A role for parasites? *Science* 218:384–87.

Hamner, C. E., L. L. Jennings, and N. J. Skojka. 1970. Cat *(Felis catus* L.) spermatozoa require capacitation. *J. Reprod. Fert.* 23:477–80.

Heming-van Battum, K. E., and B. S. Heming. 1986. Structure, function and evolution of the reproductive system in females of *Hebrus pusillus* and *H. ruficeps* (Hemiptera, Gerromorpha, Hebridae). *J. Morph.* 190:121–67.

Hershkovitz, P. 1979. *Living New World Monkeys (Platyrrhini) with an Introduction to Primates,* vol. 1. Univ. Chicago Press.

Hinton, H. E. 1964. Sperm transfer in insects and the evolution of haemocoelic insemination. In *Insect Reproduction,* ed. K. C. Highnam, pp. 95–107. Royal Entom. Soc. London.

Humphries, D. A. 1967. The mating behaviour of the hen flea *Ceratophyllus gallinae* (Shrank) (Siphonaptera: Insecta). *Animal Behav.* 15:82–90.

Inglis, W. G. 1961. The oxyurid parasites (Nematoda) of primates. *Proc. Zool. Soc. London* 136:103–22.

Lamb, R. J. 1976. Parental behavior in the Dermaptera with special reference to *Forficula auricularia* (Dermaptera: Forficulidae). *Canadian Entom.* 108:609–19.

Leopold, R. A., and M. E. Degrugillier. 1973. Sperm penetration of housefly eggs: Evidence for involvement of a female accessory secretion. *Science* 181:555–57.

Lorkovic, A. 1952. L'accouplement artificiel chez les Lépidoptères et son application dans les recherches sur la fonction de l'appareil génital des insectes. *Physiol. Comp. Oecol.* 3:313–19.

Masters, W. H., and V. E. Johnson. 1966. *Human Sexual Response.* Little Brown.

McKinney, F., K. M. Cheng, and D. J. Bruggers. 1984. Sperm competition in apparently monogamous birds. In *Sperm Competition and the Evolution of Animal Mating Systems,* ed. R. L. Smith, pp. 523–45. Academic Press.

Nichols, S. W. 1986. Early history of the use of genitalia in systematic studies of Coleoptera. *Quaest. Entom.* 22:115–41.

Obara, Y., H. Tateda, and M. Kuwabara. 1975. Mating behavior of the cabbage white butterfly, *Pieris rapae crucivora* Boisduval. V. Copulatory stimuli inducing changes of female response patterns. *Zool. Mag.* (Tokyo) 84:71–76.

Otronen, M. 1988. Studies on reproductive behavior in some carrion insects. D. Phil. Thesis, Univ. of Oxford.

Overstreet, J. W., and D. F. Katz. 1977. Sperm transport and selection in the female genital tract. In *Development in Mammals,* ed. M. H. Johnson, vol. 2, pp. 31–65. North Holland Pub. Co.

Parker, G. A. 1970. Sperm competition and its evolutionary consequences in the insects. *Biol. Rev.* 45:525–67.

Rodriguez-Sierra, J. F., W. R. Crowley, and B. R. Komisaruk. 1975. Vaginal stimulation in rats induces prolonged lordosis responsiveness and sexual receptivity. *J. Comp. Physiol. Psychol.* 89:79–85.

Roth, L. M., and B. Stay. 1961. Oocyte development in *Diploptera punctata* (Eschscholtz) (Blattaria). *J. Ins. Physiol.* 7:186–202.

Shapiro, A. M., and A. H. Porter. 1989. The lock and key hypothesis: Evolutionary and biosystematic interpretation of insect genitalia. *Ann. Rev. Entom.* 34:231–45.

Schincariol, L. A., and R. Freitag. 1986. Copulatory locus, structure and function of the flagellum of *Cicindela tranquebarica* Hergst (Doleoptera: Cicindelidae). *Int. J. Invert. Reprod. Devel.* 9:333–38.

Simmons, L. 1986. Female choice in the field cricket *Gryllus bimaculatus* (DeGreer). *Animal Behav.* 35:1463–70.

———. 1987. Sperm competition as a mechanism of female choice in the field cricket *Gryllus bimaculatus*. *Behav. Ecol. Sociobiol.* 21:197–202.

Siva-Jothy, M. 1988. Sperm repositioning in *Crocothemis erythraea*, a libellulid dragonfly with a brief copulation. *J. Ins. Behav.* 1:235–45.

Smith, R. L., ed. 1984. *Sperm Competition and the Evolution of Animal Mating Systems*. Academic Press.

Sugawara, T. 1979. Stretch reception in the bursa copulatrix of the butterfly *Pieris rapae crucivora*, and its role in behaviour. *J. Comp. Physiol.* 130:191–99.

Thornhill, R. 1983. Cryptic female choice and its implications in the scorpionfly *Harpobittacus nigriceps*. *Am. Nat.* 122:765–88.

Waage, J. K. 1979. Dual function of the damselfly penis: Sperm removal and transfer. *Science* 203:916–18.

———. 1986. Evidence for widespread sperm displacement ability amongst Zygoptera and the means for predicting its presence. *Biol. J. Linn. Soc.* 28:285–300.

Zahavi, A. 1987. The theory of signal selection and some of its implications. In *International Symposium Biological Evolution*, ed. V. P. Delfino, pp. 305–27. Adriatica Editrice.

Mating Behavior and Hermaphroditism in Coral Reef Fishes

The diverse forms of sexuality found among tropical marine fishes can be viewed as adaptations to their equally diverse mating systems

Robert R. Warner

The colorful diversity of coral-reef fishes has long been a source of fascination for both scientists and amateurs. We now know that this diversity of form and color is matched by an immense variety of social behaviors and sexual life histories, including several kinds of functional hermaphroditism. Recent observations and experimental work suggest that the sexual patterns found in fishes may best be viewed as evolutionary responses to the species' mating systems, and much of the evidence I review here bears out this idea. Since most theory in behavioral ecology has been derived from studies of terrestrial vertebrates and insects, which have strictly separate sexes, the relationships between sexual expression and mating behavior in fishes offer new insights into the role of sexuality in social evolution.

Like other vertebrates, most fish species have separate sexes, a condition known as gonochorism. However, fishes are by no means restricted to this pattern: in many species individuals are capable of changing sex, a phenomenon sometimes called sequential hermaphroditism, and in others fishes can be both sexes at the same time, displaying simultaneous hermaphroditism.

This sexual flexibility is quite widespread. At least fourteen fish families contain species that exhibit sex change from female to male, termed protogyny, as a normal part of their life histories (see Policansky 1982 for a recent review). Eleven of these families are common in coral-reef areas; and in the wrasses (Labridae), parrotfishes (Scaridae), and larger groupers (Serranidae) protogyny occurs in the great majority of the species studied (Fig. 1). Changes from female to male are also known to occur in damselfishes (Pomacentridae; Fricke and Holzberg 1974), angelfishes (Pomacanthidae), gobies (Gobiidae), porgies (Sparidae), emporers (Lethrinidae; Young and Martin 1982), soapfishes (Grammistidae), and dottybacks (Pseudochromidae; Springer et al. 1977). However, we have no indication of how common sex change might be in these families, since few species have been carefully investigated. New reports of protogynous species are constantly cropping up, and the phenomenon may be much more frequent than previously imagined.

Change of sex from male to female—protandry—appears to be less common. It is known in eight families of fishes, three of which—porgies (Sparidae), damselfishes (Pomacentridae), and moray eels (Muraenidae; Shen et al. 1979)—are found on coral reefs. The damselfish and porgy families also include species that are protogynous. Such variability within a family offers an important opportunity to test sex-change theory, and deserves further study.

Robert R. Warner is Associate Professor of Marine Biology at the University of California, Santa Barbara. He received his Ph.D. from Scripps Institution of Oceanography in 1973. Before coming to Santa Barbara in 1975, he was a postdoctoral fellow at the Smithsonian Tropical Research Institute at Balboa, Panama. His interests include the interactions of life history and behavioral characteristics, the evolution of sexual dimorphism, the adaptive significance of delayed maturity in males, and the sexual patterns of tropical marine fishes. Address: Department of Biological Sciences, University of California, Santa Barbara, CA 93106.

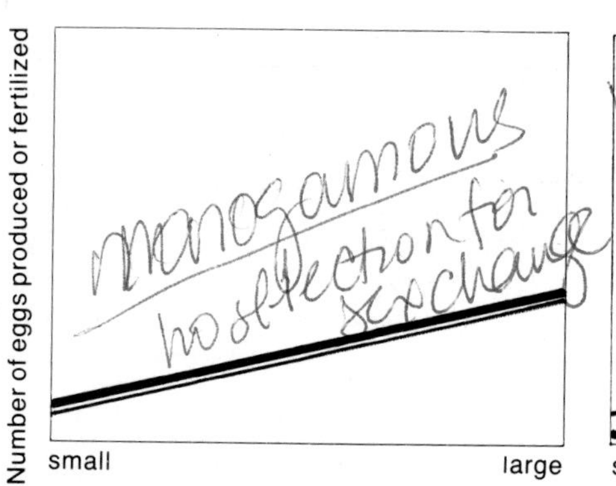

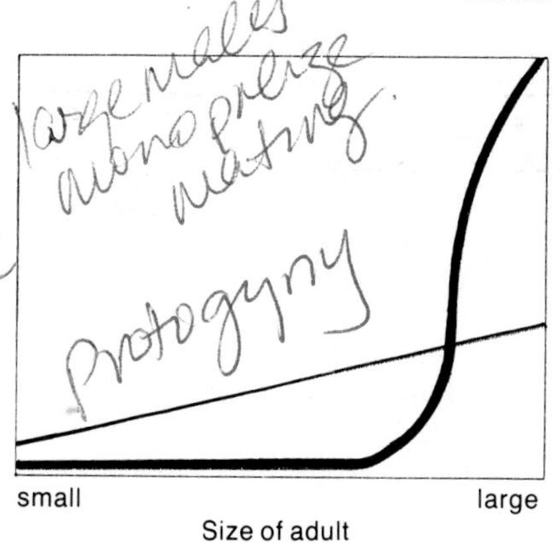

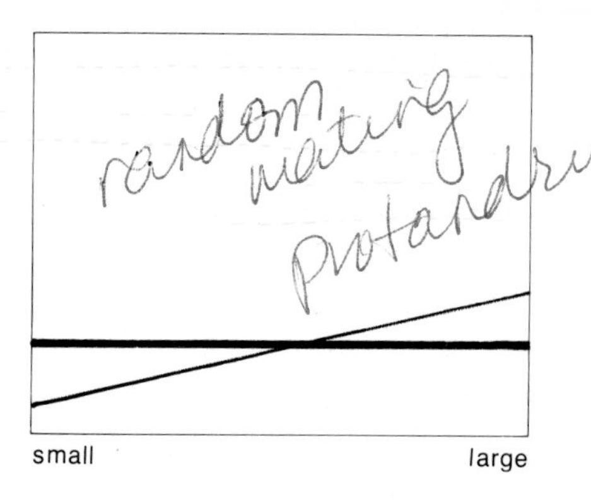

Figure 2. The expected fertility of a female fish (*thin line*), measured as the number of eggs produced, usually depends directly on size and thus shows a steady increase with growth. By contrast, a male's expected fertility (*thick line*), measured as the number of eggs fertilized, is affected by the mating system. When males and females form monogamous pairs matched by size or when males compete with each other to fertilize eggs and thus to produce the most sperm, both sexes show a similar increase of fertility with size (*left*), and no selection for sex change exists. When large males monopolize mating to the detriment of small males, however, male fertility rises dramatically at a certain point in growth (*center*), and an individual that remains a female when small and changes to a male at a large size will be selectively favored. When mating consists of random pairing (*right*), an individual would do best to function as a male while small (with a chance of fertilizing a larger female) and as a female when large (capitalizing on a high capacity for egg production). (After Warner 1975.)

Species in which individuals are simultaneous hermaphrodites, producing eggs and sperm at the same time, are common among deep-sea fishes (Mead et al. 1964) but quite rare elsewhere. However, several species of small sea basses (Serranidae) found in shallow water in tropical areas are known to be simultaneous hermaphrodites (Smith 1975). The existence of this sexual pattern in an abundant species such as the sea basses poses special evolutionary problems that are discussed below.

Because the warm, clear waters of coral reefs create conditions that are nearly ideal for observing animal behavior, the social and mating systems of tropical marine fishes have been particularly well studied. It is my intention here first to outline a central hypothesis which relates sexual patterns to social behavior, and then to test this hypothesis at several levels, using information available in the literature.

Sex change

Why should natural selection favor sexual patterns different from pure gonochorism? In other words, under what circumstances might we expect hermaphroditism to be adaptive? This question was first dealt with in a comprehensive fashion by Ghiselin (1969, 1974), who suggested the general conditions under which hermaphroditism would be expected to evolve. More recent work has related these generalities to the specific social systems of hermaphroditic species (Warner 1975; Fischer 1980; Charnov 1982). It is best to view sequential hermaphroditism, or sex change, separately from simultaneous hermaphroditism, because they are distinct phenomena under the influence of very different selective regimes.

Ghiselin proposed the "size-advantage model" to account for many cases of sex change. The concept is simple: if the expected number of offspring (measured, say, as the number of eggs produced or fertilized) differs between the sexes with size, then an individual that changes sex at the right size or age will have more offspring than one that remains exclusively male or female.

What might cause such sexual differences in the distribution of expected fertility? Two factors are important here: the relative number of male and female gametes produced and the characteristic mating behavior of the species. In many cases, a female's fertility is limited by the number of eggs she can hold or manufacture, which in turn is controlled by her size, her store of energy, or both. Thus it makes little difference whether she mates with one male or with many. Male fertility, on the other hand, is often limited not by the number of sperm an individual can produce but rather by the number of females with whom he mates and their fertility. Because of this, the fertility of males is potentially much more variable than that of females, and can reach very high levels in certain circumstances (Williams 1966; Trivers 1972).

While size of gamete production sets the stage for potential differences in the reproductive success of males and females of various sizes, it is often the mating system that determines the actual values (Fig. 2). For example, in monogamous species where both members of a pair are normally about the same size, the fertility of males and females is approximately the same over their entire size range, and changing sex conveys no advantage.

By contrast, many coral-reef fishes have mating systems in which larger males monopolize the spawning of females. In this situation smaller males may not spawn at all, while females of equivalent size have little trouble finding a mate. The spawning rate of small males is thus lower than that of small females, but large males expe-

Figure 1. The parrotfish family Scaridae is one of several common families of coral-reef fishes in which the majority of species include individuals that change from female to male, a phenomenon known as protogyny. In the striped parrotfish, *Scarus iserti*, small females are boldly striped (*top*); as females grow larger, they change sex and adopt the brilliant blue and yellow coloration found in older males (*bottom*). Groups of young females defend territories along the edges of reefs; males and sex-changed females have larger territories which usually encompass several of these harems. (Photos by D. R. Robertson.)

rience relatively high mating success. Since the distribution of fertility differs between the sexes with size, we would expect sex change to be adaptive: an individual that functioned as a female when small and as a male after attaining a large size would have more offspring over its lifetime than one that remained either male or female, and thus protogyny should be favored by natural selection (Warner 1975).

Other mating systems lead to selection for protandry. Males usually produce millions of sperm, and small individuals are physically capable of fertilizing females of almost any size. Thus in mating systems where no monopolization occurs and where mating consists of random pairing, it should be advantageous to be a male when small (since it is probable that any mating will be with larger individuals) and a female when large (thereby taking advantage of a high capacity for egg production).

Because the fertilization of eggs occurs outside the body in many fish species, spawning is not limited to simple pairs: numerous individuals can mate simultaneously in large groups. Although mating occurs more or less at random in such spawning groups, protandry is not necessarily adaptive if many males release sperm simultaneously. In this case, competition among sperm from several males to fertilize eggs creates a situation in which male fertility is limited by the number of sperm produced. Such production should increase with size in a fashion similar to egg production by females, and thus no fertility differential between the sexes exists.

The size-advantage model has been refined over the years to allow for sexual differences both in mortality and in the rate at which fertility changes with size (Warner et al. 1975; Leigh et al. 1976; Jones 1980; Charnov 1982; Goodman 1982). Individuals should change sex when the other sex has a higher reproductive value—that is, higher future expected reproduction taking into account the probability of death. This means that individuals may (and do) change sex and suffer an initial drop in reproductive success, but by making the change they increase the probability of attaining a high level of success in the future. These are complications we need not consider here, since they do not affect the general idea that the mating system can determine the adaptive value of various forms of sex change.

Testing an evolutionary idea such as this is difficult, since experimental manipulations are often impossible or exceedingly time-consuming. Typically, one must rely instead on a search for correlations between the hypothesized cause and effect. As long as sufficient variation exists in the traits in question, the search may take place among unrelated species, within a related group, or even within a single species. The wide diversity of sexual patterns and behaviors among coral-reef fishes allows investigation on all these levels. In addition, the fact that sex changes are often direct responses to external cues makes possible experimental study as well.

Mating in sex-changers

In general, the mating systems of sex-changing species are those in which reproductive success varies with sex and size. Larger males tend to monopolize mating, either by defending spawning sites that females visit or by controlling a harem of females. In most of the species, eggs are simply released into the water, and males are free to devote a large amount of time to courtship, spawning, and defense of mating sites.

Figure 3. The sexual pattern and mating system of the bluehead wrasse, *Thalassoma bifasciatium*, is typical of many wrasses. As in most protogynous species, large males play a dominant role. These older individuals – both primary males and sex-changed females – defend spawning grounds, pair-mating with as many as 150 visiting females daily (*left*). Young primary males engage in group spawning (*right*) and in "sneaking," the practice of interfering with the mating of older males by rushing in as sperm is released. Young females and young primary males such as those shown group-spawning are characterized by greenish-black lateral markings, whereas both older primary males and sex-changed females display a distinctive white band bordered with black, like the larger individual at the left. (Photos by S. G. Hoffman.)

Figure 4. The study of three closely related species of the wrasse genus *Bodianus* seems to support the idea that sex change is less common in species in which large males have less opportunity to monopolize mating. Both *Bodianus rufus* (*top*), a species in which large males defend harems of females, and *B. diplotaenia* (*center*), one in which large males defend temporary spawning sites, are protogynous, as might be expected in mating systems where large males dominate. By contrast, *B. eclancheri* (*bottom*), which spawns in groups with no pattern of domination by large males, is functionally gonochoric, with the sexes existing in equal ratios. (Photos by S. G. Hoffman.)

A good example of such a mating system is found in the wrasse *Labrides dimidiatus*, which feeds by cleaning the skin, mouth, and gills of other fishes at specific "cleaning stations" on the reef. The cleaner-wrasses at a station live in a group consisting of a single male and a harem of five or six females. The male actively defends these females and mates with each one every day. This appears to be a system in which there is no advantage to being a small male, and indeed the species is totally protogynous (Robertson 1972).

In the last decade, marine biologists have begun to study the mating behavior of a wide variety of protogynous species in their coral-reef habitats. It is striking that virtually all these species exhibit some form of monopolization of mating by large males, even though they are found in a diverse array of families such as wrasses, parrotfishes, damselfishes, angelfishes, basses, and gobies (see, for example, Moyer and Nakazono 1978a; Robertson and Warner 1978; Cole 1982; and Thresher and Moyer 1983). Haremic mating systems appear to be most common among these protogynous species. Coral-reef fishes are often quite sedentary, and it is not surprising that large males have come to dominate and defend a local group of females in many cases. It is just this kind of situation that evolutionarily favors sex change from female to male.

There are also some apparent exceptions to the trend toward the dominance of the large male in protogynous species. In a number of species of gobies (Lassig 1977), a small bass (Jones 1980), and a wrasse (Larsson 1976), the social system appears to be monogamous, and thus protogyny would not be adaptive. Lassig (1977) suggests that sex change in the gobies he studied is an adaptation to allow reconstitution of a mated pair in case of death. This explanation probably would not apply to the more active wrasses and basses, however.

For other families in which protogyny occurs, such as the groupers and the emporers, we simply lack sufficient knowledge of the mating behavior to state whether the predictions of the size-advantage model hold.

Our knowledge of the mating habits of protandrous fishes is also incomplete. Many of the species known to be protandrous live in large schools not closely associated with the substrate; however, the details of their mating behavior have not been reported. The size-advantage model suggests that mating might consist of haphazard pairing, but this prediction remains to be tested. The anemonefishes, the one group of protandrous species whose mating system is well known, fit the size-advantage model in a precise but unexpected way, as is discussed below.

While most studies of fishes known to change sex lend support to the size-advantage model, approaches from the opposite direction are less satisfactory. For example, sex change is not found in every species in which mating is monopolized by large males. Perhaps this is asking too much of evolution, since an adaptive situation does not guarantee the appearance of a trait. It may simply be that the capacity for sexual flexibility has not yet evolved in some species, or that unknown factors reduce the advantage of sex change. Unfortunately, like many evolutionary arguments, this one is virtually untestable.

Comparisons within families

We can avoid some of the uncertainty inherent in broad comparisons among families by examining the sexual patterns and mating behaviors of a group of species within a family in which we know sex change is widespread. Using this approach, the absence of sex change where it is theoretically adaptive is less easily dismissed as evolutionary lag.

The wrasses (Labridae) and the parrotfishes (Scaridae) are large and well-known families of coral-reef fishes that include many species made up of both primary males—that is, fish that remain males for their entire lives—and protogynous individuals. Both primary males and females can become dominant, territorial males if they grow large enough. The proportion of smaller males is a measure of the degree of sex change present in the species: in cases where small males are absent, sex change is at a maximum, and when they form half the population, the species is essentially gonochoristic.

The diversity of sexual types within the wrasses and parrotfishes is reflected in a diversity of mating behavior (Fig. 3). Small males either interfere with the mating activities of larger males by darting in to join the spawning couple at the moment sperm is released, a practice called "sneaking," or they take over a whole spawning site en masse and group-spawn with the females that appear there. In group spawning a single female releases her eggs in the midst of an aggregation of males, all of whom participate in fertilization. Spawning groups can contain from two to over a hundred males.

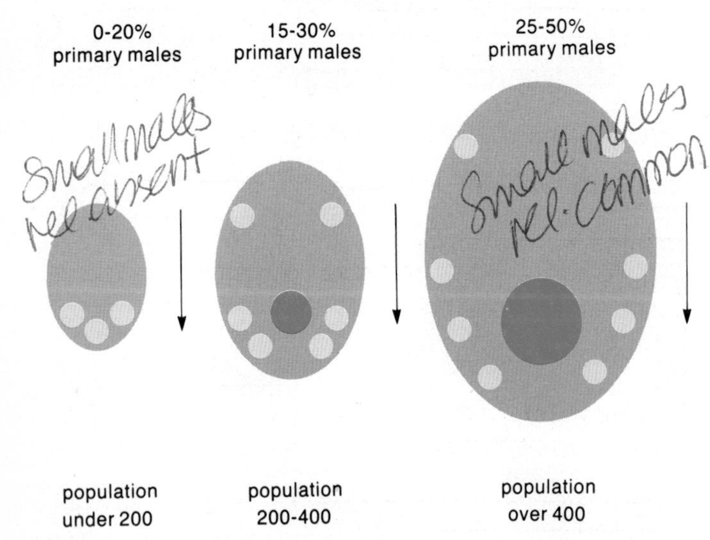

Figure 5. The mating system of the bluehead wrasse, which spends its adult life on a single reef, varies dramatically with the size of the reef. In this diagram of typical mating configurations, the predominant direction of the current is indicated by arrows; sites at the downcurrent end of the reef are preferred by spawning females. On small reefs (2 to 600 m^2), mating occurs exclusively with large males in their territories (*light blue areas*), and primary males are relatively rare. On intermediate-sized reefs (600 to 1,000 m^2), territorial males continue to occupy the prime spawning area, but a spawning group of smaller males is active just upcurrent (*dark blue area*). On the largest reefs (above 1,000 m^2), a large spawning group of small males occupies the major downcurrent spawning site, and territorial males are relegated to less productive upcurrent sites. Primary males reach their highest concentrations on such reefs, constituting up to 50% of the population.

For the size-advantage model to hold, variation in the degree of sex change should correspond to differences in the mating systems. Specifically, sex change should be less common or absent in species where large males have less opportunity to monopolize mating. Hoffman's recent work (1980 diss., 1983) on three closely related wrasses of the genus *Bodianus* provides a clear demonstration of this relationship between sexual expression and mating system (Fig. 4). *Bodianus rufus* of the Caribbean is haremic, whereas *B. diplotaenia* of the eastern tropical Pacific defends a spawning site visited by females; thus large males monopolize mating in both species. Correspondingly, small males are absent in both species, and all males are the result of sex change in functional females. On the other hand, the multicolored *B. eclancheri* of the Galapagos Islands is a group-spawner with no apparent pattern of dominance related to size or sex. Hoffman could find no evidence of sex change during adult life in this species. Individuals appear instead to be functional gonochores: change from female to male occurs before maturation, and males are equally common in all size classes of the population. The production of small males through prematurational sex change has also been noted in some species of parrotfishes (Robertson and Warner 1978).

What factors lead to changes in the monopolization of mating by large males? We have found that increased population density around spawning sites plays a role in lowering the ability of a male to defend his harem or territory adequately against smaller males (Warner and Hoffman 1980a). In extreme cases, some spawning sites can be undefendable and may be entirely abandoned to group-spawners (Warner and Hoffman 1980b).

A recent study of the wrasses of the Caribbean (genera *Thalassoma*, *Halichoeres*, *Bodianus*, and *Clepticus*) revealed that among species living in similar habitats, those with low population densities tended to have few or no small males (Warner and Robertson 1978). Regardless of whether the mating system was characterized by harems or spawning-site defense, larger males successfully monopolized mating in these species. In species living at greater densities, the proportion of primary males rose as high as 35%. Among these densely distributed species, group spawning as well as territorial mating was seen, with larger males subject to varying amounts of interference from small primary males.

The most thoroughly studied species in this group is the bluehead wrasse, *Thalassoma bifasciatum*. In this species, large males normally control the spawning sites on smaller reefs where the density of the mating population is low, and small males are nearly absent from these local populations (Fig. 5). On large reefs, where spawning sites are much more crowded, group-spawning aggregations occupy the major sites and small males are relatively common (Warner and Hoffman 1980b). Since individuals arrive on reefs as drifting planktonic larvae, the precise mechanisms leading to the distribution of small males are not known, but this example serves as a useful illustration of how density can affect the sex-changing strategy.

Not surprisingly, the effect on monopolization of mating is most pronounced at extreme densities. *T. lucasanum* of the eastern tropical Pacific is the most densely

Figure 6. The most adaptive direction for sex change can depend on the size of the social group. When the group consists of a single pair, both individuals profit if the larger member of the pair is a female, since she could produce more eggs than a smaller individual. Protandry is most adaptive in this case, and is found in the strictly monogamous anemonefishes of the genus *Amphiprion* (*top*). In larger social groups, the combined egg production of smaller members can easily exceed the egg production of the largest individual, and thus his output is maximized by functioning as a male. Protogyny would be expected here, and is found in the group-living damselfishes of the genus *Dascyllus* (*bottom*), which are closely related to anemonefishes. (Photo at top by H. Fricke; photo at bottom by F. Bam.)

distributed wrasse thus far studied; its population is essentially gonochoristic, with about 50% primary males (Warner 1982). Large territorial males are rather rare and only moderately successful in this species, and nearly all mating takes place in groups.

Certain characteristics of the habitat that allow access to spawning sites by small males apt to engage in "sneaking" should also affect mate monopolization. These characteristics are difficult to measure in a quantitative fashion, but some trends are evident. For example, small parrotfishes that live in beds of sea grass near coral reefs have a higher proportion of small males than species that exist in similar densities on the reefs themselves (Robertson and Warner 1978). In one grass-dwelling species, sex change appears to be entirely absent (Robertson et al. 1982). Sea grasses offer abundant hiding places for small fishes, and dominant males in these habitats suffer interference from smaller males in a high proportion of their matings.

Perhaps the most telling variation within a family occurs in the damselfishes (Pomacentridae), where sex change was only recently discovered (Fricke and Fricke 1977). Small damselfishes called clownfishes or anemonefishes (genus *Amphiprion*) live in or near large stinging anemones in reef areas and thus have extremely limited home ranges. They appear to be unaffected by the stinging cells of the anemone, and may enjoy a certain amount of protection from the close association (Allen 1972). An anemonefish society consists of two mature individuals and a variable number of juveniles. The species are protandrous; the largest individual is a female, the smaller adult a male (Fricke and Fricke 1977; Moyer and Nakazono 1978b). The per capita production of fertilized eggs is higher when the larger individual of a mating pair is the female, and protandry is thus advantageous to both adults (Warner 1978).

Note that the advantage of protandry in this case depends on the fact that the social group is rigidly limited to two adults. If more adults were present, the most adaptive sexual pattern could instead be protogyny. This is because the largest individual, as a male, might be able to fertilize more eggs than it could produce as a female. In accordance with this, protogyny appears in some related damselfishes (genus *Dascyllus*) in which the social groups of adults are larger (Fig. 6; Fricke and Holzberg 1974; Swarz 1980 and pers. com.; Coates 1982).

Social control of sex change

Another way of testing the size-advantage model is through an investigation of the dynamics of sex change within a species. So far, I have stressed the importance of the mating system in determining the advantage of a given sex and size. Within a mating system, it is often relative rather than absolute size that determines reproductive expectations. For example, when dominance depends on size, the probable mating success of a particular male is determined by the sizes of the other males in the local population. It would be most adaptive for individuals to be able to change from female to male when their expectations of successful reproduction as a male increase considerably. Thus the removal of a large, dominant male from a population should result in a change of sex in the next largest individual, but no change should be expected in the rest of the local population.

Such social control of sex change has been noted in several species of protogynous coral-reef fishes. Because haremic species exist in small, localized groups, they have proved to be exceptionally good candidates for studies of this kind. In the cleaner-wrasse *L. dimidiatus*, Robertson (1972) found that if the male is removed from the harem, the largest female rapidly changes sex and takes over the role of harem-master. Within a few hours she adopts male behaviors, including spawning with the females. Within ten days this new male is producing active sperm. By contrast, the other females in the harem remain unchanged.

Social control of sex change has also been found in other haremic species (Moyer and Nakazono 1978a; Hoffman 1980; Coates 1982), as well as in species that live in bigger groups with several large males present (Fishelson 1970; Warner et al. 1975; Shapiro 1979; Warner 1982; Ross et al. 1983). In all cases, it is always the largest remaining individuals that undergo sex change when the opportunity presents itself. Even when experimental groups consist entirely of small individuals, sex change can still be induced in the largest individuals present, in spite of the fact that they may be far smaller than the size at which sex change normally occurs (Hoffman 1980; Warner 1982; Ross et al. 1983).

The exact behavioral cues used to trigger sex change appear to differ among species. Ross and his co-workers have shown experimentally that the sex-change response in the Hawaiian wrasse *T. duperrey* depends solely on relative size and is independent of the sex and coloration of the other individuals in a group, whereas Shapiro and Lubbock (1980) have suggested that the local sex ratio is the critical factor in the bass *Anthias squamipinnis*. While it is still unclear how sex change is regulated in fishes that live in large groups, the mechanisms appear to operate with some precision. Shapiro (1980) found that the simultaneous removal of up to nine male *Anthias* from a group led to a change of sex in an equivalent number of females.

Social control of sex change occurs in protandrous fishes as well, and in a pattern consistent with the size-advantage model. A resident male anemonefish will change sex if the female is removed (Fricke and Fricke 1977; Moyer and Nakazono 1978b). One of the juveniles—who apparently are otherwise repressed from maturing—then becomes a functional male and the adult couple is reconstituted.

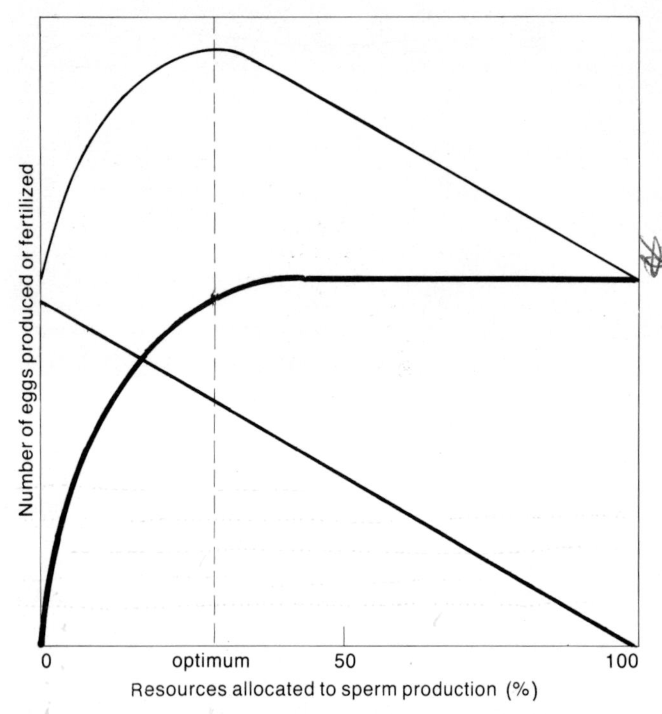

Figure 7. The number of eggs produced by an individual is normally directly related to the amount of energy devoted to their manufacture (*thin line*). In the case of sperm, however, a relatively low output can often produce maximum success (*thick line*) – that is, a small amount of sperm can fertilize all of a partner's eggs, and further investments are superfluous. Natural selection favors the individual with the highest overall reproductive rate combining male and female functions (*black line*), and in this case the optimum result is obtained by putting most of the energy into egg production. A simultaneous hermaphrodite following this strategy has a much higher reproductive success than an individual that is exclusively male or female. (After Fischer 1981.)

Simultaneous hermaphroditism

In one sense, the adaptive significance of simultaneous hermaphroditism is obvious: by putting most of their energy into egg production and producing just enough sperm to ensure fertilization, a hermaphroditic mating couple can achieve a much higher output of young than a male-female pair (Fig. 7; Leigh 1977; Fischer 1981). The problem, however, rests with the maintenance of simultaneous hermaphroditism in the face of an alternative male strategy. Consider an individual that fertilizes the eggs of a hermaphrodite, but does not reciprocate by producing eggs of its own. Instead, this individual uses the energy thus saved to find and fertilize other hermaphrodites. This strategy would spread rapidly in a purely hermaphroditic population, effectively forcing it to become gonochoristic. It would therefore appear that where simultaneous hermaphroditism is present, there should exist some means of preventing this kind of "cheating" (Leigh 1977; Fischer 1981).

Among the small coral-reef basses (Serranidae) that are known to be simultaneous hermaphrodites, two types of possible anticheating behavior have been observed. The hamlets, small basses common on Caribbean coral reefs (genus *Hypoplectrus*), appear to ensure that investments in eggs are kept nearly even between the members of a spawning pair by what Fischer (1980) has called "egg trading." In this behavior, a pair alternates sex roles over the course of mating (Fig. 8). Each time an individual functions as a female, it extrudes some, but not all, of its eggs. As a male, it fertilizes the eggs of its partner, who also parcels out eggs in several batches. Thus both individuals are forced to demonstrate their commitment to egg production, and neither has the chance for an unreciprocated fertilization of a large batch of eggs.

Figure 8. Simultaneous hermaphrodites among fishes include the hamlet (genus *Hypoplectrus*), a small bass that may alternate sexual roles as many as four times in the course of a single mating, by turns offering eggs to be fertilized and fertilizing its partner's eggs. Here the fish acting as the male curves its body around the relatively motionless female, cupping the upward-floating eggs as he fertilizes them. The strategy of parceling out eggs in a number of batches means that the egg contributions of the two partners are roughly equal, and reduces the rewards of "cheating" by fertilizing a large batch of eggs and then refusing to offer eggs for fertilization in return. (Photo by S. G. Hoffman.)

Another method of preventing desertion is to reduce the opportunities of your partner to find another mate. Some simultaneously hermaphroditic species of the genus *Serranus* delay their mating until late dusk, just before nightfall. These species do not engage in egg-trading, but presumably the onset of darkness means that time is quite limited before shelter must be taken for the night, and thus further mating is impossible (Pressley 1981).

Although these anticheating behaviors are fascinating in their own right, they give us little insight into the origin of the sexual pattern itself. Simultaneous hermaphroditism has been viewed as an adaptation to extremely low population density: if finding mates is difficult, it helps a great deal to be able to mate with whomever you meet (Tomlinson 1966; Ghiselin 1969). Thus many deep-sea fishes, sparsely distributed in their habitat, are simultaneous hermaphrodites (Mead et al. 1964). Perhaps the small serranids evolved from ancestors who lived at low densities, and developed their anticheating behaviors at a later stage when densities were higher. Alternatively, and perhaps more likely, the existence of late-dusk spawning behavior could have allowed the development of hermaphroditism. Unfortunately, we again run up against the problem of untestability and limited predictive power: several other coral-reef fishes mate late in the day, but they are not simultaneous hermaphrodites.

Broader patterns

The contrast between sex change and simultaneous hermaphroditism is intriguing: sex change, particularly protogyny, appears to be a specific adaptation to certain mating systems that happen to be common on coral reefs. These mating systems may be prevalent in coral habitats because clear water, relatively low mobility, and the absence of paternal care allow a greater degree of dominance by large males. Simultaneous hermaphroditism, on the other hand, is theoretically adaptive in a wider variety of circumstances, but is evolutionarily unstable unless male-type cheating can be prevented. While the wide dispersion of deep-sea fishes automatically works against cheating, I can see no reason why coral reefs are particularly good places for such prevention to come about.

Among hermaphroditic groups other than tropical marine fishes, our knowledge of behavior and ecology is generally insufficient to carry out a similar analysis of the relationship of sexual pattern to mating system. Broad surveys of vertebrates (Warner 1978) or organisms in general (Policansky 1982) have shown large-scale tendencies toward hermaphroditism in some groups, sporadic appearance of the phenomenon in others, and a total lack of sexual flexibility in still others. While major features such as the greater complexity of terrestrial reproduction may help to explain the lack of hermaphroditism in some large groups (Warner 1978), many others must await more thorough investigation of the life histories and behavior of the organisms in question.

On this point, Policansky (1982) has suggested that a major problem of sex-change theory is that among closely related species with similar life histories, some change sex and some do not. In light of this review, I am not yet ready to take such a dim view of the size-advantage model. Sexual expression in fishes is extraordinarily adaptable, and closely related species can have quite different sexual patterns that appear to be predictable from their different mating systems. For the fishes, at least, divergent sexual expressions may be no more surprising than differences in coloration. Detailed considerations of the mating systems and life histories of other sexually labile groups are needed to test the hypothesis further.

References

Allen, G. R., 1972. *The Anemonefishes, Their Classification and Biology.* Neptune City, NJ: T. F. H. Publications.

Charnov, E. L. 1982. *The Theory of Sex Allocation.* Princeton Univ. Press.

Coates, D. 1982. Some observations on the sexuality of humbug damselfish, *Dascyllus aruanus* (Pisces, Pomacentridae) in the field. *Z. für Tierpsychol.* 59:7–18.

Cole, K. S., 1982. Male reproductive behavior and spawning success in a temperate zone goby, *Coryphopterus nicholsi. Can. J. Zool.* 10: 2309–16.

Fischer, E. A. 1980. The relationship between mating system and simultaneous hermaphroditism in the coral reef fish *Hypoplectrus nigricans* (Seranidae). *Anim. Beh.* 28:620–33.

______. 1981. Sexual allocation in a simultaneously hermaphroditic coral reef fish. *Am. Nat.* 117:64–82.

Fishelson, L. 1970. Protogynous sex reversal in the fish *Anthias squamipinnis* (Teleostei, Anthiidae) regulated by presence or absence of male fish. *Nature* 227:90–91.

Fricke, H. W., and S. Fricke. 1977. Monogamy and sex change by aggressive dominance in coral reef fish. *Nature* 266:830–32.

Fricke, H. W., and S. Holzberg. 1974. Social units and hermaphroditism in a pomacentrid fish. *Naturwissensch.* 61:367–68.

Ghiselin, M. T. 1969. The evolution of hermaphroditism among animals. *Quart. Rev. Bio.* 44:189–208.

______. 1974. *The Economy of Nature and the Evolution of Sex.* Univ. of California Press.

Goodman, D. 1982. Optimal life histories, optimal notation, and the value of reproductive value. *Am. Nat.* 119:803–23.

Hoffman, S. G. 1980. Sex-related social, mating, and foraging behavior in some sequentially hermaphroditic reef fishes. Ph.D. diss., Univ. of California, Santa Barbara.

______. 1983. Sex-related foraging behavior in sequentially hermaphroditic hogfishes (*Bodianus* spp.). *Ecology* 64:798–808.

Jones, G. P. 1980. Contribution to the biology of the redbanded perch *Ellerkeldia huntii* (Hector), with a discussion on hermaphroditism. *J. Fish Biol.* 17:197–207.

Larsson, H. O. 1976. Field observations of some labrid fishes (Pisces: Labridae). In *Underwater 75*, vol. 1, ed. John Adolfson, pp. 211–20. Stockholm: SMR.

Lassig, B. R. 1977. Socioecological strategies adapted by obligate coral-dwelling fishes. *Proc. 3rd Int. Symp. Coral Reefs* 1: 565–70.

Leigh, E. G., Jr. 1977. How does selection reconcile individual advantage with the good of the group? PNAS 74:4542–46.

Leigh, E. G., Jr., E. L. Charnov, and R. R. Warner. 1976. Sex ratio, sex change, and natural selection. PNAS 73:3656–60.

Mead, G. W., E. Bertelson, and D. M. Cohen, 1964. Reproduction among deep-sea fishes. *Deep Sea Res.* 11:569–96.

Moyer, J. T., and A. Nakazono. 1978a. Population structure, reproductive behavior and protogynous hermaphroditism in the angelfish *Centropyge interruptus* at Miyake-jima, Japan. *Japan. J. Ichthyol.* 25:25–39.

______. 1978b. Protandrous hermaphroditism in six species of the anemonefish genus *Amphiprion* in Japan. *Japan. J. Ichthyol.* 25: 101–6.

Pressley, P. H. 1981. Pair formation and joint territoriality in a simultaneous hermaphrodite: The coral reef fish *Serranus tigrinus. Z. für Tierpsychol.* 56:33–46.

Policansky, D. 1982. Sex change in plants and animals. *Ann. Rev. Ecol. Syst.* 13:471–95.

Robertson, D. R. 1972. Social control of sex reversal in a coral reef fish. *Science* 1977:1007–9.

Robertson, D. R., R. Reinboth, and R. W. Bruce. 1982. Gonochorism, protogynous sex-change, and spawning in three sparasomatinine parrotfishes from the western Indian Ocean. *Bull. Mar. Sci.* 32: 868–79.

Robertson, D. R., and R. R. Warner. 1978. Sexual patterns in the labroid fishes of the western Caribbean, II: The parrotfishes (Scaridae). *Smithsonian Contributions to Zoology* 255:1–26.

Ross, R. M., G. S. Losey, and M. Diamond. 1983. Sex change in a coral-reef fish: Dependence of stimulation and inhibition on relative size. *Science* 221:574–75.

Shapiro, D. Y. 1979. Social behavior, group structure, and the control of sex reversal in hermaphroditic fish. *Adv. Study Beh.* 10:43–102.

______. 1980. Serial female sex changes after simultaneous removal of males from social groups of a coral reef fish. *Science* 209:1136–37.

Shapiro, D. Y., and R. Lubbock. 1980. Group sex ratio and sex reversal. *J. Theor. Biol.* 82:411–26.

Shen, S.-C., R-P. Lin, and F. C-C. Liu. 1979. Redescription of a protandrous hermaphroditic moray eel (*Rhinomuraena quaesita* Garman). *Bull. Instit. Zool. Acad. Sinica* 18(2):79–87.

Smith, C. L. 1975. The evolution of hermaphroditism in fishes. In *Intersexuality in the Animal Kingdom*, ed. R. Reinboth, pp. 295–310. Springer-Verlag.

Springer, V. G., C. L. Smith, and T. H. Fraser. 1977. *Anisochromis straussi*, new species of protogynous hermaphroditic fish, and synonomy of Anisochromidae, Pseudoplesiopidae, and Pseudochromidae. *Smithsonian Contributions to Zoology* 252:1–15.

Swarz, A. L. 1980. Almost all *Dascyllus reticulatus* are girls! *Bull. Mar. Sci.* 30:328.

Thresher, R. E., and J. T. Moyer. 1983. Male success, courtship complexity, and patterns of sexual selection in three congeneric species of sexually monochromatic and dichromatic damselfishes (Pisces: Pomacentridae). *Anim. Beh.* 31:113–27.

Tomlinson, N. 1966. The advantages of hermaphroditism and parthenogenisis. *J. Theoret. Biol.* 11:54–58.

Trivers, R. L. 1972. Parental investment and sexual selection. In *Sexual Selection and the Descent of Man, 1871–1971*, ed. B. Campbell, pp. 136–79. Chicago: Aldine Publishing Co.

Warner, R. R. 1975. The adaptive significance of sequential hermaphroditism in animals. *Am. Nat.* 109:61–82.

______. 1978. The evolution of hermaphroditism and unisexuality in aquatic and terrestrial vertebrates. In *Contrasts in Behavior*, ed. E. S. Reese and F. J. Lighter, pp. 77–101. Wiley.

______. 1982. Mating systems, sex change, and sexual demography in the rainbow wrasse, *Thalassoma lucasanum. Copeia* 1982:653–61.

Warner, R. R., and S. G. Hoffman, 1980a. Population density and the economics of territorial defense in a coral reef fish. *Ecology* 61: 772–80.

______. 1980b. Local population size as a determinant of a mating system and sexual composition in two tropical reef fishes (*Thalassoma* spp.). *Evolution* 34:508–18.

Warner, R. R., and D. R. Robertson. 1978. Sexual patterns in the labroid fishes of the western Caribbean, I: The wrasses (Labridae). *Smithsonian Contributions to Zoology* 254:1–27.

Warner, R. R., D. R. Robertson, and E. G. Leigh, Jr. 1975. Sex change and sexual selection. *Science* 190:633–38.

Williams, G. C. 1966. *Adaptation and Natural Selection.* Princeton Univ. Press.

Young, P. C., and R. B. Martin. 1982. Evidence for protogynous hermaphroditism in some lethrinid fishes. *J. Fish Biol.* 21:475–84.

Avian Siblicide

Killing a brother or a sister may be a common adaptive strategy among nestling birds, benefiting both the surviving offspring and the parents

Douglas W. Mock, Hugh Drummond and Christopher H. Stinson

Occasionally, the pen of natural selection writes a murder mystery onto the pages of evolution. But unlike a typical Agatha Christie novel, this story reveals the identity of the murderer in the first scene. The mystery lies not in "whodunit," but in why.

The case at hand involves the murder of nestling birds by their older siblings. Observers in the field have frequently noted brutal assaults by elder nestmates on their siblings, and the subsequent deaths of the younger birds. The method of execution varies among different species, ranging from a simple push out of the nest to a daily barrage of pecks to the head of the younger, smaller chick. Such killings present a challenge to the student of evolutionary biology: Does siblicide promote the fitness of the individuals that practice it, or is such behavior pathological? In other words, are there certain environmental conditions under which killing a close relative is an adaptive behavior? Moreover, are there other behaviors or biological features common to siblicidal birds that distinguish them from nonsiblicidal species?

Avian siblicide holds a special interest for several reasons. First, because nestling birds are relatively easy to observe, a rich descriptive literature exists based on field studies of many species. Second, because birds tend to be monogamous, siblicide is likely to involve full siblings. (Although recent DNA studies suggest that birds may not be as monogamous as previously thought, most nestmates are still likely to be full siblings.) Third, young birds require a large amount of food during their first few weeks of development, and this results in high levels of competition among nestlings. The competitive squeeze is exacerbated for most species because the parents act as a bottleneck through which all resources arrive. Fourth, some avian parents may not be expending their maximum possible effort toward their current brood's survival (Drent and Daan 1980, Nur 1984, Houston and Davies 1984, Gustafsson and Sutherland 1988, Mock and Lamey in press). Parental restraint may be especially common in long-lived species, in which a given season's reproductive output makes only a modest contribution to the parents' lifetime success (Williams 1966).

Siblicide—or juvenile mortality resulting from the overt aggression of siblings—is not unique to birds. It is also observed, for example, among certain insects and amphibians; in those groups, however, the behavioral pattern is rather different. Most siblicidal insects and amphibians immediately consume their victims as food, whereas in birds (and mammals) siblicide rarely leads to cannibalism. For example, tadpoles of the spadefoot toad acquire massive dentition (the so-called "cannibal morph") with which they consume their broodmates (Bragg 1954), and fig wasps use large, sharp mandibles to kill and devour their brothers (Hamilton 1979). In contrast, among pronghorn antelopes, one of the embryos develops a necrotic tip on its tail with which it skewers the embryo behind it (O'Gara 1969), and piglet littermates use deciduous eyeteeth to battle for the sow's most productive teats (Fraser 1990). Among birds and mammals it seems that the goal is to secure a greater share of critical parental care.

Although biologists have known of avian siblicide for many years, only recently have quantitative field studies been conducted. The current wave of such work is due largely to the realization that siblicide occurs routinely in some species that breed in dense colonies; such populations provide the large sample sizes needed for formal testing of hypotheses.

Models of Nestling Aggression

Our examination of siblicidal aggression focuses on five species of birds. Two of these, the black eagle (*Aquila verreauxi*) and the osprey (*Pandion haliaetus*), are raptors that belong to the family Accipitridae. A third species, the blue-footed booby (*Sula nebouxii*) is a seabird belonging to the family Sulidae. We also present studies of the great egret (*Casmerodius albus*) and the cattle egret (*Bubulcus ibis*), both of which belong to the family Ardeidae. Each of these species exhibits a distinct behavioral pattern; the range of variation is important to an understanding of siblicide.

The black eagle is one of the first birds in which siblicide was described. This species, also called Verreaux's eagle, lives in the mountainous terrain of southern and northeastern Africa, as well as the western parts of the Middle East. Black eagles generally build their nests on cliff ledges and lay two eggs between April and June. The eaglets hatch about three days apart, and so the older chick is significantly larger than the younger one. The black eagle is of particular interest for the study of

Figure 1 (overleaf). Two cattle egrets peer down at their recently evicted younger sibling. For several days before the eviction, the elder siblings pecked at the head of their smaller nestmate. Here the younger bird holds its bald and bloodied head out of reach. Soon after the photograph was made, the bird was driven to the ground and perished. (Photograph by the authors.)

Douglas W. Mock is associate professor of zoology at the University of Oklahoma. He was educated at Cornell University and the University of Minnesota, where he received his Ph. D. in ecology and behavioral biology in 1976. Address: Department of Zoology, University of Oklahoma, Norman, OK 73019. Hugh Drummond is a researcher in animal behavior at the Universidad Nacional Autònoma de México. He was educated at Bristol University, the University of Leeds and the University of Tennessee, where he received his Ph. D. in psychology in 1980. Address: Centro de Ecologia, Universidad Nacional Autònoma de México, AP 70-275, 04510 México, D.F. Christopher H. Stinson was educated at Swarthmore College, the College of William and Mary and the University of Washington, where he received his Ph. D. in 1982. Address: 4005 NE 60th Street, Seattle, WA 98115.

siblicide because the elder eaglet launches a relentless attack upon its sibling from the moment the younger eaglet hatches. In one well-documented case, the senior eaglet pecked its sibling 1,569 times during the three-day lifespan of the younger nestling (Gargett 1978).

Among ospreys, sibling aggression is neither so severe nor so persistent as it is among black eagles. Ospreys are widely distributed throughout the world, including the coastal and lacustrine regions of North America. The nests are generally built high in trees or on other structures near water. A brood typically consists of three chicks, which usually live in relative harmony. Nevertheless, combative exchanges between siblings do occur in this species; comparisons between the fighting and the pacifist populations offer insights into the significance of aggression.

The blue-footed booby lives exclusively on oceanic islands along the Pacific coast from Baja California to the northern coast of Peru. Blue-footed boobies are relatively large, ground-nesting birds that typically form dense colonies near a shoreline. Two or three chicks hatch about four days apart, and this results in a considerable size disparity between the siblings. As in many other siblicidal species, the size disparity predicts the direction of the aggression between siblings.

Young nestmates also differ in size in the two egret species we have studied. The larger of these, the great egret, is distributed throughout the middle latitudes of the world, and also throughout most of the Southern Hemisphere. Great egrets make their nests in trees or reed beds in colonies located near shallow water. The cattle egret also nests in colonies, but not necessarily close to water. Cattle egrets live in the middle latitudes of Asia, Africa and the Americas. As their name suggests, they are almost always found in the company of grazing cattle or other large mammals, riding on their backs and feeding on grasshoppers stirred up by the movement of the animals. Despite their differences in habitat, great egrets and cattle egrets have a number of behaviors in common. Typically, three or four egret nestlings hatch at one- to two-day intervals, and fighting starts almost as

Peter Steyn

Figure 2. Aggression in black eagle nestlings almost always results in the death of the younger sibling. Here a six-day-old black eagle chick tears at a wound it has opened on the back of its day-old sibling.

soon as the second sibling has hatched. Aggressive attacks lead to a "pecking order" that translates into feeding advantages for the elder siblings (Fujioka 1985a, 1985b; Mock 1985; Ploger and Mock 1986). In about a third of the nests, the attacks culminate in siblicide through socially enforced starvation and injury or eviction from the nest.

Obligate and Facultative Siblicide

It is useful to distinguish those species in which one chick almost always kills its sibling from those in which the incidence of siblicide varies with environmental circumstances. Species that practice obligate siblicide typically lay two eggs, and it is usually the older, more powerful chick that kills its nestmate. The black eagle is a good example of an obligate siblicide species. In 200 records from black eagle nests in which both chicks hatched, only one case exists where two chicks fledged (Simmons 1988). Similar patterns of obligate siblicide have been reported for other species that lay two eggs, including certain boobies, pelicans and other eagles (Kepler 1969; Woodward 1972; Stinson 1979; Edwards and Collopy 1983; Cash and Evans 1986; Evans and McMahon 1987; Drummond 1987; Simmons 1988; Anderson 1989, 1990).

A far greater number of birds are facultatively siblicidal. Fighting is frequent among siblings in these species, but it does not always lead to the death of the younger nestling. There are various patterns of facultative siblicide. For example, in species such as the osprey, aggression is entirely absent in some populations, and yet present in others (Stinson 1977; Poole 1979, 1982; Jamieson et al. 1983). In other species aggression occurs at all nests but differs in form and effect. In the case of the blue-footed booby a chick may hit its sibling only a few times per day for several weeks, and then rapidly escalate to a lethal rate of attack (Drummond, Gonzalez and Osorno 1986). Egret broods tend to have frequent sibling fights—there are usually several multiple-blow exchanges per day—but the birds do not always kill each other (Mock 1985, Ploger and Mock 1986).

Figure 3. Blue-footed booby nestlings maintain dominance over their younger siblings through a combination of aggression and threats *(upper photograph)*. The assaults do not escalate to the point of eviction unless the food supply is inadequate. An evicted chick has little chance of survival in the face of attacks from neighboring adults *(lower photograph)*. (Photographs courtesy of the authors.)

Traits of Siblicidal Species

Five characteristics are common to virtually all siblicidal birds: resource competition, the provision of food to the nestlings in small units, weaponry, spatial confinement and competitive disparities between siblings. The first four traits are considered essential preconditions for the evolution of sibling aggression; the study of their occurrence may shed some light on the origin of siblicidal behavior. The fifth trait—competitive disparities among nestmates resulting from differences in size and age—is also ubiquitous and important, but it is probably not essential for the evolution of siblicide. In fact, competitive disparities may be a consequence rather than a cause of siblicidal behavior; having one bird appreciably stronger than the other reduces the cost of fighting, since asymmetrical fights tend to be brief and it is less likely that both siblings will be hurt during combat (Hahn 1981, Fujioka 1985b, Mock and Ploger 1987).

Of the five traits common to siblicidal species, the competition for resources is probably the most fundamental. Among birds, the competition is primarily for food. Experiments have shown that the provision of additional food often diminishes nestling mortality (Mock, Lamey and Ploger 1987a; Magrath 1989). But "brood reduction"—the general term for nestling deaths brought about by the competition for food—does not necessarily entail direct aggression. Nestlings die even in nonsiblicidal species, but the usual cause of death is starvation; weaker chicks continually lose to their more robust siblings in the scramble for food. What distinguishes siblicidal species is that the competition for food is intensified to the point of overt attack. (In nonavian species, the competition may be over reproductive opportunity. For example, male fig wasps and female "proto-queen" honeybees kill all of their same-sex siblings immediately after hatching in order to gain the breeding unit's single mating slot. In certain species of mammalian social carnivores, one female dominates her sisters, rendering them effectively sterile.)

In avian species, if the source of food cannot be defended, aggression does not appear to be advantageous. The food must come in morsels small enough to be monopolized through combat. In all known species of siblicidal birds, food is presented to the young in small units through direct transfer from parent to chick (Mock 1985). For example, very young raptor chicks take small morsels held in the mother's bill, whereas boobies either reach inside the parent's throat or use their own bills to form a tube with the parent's bill, and egrets scissor the parent's bill crosswise so as to intercept the food as it emerges.

Figure 4. Great egret chicks fight frequently, regardless of food levels. Siblicide occurs in about a third of the nests, through socially enforced starvation and injury or as a result of eviction. As in other species of siblicidal birds, the parents do not interfere with the fights and evictions among their offspring. (Photograph courtesy of the authors.)

The link between the size of the food and sibling aggression lies in the relation between intimidation and monopolization. From the chick's perspective, food descends from the inaccessible heights of its parent's bill, becoming potentially available only at the moment it arrives within reach. A sibling's share depends primarily on its position relative to its competitors; that position can be enhanced through physical aggression or threat (much as the use of elbows can enhance a basketball player's chance of catching a rebound). For food items that can be taken directly from the parent's bill, the sibling's share should rise in relation to the degree of intimidation achieved. Thus, small food items create incremental rewards for aggression.

A diet of large, cumbersome items that cannot be intercepted by the chicks generally does not give rise to sibling aggression. Although killing all of its siblings would enable a chick to monopolize large items, the rewards for mild forms of aggression are sharply reduced. Thus, when food units are large, sublethal fighting may be less effective than simply eating as fast as possible. The great blue heron (*Ardea herodias*) is developmentally flexible with respect to prey size and aggression. These birds express siblicidal aggression only when the food is small enough to be taken directly from the parent. Great blue heron nestlings in Quebec fight vigorously over small units of food that can be intercepted by aggressive actions (Mock et al. 1987). In contrast, nestlings of the same species in Texas typically receive very large morsels and seldom fight (Mock 1985). Moreover, if the normally nonaggressive Texas herons are raised by great egrets, which feed their young small morsels of food, the herons quickly adopt the direct feeding method and exhibit siblicidal aggression (Mock 1984).

Figure 5. Five characteristics are common to virtually all siblicidal birds (*from top left to bottom right*): competition for food, provision of food to the nestlings in small units, weaponry, competitive disparities between siblings and spatial confinement. Four of the traits are considered essential preconditions for the evolution of sibling aggression, whereas competitive disparities between siblings may be a consequence rather than a cause of siblicidal behavior.

A shortage of food, and the ability to defend each unit of food, set the stage for siblicide, but the nestling must also possess some means of carrying out a lethal attack. In this regard it is notable that most siblicidal birds are predatory and have hooked or pointed beaks capable of inflicting serious damage on nestmates. Even so, where obvious weaponry is lacking, other means of siblicide may be possible—such as simply rolling eggs out of the nest cup.

Weaponry aside, effective aggression among nestling birds is also correlated with small nests or nesting territories. Chicks assaulted by their senior siblings do not necessarily have the option of escaping the nest. In tree-nesting species, a chick that leaves its nest risks falling from a narrow limb. Dominant chicks of the cliff-nesting kittiwake (*Rissa tridactyla*) simply drive their siblings off the nest ledge (Braun and Hunt 1983). In the dense colonies of the blue-footed booby, young chicks oppressed at home by their siblings may face even greater persecution from adult neighbors if they leave their natal territory. In a tunnel-nesting bee-eater species, the nestlings have a special hook on their beaks with which they defend the opening to the nest (and the source of the food) against their younger siblings (Bryant and Tatner 1990). In each of these cases, the lack of suitable space (either for escape or as an alternative route to food) contributes directly to the victim's death.

The competitive disparities commonly observed among nestlings of siblicidal birds may hold an important clue to the evolution of sibling aggression. Parents usually create such disparities by starting to incubate one egg at some point prior to laying the final egg in the clutch. Because eggs are produced at intervals of one or more days, the chick hatched from the egg laid first has an important head start. (Parents may also initiate competitive asymmetries by laying different-sized eggs within a clutch or by feeding certain young preferentially, but these mechanisms are less common than asynchronous hatching.)

The Oxford ornithologist David Lack proposed that asynchronous hatching is a behavioral adaptation that allows for a secondary adjustment in brood size to match resource levels

Richard Estes (Photo Researchers, Inc.)

Michael Dick (Animals Animals)

Charles J. Lamphiear (DRK Photo)

Figure 6. Food is presented directly to the chick in small units in all known species of siblicidal birds. This direct method of feeding means that a chick may increase its share of food by physically intimidating, and not just by killing, its competing siblings. The young black eagle (*top*) is fed a piece of hyrax meat by the direct-transfer method, even though the bird is well into the fledgling stage. In the blue-footed booby (*lower left*), the parent transfers small pieces of fish from its mouth directly into the mouth of a chick. An osprey chick (*lower right*) receives a piece of meat from its parent while its sibling waits. Osprey chicks take turns feeding, and will fight only if food becomes scarce.

(Lack 1954). Parents must commit themselves to a fixed number of eggs early in the nesting cycle, before the season's bounty or shortcomings can be assessed. Thus, it is often advantageous for parents to produce an additional egg or two, in case later conditions are beneficent, while reserving the small-brood option by making the "bonus" offspring competitively inferior, in case the season's resources are poor. The production of an inferior sibling may be advantageous, since the senior sibling can then eliminate its younger nestmate with greater ease. In fact, experimentally synchronizing the hatchings of cattle egrets results in an increase in fighting, which reduces the reproductive efficiency of the parents (Fujioka 1985b, Mock and Ploger 1987).

Siblicide as an Adaptation

To understand siblicide, we must understand how the killing of a close relative can be favored by natural selection. At first this may seem a simple matter. Eliminating a competitor improves one's own chance of survival, and thereby increases the likelihood that genes promoting such behavior will be represented in the next generation. According to this simple analysis, natural selection should always reward the most selfish act, and siblicide is arguably the epitome of selfishness.

The trouble with this formulation is that it implies that all organisms should be as selfish as possible, which is contrary to observation. (Siblicide is fairly common, but certainly not universal.) A more sophisticated analysis was provided in the 1960s by the British theoretical biologist William D. Hamilton. In Hamilton's view, the fitness of a gene is more than its contribution to the reproduction of the individual. A gene's fitness also depends on the way it influences the reproductive prospects of close genetic relatives.

This expanded definition of evolutionary success, called inclusive fitness, is a property of individual organisms. An organism's inclusive fitness is a measure of its own reproductive success plus the incremental or decremental influences it has on the reproductive success of its kin, multiplied by the degree of relatedness to those kin (Hamilton 1964). Hamilton's theory is generally invoked to explain apparently altruistic behavior, but the theory also specifies the evolutionary limits of selfishness.

An example will help to clarify Hamilton's idea. Suppose a particular gene predisposes its bearer, *X*, to help a sibling. Since the laws of Mendelian inheritance state that *X* and its sibling share, on average, half of their genes, *X*'s sibling has a one-half probability of carrying the gene. From the gene's point of view, it is useful for *X* to promote the reproductive success of a sibling because such an action contributes to the gene's numerical increase. Therefore, helping a sibling should be of selective advantage. It is in this light that we must understand and explain siblicide. Since selection favors genes that promote their own numerical increase, what advantage might there be in destroying a sibling—an organism with a high probability of carrying one's own genes? The solution to the problem lies in the role played by the "marginal" offspring, which may be the victim of siblicide.

In all siblicidal species studied to date there is a striking tendency for the victim to be the youngest member of the brood (Mock and Parker 1986). The youngest sibling is marginal in the sense that its reproductive value can be assessed in terms of what it adds to or subtracts from the success of other family members. Specifically, the marginal individual can embody two kinds of reproductive value. First, if the marginal individual survives in addition to all its siblings, it represents an extra unit of parental success, or extra reproductive value. Such an event is most likely during an especially favorable season, when the needs of the entire brood can be satisfied. Alternatively, the marginal offspring may serve as a replacement for an elder sibling that dies prematurely. In such instances the marginal individual represents a form of insurance against the loss of a senior sibling. The magnitude of this insurance value depends on the probability that the senior sibling will die.

Among species that practice obligate siblicide, the marginal individual offers no extra reproductive value; marginal chicks serve only as insurance against the early loss or infirmity of the senior chick. In these species, if the senior chick is alive but weakened and in-

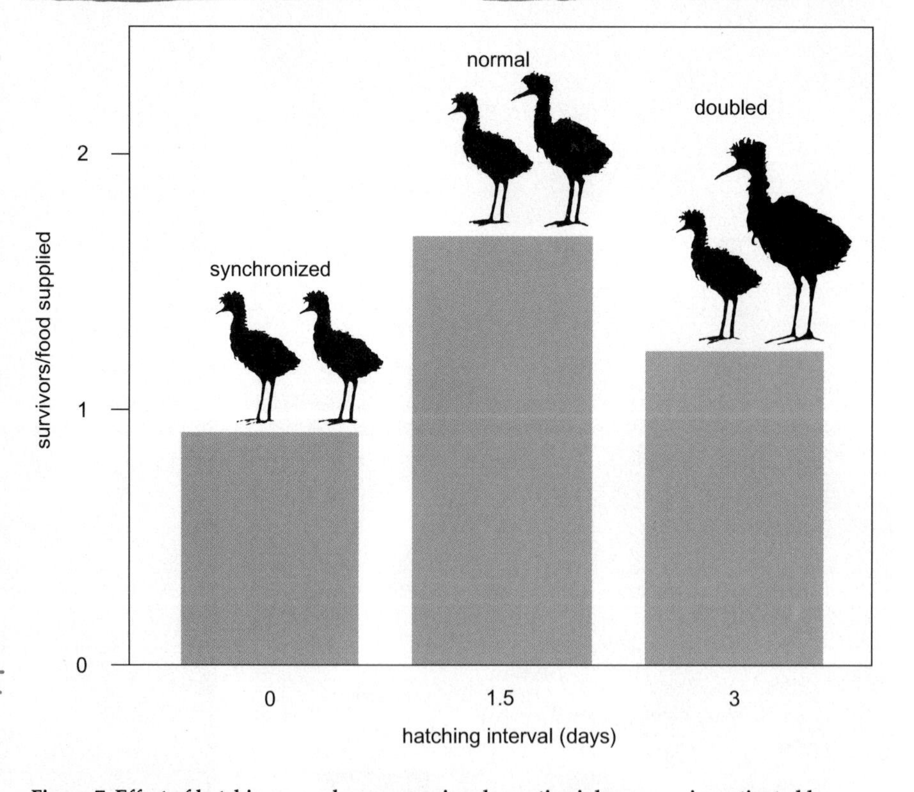

Figure 7. Effect of hatching asynchrony on avian domestic violence was investigated by switching eggs in the nests of cattle egrets. Reproductive efficiency is maximal when chicks hatch at an interval of one and one-half days (as they do under normal conditions). Synchronized hatching (an interval of zero days) increases the amount of fighting between chicks, which results in greater chick mortality. The normal one-and-a-half day interval reduces the amount of fighting since the older chick is able to intimidate the younger chick. Doubling the asynchrony, so that the eggs hatch three days apart, greatly reduces the amount of fighting but exaggerates the competitive asymmetries, so that the youngest nestmates receive little food. The experiments were performed by Douglas Mock and Bonnie Ploger at the University of Oklahoma.

capable of killing the younger chick, the latter may be able to reverse the dominance and kill the senior chick. Such scenarios appear to be played out regularly: In a sample of 22 black eagle nests in which both chicks hatched, the junior chick alone fledged in five of the nests, and the senior chick alone fledged in the remaining 17 cases (Gargett 1977). Similarly, in a sample of 59 nests of the masked booby, the junior chick was the sole fledgling in 13 nests, and the senior chick the sole fledgling in the other 46 nests (Kepler 1969). In both of these species, the junior chick's chance of being the sole survivor—its insurance reproductive value to the parents—is about 22 percent. Removing the "insurance" eggs results in a reduction in the mean number of fledglings per nest (Cash and Evans 1986). Consequently, the insurance value of the marginal offspring should improve parental fitness if the cost of producing that offspring is reasonable. (In fact, the cost of producing one additional egg seems fairly modest: approximately 2.5 percent of the body weight of the black eagle female.)

Among species that practice facultative siblicide, the marginal offspring may be a source of insurance but may also provide extra reproductive value. The relative contribution of the marginal offspring to the reproductive success of the parents appears to vary considerably within and between species. For example, among great egrets the proportion of nests in which all nestlings survive—the extra reproductive value—varies from 15 to 23 percent, whereas the proportion of the nests in which at least one senior sibling dies and the youngest sibling lives—the insurance reproductive value—may vary from 0 to 48 percent (Mock and Parker 1986). The blue-footed booby shows great variation in the extra reproductive value provided by the marginal offspring (5 to 67 percent), whereas the insurance reproductive value is generally quite low (5 to 6 percent). In both of these species the magnitudes of the total reproductive values depend on the size of the brood. In general, the marginal offspring provides a greater total reproductive value to the parents when the brood size is smaller.

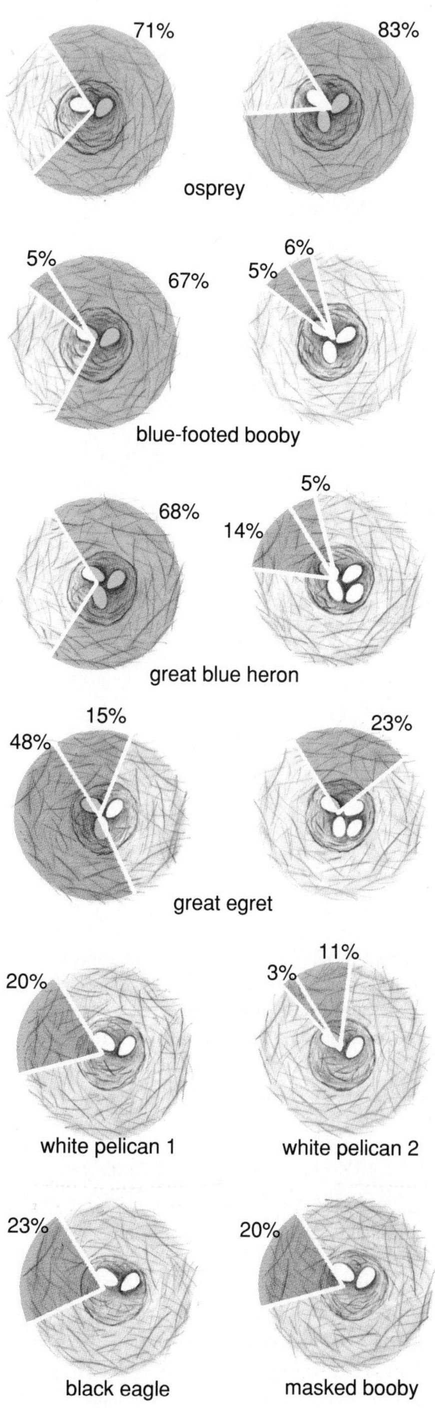

Figure 8. Reproductive value of the youngest member of a brood (the usual victim of siblicide) varies across species and brood size. The reproductive value is represented as the proportion of nests (in broods of two, three or four eggs) in which the youngest chick survives. If the youngest chick survives in addition to its elder siblings, it contributes "extra reproductive value" (*blue sections*); when the youngest chick survives as a replacement for an elder sibling that dies early, the junior bird provides "insurance reproductive value" (*red sections*). Among birds that almost always commit siblicide, such as the black eagle and the masked booby, the youngest chick's reproductive value is entirely due to its role as an "insurance policy." In species where siblicide is more occasional, such as the great egret, the youngest chick may provide either form of reproductive value. These estimates of reproductive value are maxima, since they represent survival only part way through the prefledgling period and not recruitment into the breeding population. These data are derived from studies by: Gargett 1977 (black eagle), Stinson 1977 (osprey), Cash and Evans 1986 (white pelican 1), Evans and McMahon 1987 (white pelican 2), Mock and Parker 1986 (great egret and great blue heron), and Drummond (unpublished data on the blue-footed booby). The data on the masked booby are combined from studies by Kepler 1969 and Anderson 1989.

The Timing of the Deed

A senior sibling should kill its younger sibling as soon as two conditions are met: (1) the senior sibling's own viability seems secure; and (2) the resources are inadequate for the survival of both siblings. Killing the junior sibling before these conditions are met would waste the potential fitness the junior sibling could offer in the form of extra reproductive value or insurance reproductive value. Delaying much beyond the point at which the conditions are met also has a cost. First, the food eaten by the victim is a loss of resources, and, second, the cost of execution may increase as the victim gains strength and is more likely to defeat the senior sibling.

In obligate siblicide species, the average food supply is presumably inadequate for supporting two chicks at reasonable levels of parental effort, and as a result the second chick is dispatched as soon as possible after it hatches. For example, the mean longevity of the victim in the case of the masked booby is 3.3 days (Kepler 1969), and only 1.75 days for brown boobies (Cohen et al. in preparation).

Among facultative siblicide species, the mean longevity of the victim is usually greater; in the blue-footed booby it is 18 days (Drummond, Gonzalez and Osorno 1986). Although the senior blue-footed booby chick may peck at the head or wrench the skin of its nestmate, the younger sibling is seldom killed by these direct physical assaults. Instead, death typically results from starvation or violent pecking by adult neighbors when the junior chick is routed from the home nest (Drummond and Garcia Chavelas 1989).

The Causes of Siblicide

The evolutionary difference between the obligate and the facultative forms of siblicide may be a function of the risk that a junior chick poses to the welfare of its senior sibling. That risk can be defined both in terms of resource consumption and in terms of the potential for bodily harm. If the resources are adequate only for the survival of a single chick, or if a young chick poses a significant physical threat to an older chick, then the senior sibling might be expected to destroy the younger one. On the other hand, if there is enough food for both chicks, and if the younger sibling can be subjugated so that it does not present a threat, then the survival of the younger sibling is beneficial because it increases the inclusive fitness of the senior sibling. In such circumstances, natural selection should favor a measure of clemency on the part of the senior sibling. Accordingly, we would expect obligate siblicide to evolve in circumstances in which resources are routinely limited and siblings tend to pose a physical threat to one another. In contrast, facultative siblicide should arise in circumstances in which resources are not always limited.

The analysis offered above concerns the inheritance of a long-term predisposition to siblicide. Recent studies suggest that food shortages also act as an immediate stimulus to, or proximal cause of, sibling fighting. A link between the food supply and siblicide was suggested by the finding that brood reductions in the blue-footed booby tend to occur soon after the weight of the senior chick drops about 20 percent below the weight expected at its current age in a good year (Drummond, Gonzalez and Osorno 1986). The relationship between food deprivation and aggression was confirmed by experiments in which the senior chick's neck was taped to prevent it from swallowing food. The experimentally deprived senior chicks pecked their nestmates about three to four times more frequently with the tape in place than without the tape, and they subsequently received a greater share of the food (Drummond and Garcia Chavelas 1989).

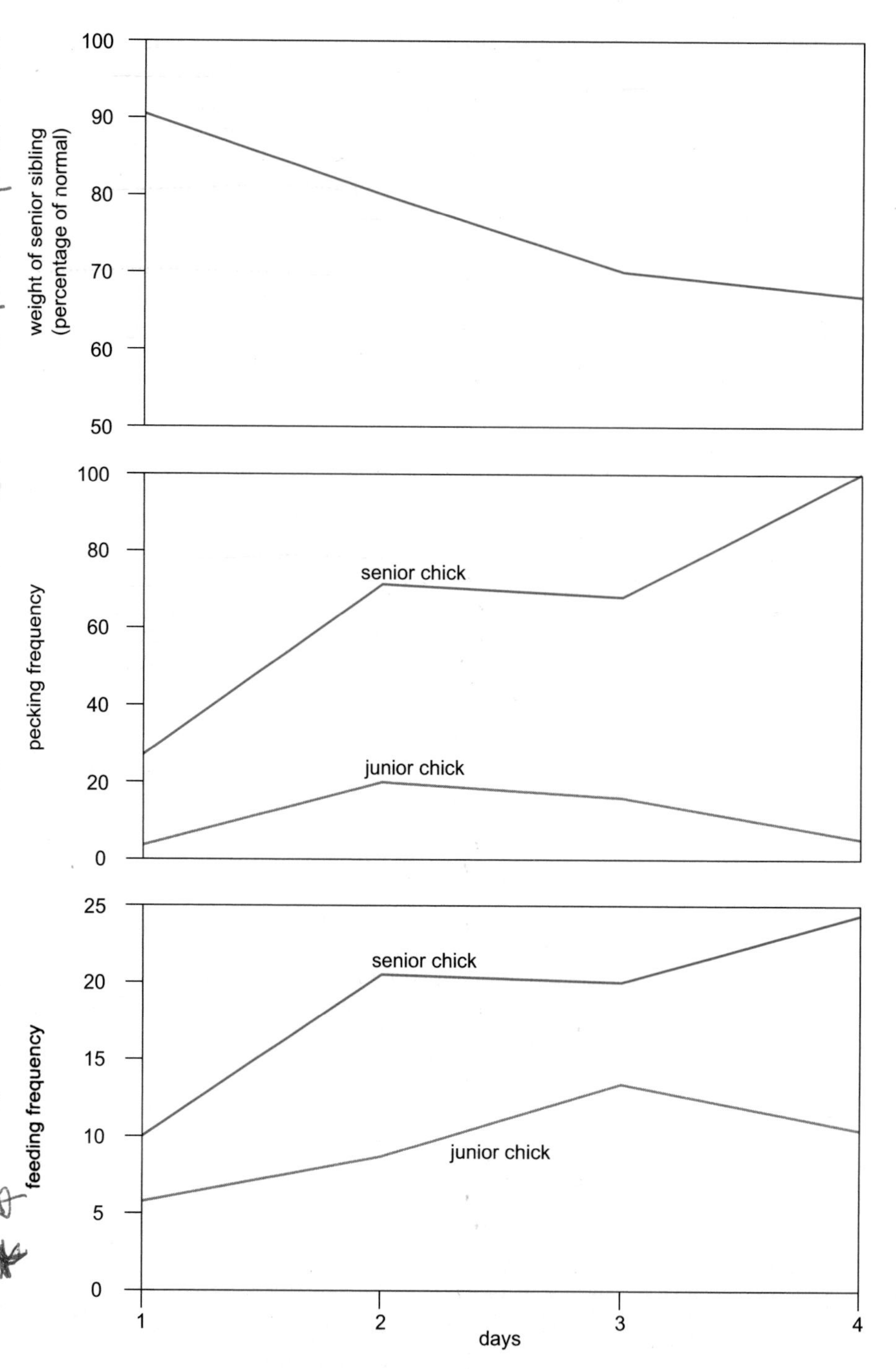

Figure 9. Effect of food deprivation on aggression and food distribution in blue-footed booby nestlings was investigated by taping the senior chick's neck to prevent it from swallowing food. As the weight of the senior chick drops more than 20 percent below normal (*top*), the rate at which it pecks its younger sibling increases more than three-fold (*middle*). The escalating aggression of the elder chick brings it a greater share of food (*bottom*). The experiments were performed by Hugh Drummond and Cecilia Garcia Chavelas at the Universidad Nacional Autònoma de México.

In older booby broods, the increase in the amount of aggressive pecking was delayed by about a day after the chick's neck was taped, suggesting that aggression is controlled by a factor that changes progressively over time, such as hunger or growth status. In fact, the increased pecking rate coincided with a 20 percent weight loss by the senior chick. When the tape was removed, the aggressive pecking rate returned toward the baseline level. These results suggest that nestling aggression among certain facultative species is a reversible response that is sensitive to the weight level of the senior chick.

There is also suggestive evidence in other species that practice facultative siblicide that the amount of food available to the nestlings may affect sibling aggression. For example, junior kittiwake siblings are lost from the nest at higher rates following prolonged periods of bad weather, when parental foraging is reduced (Braun and Hunt 1983). Among osprey populations in which there is a high rate of prey delivery to the offspring, the nestlings are amicable and may even take turns feeding (Stinson 1977). In populations where the food delivery rate is lower, the older nestlings frequently attack their younger siblings, although they do not kill them outright (Henny 1988, Poole 1982).

In contrast, the relative abundance of food does not appear to affect the level of aggression in obligate siblicide species. Black eagle nestlings kill their siblings even in the midst of several kilograms of prey, and even while the mother eagle is offering food to the senior sibling. There does not appear to be the same direct relationship between the immediate availability of food and the level of sibling aggression. Since black eagle nestlings require large amounts of food over a period of many weeks, short-term abundance of food may not be an accurate indicator of long-term food levels. As a consequence, aggression and siblicide might be favored in order to obviate any future competition (Anderson 1990, Stinson 1979).

Perhaps the appropriate "sibling aggression policy" is obtained from simple cues available to the chicks from the outset. Assuming that parents deliver food at some optimal rate, then a chick may be able to estimate in advance whether sufficient levels will be available for its own growth. It is interesting to note that the facultatively siblicidal golden eagle (*Aquila chrysaetos*) provides the same amount of food regardless of the number of chicks in the brood (Collopy 1984). If this is typical, then the senior chick may be able to detect whether the food will be enough to support all nestmates. Eagles that practice obligate siblicide generally deliver less food to the nest than facultative species, and consequently no assessment by the chicks is necessary (Bortolotti 1986). The average amount of food provided by the parents may be consistently low enough for natural selection to favor preemptive killing—a system that benefits both the senior chick and its parents. In other words, the insurance policy is canceled.

Even in species that practice facultative siblicide, aggression is sometimes insensitive to food supply. For example, the level of fighting among heron and egret chicks appears to be independent of the amount of food available (Mock, Lamey and Ploger 1987). It may be that the current food level acts as a proximate cue for sibling aggression only in those species where the current level accurately predicts future food levels. This hypothesis is consistent with the observation that daily food levels are unpredictable and unstable among egrets (Mock, Lamey and Ploger 1987). Interestingly, fighting between egrets ceases when the brood size drops from three to two (thus reducing future food demands) and may be reinstated by restoring the third chick (Mock and Lamey in press). Further studies of other species are necessary to determine whether the degree to which food levels fluctuate is related to aggressive behavior.

Future Directions

The study of siblicide as an adaptive strategy is still in its infancy. Much of the work to date has been devoted to identifying the proximate causes of aggressive behavior and documenting its utility for controlling resources. Less is known about the effects of siblicide on the inclusive fitness of the perpetrators. Although many theoretical models of avian siblicide have been proposed (O'Connor 1978; Stinson 1979; Mock and Parker 1986; Parker, Mock and Lamey 1989, Godfray and Harper 1990), the field data are limited.

Several areas of research need to be explored further. We would like to determine the short-term costs of sibling rivalry, perhaps by comparing the energetics of competitive begging and fighting. Likewise we need to know the long-term costs of temporary food shortages; there is particular interest in the relationship between the development of the chick and the amount of food available. Similarly, what is the relation between the amount of effort parents put into supplying food, the resulting chick survival rate and the long-term costs of reproduction among brood-reducing species? Is there any relation between chick gender, hatching order and siblicide—particularly in siblicidal species that have a large degree of sexual dimorphism? Another area of interest is the role of extra-pair copulations, which reduce the relatedness of nestmates and thereby increase the potential benefits of selfishness; it would be useful to know whether chicks have the ability to discriminate half-siblings from full siblings. Finally, why is it that parents appear not to interfere with the execution process in siblicidal species (O'Connor 1978; Drummond, Gonzalez and Osorno 1986; Mock 1987)? Answers to these questions can give us a better understanding of how siblicidal behavior may have evolved.

Bibliography

Anderson, D. J. 1989. Adaptive adjustment of hatching asynchrony in two siblicidal booby species. *Behavioral Ecology and Sociobiology* 25:363-368.

Anderson, D. J. 1990. Evolution of obligate siblicide in boobies. I: A test of the insurance egg hypothesis. *American Naturalist* 135:334-350.

Bortolotti, G. R. 1986. Evolution of growth rates in eagles: sibling competition vs. energy considerations. *Ecology* 67:182-194.

Bragg, A. N. 1954. Further study of predation and cannibalism in spadefoot tadpoles. *Herpetologica* 20:17-24.

Braun, B. M., and G. L. Hunt, Jr. 1983. Brood reduction in black-legged kittiwakes. *The Auk* 100:469-476.

Bryant, D. M., and P. Tatner. 1990. Hatching asynchrony, sibling competition and siblicide in nestling birds: studies of swiftlets and bee-eaters. *Animal Behaviour* 39:657-671.

Cash, K., and R. M. Evans. 1986. Brood reduction in the American white pelican, *Pelecanus erythrorhynchos*. *Behavioral Ecology and Sociobiology* 18:413-418.

Collopy, M. 1984. Parental care and feeding ecology of golden eagle nestlings. *The Auk* 101:753-760.

Drent, R. H., and S. Daan. 1980. The prudent parent: energetic adjustments in avian breeding. *Ardea* 68:225-252.

Drummond, H. 1987. Parent-offspring conflict and brood reduction in the Pelecaniformes. *Colonial Waterbirds* 10:1-15.

Drummond, H., E. Gonzalez and J. Osorno. 1986. Parent-offspring cooperation in the blue-footed booby, *Sula nebouxii*. *Behavioral Ecology and Sociobiology* 19:365-392.

Drummond, H., and C. Garcia Chavelas. 1989. Food shortage influences sibling aggression in the blue-footed booby. *Animal Behaviour* 37:806-819.

Edwards, T. C., Jr., and M. W. Collopy. 1983. Obligate and facultative brood reduction in eagles: An examination of factors that influence fratricide. *The Auk* 100:630-635.

Evans, R. M., and B. McMahon. 1987. Within-brood variation in growth and conditions in relation to brood reduction in the American white pelican. *Wilson Bulletin* 99:190-201.

Fraser, D. 1990. Behavioural perspectives on piglet survival. *Journal of Reproduction and Fertility, Supplement* 40:355-370.

Fujioka, M. 1985a. Sibling competition and siblicide in asynchronously-hatching broods of the cattle egret, *Bubulcus ibis*. *Animal Behaviour* 33:1228-1242.

Fujioka, M. 1985b. Food delivery and sibling competition in experimentally even-aged broods of the cattle egret. *Behavioral Ecology and Sociobiology* 17:67-74.

Gargett, V. 1977. A 13-year population study of the black eagles in the Matopos, Rhodesia, 1964-1976. *Ostrich* 48:17-27.

Gargett, V. 1978. Sibling aggression in the black eagle in the Matopos, Rhodesia. *Ostrich* 49:57-63.

Godfray, H. C. J., and A. B. Harper. 1990. The evolution of brood reduction by siblicide in birds. *Journal of Theoretical Biology* 145:163-175.

Gustafsson, L., and W. J. Sutherland. 1988. The costs of reproduction in the collared flycatcher *Ficedula albicollis*. *Nature* 335:813-815.

Hahn, D. C. 1981. Asynchronous hatching in the laughing gull: Cutting losses and reducing rivalry. *Animal Behaviour* 29:421-427.

Hamilton, W. D. 1964. The genetical evolution of social behaviour. *Journal of Theoretical Biology* 7:1-52.

Henny, C. J. 1988. Reproduction of the osprey. In *Handbook of North American Birds*, ed. R. E. Palmer. Yale University Press.

Houston, A. I., and N. B. Davies. 1985. The evolution of cooperation and life history in the dunnock *Prunella modularis*. In R. Sibly and R. Smith (eds.) *Behavioral Ecology: The Ecological Consequences of Adaptive Behaviour*. Blackwell: Oxford. pp. 471-487.

Jamieson, I. G., N. R. Seymour, R. P. Bancroft and R. Sullivan. 1983. Sibling aggression in nestling ospreys in Nova Scotia. *Canadian Journal of Zoology* 61:466-469.

Kepler, C. B. 1969. Breeding biology of the blue-faced booby on Green Island, Kure Atoll. *Publications of the Nuttal Ornithology Club* 8.

Lack, D. 1954. *The Natural Regulation of Animal Numbers*. Clarendon Press: Oxford.

Magrath, R. 1989. Hatch asynchrony and reproductive success in the blackbird. *Nature* 339:536-538.

Mock, D. W. 1984. Siblicidal aggression and resource monopolization in birds. *Science* 225:731-733.

Mock, D. W. 1985. Siblicidal brood reduction: The prey-size hypothesis. *American Naturalist* 125:327-343.

Mock, D. W. 1987. Siblicide, parent-offspring conflict, and unequal parental investment by egrets and herons. *Behavioral Ecology and Sociobiology* 20:247-256.

Mock, D. W., and T.C. Lamey. In press. The role of brood size in regulating egret sibling aggression. *American Naturalist*.

Mock, D. W., and G.A. Parker. 1986. Advantages and disadvantages of ardeid brood reduction. *Evolution* 40:459-470.

Mock, D. W., and B.J. Ploger. 1987. Parental manipulation of optimal hatch asynchrony in cattle egrets: An experimental study. *Animal Behaviour* 35:150160.

Mock, D. W., T. C. Lamey and B. J. Ploger. 1987. Proximate and ultimate roles of food amount in regulating egret sibling aggression. *Ecology* 68:1760-1772.

Mock, D. W., T. C. Lamey, C.F. Williams and A. Pelletier. 1987. Flexibility in the development of heron sibling aggression: An intraspecific test of the prey-size hypothesis. *Animal Behaviour* 35:1386-1393.

Nur, N. 1984. Feeding frequencies of nestling blue tits (*Parus coeruleus*): Costs, benefits, and a model of optimal feeding frequency. *Oecologia* 65:125-137.

O'Connor, R. J. 1978. Brood reduction in birds: Selection for infanticide, fratricide, and suicide? *Animal Behaviour* 26:79-96.

O'Gara, B. W. 1969. Unique aspects of reproduction in the female pronghorn, *Antilocapra americana*. *American Journal of Anatomy* 125:217-232.

Parker, G. A., D. W. Mock and T. C. Lamey. 1989. How selfish should stronger sibs be? *American Naturalist* 133:846-868.

Ploger, B. J., and D. W. Mock. 1986. Role of sibling aggression in distribution of food to nestling cattle egrets, *Bubulcus ibis*. *The Auk* 103:768-776.

Poole, A. 1979. Sibling aggression among nestling ospreys in Florida Bay. *The Auk* 96:415-417.

Poole, A. 1982. Brood reduction in temperate and sub-tropical ospreys. *Oecologia* 53:111-119.

Simmons, R. 1988. Offspring quality and the evolution of Cainism. *Ibis* 130:339-357.

Stinson, C. H. 1977. Growth and behaviour of young ospreys, *Pandion haliaetus*. *Oikos* 28:299-303.

Stinson, C. H. 1979. On the selective advantage of fratricide in raptors. *Evolution* 33:1219-1225.

Williams, G. C. 1966. Natural selection, the costs of reproduction, and a refinement of Lack's principle. *American Naturalist* 100:687-690.

Woodward, P. W. 1972. The natural history of Kure Atoll, northwestern Hawaiian Islands. *Atoll Research Bulletin* 164.

Social Organization in Jackals

Patricia D. Moehlman

The popular stereotype of jackals as skulking scavengers with base and reprehensible behavior is belied, as is so often the case, by long-term observations. Jackals are one of the few species of mammals in which males and females form long-term pair-bonds, often lasting a lifetime. They hunt together, share food, groom each other, jointly defend their territory, and provision and defend their pups together (Fig. 1). Some of the pups stay with their parents and at the age of one year help raise the next litter, their full brothers and sisters.

Complex ecological and evolutionary factors are of course involved in a social organization as elaborate as that of the jackals, and this paper will review these factors. Many of the ecological constraints that select for the jackals' monogamous mating system and their cooperative rearing of pups have been revealed in a twelve-year study of two species in Tanzania: silverbacked jackals (*Canis mesomelas*), which occupy brush woodlands peripheral to Lake Ndutu in the Serengeti Plain, and golden jackals (*C. aureus*), which live in the adjacent short-grass plains (Moehlman 1979, 1981, 1983, 1986).

The complex social system of jackals allows the successful rearing of very dependent young

Beyond these ecological constraints, there are indications that body size may provide important insights into the physiological constraints that have given rise to the jackals' cooperative breeding system. Jackals fall midway in the range of sizes found in the family Canidae, which consists of 37 or so species that are diverse in diet and distribution as well as in body size (Clutton-Brock et al. 1976; Gittleman 1984a). Within the Canidae, the pervasive mating system is long-term monogamy, a system observed in less than 3% of mammals (Kleiman 1977). As one examines the canids' feeding ecology, mating system, adult sex ratios, dispersal, and propensity for the cooperative rearing of pups, trends that correlate with body weight seem to emerge. For example, small canids (those less than about 6 kg), such as red foxes (*Vulpes vulpes*) and bat-eared foxes (*Otocyon megalotis*), are usually monogamous but have a tendency toward polygyny (one male mated with several females), with females outnumbering males, males dispersing, and some nonreproductive females helping to rear pups. Medium-sized canids (between 6 and 13 kg), including the jackals, appear to be strictly monogamous, with adult sex ratios equal, and with both males and females equally either helping to rear pups or emigrating. Most large-sized canids (greater than 13 kg), such as the African hunting dog (*Lycaon pictus*) and the timber wolf (*C. lupus*), usually have a monogamous mating system, with indications of polyandry (one female mated by several males), adult sex ratios skewed toward males, male helpers, and females emigrating.

Central among the ecological constraints on social organization are those that involve feeding ecology, as is demonstrated in the studies of the jackals near Lake Ndutu. Silverbacked and golden jackals, which have an adult body weight of 7 to 10 kg, are both highly opportunistic omnivores, and their diets can range from fruit, insect larvae, beetles, and reptiles, to birds, rodents, hares, and Thomson's gazelle (*Gazella thomsoni*). In the Serengeti study area, the habitats on which these jackals live differ in the type and size of food and in its spatial and temporal availability. This difference correlates strongly with subtle differences in pair-bonding and with more obvious differences in the cooperative breeding and survival of pups and in the dispersal of jackals at the time they reach maturity.

The critical period for both species is when they are raising and provisioning pups. Silverbacked jackals have their pups during the dry season, between June and November, when the unstriped grass rat (*Arvicanthis niloticus*), a small (60 g) diurnal rodent, is at the peak of its yearly cycle and is available as prey in large numbers (Senzota 1978). The silverbacks also depend on the *Balanites aegyptiaca* tree, whose fruit, which superficially resembles dates, is gathered daily after it drops to the ground. Foraging for food as abundant but as scattered and small as these rats and fruits is energetically costly for the silverbacks, involving forays of 6 to 8 km (see Figs. 2 and 3). In contrast, golden jackals live on smaller territories and feed on more substantial packets of food, primarily meat from large vertebrates (Fig. 4). They raise their pups during the wet season, from December to May, when they regularly kill fawns of Thomson's gazelle and scavenge on the afterbirths and carcasses of

Patricia Moehlman, a graduate of Wellesley College, the University of Texas (M.A.), and the University of Wisconsin, Madison (Ph.D.), is a staff zoologist at Wildlife Conservation International, a division of the New York Zoological Society. She is a behavioral ecologist, focusing on the evolution of social systems in Canidae and Equidae. The research reported here represents fieldwork conducted over a substantial part of the past twelve years in Tanzania and supported by the National Geographic Society, The Harry Frank Guggenheim Foundation, the Muskiwinni Foundation, and the Tanzanian Scientific Research Council. Address: Wildlife Conservation International, New York Zoological Society, Bronx, NY 10460.

Figure 1. Jackals in the Serengeti Plain form intimate family groups, with the mother and father mated monogamously for life. This social behavior is necessary, because raising pups takes so much time and energy that a mother cannot do it without assistance from the father. She is also often assisted by the pups' siblings from the previous year's litter. In this family of golden jackals (*Canis aureus*), three pups have survived to the age of about 2 months (litters usually have six pups at birth), although the pup lying in the foreground is sick and may not survive. (All photographs © by Patricia D. Moehlman.)

wildebeest (*Connochaetes taurinus*), zebra (*Equus burchelli*), Grant's gazelle (*G. granti*), and Thomson's gazelle that are killed by a hyena clan (*Crocuta crocuta*) in the study area.

Jackals are facultative cooperative hunters, meaning that they hunt in groups opportunistically, when appropriate prey is available. They have a higher success rate in killing the fawns of Thomson's gazelle when they hunt in groups of two or three than when they hunt alone (Wyman 1967; Lamprecht 1981). And as Figure 5 illustrates, a lone jackal would have less success than cooperature groups of jackals in defending and feeding on carcasses (Lamprecht 1978; Moehlman 1983). The increased success in capturing and retaining prey that comes with cooperative hunting has been considered critical to the evolution of sociality in carnivores (Kleiman and Eisenberg 1973). For example, research on another medium-sized canid, the coyote (*C. latrans*), indicates that relative prey size may be an important selective pressure for group size. In habitats where mule deer (*Odocoileus hemionus*) and elk (*Cervus elaphus*) are available in significant numbers, there was a correlation with a delay in the dispersal of pups and concurrently with larger social groups (Bowen 1981). By comparison, coyotes that feed on smaller prey, such as rodents, had smaller group sizes and dispersed earlier (Bekoff and Wells 1980). However, these correlations are confounded by studies in areas where populations of coyotes that feed on small prey are relatively high in density; in this case, because dispersal is presumably difficult, groups are relatively large (Andelt 1982). Thus, there is great flexibility in the way that medium-sized canids are able to respond through group size to differences in the type, size, and availability of food.

A comparison of medium-sized canids to their smaller and larger relatives indicates that the size of a canid species may impose limits on its feeding ecology. As body size increases within Canidae, there is a general trend from solitary to facultative cooperative to obligatory cooperative hunting, but the availability of food is a complicating factor. All the smaller canids are strictly solitary foragers, although they do form social groups; as Kruuk (1978) and Macdonald (1983) have stressed, in habitats with patchily distributed food, the minimum size of these canids' home ranges will be determined by how widely the patches are distributed, and the size of

Figure 2. A male silverbacked jackal (*C. mesomelas*) hunts for small rodents, the primary source of food during the whelping season both for himself and for feeding his pups (Fig. 3). The nutritional value of this prey is small compared to the effort required to find and capture it; therefore, if a silverback mother is to provision her litter successfully, the assistance of the father and their year-old offspring is critical.

their groups will be determined by the availability of prey in those patches. Large canids, such as the African hunting dog and the dhole (*Cuon alpinus*), form large stable groups and always hunt cooperatively, preying mostly on larger vertebrates (Frame et al. 1979; Johnsingh 1982). Timber wolves, another large canid species, are mainly cooperative hunters during the winter and form packs, but their foraging pattern can shift to small-sized prey in the spring and summer, when solitary hunting prevails and groups tend to disperse (Mech 1970; Peterson et al. 1984). The interesting exception among large canids is the maned wolf (*Chrysocyon brachyurus*), the only large canid that forages exclusively on rodents and fruits; it does not hunt or breed cooperatively and has been observed only in pairs.

Mating and territoriality

Both silverbacked and golden jackals form pair-bonds that are characterized by friendly behavior and that normally last the 6 to 8 years of their usual lifespans. They groom one another, share food, hunt cooperatively, and care for their sick or injured partners. There is little sexual dimorphism, either physically or behaviorally, and they share equally in most activities, such as marking and defending their territory, foraging, and resting.

Jackals are territorial, actively defending their territories and scent-marking them with urine when foraging. Occasionally, the presence of a large carcass can attract neighboring jackals in such numbers that the residents cannot drive them out, but the trespassers remain at the carcasses only briefly, rarely for more than 2 hours. Residents may leave their territories to drink, scavenge, and explore, but these forays are infrequent and almost always last less than 2 hours. When off their territory, the jackals do not scent-mark. Jackal pairs that forage on their territory together scent-mark twice as often as solitary foragers. This tandem marking advertises that both members of the pair are in residence, a behavior that appears to be particularly important for silverbacked jackals. For example, twice during the whelping season I observed that the territorial male disappeared and was presumably dead, and on both occasions new pairs trespassed on the territory and scent-marked it. They attacked and wounded the resident female and on one occasion ate the pups. Within days the former residents were dead, and the new pairs had taken over the territories.

Jackals are especially fierce at guarding mating exclusivity. Every instance of territorial aggression that I observed, such as the one shown in Figure 6, has been between jackals of the same sex—females do not tolerate female intruders, and males do not tolerate males. Litters with multiple sires are possible, as has been reported for domestic dogs (*C. familiaris*) (Beach, pers. com.). Thus, the maintenance of mating exclusivity may be critical for the male in all *Canis* species so that he does not invest in pups that he did not sire. From the female's perspective, mating exclusivity would prevent her mate from having other litters to invest in and reducing the aid to her pups.

Long-term pair-bonding in jackals reflects both physiological and ecological constraints. Compared to most mammals, jackals have large litters, usually six pups, with a long period of infant dependency (Kleiman and Eisenberg 1973). Among the silverbacked jackals in

Figure 3. The unstriped grass rat in the mouth of this month-old silverback pup was captured by a parent and regurgitated to the pup. Regurgitation allows most canid species to gather and transport large amounts of food back to the den efficiently.

the Ndutu study area, which forage on abundant though small items of food, paternal investment is critical to pup survival, and a pair successfully raises an average of only 1.3 pups from each litter. If a silverback male were to divide his care between several litters, probably no pups would survive and the reproductive success of both parents would decline; long-term pair-bonds therefore appear to be critical for the maintenance of territories and the rearing of pups.

Golden jackals also have large litters of dependent young, but they are more carnivorous during the whelping season than silverbacked jackals and make use of larger-sized food; cooperative hunting and defense of carcasses are a central part of their foraging behavior. Although paternal investment is important in golden jackals, it does not appear to be as important as in silverbacked jackals, and the pair-bond does not appear to be as strong. Pairs of golden jackals feed their pups twice as often as do silverbacks, and pup survival does not appear to be limited by food provisioning. The greater availability of food might enable a female to provision her litter adequately with little help and a male to provision two litters successfully.

No mate changes were observed among 11 pairs of silverbacked jackals, whereas among 14 pairs of golden jackals one mate change was observed while the resident female was still alive. She was dominant to the intruding female but became more and more peripheral to the male until she disappeared; in the subsequent three years she has not been seen again. Three other golden jackals (2 females, 1 male) have acquired new partners, although it is not known whether these mate changes were due to death or to departure. Polygyny or polyandry has not been observed in jackals, but it is possible that a sharp change in their variable foraging strategies could tip the balance that selects for equal parental investment and could thereby change their mating and rearing systems.

Monogamy is the basic mating system in the Canidae and is strongly correlated with the inability of a single female to rear a litter—generally larger and more dependent than in other mammals—without male parental investment. The degree of paternal investment that the newborns require depends on two related factors—the number of newborns and how well developed they are at birth.

Canids are unusual among mammals in that most of them regurgitate food to their young, enabling them to bring large amounts of food to the den economically

The relative development of newborns is indicated by the relationship between their weight at birth and the weight of their mothers. Among mammals, the weight of newborns generally increases with maternal body weight, but it does so as a $\frac{2}{3}$ to $\frac{3}{4}$ exponent of female weight (Western 1979). Hence, larger mammals will have newborns which are relatively smaller as a percentage of adult weight and which may require greater parental investment, a longer period of development, or

Figure 4. A golden jackal moves in for an attack on an adult Thomson's gazelle (*Gazella thomsoni*), which is too sick to escape. Because golden jackals feed on larger sized food than do silverbacks, they do not need to spend as much time and energy hunting, nor do they require as large a territory that must be patrolled and defended. Consequently, although still important, the pair-bonds in golden jackals are not as strong as in silverbacks, and the assistance of the father and siblings in raising pups is less critical.

Figure 5. A golden jackal threatens vultures (*Gyps africanus* and *G. rüppellii*) over the carcass of a Thomson's gazelle. The great difficulty that a single jackal has in competing for and defending captured food is one of the strong selective pressures for larger group sizes in jackals.

both. The canids show a similar correlation (Bekoff et al. 1981, 1984; Gittleman 1984a, b; Moehlman 1986).

However, the canids appear to be unique among mammals in showing a trend toward larger litters as maternal body size increases (Moehlman 1986); this correlation has not been recorded for any other mammalian family (Millar 1977, 1981; Western 1979). Furthermore, as canids get larger, not only do the relatively smaller and less developed pups require greater parental investment after birth, but the maternal investment before birth is also greater. Although larger females produce relatively smaller pups, the larger number of pups means that the total weight of the litter remains high and may even increase with larger females.

Sibling helpers

The energetic costs of reproduction—for gestation, lactation, and the protection and feeding of pups—are so much greater than a jackal mother could bear alone that she is assisted not only by her mate, but often also by her older offspring. One of the most intriguing aspects of the social behavior and organization in both jackal species is the tendency of some male and female offspring to remain with their parents and to help protect, provision, and socialize the next year's litter. In all 11 cases in which consecutive litters were observed and there were helpers in the second year, the helpers were offspring from the previous year's litter. Helpers are subordinate to their parents and do not scent-mark, but the parental pair share food with them, groom them, and play with them.

Pups first open their eyes at about 10 days, and they do not emerge from the underground dens where they are born until about 3 weeks of age. Mothers spend 90% of the initial 3 weeks in the den, which may be critical for preventing hypothermia in small pups. Until 7 weeks of age, pups spend most of their time in the den, and they continue to use dens until they are 14 weeks old. The use of dens reduces the threat of predation and may allow for the production of more dependent pups and for a longer period for development. The pups start to eat regurgitated food at 3 weeks; canids are unusual among mammals in that most of them regurgitate food to their young, a behavior that enables them to bring large amounts of food to the den economically and safely (Fig. 3). The pups are weaned at 8 to 9 weeks of age, and at 14 weeks they are well coordinated and start to forage with the adults. The initial 14 weeks are critical to survival, and all observed mortality occurred within this period.

A pair that raises a litter without helpers must divide its activity between staying at the den to guard the pups and foraging for food. Typically, adults do all their resting near the den, so that when they are not foraging they are guarding. The presence of a single adult can provide the pups with protection, which consists of both vocally warning the pups, who then run into the den, and threatening the potential predator; Figure 7 shows how adept jackals are at biting 55-kg hyenas in the rump and chasing them away from the den. During the first 14 weeks in families with no helpers, the pups in both jackal species were left unguarded about 20% of the time. Primarily because they make use of larger-sized and energetically less costly food resources, unaided pairs of golden jackals feed their pups at a higher rate than do unaided silverbacked pairs (0.52 versus 0.32 feedings per hour), and they are somewhat more successful at raising pups to the age of 14 weeks (1.8 versus 1.3 pups) (Moehlman 1986).

The addition of helpers in both species correlates with improved protection, provisioning, and survival of pups. Among silverbacked jackals, pups in families with one helper were left alone 16% of the time, and with two helpers were left alone only 8% of the time. More helpers also resulted in significantly higher rates of feeding by regurgitation, with helpers contributing up to 33% of the pups' food; the energetics of foraging for dispersed small packets of food appears to limit the ability of adults to feed both themselves and the pups, so that when pups outnumbered the adults in the family, they were observed to be rough coated, thin, and weak legged, and they either died or disappeared. In general, as Figure 8 shows, there is a significant correlation among silverbacked jackals between the number of adults in a family and the survival of pups to the age of 14 weeks (Moehlman 1979).

Jackals are one of the few species of mammals in which males and females form long-term pair bonds, often lasting a lifetime

Could this correlation simply reflect a variation in parental competence? That is, might a first-time mother be less efficient in caring for pups (regardless of the fact that she also would not have helpers)? Long-term observations on one pair do not support this alternative explanation and indicate that the presence of helpers may be critical. In 1976, a silverback pair that I observed had one helper, and 4 pups survived for at least six months. In 1977 this same pair had no helper, and none

Figure 6. While her male partner looks on, a female silverbacked jackal (*middle*) attacks an intruding female. The intruder is not tolerated, because she could potentially displace the resident; furthermore, in competing for the male's help in raising pups, the intruder would be a threat to the survival of the resident female's pups. Almost all aggression among jackals is between members of the same sex. The male would attack an intruding male, because he otherwise risks being displaced or investing his time and energy in raising pups he did not sire.

of the 5 pups I observed at 3 weeks of age survived to 14 weeks. Although they were experienced parents, they could not successfully raise any pups without helpers.

A second alternative explanation, that the variation in pup survival could reflect a variation in territory quality and that a lack of helpers or of surviving pups might reflect poor habitat and food resources, is also not supported by observations. In 1976 and 1977 rainfall was high at Ndutu and there was an abundance of the jackals' prey, the unstriped grass rat, which could often be observed running back and forth between the thornbushes. The silverback pair I observed did not successfully raise any pups in 1977, the year they had no helper, despite the abundance of prey (32,083 rats/km^2). In contrast, on an adjoining and similar-sized territory that same year, another pair of silverbacked jackals had one helper and succeeded in raising 3 of 6 pups (first observed at 3 weeks) to the age of 14 weeks. Although still abundant, the density of the rats was much lower (13,125 rats/km^2) on this second pair's territory (Senzota 1978; Moehlman 1983).

Among golden jackals, pups in families with helpers were seldom left alone. All adults spent more time at the den than did silverbacked jackals, which may also reflect that less time is needed for foraging. Pups are consistently provisioned at a higher rate than among silverbacked jackals, and there was no observed correlation between the number of adults and the rate of feeds. Although both male and female helpers supply at a *rate* similar to silverback helpers, their contributions are a smaller percentage of the total feeds (Moehlman 1986). In general, pup survival does correlate significantly with the addition of helpers, but the data are more variable, and the correlation is much weaker than for silverbacked jackals (Fig. 8). Nevertheless, despite the fact that golden-jackal pups are better provisioned and guarded than those of silverbacked jackals, fewer survive to 14 weeks of age. The rainy season when they are born means an abundant food supply, but it also means flooded dens, and golden pups die of exposure and illness. This density-independent factor can negate the contributions of parents and helpers.

Helping versus dispersing

Both species of jackals can become sexually mature at the age of 10 or 11 months (Taryannikov 1976; Ferguson et al. 1983). At this age, a young jackal has the option of delaying reproduction and staying on its natal territory for an extra 6 to 8 months to help rear its full siblings—which are on average as closely related to the young jackal as its own offspring would be—or dispersing and attempting to acquire a mate and territory in order to rear a litter of its own pups (Moehlman 1979; Emlen 1982a, b). Helpers do not display sexual behavior and may experience social suppression of endocrine function. By visual estimate, male helpers have smaller testes than do their fathers. A similar social control of sexuality

Figure 7. When they are not foraging for food, jackals must guard the den against potential predators, such as this hyena (*Crocuta crocuta*) being bitten by a golden jackal. Guarding pups and foraging tax virtually all the time and energy of the parents when they are unaided by their year-old offspring, a burden that constitutes a strong selective pressure for pups to delay leaving the nest in order to help raise the next year's litter.

has been reported in a study of a captive wolf pack: when the alpha female wolf died, the most dominant female pup became estrous, while her two similarly sized but subordinate sisters remained anestrous; the study concluded that wolves were physiologically capable of reproducing during their first year, but that the social environment could suppress the first seasonal estrus (Seal et al. 1979).

Retaining helpers potentially increases the parents' reproductive success—that is, it increases the parents' chances of passing on their genes to future generations.

Gestation, lactation, and the protection and feeding of pups are so difficult that a jackal mother is assisted not only by her mate, but often also by her older offspring

Parents clearly benefit more by allowing helpers to invest in their pups, the helpers' siblings, rather than by having helpers emigrate to produce their own pups, because each parent accounts for half of the genetic makeup of its pups (a relatedness of $r = \frac{1}{2}$) but only a fourth of its grandpups ($r = \frac{1}{4}$). Females may also improve their own future reproductive success by tolerating helpers which feed them during the lactation period. However, the benefits to parents of retaining helpers are limited by available resources and the energetics of pup provisioning.

To answer questions concerning the relative costs and benefits of helping or dispersing, we need data on the success of emigrating, mating, establishing a territory, and rearing pups at one year of age versus two years of age, and these data are not available. Although a few older helpers have been observed in silverbacked jackals—one 2-year-old female (Moehlman 1983) and one 3-year-old male (Ferguson et al. 1983)—these are rare occurrences. However, the existing data on the two jackal species make it possible to interpret the differences between them in terms of the costs and benefits of dispersing, because the two species are closely related and socially similar but inhabit areas with different vegetation and food resources.

The percentage of surviving pups that stay to become helpers is much smaller among silverbacked jackals (24%) than among golden jackals (70%). It is easier for silverbacked jackals to disperse, because their feeding ecology is relatively stable, with an abundant food supply throughout the year and with no bottlenecks. The density of unstriped grass rats varies sharply from between 3,130 and 7,790 rats/km^2 during the wet season to between 13,130 and 32,080 rats/km^2 during the dry season (Senzota 1978); but because the low point of the rodent's population cycle occurs during the wet season, when there is an influx of migrating herds of wildebeest, zebra, and gazelle, food remains abundant. This situation, which makes it possible for young adults to stay on the natal territory, also presumably makes it easier for them to disperse. The brush-woodland habitat, with its poorer visibility conditions, would also allow young jackals to avoid visual detection and to establish themselves initially at the edges of existing territories.

However, it is difficult for silverbacks to raise pups. Even if the opportunities exist to obtain a territory and a mate, the cost of rearing young may be prohibitive (Emlen 1982a). Pairs of silverbacked jackals raise an average of only 1.3 pups from each year's litter, and even experienced pairs lose whole litters. Jackals that stay and help derive important benefits. They acquire extended experience on familiar terrain and thus may improve their survivorship and the quality of their future parental care. Since helpers are full siblings, on average as closely related to their siblings ($r = \frac{1}{2}$) as they would be to their own pups, by delaying their reproduction for one year they potentially improve their inclusive fitness through the survival of close relatives, increasing the chances that their genes will be passed on to future generations. But even with the constraints on rearing their own offspring, most silverbacks appear to opt for dispersing and attempting reproduction at one year of age.

In contrast to silverbacked jackals, 70% of known surviving pups among golden jackals were observed helping with the next year's litter. During the dry season, when they are over 7 months old, there is a food bottleneck, and the young jackals of that year leave their natal territory and return in the wet season when food is abundant. This high rate of return by golden jackals at the age of 11 or 12 months may reflect a high density of jackals; in a saturated habitat, in which it is difficult to disperse successfully, mate, and acquire a territory, one-year-old goldens are simply forced to return home. Current data on golden jackals indicate that pairs may adequately provision their litters and that the major cause of mortality is a density-independent factor. Thus, the major ecological constraint on the dispersal of golden jackals appears to be the difficulty of acquiring a territory. In two instances, golden jackals have been observed on their natal territories in years when their parents did not have another litter and no increase in inclusive fitness could be achieved by helping younger siblings. This is perhaps another indication of the relative benefits of staying on the home territory—comparable to the benefits derived by silverbacks—and the difficulty of dispersing.

Sex ratios

Other than gestation and lactation, there is a symmetry of sexual roles among jackals. Both members of the mated pair mark and defend their territory; both feed, guard, and defend the pups at similar rates; and both groom, play with, and otherwise socialize the pups. Helpers also display symmetry of roles in defending the territory and feeding, guarding, and socializing the pups. What accounts for this symmetry? A comparison of jackals with the rest of the Canidae provides insight as to how this sexual selection might have evolved.

In small canids, since females have fewer and relatively heavier newborns that potentially require less paternal investment after birth, they are investing more than do the males and are the limiting sex. Thus, the potential arises for a male to invest in the offspring of more than one female, and polygyny is possible. Parental-investment and sexual-selection theory would then predict that males would be more likely to disperse (Fisher 1930; Trivers 1972). This set of behaviors expected in small canids has in fact been observed, particularly in several different species of foxes (Macdonald 1981, 1983; Hersteinsson 1984). The availability of food and the ability of the male to control resources can affect the relative costs and benefits of maternal and paternal investment and can thereby determine whether the mating system is monogamous or polygynous.

At the other end of the scale, large canids produce larger litters of relatively smaller pups, so that females must make a large investment in their offspring before birth, and they require substantial male investment in their pups after birth. If their pups are to survive, females cannot afford to share this male investment with other females, and competition for males can be intense. One would therefore expect a significant bias in the pup and adult sex ratios toward males, fierce competition between females for dominance, females emigrating, and males helping.

This pattern is observed in African hunting dogs, which are predominantly monogamous, with limited observations of polyandrous matings (van Lawick 1973). Data indicate that a pair of African hunting dogs cannot successfully raise a litter without helpers (Malcolm and Marten 1982). They always hunt cooperatively, concentrating on larger vertebrate prey, and larger packs tend to be more successful in hunting and in defending carcasses (Kruuk 1975; Frame et al. 1979). Thus, African hunting dogs tend to form stable patrilineal social groups that rear the large litters cooperatively. The other large canids, dholes and timber wolves, have similar constraints of body size and feeding ecology, and they exhibit similar mating and breeding strategies (Harrington et al. 1982, 1983; Johnsingh 1982).

As would be expected, sex ratios in African hunting dogs are biased toward males. Frame and her colleagues (1979) have observed that females will compete to exclude other reproductive efforts in the group, because a second litter at the same time will reduce nutrition provided to each pup, and because a second litter temporally overlapping the first will restrict pack movements for a longer period of time and thus potentially

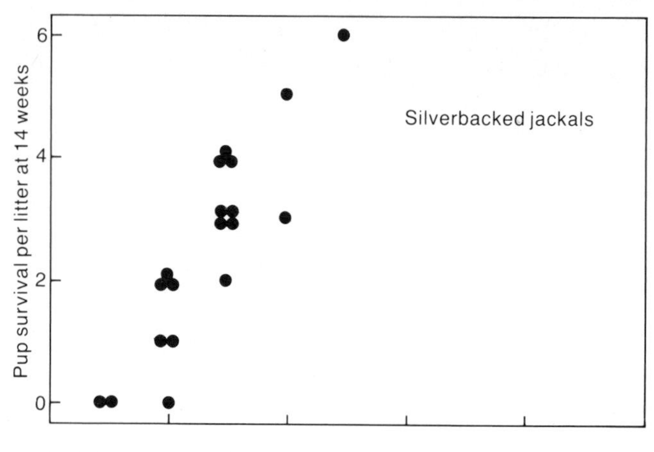

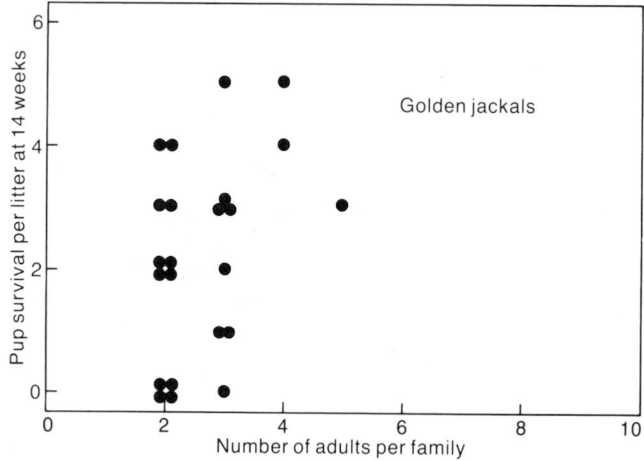

Figure 8. The chances of survival of jackal pups increase significantly with the number of adults—parents plus older siblings—that share the tasks of raising them. The correlation between the number of adults and the number of pups that survive (from a typical litter of six pups at birth) is stronger for silverbacked than for golden jackals, primarily because of their different feeding ecologies. More time and energy is needed to forage for the small fruits and rodents that the silverbacks feed their pups than to hunt and scavenge the large vertebrates that are the golden's main source of food. Therefore, the addition of helpers is more consequential to the silverbacks. (After Moehlman 1986.)

Figure 9. Jackal males, such as this golden jackal father with his 3-month-old pups, invest a similar amount of care in their offspring as do females. This largely explains why they form lifelong monogamous relationships—the full-time parental investment of both parents (without sibling helpers) is just sufficient to raise an average of one to two pups per litter. In contrast, smaller canids have a tendency to be polygynous, because the mother does not require a paternal investment equal to her own to raise her relatively larger, better developed pups; and larger canids tend to be polyandrous, because the mother requires more help than just one male can provide to raise her pups, which are even more dependent on parental care than in jackals.

reduce hunting success. Competition between females can be so intense that dominant females have been observed killing the pups of subordinate females.

Because subordinate females among African hunting dogs are unlikely to breed successfully, they tend to emigrate (Frame and Frame 1976). By comparison, subordinate males are more likely to have breeding success within the pack for several reasons: the potential for mixed paternity resulting from the multiple insemination of the female allows subordinate males some increase in reproductive fitness (Emlen 1982b); by staying in their natal pack, male helpers can derive the inclusive fitness benefits of investing in pups that are close relatives; and there is the potential for a change in dominance. Data indicate that individual male members of a pack tend to feed pups by regurgitation more than their mother does (Malcolm and Marten 1982). Thus, the limiting sex is male, and females will compete for males. In addition, sons may ultimately be cheaper to raise than daughters, since they will stay and help, and this might explain the bias in sex ratio toward males at birth (Trivers and Hare 1976; Heerdon and Kuhn 1985; Emlen et al. 1986).

The apparent correlations of mating and breeding strategies with the possible physiological constraints of body size and the dependency of pups is complicated by ecological factors. Western (1979) has rightly stressed the importance of distinguishing between "first order" life-history strategies that are a consequence of size and "second order" strategies that are in response to ecological constraints. Ecological parameters, particularly the size and the spatial and temporal availability of food, can affect territory size, group size, and reproductive strategies, and this effect can be seen in the several canid species that have litter sizes inconsistent with other species of similar weight.

Arctic foxes, for example, have unusually large litters for their weight category, and ecological parameters appear to have a major impact on their reproductive effort. In northwestern Canada, where populations of two lemming species, the foxes' basic prey, fluctuate widely and can be very abundant during whelping season, the mean litter size of arctic foxes is 10.6 (MacPherson 1969). By contrast, in coastal areas of Iceland, where food availability is patchy but relatively steady in abundance, the mean litter size is 4.0 (Hersteinsson 1984). Arctic foxes are capable of exploiting periodically abundant food resources by dramatically increasing their reproductive rate. Compared to canids of similar body size, they have twice as many teats (6 or 7 versus 3 or 4 pairs) and are morphologically equipped to handle large litters (Ewer 1973).

Another anomalous species is the maned wolf, a large canid (about 23 kg) whose mean litter size of two pups is the lowest in the Canidae. The only large canid that is omnivorous, it forages primarily on rodents and fruit (Dietz 1984). Feeding on items of food that are so small may impose energetic constraints on the female and on group formation.

The comparison to larger and smaller canids suggests that jackals, as well as coyotes, may be monogamous and have a symmetry of sexual roles in large part because they are medium-sized canids. As such, their body weight and the associated weight and number of their newborn pups may be at a fulcrum point in the balance between female investment in pups at birth and the relative investment needed from the male to ensure the survival of their offspring (Fig. 9). The mating system and the need for cooperative investment in pups would then determine the relative sexual balance among the helpers who remain and, by elimination, the relative dispersal of the two sexes.

This sexual balance could tip toward polyandry or polygyny if the type and availability of food were significantly different and changed the parental investment needed from the male. Field studies of medium-sized canids living under environmental conditions substantially different from those found in this study are needed to understand the extent to which ecological constraints, as opposed to physiological limitations, determine the reproductive strategies and sex ratios of jackal societies.

References

Andelt, W. F. 1982. Behavioral ecology of coyotes on Welder Wildlife Refuge, South Texas. Ph.D. diss., Colorado State Univ., Ft. Collins.

Bekoff, M., T. J. Daniels, and J. L. Gittleman. 1984. Life-history patterns and the comparative social ecology of carnivores. *Ann. Rev. Ecol. Syst.* 15:191–232.

Bekoff, M., J. Diamond, and J. B. Mitton. 1981. Life-history patterns and sociality in canids: Body size, reproduction, and behavior. *Oecologia* 50:386–90.

Bowen, W. D. 1981. Coyote social organization and prey size. *Can. J. Zool.* 59:639–52.

Clutton-Brock, J., G. B. Corbett, and M. Hills. 1976. A review of the family Canidae, with a classification by numerical methods. *Bull. Mus. Nat. Hist.* 29:117–99.

Dietz, J. M. 1984. Ecology and social organization of the Maned Wolf (*Chrysocyon brachyurus*). *Smithsonian Contrib. Zool.* 392:1–51.

Emlen, S. T. 1982a. The evolution of helping. I. An ecological-restraints model. *Am. Nat.* 119:29–39.

———. 1982b. The evolution of helping. II. The role of behavioral conflict. *Am. Nat.* 119:40–53.

Emlen, S. T., J. M. Emlen, and S. A. Levin. 1986. Sex-ratio selection in species with helpers-at-the-nest. *Am. Nat.* 127:1–8.

Ewer, R. F. 1973. *The Carnivores.* Cornell Univ. Press.

Ferguson, J. W., J. A. Nel, and M. J. DeWet. 1983. Social organization and movement patterns of black-backed jackals, *Canis mesomelas,* in South Africa. *J. Zool. London* 199:487–502.

Fisher, R. A. 1930. *The Genetic Theory of Natural Selection.* Oxford Univ. Press.

Frame, L. H., and G. W. Frame. 1976. Female African wild dogs emigrate. *Nature* 263:227–29.

Frame, L. H., J. R. Malcolm, G. W. Frame, and H. van Lawick. 1979. Social organization of African wild dogs (*Lycaon pictus*) on the Serengeti Plains, Tanzania, 1967–1978. *Z. Tierpsychol.* 50:225–49.

Gittleman, J. L. 1984a. The behavioral ecology of carnivores. Ph.D. diss., Univ. of Sussex, Brighton, England.

———. 1984b. Functions of communal care in mammals. In *Evolution: Essays in Honor of John Maynard Smith,* ed. P. J. Greenwood and M. Slatkin, pp. 187–205. Cambridge Univ. Press.

Harrington, F. H., L. D. Mech, and S. H. Fritts. 1983. Pack size and wolf pup survival: Their relationship under varying ecological conditions. *Beh. Ecol. Sociobiol.* 13:19–26.

Harrington, F. H., P. C. Pacquet, J. Ryon, and J. C. Fentress. 1982. Monogamy in wolves: A review of the evidence. In *Wolves of the World,* ed. F. H. Harrington, pp. 209–22. Noyes.

Heerden, J. van, and F. Kuhn. 1985. Reproduction in captive hunting dogs *Lycaon pictus. S. Afr. J. Wildl. Res.* 15:80–84.

Hersteinsson, P. 1984. The behavioral ecology of the Arctic Fox (*Alopex lagopus*) in Iceland. Ph.D. diss., Oxford Univ.

Johnsingh, A. J. T. 1982. Reproductive and social behavior of the dhole, *Cuon alpinus* (Canidae). *J. Zool. London* 198:443–63.

Kleiman, D. G. 1977. Monogamy in mammals. *Quart. Rev. Biol.* 52:39–69.

Kleiman, D. G., and J. F. Eisenberg. 1973. Comparisons of canid and felid social systems from an evolutionary perspective. *Anim. Beh.* 21:637–59.

Kruuk, H. 1975. Functional aspects of social hunting in carnivores. In *Function and Evolution in Behavior: Essays in Honor of Professor Niko Tinbergen,* ed. G. Baerenos et al., pp. 119–41. Oxford Univ. Press.

———. 1978. Foraging and spatial organization of the European badger *Meles meles* L. *Behav. Ecol. Sociobiol.* 4:75–89.

Lamprecht, J. 1978. On diet, foraging behavior, and interspecific food competition of jackals in the Serengeti National Park, East Africa. *Z. Saugetier.* 43:210–23.

———. 1981. The function of social hunting in larger terrestrial carnivores. *Mammal. Rev.* 11:169–79.

Lawick, H. van. 1973. *Solo.* London: Collins.

Macdonald, D. W. 1981. Resource dispersion and the social organization of the red fox (*Vulpes vulpes*). In *Worldwide Furbearer Conference Proceedings,* ed. J. A. Chapman and D. Pursley, pp. 918–49. Frostburg, Md.

———. 1983. The ecology of carnivore social behavior. *Nature* 301:379–84.

MacPherson, A. H. 1969. The dynamics of Canadian arctic fox populations. *Can. Wildl. Serv. Repr. Ser.* 8:1–49.

Malcolm, J. R., and K. Marten. 1982. Natural selection and the communal rearing of pups in African wild dogs (*Lycaon pictus*). *Beh. Ecol. Sociobiol.* 10:1–13.

Mech, L. D. 1970. *The Wolf: Ecology and Social Behavior of an Endangered Species.* Nat. Hist. Press.

Millar, J. S. 1977. Adaptive features of mammalian reproduction. *Evolution* 31:370–86.

———. 1981. Prepartum reproductive characteristics of eutherian mammals. *Evolution* 35:1149–63.

Moehlman, P. D. 1979. Jackal helpers and pup survival. *Nature* 277:382–83.

———. 1981. Why do jackals help their parents? Reply to Montgomerie. *Nature* 289: 824–25.

———. 1983. Sociobiology of silverbacked and golden jackals (*Canis mesomelas, C. aureus*). In *Recent Advances in the Study of Mammalian Behavior,* ed. J. F. Eisenberg and D. G. Kleiman, pp. 423–53. Special Publ. No. 7, the American Society of Mammalogists.

———. 1986. Ecology of cooperation in Canidae. In *Ecological Aspects of Social Evolution,* ed. D. Rubenstein and R. Wrangham, pp. 64–86. Princeton Univ. Press.

Peterson, R. O., J. D. Woolington, and T. N. Bailey. 1984. Wolves of the Kenai Peninsula, Alaska. *Wildl. Monogr.* 88:1–52.

Seal, U. S., E. D. Plotka, J. M. Packard, and L. D. Mech. 1979. Endocrine correlates of reproduction in the wolf. *Biol Reprod.* 21:1057–66.

Senzota, R. B. M. 1978. Some aspects of the ecology of two dominant rodents in the Serengeti ecosystem. M.Sc. Thesis, Univ. Dar-es-Salaam.

Taryannikov, V. I. 1976. Reproduction of the jackal (*Canis aureus* L.) in Central Asia. *Ekologiya* 2:107.

Trivers, R. L. 1972. Parental investment and sexual selection. In *Sexual Selection and the Descent of Man, 1871–1971,* ed. B. Campbell, pp. 136–79. Aldine.

Trivers, R. L., and H. Hare. 1976. Haplodiploidy and the evolution of the social insects. *Science* 191:249–61.

Western, D. 1979. Size, life history, and ecology in mammals. *Afr. J. Ecol.* 17:185–204.

Wyman, J. 1967. The jackals of the Serengeti. *Animals* 10:79–83.

Naked Mole-Rats

Like bees and termites, they cooperate in defense, food gathering and even breeding. How could altruistic behavior evolve in a mammalian species?

Rodney L. Honeycutt

Biological evolution is generally seen as a competition, a contest among individuals struggling to survive and reproduce. At first glance, it appears that natural selection strongly favors those who act in self-interest. But in human society, and among other animal species, there are many kinds of behavior that do not fit the competitive model. Individuals often cooperate, forming associations for their mutual benefit and protection; sometimes they even appear to sacrifice their own opportunities to survive and reproduce for the good of others. In fact, apparent acts of altruism are common in many animal species.

It is easy to admire altruism, charity and philanthropy, but it is hard to understand how self-sacrificing behavior could evolve. The evolutionary process is based on differences in individual fitness—that is, in reproductive success. If each organism strives to increase its own fitness, how could natural selection ever favor selfless devotion to the welfare of others? This question has perplexed evolutionary biologists ever since Charles Darwin put forth the concepts of natural selection and individual fitness. An altruistic act—one that benefits the recipient at the expense of the individual performing the act—represents one of the central paradoxes of the theory of evolution.

In seeking to explain this paradox, biologists have focused their attention on the social insects—ants, bees, wasps and termites. These species exhibit an extreme form of what has been called reproductive altruism, whereby individuals forgo reproduction entirely and actually help other individuals reproduce, forming entire castes of sterile workers. Since reproductive success is the ultimate goal of each player in the game of natural selection, reproductive altruism is a remarkable type of self-sacrifice.

Helping behavior is common in vertebrate societies as well, and some species cooperate in breeding. But until recently there did not appear to be a close vertebrate analogue to the extreme form of altruism observed in social insects. Such a society may now have been found in the arid Horn of Africa, where biologists have been studying underground colonies of a singularly unattractive but highly social rodent.

The naked mole-rat, *Heterocephalus glaber*, appears to be a eusocial, or truly social, mammal. It fits the classical definition of eusociality developed by Charles Michener (1969) and E. O. Wilson (1971), who extensively studied the social insects. In the burrow colonies of naked mole-rats there are overlapping adult generations, and as in insect societies brood care and other duties are performed cooperatively by workers or helpers that are more or less nonreproductive. A naked mole-rat colony is ruled, as is a beehive, by a queen who breeds with a few select males. Furthermore, the other tasks necessary to underground life—food gathering, transporting of nest material, tunnel expansion and cleaning and defense against predators—appear to be divided among nonreproductive individuals based on size, much as labor in insect societies is performed by the sterile worker castes.

The naked mole-rat is not the only vertebrate that can be described as eusocial, but no other vertebrate society mimics the behavior of the eusocial insects so closely. The fact that highly social behavior could evolve in a rodent population suggests that it is time to reexamine some old theories about how eusocial behavior could come into being—theories that were based on the characteristics of certain insects and their societies. In the past decade, since Jennifer U. M. Jarvis first revealed the unusual social structure of a naked mole-rat colony, a number of biologists have been at work considering how a eusocial rodent could evolve. I shall discuss the state of that work briefly here, examining what is known about the naked mole-rat's ecology, behavior and evolution and about altruistic animal societies.

Introducing the Naked Mole-Rat

The naked mole-rat is a member of the family Bathyergidae, the African mole-rats—so named because they resemble rats but live like moles. Many rodents burrow and spend at least part of their life underground; all 12 species of Bathyergidae live exclusively underground, and they share a set of features that reflect their subterranean lifestyle and that demonstrate evolutionary convergence, the independent development of similar characteristics. Like the more familiar garden mole, a mole-rat has a stout, cylindrical body, a robust skull, eyes that are small or absent, reduced external ears, short limbs, powerful incisors and sometimes claws for digging, and a somewhat unusual physiology adapted to the difficulties of life underground, including a burrow

Rodney L. Honeycutt is an associate professor in the Department of Wildlife and Fisheries Sciences and a member of the Faculty of Genetics at Texas A&M University. He began his research on the genetics and systematics of African mole-rats in 1983 during his tenure as assistant professor of biology at Harvard University and assistant curator of mammals at Harvard's Museum of Comparative Zoology. He is interested in the evolution and systematics of mammals and has taught courses in mammalian biology for the past seven years. His research has taken him to regions of Africa, South America, Central America and Australia. Address: Department of Wildlife and Fisheries Sciences, Texas A&M University, 210 Nagle Hall, College Station, TX 77843.

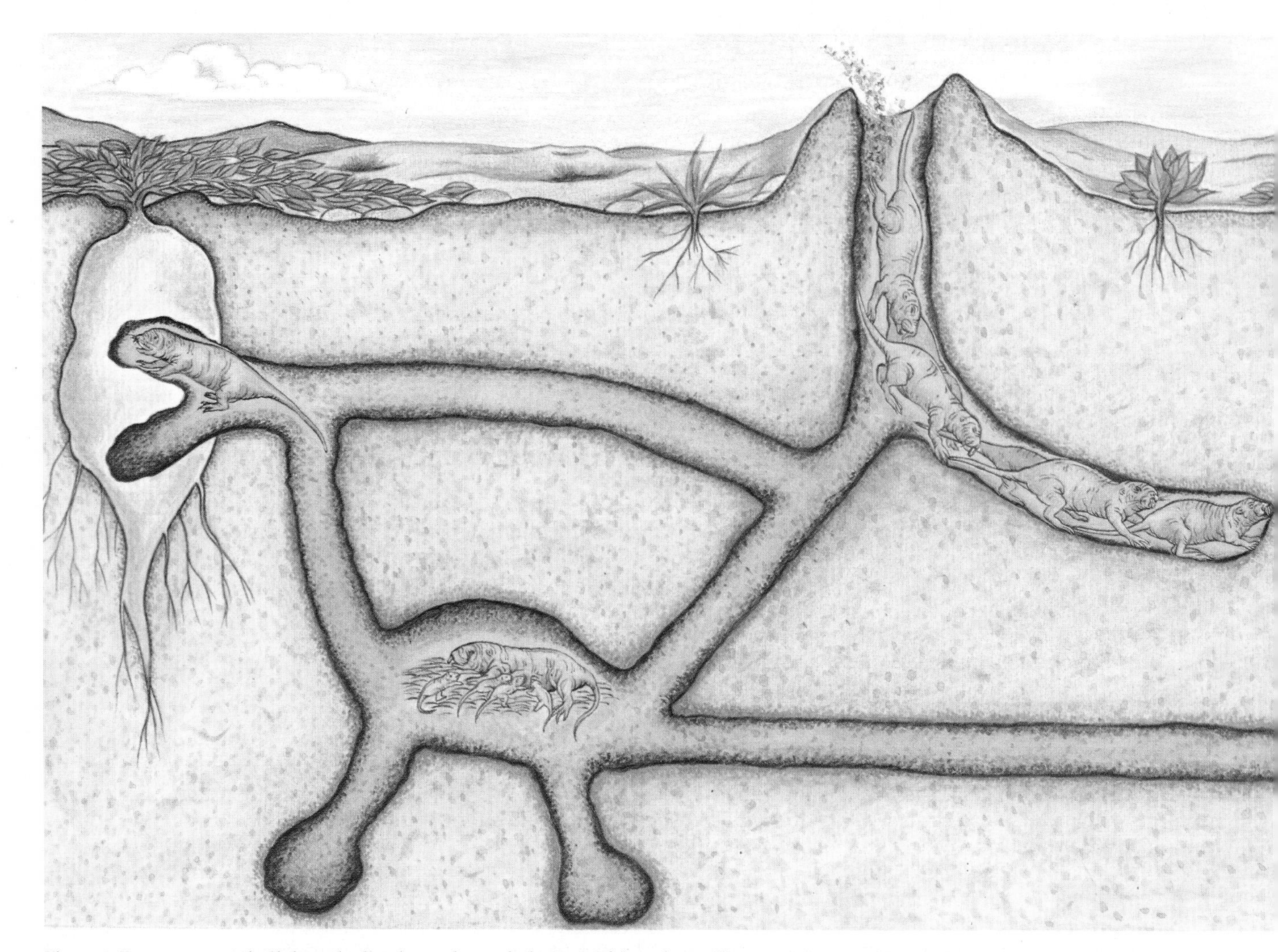

Figure 1. Burrow system built by naked mole-rats beneath the East African desert illustrates the complex social organization that makes the subterranean species unusual. Reproduction in a naked mole-rat colony, which usually has 70 to 80 members, is controlled by a queen, the only breeding female, shown here nursing newborns in a nest chamber. Digging tunnels to forage for food is one of the functions of

atmosphere high in carbon dioxide. All Bathyergidae species are herbivorous, and all but one sport fur coats.

Field biologists who encountered naked mole-rats in the 19th century thought that these small rodents—only three to six inches long at maturity, with weights averaging 20 to 30 grams—were the young of a haired adult. But subsequent expeditions showed that adult members of the species are hairless except for a sparse covering of tactile hairs. Oldfield Thomas, noting wide variations in the morphological characteristics of the naked mole-rats, identified what he thought were several species. *H. glaber* is currently considered a single species, within which there is great variation in adult body size.

Naked mole-rats inhabit the hot, dry regions of Ethiopia, Somalia and Kenya. Like most of the Bathyergidae species, they build elaborate tunnel systems. The tunnels form a sealed, compartmentalized system interconnecting nest sites, toilets, food stores, retreat routes and an elaborate tunnel system allowing underground foraging for tubers *(Figure 1)*. Like the morphology of the animals, the tunnel system is an example of convergent evolution, being similar to those of the other mole-rats in its compartmentalization, atmosphere and more or less constant temperature and humidity. Naked mole-rats subsist primarily on geophytic plants (perennials that overwinter in the form of bulbs or tubers), which are randomly and patchily distributed. The mole-rats forage broadly by expanding their burrows, but their distribution is limited by food supply and soil types. Like most rodents that live underground, they are not able to disperse over long distances.

The tunnel systems of naked mole-rats can be quite large, containing as many as two miles of burrows. The average colony is thought to have 70 to 80 members. In order to study the social organization of the naked mole-rats, biologists have had to devise ways to capture whole colonies and recreate their burrow systems in the laboratory. This is not an easy task, but it is possible because the rodents have a habit of investigating opened sections of their burrow systems and then blocking them. One can create an opening, then capture the naked mole-rats as they come to seal it. Cutting off their retreat requires quick work with a spade, hoe or knife, and the procedure must be repeated in various parts of the tunnel system in order to retrieve an entire colony. A carefully reconstructed colony can survive quite well in captivity, and naked mole-rats are beginning to become an attraction at zoos.

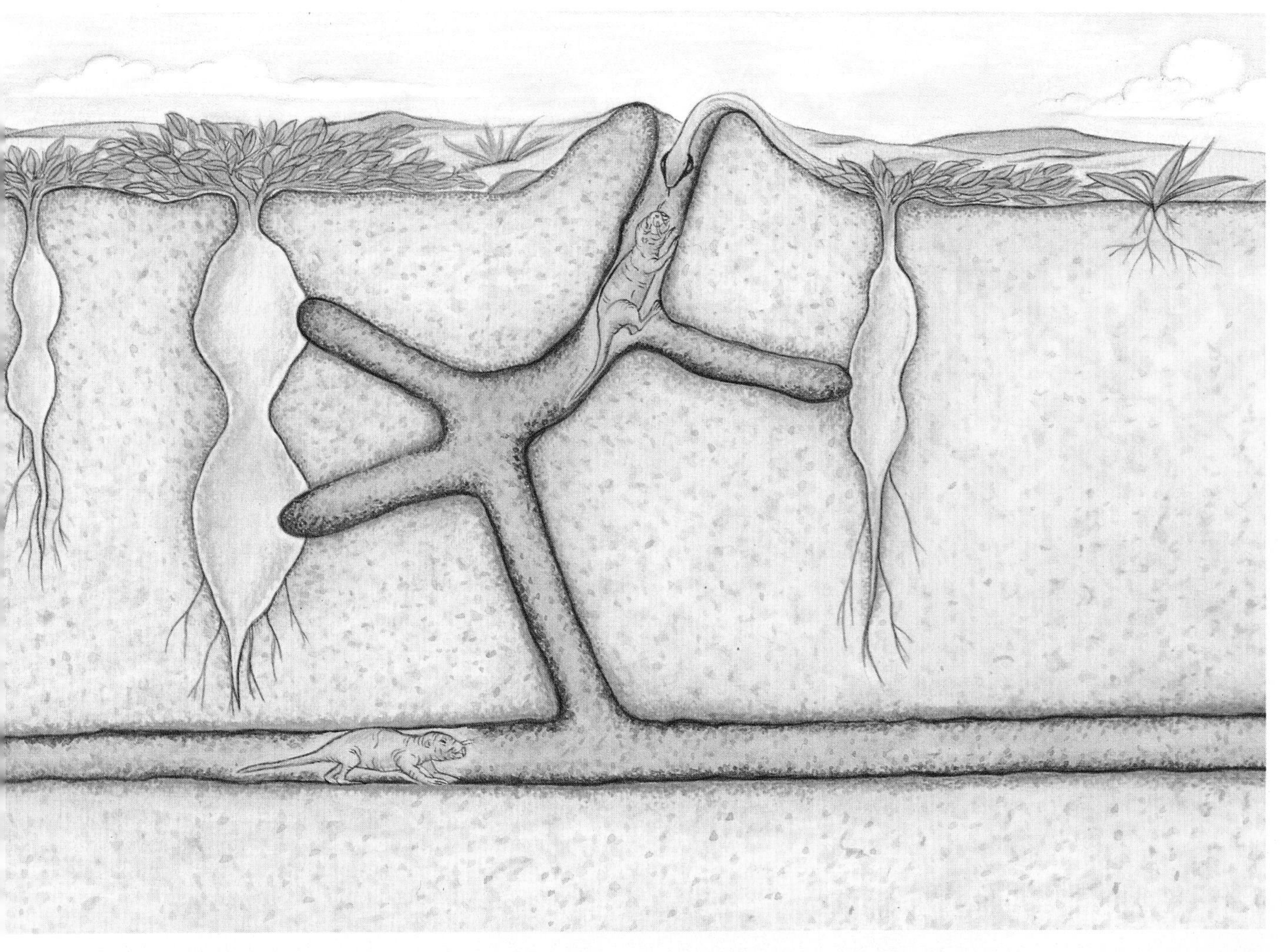

nonreproductive workers, which often form digging teams; one individual digs with its incisors while others kick the dirt backward to a mole-rat that kicks it out of the tunnel. The molehills or "volcanoes" formed in this way are plugged to create a closed environment and deter predators such as the rufous-beaked snake. Tubers and bulbs are the naked mole-rats' food source.

Most African mole-rats excavate by digging with their large incisors, removing the dirt from the burrow with their hind feet. The digging behavior of naked mole-rats, which are most active during periods when the soil in their arid habitat is moist, appears to be unlike that of the other mole-rats in two respects. First, instead of plugging the surface opening to a tunnel during excavation, the naked mole-rats "volcano," kicking soil through an open hole to form a tiny volcano-shaped mound. When excavation is complete, the tunnel is plugged to form a relatively airtight, watertight and predator-proof seal *(Figure 4)*. Second, naked mole-rats have been observed digging cooperatively in a wonderfully efficient arrangement that resembles a bucket brigade. One animal digs while a chain of animals behind move the dirt backward to an

Figure 2. Wrinkled, squinty-eyed and nearly hairless, the first naked mole-rats found by biologists were thought to be the young of a haired adult. The rodents are just three to six inches long at maturity, although there is great variation in body size within each colony. Other morphological features reflect the fact that the naked mole-rats live entirely underground: small eyes, two pairs of large incisors for digging, and reduced external ears. (Except where noted, photographs courtesy of the author.)

Figure 3. Habitat of the naked mole-rats is hot, dry and dotted with patches of vegetation. Visible in the foreground of this photograph, taken in Kenya, are the molehills formed by the rodents.

Figure 4. "Volcanoes" formed when naked mole-rats kick sand out of a tunnel, then plug the opening, make the animals' burrows easy to find. Naked mole-rats are most vulnerable to predators while forming volcanoes; the activity often attracts the attention of snakes.

animal at the end, which kicks the dirt from the burrow. One 87-member colony was seen to remove about 500 kilograms of soil per month by this process. Another colony of similar size moved an estimated 13.5 kilograms in an hour—about 380 times the mean body weight of a naked mole-rat. A team kicking dirt through a surface opening is vulnerable to attack from snakes; the mounds also make *H. glaber*'s colonies easy for scientists to find.

Naked mole-rats are long-lived animals and prolific breeders. Several individuals caught in the wild are surviving after 16 years in captivity; two of these are females that still breed. In captive colonies females have produced litters as large as 27, and in wild populations litter sizes can be as high as 12. The naked mole-rat breeds year-round, giving birth about every 70 to 80 days. This fecundity is unusual among the Bathyergidae. The other highly social species of African mole-rat, *Cryptomys damarensis*, is also a year-round breeder but produces smaller litters, with an average size of five.

The major threat to the longevity of a naked mole-rat, and probably to all of the mole-rats, is predation. On at least two occasions I have encountered the rufous-beaked snake in a mole-rat burrow; one snake had three mole-rats in its stomach. Similar field observations have been made by other investigators. Encounters between mole-rats and snakes in the laboratory suggest that avoidance may not be the mole-rat's only strategy against predators; individuals have also been seen attacking the predator in their defense of the colony.

The naked mole-rat's closest relatives are the 11 other species in the Bathyergidae, which are all of exclusively African origin and distribution *(Figure 5)*. It has been difficult to determine which of the 32 other rodent families shares a common ancestry with the Bathyergidae, but a consensus arising from recent studies places the family in the rodent suborder Hystricognathi, which includes caviomorph rodents from the New World—porcupines, guinea pigs and chinchillas—and porcupines and cane rats from the Old World. The naked mole-rat is the most divergent species within the Bathyergidae, its evolutionary branch splitting off at the base of the family's phylogenetic tree *(Figure 6)*.

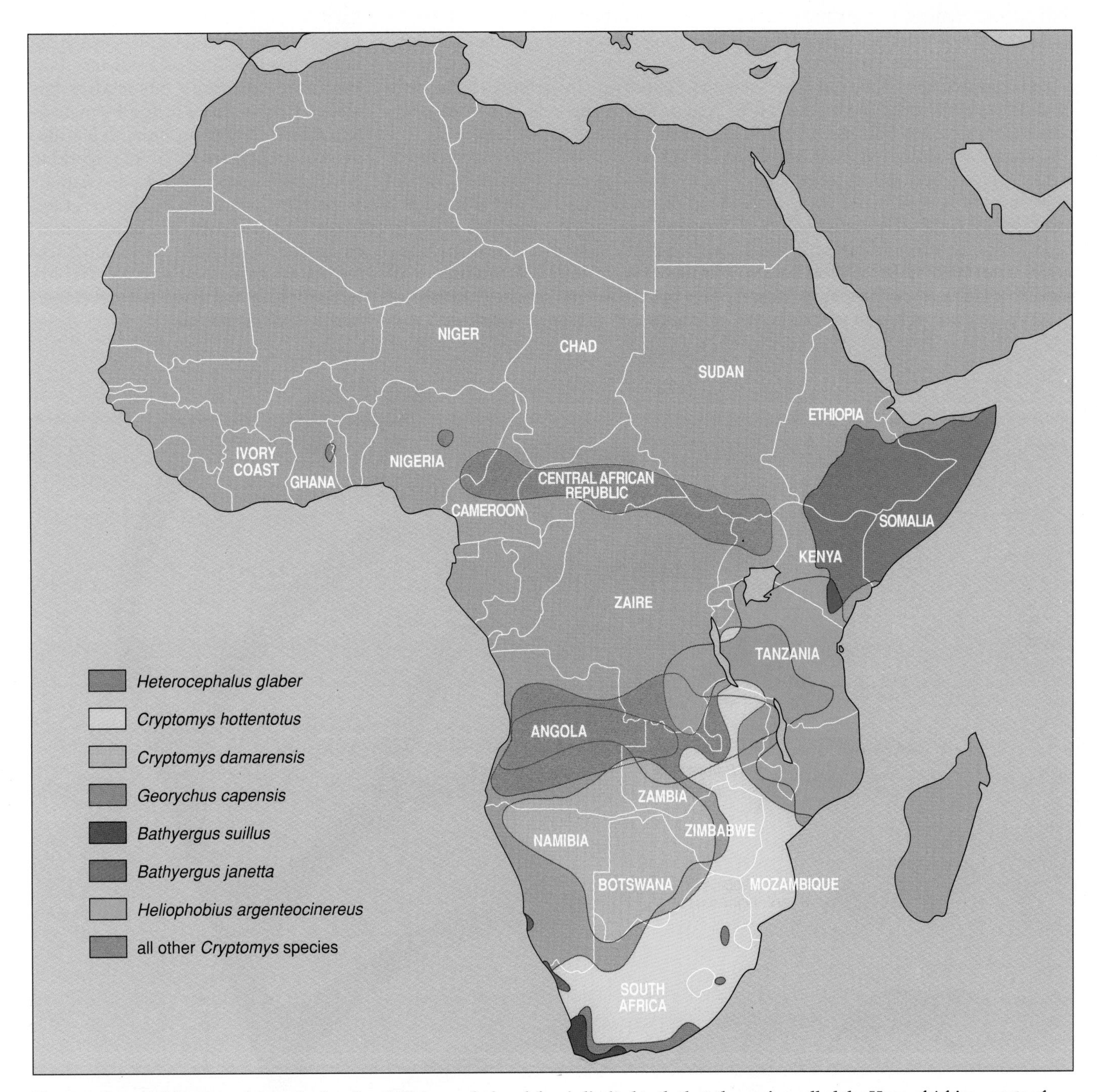

Figure 5. Geographic range of the naked mole-rat, *Heterocephalus glaber*, is limited to the hot, dry region called the Horn of Africa—parts of Ethiopia, Kenya and Somalia. On the map are shown the areas inhabited by other species of African mole-rats. All species in the family Bathyergidae live entirely underground. Most are solitary or colonial; the other species with a highly developed social structure, *Cryptomys damarensis*, is found in Southern Africa.

How Do Altruistic Societies Evolve?

Darwin called the development of sterile castes in insect societies a "special difficulty" that initially threatened to be fatal to his theory of natural selection. His solution to the problem was surprisingly close to current hypotheses based on genetic relatedness, even though he did not have a knowledge of genetics. Darwin suggested that traits, such as helping, that were observed in sterile form could survive if individuals that expressed the traits contributed to the reproductive success of those individuals that had the trait but did not express it.

Today the notion of *inclusive fitness* forms the foundation for theories about how reproductive altruism might evolve. The idea arose in 1964 from William Hamilton's remarkable genetic studies of the Hymenoptera, the insect order that includes the social ants, bees and wasps. Hamilton showed that if the genetic ties within a generation are closer than the ties between generations, each member of the generation might be motivated to invest in a parent's reproductive success rather than his or her own. Inclusive fitness is a combination of one's own reproductive success and that of close relatives.

In the Hymenoptera, Hamilton found an asymmetric genetic system that could contribute to the development of reproductive altruism by giving individuals chances to maximize their inclusive fitness without reproducing. Hymenopteran males arise from unfertilized eggs and thus have only one set of chromosomes (from the mother); females have one set from each parent. The males are called haploid, the females diploid, and this system of sex determination is referred to as *haplodiploidy (Figure 9)*. The daughters of a monogamous mother share identical genes from their father and half their mother's genes; they thus have three-quarters of their genes in common. A female who is more closely related to her sister than to her mother or her offspring can propagate her own genes most effectively by helping create more sisters. Sterile workers in hymenopteran insect colonies are all female.

Hamilton's work prompted a flurry of interest in genetic asymmetry, but he and others recognized that it was not a general explanation for how eusocial societies might evolve. There are many limitations; for instance, multiple matings by females reduce the closeness of relationships between sisters, and it is hard to explain the incentives for females to tend juvenile males, which are not as closely related as are sisters. Furthermore, although eusociality has evolved more times in the Hymenoptera than in any other order, it has also evolved in parts of the animal world in which both sexes are diploid— namely Isoptera, which includes the social termites, and Rodentia, the order that includes the naked mole-rat. Finally, there are many arthropod species that are haplodiploid and have not developed highly social behavior.

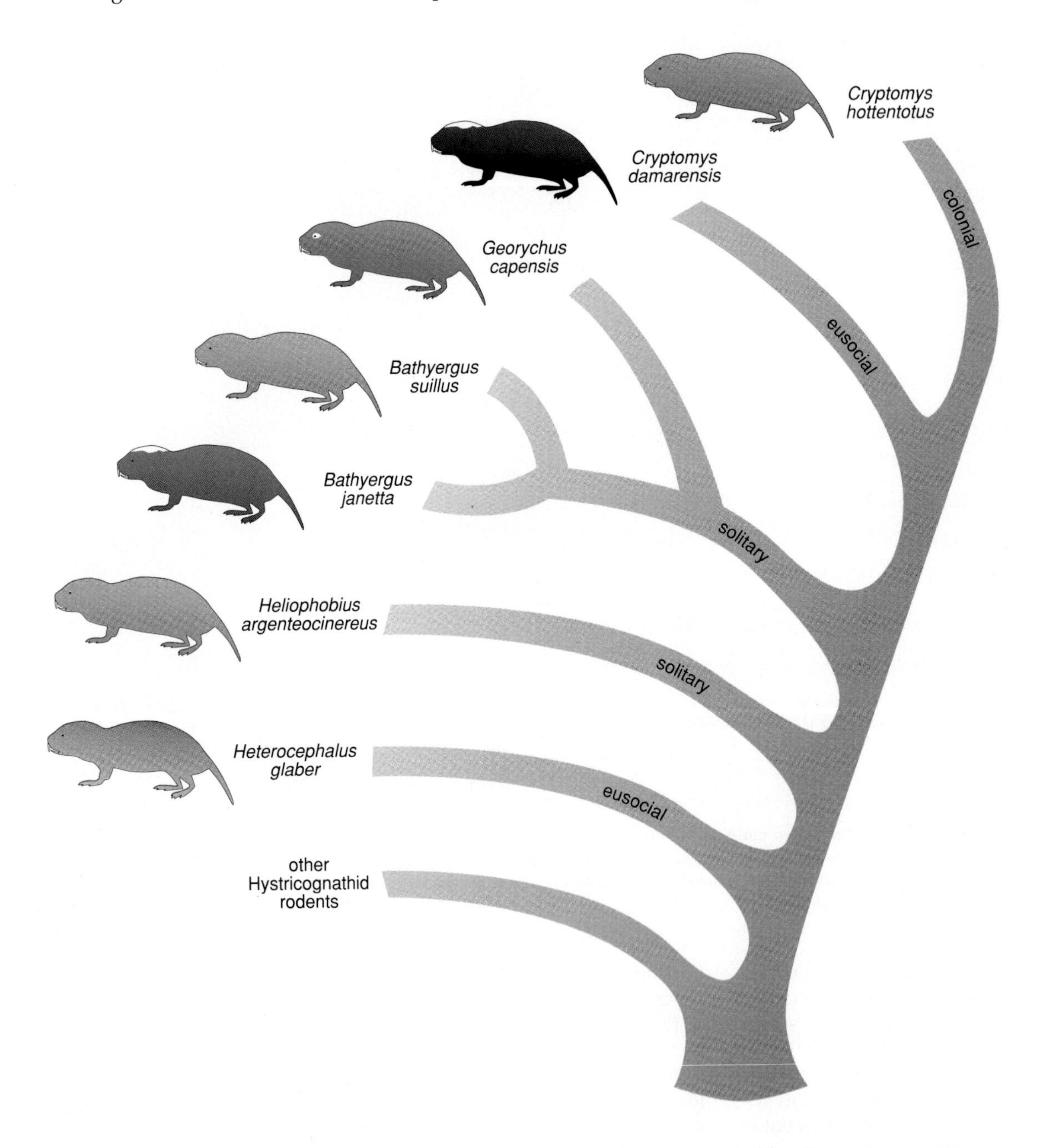

Figure 6. Phylogenetic tree for the family Bathyergidae, the African mole-rats, shows that the two eusocial, or truly social, species are quite divergent. Among other rodents, the suborder Hystricognathi, which includes porcupines, guinea pigs and chinchillas, appears to have the closest genetic link with the African mole-rats. Although there is much similarity among the Bathyergidae species in their physiological characteristics and their subterranean lifestyle, the phylogenetic distance between the eusocial species of mole-rats suggests that complex social behavior evolved separately in the two cases.

There is another way that close kinship might develop among the members of a generation, and it is considered a possible explanation for the evolution of the termite and naked mole-rat societies. Several generations of inbreeding could result in a higher degree of relatedness among siblings than between parents and offspring *(Figure 10)*. When male and female mates are unrelated, but each is the product of intense inbreeding, their offspring can be genetically identical and might be expected to stay and assist their parents for the same reasons set forth in the haplodiploid model. The inbreeding model was developed by Stephen Bartz in 1979 to explain the development of eusocial behavior in termites, which live in a contained and protected nest site conducive to multigenerational breeding.

Genetics alone cannot provide a comprehensive explanation for the evolution of eusociality. Other possible explanations, especially relevant to termites and vertebrate helpers, lie in combinations of ecological and behavioral factors. These factors perhaps provided preconditions or starting points for the eventual evolution of a eusocial lineage or species. The best way to understand the development of eusociality may be to consider the costs and benefits associated with remaining in the natal group and helping, as compared to the costs and benefits of dispersing and breeding.

Probably one of the most important preconditions for the development of eusociality is parental care in a protected nest, where offspring are defended against predators and provided with food. If there is a high cost associated with dispersal—in terms of restricted access to food, lack of breeding success or increased vulnerability to predators—then there may be an incentive for juveniles to remain in the protected nest and become helpers. Helpers that remain in the nest for multiple generations may forgo reproduction indefinitely as a consequence of maternal manipulation.

The short-term benefits of group living seem to accrue mainly to those individuals who are reproducing, since they benefit from the help others provide with defense and obtaining food. In fact, there is a correlation between the breeder's reproductive fitness and the number of helpers in cooperatively breeding vertebrate species. Thus the long-term effect of helping may be an

Figure 7. Catching naked mole-rats requires some understanding of their behavior. Mole-rat catchers create an opening from the surface to a burrow, which is normally kept sealed by the animals, and wait quietly for a mole-rat to investigate. A spade, hoe, pick or knife blade is driven quickly into the tunnel to block the mole-rat's escape. (Photograph courtesy of Stan Braude, University of Missouri at St. Louis.)

Figure 8. Captive naked mole-rats, carrying identifying tattoos, adapt well to being placed together in bins, apparently because the highly social animals tend to huddle together for warmth in their burrows in the wild. These rodents are part of Jennifer U. M. Jarvis's collection at the University of Cape Town.

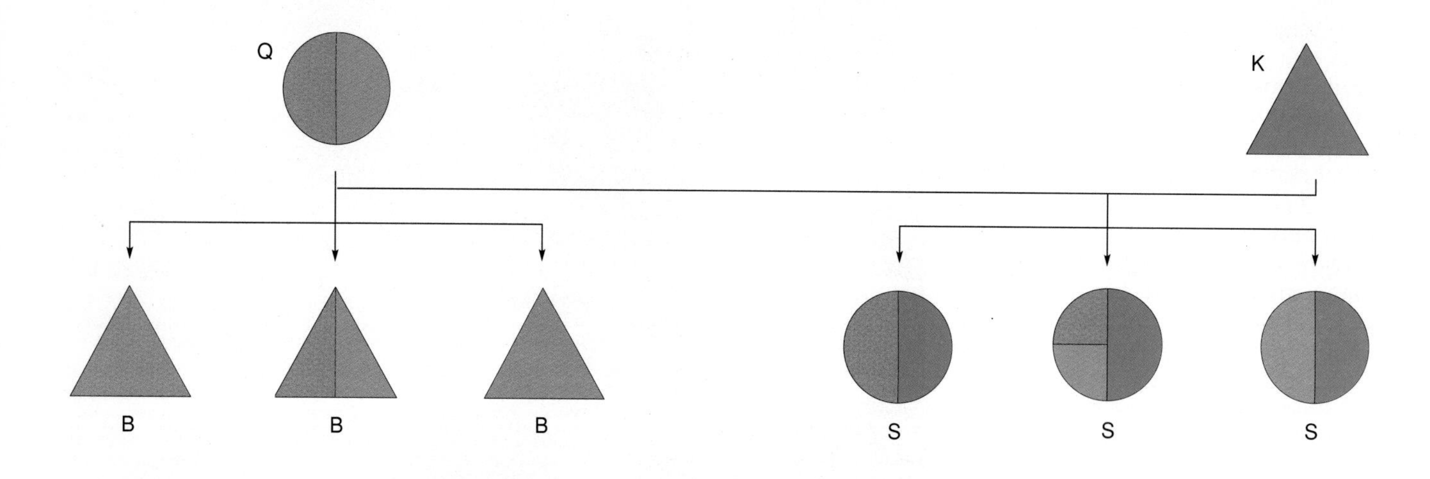

degrees of relatedness in haplodiploid species

	daughter	son	mother	father	sister	brother
female	$\frac{1}{2}$	$\frac{1}{2}$	$\frac{1}{2}$	$\frac{1}{2}$	$\frac{3}{4}$	$\frac{1}{4}$
male	1	0	1	0	$\frac{1}{2}$	$\frac{1}{2}$

Figure 9. Haplodiploidy, an asymmetric genetic system, is thought to contribute to the development of reproductive altruism in ants, bees and wasps—species with intricate social systems that include sterile castes of workers. In a haplodiploid species, males *(triangles)* arise from unfertilized eggs and have only one set of chromosomes, whereas females *(circles)* have one set of chromosomes from each parent. The relatedness between sisters—the fraction of their genes that are shared—is thus greater than the relatedness of mother and daughter *(bottom panel)*. William D. Hamilton hypothesized that females seeking to increase their inclusive fitness—a combination of their own reproductive success and that of close relatives—might in a haplodiploid species become helpers, advancing the continuation of their own genetic heritage by helping with the reproduction of sisters rather than their own offspring. Although haplodiploidy is not considered a full explanation of how eusocial behavior would evolve in ants, bees and wasps, it is notable that most species in which reproductive altruism has evolved are haplodiploid, and that the sterile workers among the haplodiploid insects are all female. In this illustration, the parents are labeled *Q* and *K* and the offspring *S* and *B*, following the scheme in Figure 10; for simplicity, the effects of any recombination of genes are not depicted.

increase in inclusive fitness for the helpers. This may prove to be a very important consideration in species where the probability of a dispersing individual procuring a nest site and eventually breeding is extremely low.

Naked Mole-Rat Society

In some ways the social organization observed in naked mole-rat colonies is more akin to the societies of the social insects than to the social organization of any other vertebrate species. In other respects, mole-rats are unique and may always remain a bit of a mystery.

Some similarities between naked mole-rat societies and the insect societies are striking. A naked mole-rat colony, like a beehive, wasp's nest or termite mound, is ruled by its queen or reproducing female. Other adult female mole-rats neither ovulate nor breed. The queen is the largest member of the colony, and she maintains her breeding status through a mixture of behavioral and, presumably, chemical control. She is aggressive and domineering; queenly behavior in a naked mole-rat includes facing a subordinate and shoving it along a burrow for a distance. Queens have been long-lived in captivity, and when they die or are removed from a colony one sees violent fighting among the larger remaining females, leading to a takeover by a new queen.

Most adult males produce sperm, but only one to three of the larger males in a colony breed with the queen, who initiates courtship. There is little aggression between breeding males, even upon removal of the queen. The queen and breeding males do not participate in the defense or maintenance of the colony; instead, they concern themselves with the handling, grooming and care of newborns.

Eusocial insect societies have a rigid caste system, defined on the basis of distinctions in behavior, morphology and physiology. Mole-rat societies, on the other hand, demonstrate behavioral asymmetries related primarily to reproductive status (reproduction being limited to the queen and a few males), body size and perhaps age. Smaller nonbreeding members, both male and female, seem to participate more in gathering food, transporting nest material and clearing tunnels. Larger nonbreeders are more active in defending the colony and perhaps in removing dirt from the tunnels. Jarvis has suggested that differences in growth rates may influence the length of time that an individual performs a task, regardless of its age.

Naked mole-rats, being diploid in both sexes, do not have an asymmetric genetic system such as haplodiploidy. As Bartz has proposed for termites, inbreeding in naked mole-rats may create a genetic asymmetry that mimics the result of haplodiploidy. There is genetic evidence suggesting that naked mole-rats are highly inbred within colonies and even between colonies in a local area. An important part of the question about breeding within and between colonial groups cannot be answered, however, since there is very little information on how mole-rat colonies are established. This makes it difficult to evaluate the naked mole-rats using Bartz's model of inbreeding and eusociality in termites.

Still, among the eusocial insects termites offer the closest comparison with the naked mole-rats. Termites are the only eusocial insects outside the Hymenoptera, and all termites are diploid, with two sets of chromosomes. Worker groups include nonreproductive males and females, and they perform primarily tasks associated with maintaining and defending the colony. The queen termite is more passive than a naked mole-rat queen and uses chemical control. Termite colonies are much larger, sometimes having more than 10,000 workers, and the definition of castes is more rigid.

The naked mole-rat cannot be considered the only eusocial vertebrate species, but it does represent the most advanced form of vertebrate eusociality and the one most analogous to eusociality in insects. Helping or cooperative breeding has evolved many times in vertebrates, and in many of those species the social system includes both a small number of reproducing individuals (usually a dominant breeding pair) and several nonbreeding individuals (males and females), representing offspring from previous years, that serve as helpers or alloparents. As in naked mole-rats, these nonbreeders participate in foraging for food, care of young and defense against predators. Unlike naked mole-rats, most cooperatively breeding vertebrates (an exception being the wild dog, *Lycaon pictus*) are dominated by a pair of breeders rather than by a single breeding female. The division of labor within a social group is not as pronounced in other vertebrates, and the colony size is much smaller. In addition, mating by subordinate females in many social vertebrates is not totally suppressed, whereas in naked mole-rat colonies subordinates are not sexually active, and many may never breed.

Several ecological and behavior factors may have facilitated the evolution of eusociality in naked mole-rats. Richard Alexander, Katharine Noonan and Bernard Crespi (1991) have sug-

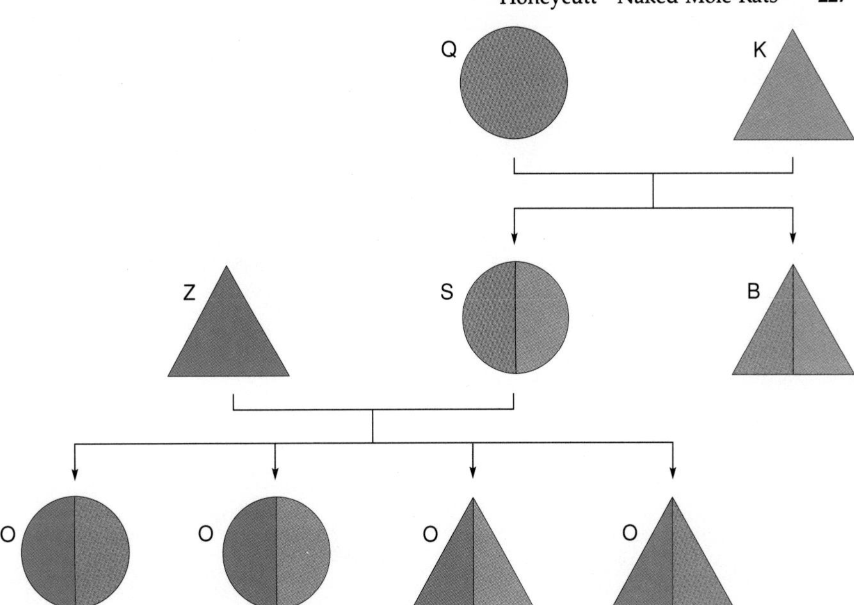

Figure 10. Genetic asymmetry can be produced by cycles of inbreeding and outbreeding in a way that may encourage the evolution of reproductive altruism. Stephen Bartz developed a genetic model to explain how complex social behavior could have evolved in termites living within the confines of a bark-covered chunk of rotting wood. Bartz's hypothesis begins with the mating of a male and a female who are unrelated but are each the product of intense inbreeding (the "queen" and "king," or *Q* and *K, above*), so that for each, both halves of the genotype are essentially identical. The products of this union (*S* and *B*) are essentially identical and therefore more related to one another than to their parents; this genetic asymmetry is thought to encourage helping behavior in both sexes because each sibling can increase its inclusive fitness by assisting in the creation of brothers and sisters. If one of the offspring mates with a similarly inbred but unrelated individual, as in the case of *S* and *Z*, the new parents and the new offspring (*O*) are less closely related than than are the original siblings, *S* and *B*. The result mimics the close ties between siblings that are produced by haplodiploidy (*Figure 9*), but the genetic asymmetry disappears in subsequent generations unless specific patterns of inbreeding and outbreeding are followed. (Adapted from Bartz 1979.)

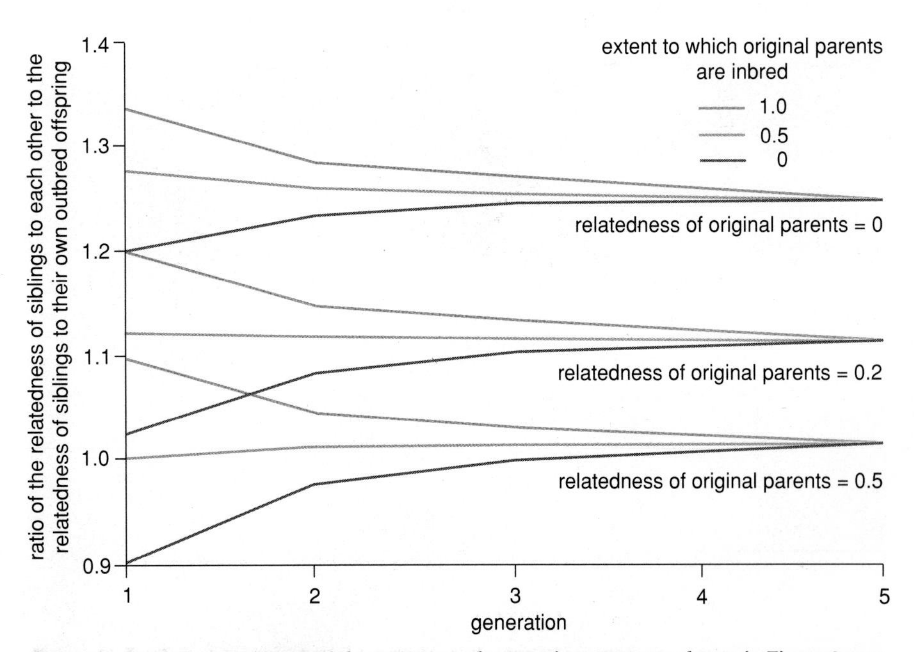

Figure 11. Brother-sister incest might perpetuate the genetic asymmetry shown in Figure 9 over several generations. On this graph, a ratio of relatedness greater than 1.0 means that siblings are more related to one another than to their offspring and are therefore encouraged to become helpers rather than breeders. It is evident that helping behavior is most encouraged when the original parents are highly inbred but unrelated; brother-sister mating makes the inbreeding of the parents unimportant after a few generations, but the importance of the relatedness of the original parents persists. A similar pattern might have contributed to the evolution of social behavior in the confined quarters in which naked mole-rats breed. (From Bartz 1979.)

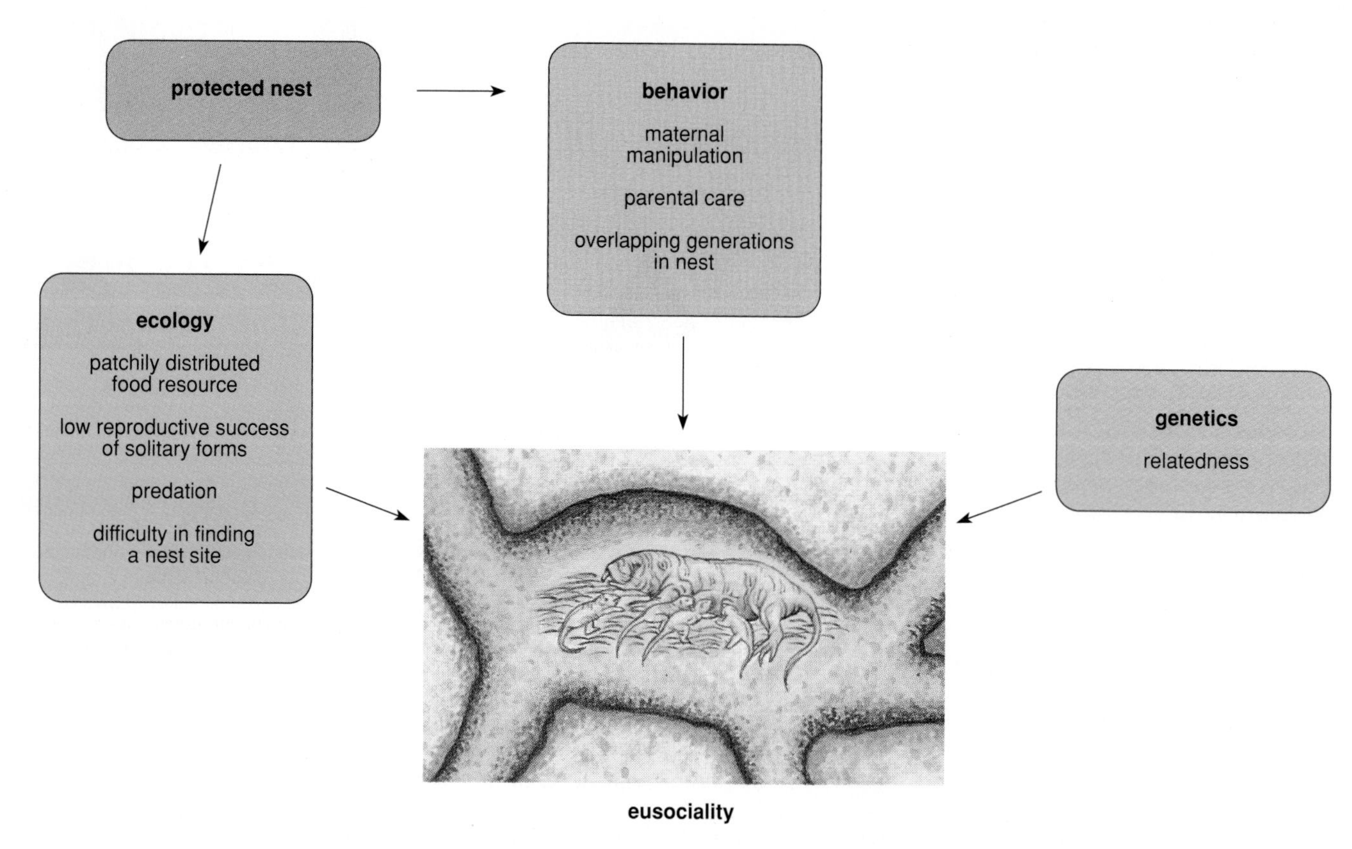

Figure 12. Genetic, ecological and behavioral factors probably combine to promote the development of eusociality in an animal species. All of the factors shown above may have been important in the evolution of the complex social organization of naked mole-rat colonies. Inbreeding in the rodents' underground burrows may have created a high degree of genetic inbreeding within generations, promoting helping behavior. In the ecosystem inhabited by the naked mole-rats, the costs associated with leaving the nest may be high compared to the benefits of group living. Behavioral patterns such as parental care may also have predisposed the species to large-scale group living. A closed burrow system that provides a protected nest environment might be crucial in tipping the ecological and behavioral balance toward organized group living and cooperative breeding.

Figure 13. Ruler of a famous group of naked mole-rats, the queen of Jennifer U. M. Jarvis's captive colony at the University of Cape Town in South Africa is distinguished from her subjects by her large size. She is still breeding after 16 years of captivity. Jarvis's colony served as the basis for the first description of eusocial behavior in a mammal.

gested that the subterranean niche shared by termites and naked mole-rats may be an important precursor for the evolution of eusociality. Life underground provides relative safety from predators and access to a readily available food source that does not require exit from the underground chamber. It also offers an expandable living place that can accommodate a large group.

Since the naked mole-rats share their subterranean niche with the other mole-rat species, it is interesting to speculate about why eusocial behavior has or has not evolved among the other Bathyergidae. Of the other 11 species, all are solitary but two—one of which, *Cryptomys damarensis*, may be termed eusocial. *C. damarensis* is not a particularly close relative of the naked mole-rat; whereas *H. glaber* diverged at the base of the family phylogenetic tree, *C. damarensis* is a distantly related and much more recent species, suggesting that complex social behavior in the two species evolved quite separately *(Figure 6)*. Another species in the *Cryptomys* genus, *C. hottentotus*, is the only other social member of the family; its small colonies (two to 14 members) have less well-developed social structures and vary in size and organization.

C. damarensis colonies are somewhat smaller than those of the naked mole-rat, having eight to 25 members. They also include a single reproductive female and one or more reproductive males and exhibit a division of labor among reproductive individuals based on size. One important difference between the two species has been suggested: *C. damarensis* colonies appear to be less stable over time, and the effects of multigenerational group living and inbreeding may be less pronounced. The significance of the presumed differences is a matter that will require further study because the dynamics associated with the duration of colonies and the founding of new colonies in wild naked mole-rats are not well understood. For instance, new colonies presumably are formed from existing colonies by budding, or fissioning, but the frequency of this event and its causes are not known.

Figure 14. *Cryptomys damarensis* is an African mole-rat distantly related to the naked mole-rats but sharing many kinds of eusocial behavior. Its somewhat larger colonies appear to be less stable over time. The species is found in Southern Africa.

The features of the subterranean niche may supply part of the explanation of social living in mole-rats, even though many solitary, non-social species of rodents in the Bathyergidae and other families occupy a similar niche. Restricted access to food and an unpredictable environment may also provide clues to the evolution of eusociality in both naked mole-rats and *C. damarensis* because as resources become more difficult to find, the energetic cost associated with finding them increases. Several authors have suggested that cooperation in food foraging and communal living might be promoted by the patchy distribution of the food source.

There is no simple explanation for the evolution of eusociality, and the hypotheses that fit the naked mole-rats and the other social species should not be considered mutually exclusive. Reproductive altruism is more likely to occur among genetically related individuals, but relatedness is not a sufficient explanation. Each eusocial species has a unique combination of life-history characteristics associated with both its ecology and its behavior, and some or perhaps all of these characteristics may have predisposed a particular species for group living and cooperative breeding. The fact that various factors can work together in the development of eusociality may provide the ultimate explanation for the novelty, and therefore the mystery, of each example of eusocial behavior.

Bibliography

Alexander, R. D., K. M. Noonan and B. J. Crespi. 1991. The evolution of eusociality. In *The Biology of the Naked Mole-Rat*, ed. P. W. Sherman, J. U. M. Jarvis and R. D. Alexander, 3–44. Princeton, N.J.: Princeton University Press.

Allard, M. W., and R. L. Honeycutt. 1992. Nucleotide sequence variation in the mitochondrial 12*S* rRNA gene and the phylogeny of African mole-rats (Rodentia: Bathyergidae). *Molecular Biology and Evolution* 9 (in press).

Andersson, M. 1984. The evolution of eusociality. *Annual Review of Ecology and Systematics*. 15:165–189.

Bartz, S. H. 1979. Evolution of eusociality in termites. *Proceedings of the National Academy of Sciences (U.S.A.)* 76:5764–5768.

Bennett, N. C., and J. U. M. Jarvis. 1988. The social substructure and reproductive biology of colonies of the mole-rat, *Cryptomys damarensis* (Rodentia, Bathyergidae). *Journal of Mammalogy*. 69:293–302.

Brown, J. L. 1987. *Helping and Communal Breeding in Birds*. Princeton, N. J.: Princeton University Press.

Emlen, S. T. 1991. Evolution of cooperative breeding in birds and mammals. In *Behavioral Ecology: An Evolutionary Approach*, 3rd edition, ed. J. R. Krebs and N. B. Davies, 301–337. Palo Alto, Calif.: Blackwell Scientific Publications.

Genelly, R. E. 1965. Ecology of the common mole-rat (*Cryptomys hottentotus*) in Rhodesia. *Journal of Mammalogy* 46:647–665.

Hamilton, W. D. 1964. The genetical evolution of social behavior. *Journal of Theoretical Biology* 7:1–52.

Jarvis, J. U. M. 1981. Eusociality in a mammal: Cooperative breeding in naked mole-rat colonies. *Science* 212:571–573.

Macdonald, D. W., and P. D. Moehlman. 1982. Cooperation, altruism, and restraint in the reproduction of carnivores. In *Perspective in Ethology*, Vol. 5, ed. P. P. G. Bateson and P. H. Klopfer, 433–467. New York: Plenum Press.

Michener, C. D. 1969. Comparative social behavior of bees. *Annual Review of Entomology* 14:277–342.

Reeve, H. K., D. F. Westneat, W. A. Noon, P. W. Sherman and C. F. Aquadro. 1990. DNA "fingerprinting" reveals high levels of inbreeding in colonies of the eusocial naked mole-rat. *Proceedings of the National Academy of Sciences (U.S.A.)*. 87:2496–2500.

Trivers, R. 1985. *Social Evolution*. Menlo Park, Calif.: The Benjamin/Cummings Publishing Company, Inc.

Wilson, E. O. 1971. *The Insect Societies*. Cambridge, Mass.: Harvard University Press.

David M. Buss

Human Mate Selection

Opposites are sometimes said to attract, but in fact we are likely to marry someone who is similar to us in almost every variable

Historically, human mating systems have deviated from randomness in nearly every way imaginable. Major variants include polygyny, in which men take multiple wives; polyandry, in which women take multiple husbands; endogamy, or inbreeding, in which close genetic relatives mate; exogamy, or outbreeding, in which mating with genetic relatives is avoided; and hypergamy, usually paired with polygyny, in which women marry upward in the socioeconomic hierarchy. One deviation from randomness that has never been reliably demonstrated, however, is the tendency of opposites to marry or mate. On the contrary, assortative mating, which can be defined as the coupling of individuals based on their similarity on one or more characteristics, is the most common deviation from random mating in Western societies.

Who mates with whom has been a subject of intense interest among scientists ranging from biologists and geneticists to psychologists and sociologists. Part of the intrigue of human mating patterns lies in their range of impact, which transcends disciplinary boundaries. Sociologists study mate selection because more than 90% of all people are married at some point in their lives (Price and Vandenberg 1980), and these marriages affect social trends such as the distribution of wealth. Cultural historians are interested in it because institutions such as colleges and universities promote assortative mating by placing similar individuals of mating age into close proximity. Social psychologists have long been concerned with attraction, which is usually a prerequisite for mating; personality psychologists work with enduring dispositions of individuals, which often affect mate selection. Biologists focus on the evolutionary change produced by mating patterns. And behaviorial geneticists are interested in assortative mating because it can affect heritability estimates, create correlations among traits that were initially unrelated (e.g., between intelligence and physical attractiveness), and increase both genotypic variance in subsequent generations and the correlations between biological relatives on those traits for which assortative mating occurs.

Assortative mating has been examined with respect to a wide array of variables, including physical characteristics, age, ethnic origin, religion, socioeconomic status, intellectual and cognitive variables, personality traits, and social attitudes. In general, the tendency to choose someone similar to oneself as a mate is so pervasive that Thiessen (1979) prefers the term "assortative narcissism." Indeed, negative assortative mating in human populations has never been reliably demonstrated, with the single exception of sex. This section summarizes the major empirical findings pertaining to human assortative mating, with special emphasis given to recent data. The following section discusses one causal mechanism that may partially account for existing patterns of assortment, and the final section focuses on several important consequences of assortative mating in human populations.

Age is probably the variable for which assortment—or similarity with one's mate—is the strongest. Correlations between spouses for age typically range between 0.7 and 0.9, with a mean of about 0.8; in this context, more than 0.5 is a high degree of correlation. It should be noted, however, that younger couples tend to be more similar in age than older couples, a finding that reflects a larger age gap between spouses in second marriages (Secord 1983). Husbands and wives are also similar with respect to race, religion, ethnic background, and socioeconomic status. For example, Burgess and Wallin (1953) found that 79.4% of a sample of couples had married someone of the same religious faith, while only 37.1% congruence would have been expected on the basis of chance alone. Similar levels of assortment on the basis of religion have been found in recent data (Buss 1984b). Warren (1966) reported correlations of approximately 0.6 for educational level, 0.3 for socioeconomic status, and 0.2 for the number of siblings each spouse had. Controlling for the level of education reduces the correlation for socioeconomic status, but status remains statistically significant.

Physical location, another variable that shows strong marital assortment, has two components: neighborhood and geographical region (Vandenberg 1972). Hollingshead (1950) reported a contingency coefficient of 0.71 between spouses for the social class of the neighborhoods where they lived prior to marriage. Geographically, Spuhler and Clark (1961) found that the median distance between the birthplaces of husband and wife and the place where they were married was only 177 km. Conceptions of ro-

Assistant Professor in the Department of Psychology and Social Relations at Harvard University, David M. Buss received his Ph.D. in psychology from the University of California, Berkeley, in 1981. He is the author of over 30 scientific publications in the areas of mate selection, personality, and social psychology. His current research interests include the development of a new approach to the study of personality and social psychology, and an investigation of cross-cultural patterns of human mate selection. Address: Department of Psychology and Social Relations, Harvard University, Cambridge, MA 02138.

mantic love aside, the "one and only" typically lives within driving distance: it is naturally easier to become intimate with someone who is close-by. These measures of propinquity, of course, are related to other variables such as socioeconomic status. For example, economic considerations affect where one lives. Correlations between variables illustrate both the complexity of choosing a mate and the difficulty of separating causal from concomitant variables in the selection process.

Spuhler (1968) summarized studies on assortment for a variety of physical characteristics, ranging from height, weight, and eye color to less obvious traits such as lung volume, nose breadth, and earlobe length. Coefficients of assortment for these characteristics typically average between 0.1 and 0.2, although figures as high as 0.3 and 0.4 are not uncommon. Individuals probably do not choose their mates on the basis of nose breadth or earlobe length, but selection for other variables such as height and race will cause auxiliary correlations on those characteristics that covary with them.

In addition to specific physical characteristics, spouses seem to select one another on the basis of overall physical attractiveness. An early study by Schooley (1936), for example, found a correlation among spouses of 0.41 for physical appearance. Because attractiveness can vary with age, more recent studies have controlled for age; this research has also reported positive correlations, however. For instance, Price and Vandenberg (1979) discovered correlations of 0.3 and 0.25 in two samples of couples.

Among psychological characteristics, attitudes, opinions, and world views have the strongest assortative mating coefficients. Early studies showed strong correlations between spouses on attitudes about such topics as war, birth control, and contemporary political issues (see Richardson 1939). More recently, Hill and his colleagues (1976) found a correlation of 0.5 for opinions on sex roles, and I found correlations from 0.37 to 0.5 for attitudes about technological growth, societal goals, and so forth (Buss, unpubl.).

Marital assortment for cognitive abilities is consistently moderate across a large number of studies. Jensen (1978) computed a median correlation for this research of 0.44. Recent studies, however, have suggested that the correlations between spouses for mental abilities may be closer to 0.35 (Johnson et al. 1980). Significant positive correlations are found even after controlling for socioeconomic status and education (Watkins and Meredith 1981). Several studies have reported greater assortment for some specific cognitive abilities than for others. Perhaps the most consistent finding is that verbal abilities tend to show higher correlations than spatial abilities, perceptual speed and accuracy, and visual memory (Watkins and Meredith 1981; Zonderman et al. 1977). Thus, spouses do appear to be moderately assorted for cognitive abilities, and there is no evidence that this is a result of any phenotypic convergence over the course of the marriage (Zonderman et al. 1977).

Studies examining assortative mating for personality variables have typically involved subjects' evaluating themselves on scales and inventories, with small (approximately 0.2) but consistently positive correlations between spouses. More recently, I examined correlations for a set of 16 personality traits such as dominance, extraversion, and quarrelsomeness, using three separate sources of data: self-evaluation, and ratings of the subjects by their spouses and by independent interviewers (Buss 1984a). The results from all the sources generally supported the previously obtained low positive correlations. An interesting exception, however, was found for dominance and submissiveness in the spouse ratings and the interviewer ratings, both of which yielded negative correlations between spouses. It may be that the spouses and the interviewers rated husbands and wives by implicitly comparing them with each other rather than with other reference groups such as peers; another possible explanation is that spouses become increasingly complementary in dominance and submissiveness within their marriage, in spite of their overall similarity with respect to the larger peer group. Finally, statistical analysis showed that older couples tended to be less, rather than more, similar to each other, again suggesting that spouses do not converge phenotypically during marriage.

Other recent research has examined correlations between spouses on the frequency with which they perform specific acts and classes of acts (Buss 1984b). In particular, correlations were computed for each of 800 acts from eight categories of personality traits drawn from Wiggins's (1979) model of interpersonal behavior: dominance (one act in this category was "I directed the conversation around to myself"), submissiveness (a sample act was "I accepted verbal abuse without defending myself"), quarrelsomeness ("I made belittling comments about the people who walked by"), agreeableness ("I helped a friend with a difficult assignment"), extraversion ("I told several jokes at the party"), introversion ("I stayed at home to watch TV rather than speak to the person"), being calculating ("I made a friend in order to obtain a favor"), and ingenuousness ("I told a secret to

Table 1. Characteristics commonly sought in a mate

Rank	*Characteristics preferred by males*	*Characteristics preferred by females*
1	kindness and understanding	kindness and understanding
2	intelligence	intelligence
3	*physical attractiveness*[a]	exciting personality
4	exciting personality	good health
5	good health	adaptability
6	adaptability	*physical attractiveness*
7	creativity	creativity
8	desire for children	*good earning capacity*
9	college graduate	college graduate
10	good heredity	desire for children
11	*good earning capacity*	good heredity
12	good housekeeper	good housekeeper
13	religious orientation	religious orientation

[a] The sex differences in ranking are significant beyond the 0.001 level (n = 162) for characteristics in italics.

someone who had previously betrayed my trust"). After the 25 most prototypical acts within each category were composited, spouses showed an average correlation of 0.2 for self-evaluations of how often they performed the acts and 0.31 for the reports of the spouse (Buss and Craik 1983, 1984). The categories of extraversion, quarrelsomeness, and ingenuousness showed particularly strong correlations between spouses. As was the case with the personality variables, the couples who had been married longer were less similar to each other than were couples who had been married for a shorter period of time.

In summary, there is a rough hierarchy of characteristics based on how high correlations are between spouses on these variables. In general, age, education, race, religion, and ethnic background show the strongest assortment. These are followed by attitudes and opinions (0.5) and then by mental abilities (0.4), socioeconomic status (0.3), height, weight, and eye color (0.25–0.3), classes of acts and personality variables (0.2–0.25), number of siblings (0.2), and a host of physical characteristics (0.15). The specific studies are thoroughly reviewed in articles by Vandenberg (1972), Jensen (1978), and Thiessen and Gregg (1980).

Preferences in mate selection

Evolutionary considerations of mate selection date back to Darwin (1871). After completing *On the Origin of Species . . .* in 1859, Darwin observed that many sex differences in characteristics such as the plumage of peacocks seemed to have no survival value and therefore appeared not to be part of natural selection. To account for these findings, he proposed the concept of "sexual selection" as a second process causing evolutionary change. Sexual selection, Darwin thought, would account for findings that he believed could not be explained by natural selection alone.

Darwin's concept of sexual selection subsumes two closely related processes. The first, called intrasexual selection, is the tendency of members of one sex to compete with each other for access to members of the opposite sex. The second, intersexual selection, is the preferential choice members of one sex express for certain members of the opposite sex. Darwin called intersexual selection "female choice" because he observed that, throughout the animal world, females tend to be more discriminating in their choice of mates. Patterns of sexual selection do not immediately involve environmental or ecological adaptations. Instead, they primarily concern the behavioral interactions of the members of a species, which are not necessarily affected by the prevailing demands of the physical environment. If females prefer males with certain characteristics, then the preferred male characteristics will be increasingly represented in subsequent generations.

It is now recognized that sexual selection operates through differential reproductive success (see, for example, Campbell 1972). Natural selection subsumes sexual selection. There is one basic evolutionary mechanism, not two, and the proximate processes of evolutionary change reduce to the differential replication of genes. Nonetheless, sexual selection describes a central process by which genes are differentially reproduced—a process that may be more relevant among humans today than variance in life expectancy or fertility.

There are three levels of preferences in sexual selection: those that are shared by most individuals, those that vary according to sex, and those that vary among individuals. Each level of preferences has distinct consequences for assortative mating.

Characteristics in a mate that are commonly desired are, unfortunately, not possessed by all potential spouses. In a relatively monogamous mating system, it follows that some individuals must settle for a mate who is less than ideal. In addition, if any individuals have to make do with no mates at all, it will probably be those who lack desired characteristics. Those who do possess the valued traits typically marry others with the same or with equally sought-after characteristics. For instance, someone who is dependable might marry someone who is intelligent in what is called cross-character assortment (Buss and Barnes, unpubl.). The previously uncorrelated traits of dependability and intelligence may then covary in the children of such marriages.

Sex differences in the characteristics that are desired in a potential mate can also produce cross-character assortment. If females generally prefer intelligent males because they typically have higher incomes and status, and if most males prefer physically attractive females, then over time these two characteristics will tend to covary (Vandenberg 1972). Indeed, one large-scale longitudinal study has found that physically attractive females do marry males of higher socioeconomic status (Elder 1969). Cross-assortment based on sex differences in preferred characteristics remains an important, but little-examined, aspect of human mating patterns. What specific sex differences occur will be discussed below.

Not all individuals value the same characteristics in a potential mate, and these differences too can produce assortative mating. Thiessen and Gregg (Thiessen 1979, Thiessen and Gregg 1980) have proposed that the tendency to seek in mates the phenotypic characteristics reflecting similar genetic material is a reproductive strategy that represents a compromise between endogamy and exogamy. Positive assortment results in offspring related to each parent by 50% plus the degree to which couples who mate possess genes in common. Thus, an individual's genetic reproduction is enhanced by mating with someone who shares at least some of his genes. If Thiessen's argument is correct, it follows that individuals will differ in their selection preferences, seeking in mates those characteristics that they themselves possess.

In establishing links between selection preferences and assortative narcissism, Thiessen and Gregg (1980) identified an important gap in our current knowledge: our inability to pinpoint exactly what traits prospective mates consider important. A recent series of studies using several samples of individuals, some with mates and some without, was undertaken to fill this gap (Buss and Barnes, unpubl.). In the first study, we asked 93 married couples to rate the desirability of 76 characteristics in a potential mate, and we identified which traits were commonly desired, which reflected sex differences in preferences, and which were the result of individual preferences.

The 15 characteristics that received the highest desirability ratings for the sample as a whole were, in order: providing good companionship, honesty, consideration, having a good disposition, affectionateness, dependability, intelligence, kindness, understanding, being interesting to talk to, loyalty, faithfulness, having a good sense of humor, adaptability, and gentleness. Statistical analyses yielded 25 significant sex differences; only about 4 would have been expected to occur by chance alone. Females valued the characteristics of good earning capacity, good family background, professional status, kindness, gentleness, and considerateness more than males did. Males rated the attributes of physical attractiveness, beauty, frugality, and being a good housekeeper more highly than females did.

We identified nine composite characteristics that reflected individual preferences, which we used as the basis for subsequent studies: kindness or considerateness, being socially exciting, being cultured or intelligent, religious orientation, interest in domesticity, professional status, liking children, political conservatism, and being easygoing or adaptable. Individuals' ratings of these factors were found to relate substantially to the characteristics of their spouses, suggesting that their preferences influenced their choice of mate.

A second study, using subjects of different age, geographic location, marital status, and education, was designed both to examine the ordering of the most desirable factors of the first study and to test the replicability and generality of the sex differences found in that study (Buss and Barnes, unpubl.). The subjects in the second study ranked 13 characteristics from most to least desired in a potential mate.

Table 1 shows how the males and females in the second study rated these characteristics. As in the first study, kindness and intelligence were strongly preferred characteristics, while being a good housekeeper and religious orientation were not highly ranked by the sample as a whole. The two most striking sex differences concerned physical attractiveness and good earning capacity; these differences occurred regardless of the subjects' age, education, geographic location, or marital status, and they replicated the sex differences found in the first study.

Are couples assortatively mated on the basis of how they rate the characteristics they value in a spouse? To address this question, we computed the correlations between spouses for each of the nine composite traits in the first study. The results showed strong positive assortment for religious orientation (0.65) and liking children (0.52); moderate positive correlations for being cultured or intelligent (0.39), being socially exciting (0.37), political conservatism (0.36), and being easygoing or adaptable (0.35); and small, nonsignificant correlations for professional status (0.22) and for kindness or considerateness and interest in domesticity (0.16 in both cases). Thus, spouses appear to have similar selection preferences, but the magnitude of their similarity varies greatly with the particular composite or individual characteristic under consideration.

Four general conclusions can be reached from these studies. First, at least in the United States, there is a moderate consensus about which attributes are preferred in potential mates (e.g., kindness and intelligence), and this consensus transcends differences in age, education, marital status, and geographic location. Second, sex differences were found within each sample, and these differences also transcend variations in the samples. Third, individuals differ in their selection preferences, and their mate selection is affected by these differences. Fourth, couples show positive assortment for individual selection preferences.

Consequences of assortative mating

A complete discussion of the consequences of mate selection would surely include such considerations as individual spouses' happiness, personality change over time as a function of choosing a mate, the distribution of wealth in society, and genetic changes in subsequent generations. The present discussion will focus more narrowly on three known genetic consequences of assortative mating for characteristics that show significant heritability: increased genotypic variance, correlations among traits that were initially unrelated, and effects on variance within and between families.

Assortative mating for traits showing significant heritability has no effect on the frequency with which a gene occurs unless assortment is linked with selection. The population mean remains unchanged. But assortative mating does increase the frequency of genotypes—combinations of genes—that produce extreme phenotypes, and it decreases the frequency of genotypes that create average phenotypes. Overall, the net effect is to increase the amount of variation within the population for the traits for which assortment occurs. Thus, using height as an example of these traits, a greater percentage of individuals in subsequent generations will be quite tall or quite short, while a smaller percentage will be of medium height. Although the increase in genotypic variance resulting from positive assortative mating is small for many characteristics, it accumulates over time.

The results of increasing the amount of variation for certain characteristics have not yet been examined empirically, but several possible consequences could have a significant effect. First, for those heritable traits that are commonly preferred in mates such as intelligence and physical attractiveness, there will be an increasing difference between the haves and the have-nots in subsequent generations. Second, societal institutions may become increasingly strained or may require modification in order to accommodate the increasing variance. Third, the characteristics for which assortment occurs may become even more important in evaluating an individual because such judgment is often based on the difference between individuals (Buss 1983). And fourth, an increase in variance might make it easier for an individual at one extreme to find and marry someone else at that extreme, so that a positive feedback loop is established for assortative mating (Allen 1970; Buss 1984c).

A second genetic consequence of assortative mating is the creation of new correlations among previously unrelated traits. This effect will occur most strongly in mating systems in which there is some consensus about which characteristics are preferred in a spouse, significant sex differences occur in selection preferences, and the choice of a mate is based on overall "market value," or the sum of a person's desirability in a variety of characteristics. If members of a mating population agree to some degree on preferences in a spouse, as the evidence suggests is the case (Buss and Barnes, unpubl.), and if market value can be calculated by combining several traits, then over generations the socially desirable qualities will increasingly covary. Thus, the differences between the haves and the have-nots on the preferred characteristics that can be combined will become even greater than if assortative mating operated independently on each desired characteristic.

A third important effect of assortative mating is that it generally increases the correlations among biological relatives on those characteristics for which assortment occurs. Family members become more similar to each other, which means each family is more homogeneous. Simultaneously, differences between families increase. These effects have implications for recent sociobiological theories about kin selection, reciprocal altruism, and nepotism. Familial communication and cooperation, for example, may be predicted to increase with greater homogeneity, since the benefits of altruistic and nepotistic acts would increase while the disadvantages decreased (Thiessen and Gregg 1980). In this sense, assortative mating promotes the replication of an individual's genes without incurring additional reproductive costs.

Because families provide most of children's early environment, the inequalities resulting from increasing genotypic variance between families may be even further compounded by a correlation between genotype and environment (Plomin et al. 1977). For example, the trait of extraversion can become linked with an environment in which parents talk a great deal. Assortative mating will increase the magnitude of this sort of correlation for children, since it means that parents are more similar or in closer agreement on those attributes that affect the environment they provide.

Strengthening such correlations can be expected to amplify the similarities within each family and the differences between families, thus increasing both the inequality among members of a population and the correlations among socially desirable traits. These expected consequences of current mating patterns are sufficiently important to warrant careful empirical examination.

How rapidly do these consequences of assortative mating occur? The answer depends both on the heritability of each characteristic and on the intensity of assortment, and these values can only be estimated crudely. Jensen (1978) has hypothesized that if the present level of assortative mating for intelligence has existed for several generations, it may account for more than half the individuals now alive whose IQs are greater than 130. The general effects of assortative mating are likely to be small for one generation, but because they accumulate over generations, they acquire considerable importance in the long run.

There is no evidence that the patterns of mate selection in Western societies have changed substantially over the past 50 years: current levels of assortment are comparable to those that occurred in the 1920s and 1930s. However, modern trends toward increasing geographical mobility and equality of opportunity may ultimately increase the intensity of assortment by making it easier for similar individuals of mating age to congregate.

References

Allen, G. 1970. Within group and between group variation expected in human behavioral characters. *Beh. Genetics* 1:175–94.

Burgess, E. W., and P. Wallin. 1953. *Engagement and Marriage.* Lippincott.

Buss, D. M. 1983. Evolutionary biology and personality psychology: Implications of genetic variability. *Person. and Individual Differences* 4:51–63.

———. 1984a. Marital assortment for personality dispositions: Assessment with three different data sources. *Beh. Genetics* 14: 111–23.

———. 1984b. Toward a psychology of person-environment (PE) correlation: The role of spouse selection. *J. Person. and Soc. Psychol.* 47:361–77.

———. 1984c. Evolutionary biology and personality psychology: Toward a conception of human nature and individual differences. *Am. Psychol.* 39:1135–47.

———. Unpubl. Marital congruence for contemporary world views: Initial assortment or phenotypic convergence?

Buss, D. M., and M. Barnes. Unpubl. Preferences in mate selection.

Buss, D. M., and K. H. Craik. 1983. The act frequency approach to personality. *Psychol. Rev.* 90:105–26.

———. 1984. Acts, dispositions, and personality. In *Progress in Experimental Personality Research*, vol. 13, ed. B. A. Maher and W. B. Maher. Academic Press.

Campbell, B., ed. 1972. *Sexual Selection and the Descent of Man: 1871–1971.* Aldine.

Darwin, C. 1859. *On the Origin of Species by Means of Natural Selection.* London: John Murray.

———. 1871. *The Descent of Man and Selection in Relation to Sex.* London: John Murray.

Elder, G. H. 1969. Appearance and education in marriage mobility. *Am. Sociol. Rev.* 34: 519–33.

Hill, C. T., Z. Rubin, and L. A. Peplau. 1976. Breakups before marriage: The end of 103 affairs. *J. Social Issues* 32:147–68.

Hollingshead, A. B. 1950. Cultural factors in the selection of marriage mates. *Am. Sociol. Rev.* 15:619–27.

Jensen, A. R. 1978. Genetic and behavioral effects of nonrandom mating. In *Human Variation: Biopsychology of Age, Race, and Sex*, ed. R. T. Osborne, C. E. Noble, and N. J. Wey. Academic Press.

Johnson, R. C., F. M. Ahern, and R. E. Cole. 1980. Secular changes in degree of assortative mating for ability? *Beh. Genetics* 10:1–7.

Plomin, R., J. C. DeFries, and J. C. Loehlin. 1977. Genotype-environment interaction and correlation in the analysis of human behavior. *Psychol. Bull.* 84:309–22.

Price, R. A., and S. G. Vandenberg. 1979. Matching for physical attractiveness in married couples. *Person. and Social Psychol. Bull.* 5:398–400.

———. 1980. Spouse similarity in American and Swedish couples. *Beh. Genetics* 10:59–71.

Richardson, H. M. 1939. Studies of mental resemblance between husbands and wives and between friends. *Psychol. Bull.* 36: 104–20.

Schooley, M. 1936. Personality resemblances among married couples. *J. Abnormal and Social Psychol.* 31:340–47.

Secord, P. F. 1983. Imbalanced sex ratios: The social consequences. *Person. and Social Psychol. Bull.* 9:525–43.

Spuhler, J. N. 1968. Assortative mating with respect to physical characteristics. *Eugenics Quart.* 15:128–40.

Spuhler, J. N., and P. J. Clark. 1961. Migration into the human breeding population of Ann Arbor, Michigan 1900–1950. *Human Biol.* 33:222–36.

Thiessen, D. D. 1979. Biological trends in behavior genetics. In *Theoretical Advances in Behavior Genetics*, ed. J. R. Royce and L. P. Mos. Alphen aan den Rijn: Sijthoff and Noordhoff.

Thiessen, D. D., and B. Gregg. 1980. Human assortative mating and genetic equilibrium: An evolutionary perspective. *Ethol. and Sociobiol.* 1:111–40.

Vandenberg, S. G. 1972. Assortative mating, or who marries whom? *Beh. Genetics* 2:127–58.

Warren, B. L. 1966. A multiple variable approach to the assortative mating phenomenon. *Eugenics Quart.* 13:285–90.

Watkins, M. P., and W. Meredith. 1981. Spouse similarity in newlyweds with respect to specific cognitive abilities, socioeconomic status, and education. *Beh. Genetics* 11:1–21.

Wiggins, J. S. 1979. A psychological taxonomy of trait descriptive terms: The interpersonal domain. *J. Person. and Social Psychol.* 37: 395–412.

Zonderman, A. B., S. G. Vandenberg, K. P. Spuhler, and P. R. Fain. 1977. Assortative marriage for cognitive abilities. *Beh. Genetics* 7:261–71.

Index

Abalone, 140–153
Abnormal behavior, 37–38
Ache tribe, 154–160
Acoustical perception, 74–86
Acoustical tuning, 89, 92–93
Adaptability, 53–54
Adaptation, 119, 168–169, 195, 204–205
Alarm calls, 55
Algae, coralline, 140–153
Africa, 219–220, 223
Altruism, 116–117, 219, 223, 226
Antipredator behavior, 53, 55, 74–86, 91–92, 94–95, 165–169
Ants, 108–117, 167–168
Assortative mating, 230–234

Baboons, 15–18
Baculum, 183
Bathyergidae, 219–220, 222, 224
Bats, 74–79, 87–95
Bees, *See* honey bees
Beekeeping, 137–138, 163
Bee nests, 162–163
Behaviorism, 50, 55–56
Belding's ground squirrel, 41–47
Bioassays, 130–133
Biological clocks, 97–104
  effects of drug treatments, 99–100
Birth interval, 33
Boobies, 200–207
Brain abnormalities, in schizophrenia, 63

Canidae, 209–211, 217–218
Cheating, in science, 3
Chemotaxis, 146
Chimpanzees, 28, 50, 52
Circadian rhythms, 97–104
Cladism, 124–126
Cleaner fish, 191, 194
Cognitive psychology, 49
Communal nest-weaving, 108–117
Communication
  dance, 52–53, 161–165
  vocal, 55–56
Comparative method, 109, 116, 166
Competition, among siblings, 200–207
Conscious awareness, in animals, 49–56
Conservation of species, 20
Convergent evolution, 103, 220
Cooperative breeding, 213–214, 227
Cooperative hunting, 155–156, 209–212
Copulation behavior, 182–184
Copulatory courtship, 182–183
Coralline red algae, 140–153
Counterstrategy, female, 29–30, 38
Cowbirds, 69
Crickets, 77–86, 176–177
Cryptomys mole-rats, 224, 228–229
Cyclic AMP, 148–150

Dance language, of bees, 52–53, 161–165
Darwin, Charles, 29, 118–127, 172, 181, 185, 219, 223, 232
Data, recording, 18
Deception, 29, 53
Density effects, 37–38
Diet, of bats, 93–94
Dispersal, 41–47, 140, 153, 214–218
  evolutionary origins, 46–47
Dolphins, 19–20, 66
Dominance, 34, 134, 191, 217
Dopamine, 60–63

Earthworms, 120
Echolocation, 74, 90–95
Ecological constraints, 209
Egg trading, 194–195
Egrets, 197–205
Entrainment, 97, 101
Eusociality, 219, 225–229
Evolution, 25, 37–38, 115, 117, 118–127, 166, 172, 191, 197, 200, 219, 225, 228–229
Experiments, 3–4

Female mate choice, 15, 181–184, 230–234
Fertilization
  external, 190–191
  internal, 180, 185
Fishes, 188–195
Fitness, 45
Flight, insect, 80–86
Food choice, 155–156
Food sharing, 156–157
Food size, and siblicide, 201, 207
Frequency-modulated calls, 89–92
Friendship, 17

GABA, 146–150
Genetic predispositions, 58
Genitalia, 180–186
Geophytic plants, 220
"Good genes" hypothesis, 185
Gorillas, 28
Gray parrot, 68
Ground squirrels, 42–47
Guarding, of a mate, 174–175

Habitat
  destruction, 37
  saturation, 216
  selection, 140, 150–152
Hangingflies, 173–174
Haplodiploidy, 224–227
Helping behavior, 213–215, 223, 225, 227
Hermaphroditism, 188–191, 194–195
Heron, great blue, 201, 207
History, importance of, 118–127
Homology, 118, 124–127
Honey, 163–165

Honey bees, 128–138, 161–169
Hormones, 43
Humans
  demography, 158–159
  fertility, 158–159
  mate choice, 230–234
Hunter-gatherers, 154–160

Identical twins, 58
Ion channels, 148–150
Inbreeding, 46, 225–227
Inclusive fitness, 204, 216–217, 223, 226
Inducer, chemical, 145–148
Infant mortality, 31, 34–35
Infanticide, 23, 25–30, 36–37
Information processing, 6, 11, 50
Insect genitalia, 180–186
Integrity, in science, 3–5
Intelligence, 19
Internal fertilization, 180, 185
Intromission, 182

Jackals, 209–218
Jet lag, 99

Katydids, 175–177
Killdeer, 53
"Killer bees," 137
Kinship, 225
!Kung, 154, 157–159

Lacewings, 76
Langurs, 21–30, 31–38
Larval settlement, 140, 150–153
Learning, 65–72
Levels of analysis, 41–47
Life histories, 188, 195, 217–218

Macaques, 51
Male takeovers, 34–35
Males
  coalitions, 19
  mate choice, 174–177
  multi-male bands, 33
  rivalry, 31, 33
Mandibular pheromone, 128–137
Manic depression, 99
Mate choice, human, 230–234
Maternal care, 32
Mating, assortative, 230–234
Mental illness
  manic depression, 99
  schizophrenia, 58–63
Messenger bees, 136–137
Mimicry, vocal, 65–72
Mistakes, in science, 4
Molecular genetics, 100
Molecular phylogeny, 126
Mole-rats, naked, 219–229

Monogamy, 174, 212
Moths, 75, 87–88, 91–95
Mozart, Wolfgang Amadeus, 65–72
Mutations, effects on biological clocks, 100

Naked mole-rats, 219–229
Nest-weaving, 108–117
Neurotransmitters, 60–63, 146–150, 153
Nuptial gifts, 174–178

Oecophylla ants, 108–111, 114–117
Ontogeny
  of behavior, 59
  of bird song, 65–72
  of dispersal, 44–45
Optimal foraging, 155–156
Orientation calls, 89

"Panda principle," 121–123
Paraguay, 155
Parental care, 174–177, 209–217
Parental investment, 173–176
Penis, 180–183
Personality, 17–18
Pheromones, 128–138
  transmission, 134–137
Philopatry, 43, 47
Phonotaxis, 78–80
Phylogenetic grade, 109, 116–117
Phylogenetic inertia, 47
Phylogeny, 26–27, 165–166
  molecular, 126
Pleiotropy, 181
Pollination, 128, 137–138, 164–165
Polyandry, 212, 216–217, 230
Polygyny, 212, 230
Polyrhachis ants, 108, 111, 112–114, 115–116
Predation, 165–169
Predator distraction, 53
Prey detection, 89–95
Process, scientific, 13–14
Prostaglandins, 142
Protandry, 188–191, 193
Protein synthesis, 100, 102–103
Protogyny, 188–191
Proximal mechanisms, 41, 47, 206–207

Queen
  honey bee, 128–138
  mole-rat, 226
  termite, 227
Queen rearing, 131–134

Reader expectations, 6, 11, 14
Receptors, 60, 62, 146–150
Reproductive suppression, 134, 214–215, 226–227

Schizophrenia, 58–63
Science
  cheating in, 3
  integrity in, 5
  mistakes in, 4
  scientific method, 122–123
  scientific process, 13–14
  writing, 6–14
Sea otters, 54–55
Sex
  change, in fishes, 188–195
  ratio, 216–217
Sexual competition, 24–25, 31, 33, 46, 172–174
Sexual receptivity, 26, 36–37
Sexual selection, 15, 29, 172–176, 181–185, 216, 232–233
Siblicide, 197–207
  facultative, 200, 206–207
  obligate, 200, 204–207
Sibling rivalry, 197–207
Silk weaving, 108–117
Size-advantage model, 189–192
"Sneaking," in fishes, 192–193
Social interactions, 17–19, 67–68
Social pathology, 22–23, 27, 37–38
Social tutors, 69
Societies
  bees, 161–169
  jackals, 214–218
  langurs, 21, 32–36
  mole-rats, 219–229
  ground squirrels, 42–47
Sociochemistry, 128
Starling, 65–72
Startle response, 94
Stress position, 8
Subject–verb separation, 9
Systematics, 124
Swarming, 133–134, 165

Temperature
  compensation, 98
  control, 163–165
Termites, 224, 226–227
Territoriality, 211–212
Thermoregulation, 165
Tool use, 52, 54–55
Topic position, 10
Transducers, 146–149

Ultimate causes, 41, 47
Ultrasound, 80–86, 90
Uniformitarianism, 119–120

Volcanoing behavior, 221

Weaver ants, 108–117
Wrasses, 188, 191–194
Writing style, 14